CW00924446

Birkhäuser

Advances in Mechanics and Mathematics

Advances in Continuum Mechanics

Volume 49

This series includes collective thematic books and monographs on advanced research topics in continuum mechanics. It is a unique forum for reference reviews discussing critically foundational issues in classical field theories under the light of up-to-date problems raised by the offspring of condensed matter physics and industrial needs, including those concerning the mechanics of microstructures and the relevant multi-field (or phase-field) representations. The series aims also to show new mathematical methods in continuum mechanics and/or recent developments of existing powerful practice. Also, the series pays attention to applications with concrete conceptual content. The series editors select editors of books on specific themes, which may split into sequential volumes.

Reuven Segev

Foundations of Geometric Continuum Mechanics

Geometry and Duality in Continuum Mechanics

Reuven Segev
Department of Mechanical Engineering
Ben-Gurion University of the Negev
Beer-Sheva, Israel

ISSN 1571-8689 ISSN 1876-9896 (electronic)
Advances in Mechanics and Mathematics
ISSN 2524-4639 ISSN 2524-4647 (electronic)
Advances in Continuum Mechanics
ISBN 978-3-031-35657-5 ISBN 978-3-031-35655-1 (eBook)
https://doi.org/10.1007/978-3-031-35655-1

Mathematics Subject Classification: 74Axx, 53-XX

This book is published under the imprint Birkhäuser, www.birkhauser-science.com by the registered company Springer Nature Switzerland AG
The registered company address is: Gewerbestrasse 11, 6330 Cham, Switzerland

Dedicated to my beloved family, Mina,
Yarin, Yehonathan, Orit, Gideon, Tamar,
Ido, Noa,
Shira, Roni, Hallel, Carmel,
Ruth, Yaara, Hadar, Neomi, and Ethan.

Preface

This book has been written with the purpose of compiling my studies in the geometric foundations of continuum mechanics. One of the objectives has been greater generality. The weaker the assumptions used in a formulation of a theory, the more convincing it is, and the more pure and credible its fundamental concepts are. As an example of this paradigm, in *The Classical Field Theories* [TT60, p. 660], Truesdell and Toupin wrote:

> A principal objective is to isolate those aspects of the theory which are independent of the assigned geometry of spacetime from those whose formulation and interpretation depend on or imply a particular spacetime geometry. For example, we regard the conservation of charge as a physical or intuitive concept logically independent of the concepts of rigid rods, uniform clocks, and inertial frames, and we have chosen to express this law in a mathematical form likewise independent of the representation of these extraneous entities.

In addition, it is hoped that greater generality will lead to greater elegance of the theory. In particular, much value is given to the intimate relationship between the physical aspects of the theory and the mathematical notions used in its formulation. It is hoped that the readers will find it of theoretical and aesthetic value.

The message of this book may be summarized by the following points.

- Continuum kinematics, flux, force, and stress theory can be formulated in the setting of general differentiable manifolds, devoid of any metric or parallelism structures.
- A global approach to the foundations of continuum mechanics, as considered in Part III, is advantageous. In particular, using the notion of duality in the formulation of the theory leads to significant results.
  - The representation of a mechanical system by its configuration manifold can be applied to continuum mechanics. Tangent vectors to the infinite-dimensional configuration manifold are generalized velocity fields, and elements of the cotangent bundle are force functionals producing mechanical power.
  - Due to the infinite-dimensional character of the configuration space, the choice of topology will determine the nature of forces. We view this fact as

the source of the significant gap between the mechanics of systems with finite numbers of degrees of freedom and continuum mechanics.
  - The principle of material impenetrability, postulated in the framework of continuum kinematics, makes the C^1-topology a natural choice.
  - Using a representation process based on the Hahn-Banach theorem, stresses and hyper-stresses emerge naturally as vector-valued measures representing forces. No other assumption of physical nature, e.g., equilibrium, is needed.
  - The representation by stress measures emphasizes the role of stresses as means for the restriction of forces from a body to its sub bodies.
  - The relation between a stress and the force it represents is a generalization of the principle of virtual work, or the equilibrium equation, in continuum mechanics. This operator is the dual of the jet extension mapping of sections of a vector bundle.
  - As another application of the notion of duality, using Whitney's geometric integration theory, it is possible to extend continuum mechanics to non-smooth fields or fractal bodies.
  - Analogous representation techniques provide applicable results for classical stress analysis of structures: optimal stress distributions, worst case loadings, and load capacity of elastic-perfectly plastic structures.

- Smooth exterior calculus on manifolds implies the following generalizations of standard classical counterparts.
  - Flux fields are represented by differential forms rather than vector fields.
  - The divergence operator, appearing in the classical balance equation for a scalar extensive property, is replaced by the exterior derivative operator.
  - The Cauchy construction for the existence of flux fields may be generalized to manifolds of arbitrary dimensions.
  - The stress field of continuum mechanics is replaced by two distinct objects: the traction stress that determines the surface forces on the boundary of a body and the boundaries of its sub bodies, and the variational stress that acts on the derivative (jet) of a velocity field to produce power. The variational stress determines the traction stress but not vice-versa.
  - The generalized Cauchy formula is simply expressed by the restriction of forms to vectors tangent to a particular hypersurface.
  - Application of exterior calculus enables one to represent the geometry of defects in a body, represented by a differential manifold of any dimension and devoid of any metric or parallelism structure. Thus, the geometry of the dislocations is invariant under deformations of the body.
  - Electromagnetism and its generalizations in a spacetime of any dimension (p-form electrodynamics) are shown to be special cases of geometric continuum mechanics.

The book is intended to serve readers either with a continuum mechanics background or with a mathematics background. It would be futile to present comprehensive, rigorous backgrounds for both disciplines. Thus, except for the analysis used in the theory, only the main mathematical objects and constructions are reviewed. Interested readers are referred to the excellent texts available for both geometry (e.g., [AMR88, Lee02]) and analysis (e.g., [Trè67]). No background on continuum mechanics is given. From my experience, mathematicians find the global geometric approach to continuum mechanics as presented here, more natural.

The continuing support for my research, granted by the H. Greenhill Chair for Theoretical and Applied Mechanics, is thankfully acknowledged.

I have been fortunate to interact with some of the best-known, and some less-known, contemporary scholars in the field of mechanics. I was inspired by all and learned a lot. My sincere gratitude is humbly offered to them.

Beer-Sheva, Israel
October, 2022

Reuven Segev

Contents

Chapter 1
Introduction

Continuum theories, or classical field theories, provide the scientific foundations for modern technology. The strengths of structures and the flow fields around the wings of aircraft are computed using equations derived from the principles of continuum mechanics. In many such analyses, concepts such as flux vector fields and stresses are taken for granted. Yet, the existence of notions such as flux vector fields and stresses follows from mathematical theorems based on hypotheses of mathematical and physical nature. For this reason, the foundations of continuum mechanics have been scrutinized for over a century.

As early as 1957, W. Noll [Nol59] defined a body in continuum mechanics as a differentiable manifold that may be covered by a single chart. The various configurations of the body in a Euclidean physical space are viewed as charts on the abstract body object. The central point is that no preferred reference configuration is available, and the differentiable structure of the body is exhibited through its configurations in space. In particular, on the abstract body manifold, notions such as straight lines, angles, and lengths are meaningless. For our setting, both the body manifold and the space manifold have no geometric structure, such as a Riemannian metric or a connection, in addition to the differentiable structure.

The following chapters present a theoretical study of the geometric foundations of continuum mechanics. Two main themes are threaded through the theory: the geometry of differentiable manifolds and duality.

On the geometric side, the Euclidean settings, where continuum theories are formulated traditionally, are forsaken. Rather, the theory is formulated here in the framework of general differentiable manifolds and, except for some particular examples, without assuming the availability of a connection or a Riemannian metric. In classical elasticity theory, it was assumed that a natural, stress-free reference configuration is given. However, such an assumption does not hold for mechanically and thermally processed structural elements, and neither for biological tissues. As a result, no natural metric is induced on a material body by such a reference configuration. Even if the physical space is considered as a Euclidean

R. Segev, *Foundations of Geometric Continuum Mechanics*, Advances in Continuum Mechanics 49, https://doi.org/10.1007/978-3-031-35655-1_1

space, in various theories, the configurations of a body assign to each material point some generalized coordinates that are sometimes referred to as internal degrees of freedom, microstructure, hidden variables, order parameters, substructure, etc. These internal degrees of freedom often lack a natural metric structure.

The absence of a Riemannian metric is typically made up for by using dual spaces—the cotangent bundle of the underlying manifold. Furthermore, as is well known from the theory of Schwartz distributions, duality enables one to obtain irregular, non-smooth objects as continuous linear functionals defined on spaces of differentiable fields. Indeed, the main message of this book is that in very general settings, significant properties of forces and stresses follow mathematically when forces are defined as elements of the cotangent bundle of an appropriately chosen configuration space.

The book is divided into three parts. The first part is algebraic in nature and contains no analysis. On the one hand, it contains the algebraic aspects of a general version of the Cauchy theory of fluxes; on the other hand, it serves as an introduction to exterior algebra. The second part presents the theory of smooth distributions of forces and stresses on manifolds. For this reason, it contains an overview of analysis on smooth manifolds.

Part III of the book studies the application to continuum mechanics of a general paradigm for mechanical systems. A mechanical system is represented mathematically by its configuration space, assumed to be a differentiable manifold. Tangent vectors to the configuration space are interpreted as generalized velocities, and generalized forces are modeled by linear functionals acting on generalized velocities. The configuration space of a continuous body in space is evidently infinite-dimensional. Therefore, notions and results from functional analysis should be used in the study of this structure in conjunction with geometry—the domain of nonlinear global analysis. A fundamental message of this book is that the gap between continuum mechanics and the mechanics of systems having finite numbers of freedom stems from this transition to infinite dimensions and the need to use tools of functional analysis. In particular, we show that the existence of stresses and their properties follow from a representation procedure for force functionals. For further details on the results obtained, see the list in the introductory paragraphs of Chap. 21.

This introduction is followed by a prelude that illustrates how some of the main ideas of the global approach to continuum mechanics, presented in Part III, apply to a mechanical system with a finite number of degrees of freedom.

Part I of the monograph lays out the algebraic-geometric foundations of flux and stress theories. In the geometric setting of an n-dimensional affine space, fluxes are defined as real-valued mappings of r-simplices, $r \leqslant n$, satisfying some natural properties such as translation invariance, homogeneity, additivity, and balance. It is shown in Chap. 4 that the flux through a simplex is represented by a completely antisymmetric multilinear tensor acting on the vectors that make up the simplex. This generalizes the standard Cauchy tetrahedron construction in two aspects: no metric is used in space, and the construction holds for any values of n and $r \leqslant n$. Rather than merely the case $r = 2$, $n = 3$, this result enables the computation of fluxes through any r-dimensional simplex in an n-dimensional space. As an additional feature, the proof is completely combinatorial.

Once flux mappings have been defined on simplices, they are extended in Chap. 5 by linearity to polyhedral chains—linear combinations of simplices. In addition, the major role played by antisymmetric tensors motivates the introduction of additional aspects of exterior algebra in Chap. 5.

In Part II we consider the smooth theory of fluxes, forces, and stresses as formulated on general n-dimensional differentiable manifolds. With some necessary modifications, the analog of Cauchy's stress theory is formulated on manifolds. An important outcome of the theory should be mentioned. When the theory is formulated in Euclidean spaces, the stress tensor object plays two roles. On the one hand, it determines the traction fields on subbodies, and in addition, it acts on the gradient of a velocity field to produce power. In the generalized theory, these two roles are played by two distinct mathematical objects. The *traction stress* is the object determining the traction on a subbody, and the *variational stress* produces power when acting on the derivative—the jet—of a velocity field. The variational stress determines the traction stress, but not vice-versa. Another significant modification is a generalized definition of the divergence of the variational stress.

Chapter 6 reviews the essentials of smooth analysis on manifolds. Deviating somewhat from the main theme, the geometry of smooth distributions of defects is presented in Chap. 7 using exterior calculus. The framework applies to manifolds of arbitrary dimensions and no metric properties are used.

Chapter 8 presents the analog, in the setting of general manifolds, of the Cauchy theory of fluxes. Specifically, the flux field is represented by a differential form rather than a vector field. The results of Chap. 4 are used in the proof. The divergence of the flux vector field, appearing in the classical form of the balance equation, is replaced by the exterior derivative of the flux form. Chapter 9 uses the properties of flux differential forms to study the structure of spacetime. The notion of a material point in continuum mechanics is motivated even for the case of volumetric growth.

Chapter 10 introduces smooth stress theory on manifolds in analogy with the traditional Cauchy construction. Now, force fields cannot be represented as vector fields and cannot be integrated over the manifold. Thus, the value of a force distribution at a point is a linear mapping that acts on generalized velocities to produce power density. The existence of traction stresses follows from the existence of fluxes. Traction stresses do not act on derivatives, or rather jets, of generalized velocities to produce power density. The power density is given using another mathematical object, which we refer to as the variational stress. The relation between the two stress objects is studied, and the differential balance equation is formulated using a generalized divergence operator. Finally, we consider the particular case where a connection is given on the space where generalized velocities assume their values.

Metric-free (pre-metric) electromagnetism is used in Chap. 11 as an example for the notions of smooth geometric stress theory. After considering the Lorentz force and metric-independent Maxwell stress tensor in a four-dimensional spacetime, electrodynamics in a spacetime of arbitrarily large dimension (p-form electrodynamics as in [HT86, HT88]) is shown to be a particular case of smooth stress theory on manifolds. Here, generalized velocity fields are taken to be differential forms (rather than vector fields) representing potential fields.

We continue with a formulation of nonlinear elasticity in Chap. 12. Here we make an additional generalization and regard the configuration of a body, not as a mapping of a body manifold into a space manifold, but rather as a section of a fiber bundle over the body manifold. This means roughly that each point in the body "sees" another version of the space manifold. Kinematics, stress theory, and constitutive relations are considered and the problem of nonlinear elasticity is formulated.

It is noted that the notion of equilibrium has not been mentioned so far as it is meaningless in the settings of general differentiable manifolds. In Chap. 13 we introduce equilibrium through an invariance condition for the power in the case where a Lie group acts on the space manifold.

Next, we consider the dynamics of a particle and a continuous body in a proto-Galilean spacetime. By a proto-Galilean spacetime, we mean that the time variable is absolute, so it is meaningful to state that two events occurred at the same time. However, spacetime has a general fiber bundle structure over the time axis, the typical fiber of which is a general manifold. Thus, a motion of a particle is a section of the fiber bundle, an embedding of the time axis in spacetime. It follows that the theory of nonlinear elasticity in fiber bundles as in Chap. 12 may be used to formulate dynamics and that the dynamics law is specified as a constitutive relation. A similar, albeit more complicated, structure is presented for the dynamics of bodies.

Part III presents the global non-smooth theory. The basic object considered is the infinite-dimensional configuration manifold of a continuous body in space. The guiding principle is that generalized velocities are elements of the tangent bundle of the configuration space and forces are elements of the cotangent bundle of the configuration space. The properties of manifolds of mappings imply that generalized velocities are represented by differentiable vector fields. Thus, forces are continuous linear functionals on spaces of differentiable vector fields. As such, the nature of forces depends on the topology used in the space of velocity fields. A natural choice of topology, the C^1-topology, is suggested by the standard assumption that configurations of a material body in the physical space are embeddings. The existence of stress, which may be as irregular as a Radon measure, follows from a mathematical representation theorem for forces. Simply put, the existence of stress and the condition for its equilibrium with external loading stem from the general framework and the kinematic assumption that configurations are embeddings. The origin of the nonunique relation between forces and stresses is also traced back to kinematics. Moreover, generalizing the topology to the C^r-topology for any $r > 1$ results in r-th order hyper-stresses and r-th order continuum mechanics.

Chapter 14 is concerned with the C^r-Banach space structure on the vector space of C^r-sections of a vector bundle. Note that the norm depends on some choices one has to make; however, other choices will lead to equivalent norms. For this reason, one often refers to such topological vector spaces as Banachable. Such a Banachable space of sections will serve later as the model space for the infinite-dimensional configuration manifold and as the tangent space to the configuration space at any given configuration.

The Banachable manifold structure on the space of sections of a fiber bundle is reviewed in Chap. 15. This includes, for the case of a trivial fiber bundle, manifolds of mapping of the body manifold into the space manifold. Configurations of the body manifold into space manifold are traditionally restricted by the principle of material impenetrability to be embeddings. This implies that in order to construct a meaningful configuration manifold for continuum mechanics, one has to endow the collection of embeddings with a suitable manifold structure. Indeed, for C^r-mappings of the body into space, with $r \geqslant 1$, the set of embeddings is open in the manifold of mapping. This implies that the case $r = 1$ has special significance in continuum mechanics, but cases where $r > 1$ are meaningful also. Indeed, standard continuum mechanics corresponds to $r = 1$, while cases where $r > 1$ lead to continuum mechanics of higher order and the representation of forces by hyper-stresses.

Before arriving at the technical details, the framework for the global analytic approach to force and stress theory in continuum mechanics is outlined in the short Chap. 16. As generalized velocity fields are modeled by C^r-sections of a vector bundle W, by our general paradigm, forces belong to the dual space, $C^r(W)^*$, of the space, $C^r(W)$, of C^r-sections of the vector bundle. Thus, Chaps. 17, 18, and 20 are concerned with various issues related to such functionals. Specifically, Chap. 17 describes how a linear functional on the space of sections of a vector bundle is represented locally for a given atlas. Particular attention is given to manifolds with corners, for which we use [Mel96].

Duality gives rise to non-smooth objects. Chapter 18 introduces de Rham currents, the geometric objects analogous to Schwartz distributions. As a simple example of the use of de Rham currents, Chap. 19 presents the singular counterpart of Chap. 7 on smooth distributions of defects in bodies. de Rham currents do not provide sufficient tools for the formulation of non-smooth stress theory in continuum mechanics. The additional algebraic structure required is that of vector-valued currents presented in Chap. 20.

Having developed the language needed, the existence of stresses and hyper-stresses as Radon measures representing force functionals is presented in Chap. 21. The relation between a force functional and a representing stress is a generalization of the principle of virtual work of continuum mechanics. Chapter 22 provides further details for the standard case of simple stresses—hyper-stresses of order one. As an example, we present the non-smooth counterpart of p-form electrodynamics, the smooth version of which has been treated in Chap. 11.

Another duality relation makes it possible to formulate a theory for non-smooth bodies and non-smooth velocity fields. Using Whitney's geometric integration theory [Whi57], bodies and velocities are modeled as flat chains in a Euclidean space. Flat chains are obtained by the completion of the vector space of polyhedral chains relative to a norm that is naturally suggested by the classical assumptions of flux theory. As a result, bodies with fractal boundaries, such as the Koch snowflake, are admissible. By duality, fluxes and forces are flat co-chains as studied by Whitney.

A representation theorem implies that co-chains are represented by forms. The applications of Whitney's geometric integration theory to continuum mechanics are presented in Chap. 23.

The representation of forces by stresses, as described above, makes use of the Hahn-Banach theorem. We recall that the extended linear functional obtained by the Hahn-Banach theorem has an optimal norm in comparison with all possible extensions. This observation suggests the following stress optimization problem for classical stress analysis in $\mathbb{R}^3$. Let a force on a body be given and consider the collection of stress distributions that equilibrate the given force where no particular constitutive relation is given. Of this collection of equilibrating stress fields, what stress field will have the least norm? This optimization problem and related issues are studied in Chap. 24. We show that the optimal norm of an equilibrating stress is equal to the norm of the given force. This result of the optimization problem raises another question. Is there a constitutive relation for the stresses that will result in an optimal stress distribution for a given force? It turns out that perfectly plastic bodies are under optimal stress fields when they are in plastic limit states. This observation implies that for a perfectly plastic body, there is a number C, to which we refer as the load capacity of the body, such that the body can support any external load F, as long as $\|F\| \leqslant Cs_Y$, where s_Y is the yield stress of the material. The load capacity ratio depends only on the geometry, and the expression for C is particularly elegant when the body is loaded by surface forces only. In such a case, C is equal to the norm of the trace mapping assigning to every vector field on the body the corresponding field on the boundary. Some examples of computations for structures, including computations of worst-case loadings, all based on the geometry of the structures only, are given in [FS09].

Chapter 2
Prelude: Finite-Dimensional Systems

It has been claimed that there is a fundamental difference between continuum mechanics and the mechanics of systems consisting of finite numbers of material points and rigid bodies. The message we advocate, which embodies the essence of the book, is that this difference is a manifestation of the transition in mathematical analysis from finite to infinite dimensionality.

A basic characteristic of continuum mechanics is static indeterminacy. Given a force acting on a body, the stress distribution in the body cannot be determined by equilibrium alone. This is the reason why material properties, represented by constitutive relations between the stress field and the kinematics of the body, are needed in order to determine the stress distribution uniquely.

The source of static indeterminacy may be traced to the kinematics of the mechanical system. This applies to both the finite-dimensional and the infinite-dimensional cases. Thus, significant insight into the geometric and algebraic structures to follow in later parts may be obtained by following the much simpler finite-dimensional analog, as the rest of this chapter considers.

2.1 The Framework for the Problem of Statics

The Physical Space Through a large part of the book, unless otherwise stated, it is assumed that there is a fixed *physical space* $\mathcal{S}$ where the mechanics of systems takes place. In a somewhat more general setting of a four-dimensional nonrelativistic spacetime $\mathscr{E}$, the physical space may be thought of as the collection $\mathscr{E}_t$ of all events occurring at some particular instant t. We will refer to this interpretation of $\mathcal{S} = \mathscr{E}_t$ as an *instantaneous fiber of spacetime*. It is assumed that $\mathcal{S}$ has the structure of a differentiable manifold. For most of what follows, the physical space will not be endowed with any additional geometric structure such as a Euclidean structure, a Riemannian structure, a connection, etc.

R. Segev, *Foundations of Geometric Continuum Mechanics*, Advances in Continuum Mechanics 49, https://doi.org/10.1007/978-3-031-35655-1_2

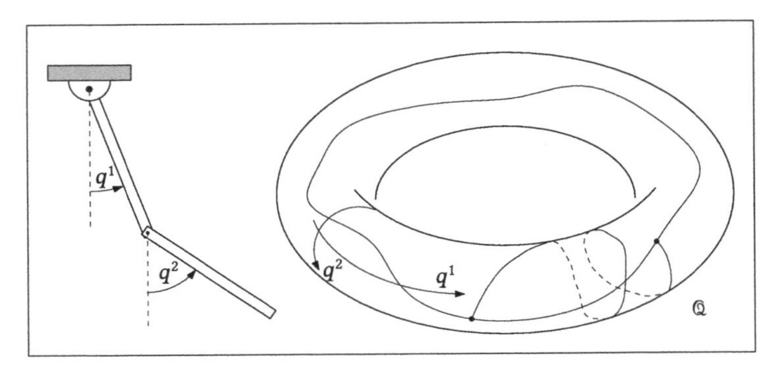

Fig. 2.1 The configuration space of the planar double pendulum

The Configuration Space The basic object in the study of the mechanics of a system is its configuration space. The configuration space contains the collection of configurations of the mechanical system in the physical space. It is viewed as a geometrical object and assumed to be a differentiable manifold $\mathbb{Q}$. For example, the configuration space of a planar double pendulum is represented by the two-dimensional torus as illustrated in Fig. 2.1.

The dimension of the configuration space, $n = \dim \mathbb{Q}$, is the number of degrees of freedom that the mechanical system has. Let $\varphi : U \to \mathbb{R}^n$ be a chart on an open subset $U \subset \mathbb{Q}$. For a configuration $q \in \mathbb{Q}$, the coordinates

$$(q^1, \ldots, q^n) = \varphi(q) \tag{2.1}$$

assigned to it by the chart are the *generalized coordinates* of q.

Motions and Generalized Velocities A curve $c : [a, b] \to \mathbb{Q}$, for a non-empty interval $[a, b] \subset \mathbb{R}$, is interpreted as a *motion* of the mechanical system.

Tangent vectors to the configuration space are referred to as *generalized velocities, virtual velocities,* or *virtual displacements.*[1] Thus, if $q \in \mathbb{Q}$ is a configuration, $T_q\mathbb{Q}$, the tangent space to the configuration manifold at q, contains the generalized velocities at that configuration.

Generalized Forces and Mechanical Power *Generalized forces* are modeled mathematically by elements of the cotangent bundle $T^*\mathbb{Q}$. Thus, a generalized force at the configuration q, $F \in T_q^*\mathbb{Q} = (T_q\mathbb{Q})^* = (T^*\mathbb{Q})_q$, is a linear mapping which acts on a generalized velocity $v \in T_q\mathbb{Q}$ to produce a real number $F(v)$ that is interpreted as mechanical power, *virtual power*, or *virtual work*. It follows that the

[1] Note that the term "velocity" is meaningful in this context only when $\mathcal{S}$ is interpreted as a fixed space manifold rather than as an instantaneous fiber of spacetime. When $\mathcal{S}$ is interpreted as an instantaneous fiber of spacetime, the terminology "virtual displacements" is more appropriate.

notion of power is fundamental and serves as the bridge between kinematics and statics—the theory of forces.

Constraints and Supported Systems A *simple constraint* for the mechanical system is specified by a differentiable mapping $h : \mathbb{Q} \to \mathbb{R}$. We will refer to h as the *kinematics mapping*. For example, if the space manifold $\mathcal{S}$ is identified with $\mathbb{R}^3$, the kinematic mapping may assign to a particular particle in the system some component of its location in space. A configuration, q, of the system, is said to be *admissible* relative to the constraint h if it is the solution of the equation $h(q) = 0$. Typically, such a constraint is defined by some support or a mechanical linkage. More generally, a *constraint* is specified by a kinematic differentiable mapping $h : \mathbb{Q} \to \mathcal{P}$, where $\mathcal{P}$ is an m-dimensional differentiable manifold. Given an element $p_0 \in \mathcal{P}$, admissible configurations are those solving the equation $h(q) = p_0$. Returning to the example where $\mathcal{S} = \mathbb{R}^3$, and setting $\mathcal{P} = \mathbb{R}^3 \times \mathbb{R}^3$, the kinematic mapping may restrict two points in the system to be fixed in two points in space.

Elements of the tangent bundle $T\mathcal{P}$ may be interpreted as generalized velocities of the constraints or the supports. In fact, a generalized velocity of the constraint is a velocity that the constraint is supposed to prevent. In analogy with the original system, elements of $T^*\mathcal{P}$ are interpreted as generalized forces acting at the constraints.

Consider a fixed configuration, $q \in \mathbb{Q}$, of the system. The tangent to the kinematic mapping at q,

$$T_q h : T_q \mathbb{Q} \longrightarrow T_{h(q)}\mathcal{P}, \tag{2.2}$$

is a linear mapping that associates with generalized velocities of the system, the corresponding generalized velocities of the supports. Elements of Image$T_q h$ are those constraint velocities that are *compatible* with the kinematics of the system. (See Fig. 2.2 for an illustration.)

Of particular importance is the case where $T_q h$ is injective. It follows immediately that $n \leqslant m$. In such a case, only the zero generalized velocity of the system is mapped to the zero generalized velocity of the constraints. In other words, if the constraints prevent generalized velocities in $T_{h(q)}\mathcal{P}$, they prevent generalized velocities of the system. Thus, if $T_q h$ is injective, the system is *fully constrained* or *kinematically determinate*.

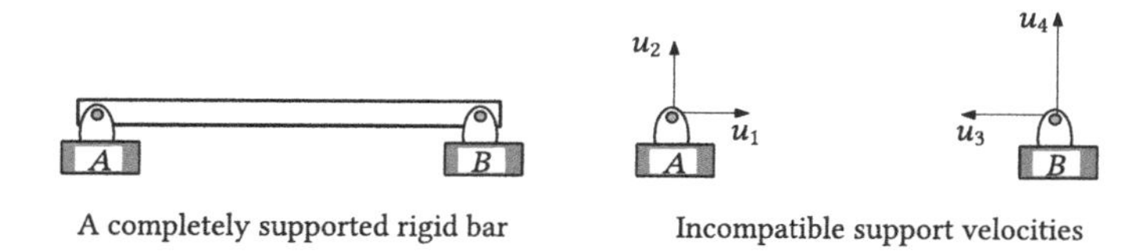

Fig. 2.2 Incompatible constraint velocities

Reactions, Equilibrium, and the Problem of Statics Traditionally, elements of $T^*\mathcal{P}$—generalized forces acting at the constraints—are referred to as *reactive forces* or *reactions*. The dual or adjoint of the tangent $T_q h$,

$$T_q^* h := (T_q h)^* : T_{h(q)}^* \mathcal{P} \longrightarrow T_q^* \mathbb{Q}, \tag{2.3}$$

maps the reactive forces to the corresponding generalized forces acting on the system. The definition of the dual mapping implies that in case $F = T_q^* h(g)$, $g \in T_{h(q)}^* \mathcal{P}$, $F \in T_q^* \mathbb{Q}$, one has

$$F(v) = g(T_q h(v)), \quad \text{for all} \quad v \in T_q \mathbb{Q}. \tag{2.4}$$

Thus, the power expended by the generalized force F when acting on a generalized velocity v is equal to the power expended by the reaction force g for the corresponding velocities of the constraints. In such a case, we will say that the reactive force g is in *equilibrium* with the generalized force F.

Now, the *problem of statics* may be formulated as follows: Given a force $F \in T_q \mathbb{Q}$, find the solutions $g \in T_{h(q)}^* \mathcal{P}$ of the equation

$$F = T_q^* h(g). \tag{2.5}$$

2.2 On the Solutions of the Problem of Statics

The properties of the solutions of the problem of statics will be outlined below. First, in order to simplify the notation, we set

$$\mathbf{V} := T_q \mathbb{Q}, \quad \mathbf{U} := T_{h(q)} \mathcal{P}, \quad J := T_q h : \mathbf{V} \longrightarrow \mathbf{U}, \tag{2.6}$$

so that the equilibrium condition (2.5) may be written as

$$F = J^*(g). \tag{2.7}$$

Only the case of kinematically determinate systems will be considered here although the analysis can be extended to the general case.

2.2.1 Existence of Solutions

We first show that for every $F \in \mathbf{V}^* := T_q^* \mathbb{Q}$, the problem of statics has some solution when $J := T_q h$ is injective.

The mapping J is decomposed in the form $J = \mathcal{J} \circ \widehat{J}$. Here

$$\widehat{J} : \mathbf{V} \longrightarrow \text{Image}J, \qquad \widehat{J}(v) = J(v), \quad \text{for all} \quad v \in \mathbf{V}$$

is the restriction of J to its image. Thus, $\widehat{J}$ is an isomorphism. Naturally,

$$\mathcal{J} : \text{Image}J \longrightarrow \mathbf{U}, \qquad \mathcal{J}(u) = u, \quad \text{for all} \quad u \in \text{Image}J \tag{2.8}$$

is the inclusion of Image$J$ in $\mathbf{U}$.

Since $\widehat{J}$ is an isomorphism, one can consider the linear functional

$$g_0 = F \circ \widehat{J}^{-1} : \text{Image}J \longrightarrow \mathbb{R},$$

an element of the dual space $(\text{Image}J)^*$. (See diagram in (2.9).)

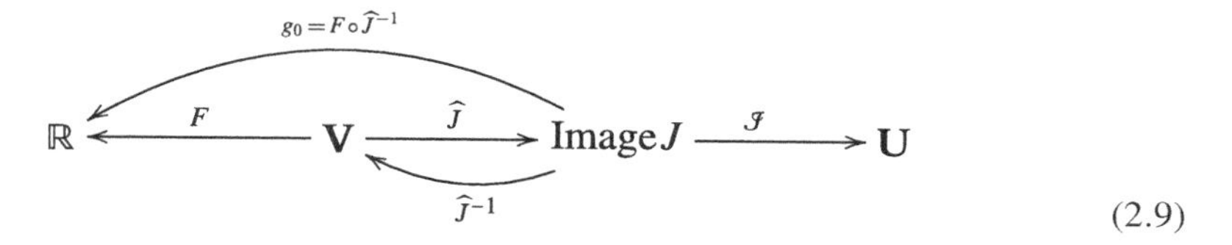

(2.9)

It follows from the definition of g_0 that for each $v \in \mathbf{V}$,

$$g_0(J(v)) = F(v), \tag{2.10}$$

which is similar to the equilibrium condition (2.7). However, g_0 is defined only on Image$J$ and not on $\mathbf{U}$, as required. In other words, we can evaluate the power only for compatible constraint velocities while an element of $\mathbf{U}^*$ associates mechanical power with constraint velocities that are not necessarily compatible. This is resolved using the Hahn-Banach theorem. It guarantees the existence of some extension $g \in \mathbf{U}^*$ of g_0, that is,

$$g(w) = g_0(w), \quad \text{for all} \quad w \in \text{Image}J. \tag{2.11}$$

Hence,

$$g(J(v)) = F(v), \quad \text{for all} \quad v \in \mathbf{V}, \qquad \text{and so,} \qquad F = J^*(g). \tag{2.12}$$

2.2.2 *Static Indeterminacy and Optimal Solutions*

Evidently, for the case where $m = \dim \mathbf{U} > n = \dim \mathbf{V}$, the equation of equilibrium does not have a unique solution. From the foregoing analysis, the nonuniqueness arises due to the fact that J is not surjective. The nonuniqueness of the solution is traditionally referred to as *static indeterminacy*. For continuum mechanics, where the stress fields play the role of the reactive forces, static indeterminacy is inherent. As will be seen in Part III, the inherent static indeterminacy of continuum mechanics also follows from the fact that the analog of the mapping J is not surjective.

Assume that a norm $\|\cdot\|$ is given on $\mathbf{U}$ and consider the dual norm on $\mathbf{U}^*$, that is,

$$\|g\| := \sup\left\{ \frac{|g(u)|}{\|u\|} \,\middle|\, u \in \mathbf{U},\ u \neq 0 \right\}. \tag{2.13}$$

Since there is a collection of reactive forces, $(J^*)^{-1}(F)$, one may consider the optimization problem

$$\Omega_F = \inf\{\|g\| \mid g \in \mathbf{U}^*,\ J^*(g) = F\}. \tag{2.14}$$

The optimization problem may be motivated by the following situation. Assume that a robotic hand grasps an object and an external force F is applied to the object. The grasping hand serves as a constraint on the object and one wishes the grasping to be as gentle as possible by minimizing the norm of the reactive forces.

To analyze the optimization problem, it is noted that as an element of $(\text{Image}J)^*$,

$$\begin{aligned} \|g_0\| &= \sup\left\{ \frac{|g_0(w)|}{\|w\|} \,\middle|\, w \in \text{Image}J,\ w \neq 0 \right\}, \\ &= \sup\left\{ \frac{|F(v)|}{\|J(v)\|} \,\middle|\, v \in \mathbf{V},\ v \neq 0 \right\}. \end{aligned} \tag{2.15}$$

As can be expected, $\|g_0\|$ is completely determined by the force F and the kinematic mapping. Any element $g \in (J^*)^{-1}(F)$—an equilibrating reactive force—is an extension of g_0 and

$$\begin{aligned} \|g\| &= \sup\left\{ \frac{|g(u)|}{\|u\|} \,\middle|\, u \in \mathbf{U},\ u \neq 0 \right\}, \\ &\geqslant \sup\left\{ \frac{|g(w)|}{\|w\|} \,\middle|\, w \in \text{Image}J,\ w \neq 0 \right\}, \\ &= \sup\left\{ \frac{|g_0(w)|}{\|w\|} \,\middle|\, w \in \text{Image}J,\ w \neq 0 \right\}. \end{aligned} \tag{2.16}$$

It follows that in general $\|g\| \geqslant \|g_0\|$ for every $g \in (J^*)^{-1}(F)$. We make use of the Hahn-Banach theorem again. It asserts that there is some extension g_{hb} of g_0 that preserves the norm, that is, $\|g_{\text{hb}}\| = \|g_0\|$. We conclude that there is an optimal $g_{\text{hb}} \in \mathbf{U}^*$ with

$$\Omega_F = \|g_{\text{hb}}\| = \sup\left\{ \frac{|F(v)|}{\|J(v)\|} \,\middle|\, v \in \mathbf{V},\ v \neq 0 \right\}. \tag{2.17}$$

2.2.3 Worst-Case Loading and Load Capacity

It is now assumed that some meaningful norm is given also on $\mathbf{V}$, and we use the dual norm on $\mathbf{V}^*$. Equation (2.17) presents the dependence of the optimum Ω_F on the force F. In particular, it implies that for any $\lambda > 0$, $\Omega_{\lambda F} = \lambda \Omega_F$. Thus, it makes sense to consider the normalized optimum

$$\Omega_{F/\|F\|} = \frac{\Omega_F}{\|F\|}, \tag{2.18}$$

which quantifies the sensitivity of the grasping to the normalized force. Let us return to the situation described above where one wishes to apply the gentlest grasping forces to support any given force F. Assume that by a force control system, the robot can apply an optimal collection of grasping forces to each external load F.[2] Assume, also, that the force control system is bounded by some maximum $\mathbf{M} > 0$, so that it can apply the required optimal grasping forces only if $\Omega_F \leqslant \mathbf{M}$.

With this limitation in mind, it makes sense to define the *worst-case loading* problem

$$K := \sup\left\{ \frac{\Omega_F}{\|F\|} \;\middle|\; F \in \mathbf{V}^*,\ F \neq 0 \right\}. \tag{2.19}$$

It follows immediately that for any $F \in \mathbf{V}^*$,

$$\text{if} \qquad \|F\| \leqslant \frac{\mathbf{M}}{K}, \qquad \text{then,} \qquad \mathbf{M} = K\frac{\mathbf{M}}{K} \geqslant K\|F\| \geqslant \Omega_F. \tag{2.20}$$

Defining the *load capacity ratio of the grasping*, C, as the inverse of K, it is concluded that the grasping mechanism can support every load F satisfying

$$\|F\| \leqslant C\mathbf{M}. \tag{2.21}$$

The Expression for the Load Capacity Ratio Substituting the expression for the optimum Ω_F into the expression for K, one has

$$\begin{aligned} \frac{1}{C} = K &= \sup_F \left\{ \frac{1}{\|F\|} \sup_v \left\{ \frac{|F(v)|}{\|J(v)\|} \right\} \right\}, \\ &= \sup_v \left\{ \frac{1}{\|J(v)\|} \sup_F \left\{ \frac{F(v)}{\|F\|} \right\} \right\}. \end{aligned} \tag{2.22}$$

[2] It is noted that optimal reactive forces are applied by Coulomb friction forces on the threshold of slippage. We will also observe in Part III that the state of stress is optimal for ideally plastic materials on the verge of plastic collapse.

We recall that duality theory implies that

$$\sup\left\{\frac{F(v)}{\|F\|}\;\middle|\; F\in\mathbf{V}^*,\; F\neq 0\right\}=\|v\|. \tag{2.23}$$

Therefore,

$$\frac{1}{C}=K=\sup\left\{\frac{\|v\|}{\|J(v)\|}\;\middle|\; v\in\mathbf{V},\; v\neq 0\right\}=\|\widehat{J}^{-1}\|. \tag{2.24}$$

The ideas and constructions of this chapter are extended to the infinite-dimensional case of continuum mechanics in Chaps. 21 and 24.

Part I
Algebraic Theory: Uniform Fluxes

Part I of the text is concerned with the algebraic aspects of flux theory. The central subject of the theory of Cauchy fluxes is the proof of the existence of a Cauchy flux object. The existence of the flux object relies on a number of assumptions, which are usually referred to as Cauchy's postulates. The Cauchy postulates are concerned with balance of the extensive property under consideration and the dependence of the flux on the geometry of sub-regions. Alternative collections of assumptions have been proposed by modern authors, e.g., [Nol59, GM75, FV89]. The geometric setting is traditionally a Euclidean three-dimensional space and two-dimensional surfaces through which the total flux is computed. The flux object at a point is shown in these versions of the Cauchy theorem to be a vector.

To lay down the algebraic foundations for the theories described in the following parts, we consider here uniform fluxes or algebraic fluxes. The geometric setting is that of an n-dimensional affine space, and no Euclidean structure is assumed. The flux through an r-dimensional polyhedral chain is shown to be represented by an alternating (or completely skew symmetric), covariant r-tensor. The proof relies on postulates of translation invariance, homogeneity, and balance.

The nature of the flux object as an alternating tensor leads us to a review of exterior algebra from the point of view of uniform flux theory.

Chapter 3
Simplices in Affine Spaces and Their Boundaries

In this chapter, we review some of the constructions involving simplices in affine spaces. These constructions will be used in the following chapter for the formulation and proof of the algebraic Cauchy flux theorem—Cauchy theorem for uniform flux fields—in an n-dimensional affine space.

3.1 Affine Spaces: Notation

Considering an affine space, $\mathscr{A}$, having a translation (or tangent) space, $\mathbf{V}$ (see, e.g., [AM77, pp. 2–6] [Whi57, pp. 349–351]), we will use the following notation. For each ordered pair (p, p') of points in $\mathscr{A}$, there is a vector $v \in \mathbf{V}$, which we regard as pointing from p to p', and we write $v = p' - p$ and $p' = p + v$. The operation $p + v$ will be referred to as the *translation* of p by v. Note that the minus sign and the plus sign, which we use in this context in order to simplify the notation, are not those used for operations in vector spaces. A choice of a particular point O, to which we will refer as an *origin*, induces a bijection of $\mathscr{A}$ and $\mathbf{V}$ by $p \mapsto p - O$. This bijection is meaningful as for any triplet of points $p, p', p'' \in \mathscr{A}$, it is postulated that

$$p'' - p = (p'' - p') + (p' - p). \tag{3.1}$$

It is noted the plus sign above is the addition in the vector space $\mathbf{V}$ while the minus signs indicate the basic operation on pairs of points in an affine space. In particular, it follows from Eq. (3.1) that

$$p - p = 0, \quad p - p' = -(p' - p), \quad \text{and} \quad (p - q) - (p' - q) = p - p'. \tag{3.2}$$

The affine space $\mathscr{A}$ is said to be n-dimensional if $\mathbf{V}$ is n-dimensional.

R. Segev, *Foundations of Geometric Continuum Mechanics*, Advances in Continuum Mechanics 49, https://doi.org/10.1007/978-3-031-35655-1_3

It is emphasized that the difference between an affine space and a Euclidean space is that an affine space does not have a metric structure. Notions such as distance, length, and angle are meaningless.

3.2 Simplices

Let $\mathscr{A}$ be an n-dimensional affine space, and let $p_0, \ldots, p_r \in \mathscr{A}$ be $r + 1$ points such that the vectors $v_i = p_i - p_0 \in \mathbf{V}$, $i = 1, \ldots, r$, are linearly independent. The collection of vectors $\{v_i\}$, $i = 1, \ldots, r$, may serve as a basis for an r-dimensional subspace of $\mathbf{V}$. Translating p_0 by all vectors in this subspace, one obtains an r-dimensional affine subspace, or a linear variety, of $\mathscr{A}$.

The r-dimensional *simplex* generated by the sequence of points $p_0, \ldots, p_r$ is the subset of $\mathscr{A}$ defined by

$$s = \left\{ p_0 + \textstyle\sum_{i=1}^{r} x^i (p_i - p_0) \mid x^i \geqslant 0,\ \textstyle\sum_{i=1}^{r} x^i \leqslant 1 \right\}. \tag{3.3}$$

Let O be some origin and $w_i = p_i - O$, $i = 0, \ldots, r$. Then,

$$\begin{aligned} p_0 + \textstyle\sum_{i=1}^{r} x^i (p_i - p_0) &= O + w_0 + \textstyle\sum_{i=1}^{r} x^i (p_i - p_0), \\ &= O + w_0 + \textstyle\sum_{i=1}^{r} x^i (w_i - w_0), \\ &= O + \left(1 - \textstyle\sum_{i=1}^{r} x^i\right) w_0 + \textstyle\sum_{i=1}^{r} x^i w_i, \end{aligned} \tag{3.4}$$

where the second line follows from Eq. (3.1). Setting $x^0 = 1 - \sum_{i=1}^{r} x^i$, we obtain

$$p_0 + \textstyle\sum_{i=1}^{r} x^i (p_i - p_0) = O + \textstyle\sum_{j=0}^{r} x^j w_j. \tag{3.5}$$

Since the left-hand side of the last equation is independent of the choice of O, the same holds for the right-hand side, and we use the notation

$$\textstyle\sum_{j=0}^{r} x^j p_j, \tag{3.6}$$

for both. Thus, one may define a simplex alternatively as

$$s = \left\{ \textstyle\sum_{i=0}^{r} x^i p_i \mid x^i \geqslant 0,\ \textstyle\sum_{i=0}^{r} x^i = 1 \right\}. \tag{3.7}$$

For a point $p \in s$ such that $p = \sum_{i=0}^{r} x^i p_i$, the numbers x^i are the *barycentric coordinates* of p.

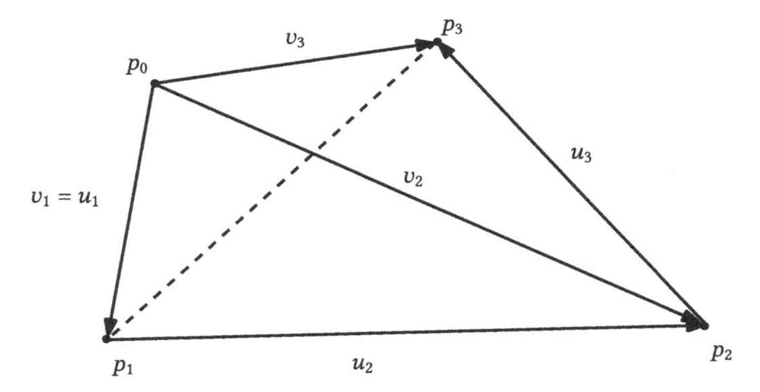

Fig. 3.1 Illustrating a 3-simplex

Clearly, the simplex s is defined equivalently by either (1) the collection of points $p_0, \ldots, p_r$, (2) the point p_0 and the collection of vectors $v_i = p_i - p_0, i = 1, \ldots, r$, and (3) the point p_0 together with the collection of vectors $u_i = p_i - p_{i-1}, i = 1, \ldots, r$. Using the convention $v_0 = 0$, one has $u_i = v_i - v_{i-1}$, $v_i = \sum_{j=1}^{i} u_j$ for $i = 1, \ldots, r$. (See Fig. 3.1 for illustration.) Since

$$\begin{aligned} s &= \left\{p_0 + \textstyle\sum_{i=1}^{r} x^i v_i \mid x^i \geqslant 0,\ \sum_{i=1}^{r} x^i \leqslant 1\right\}, \\ &= \left\{p_0 + \textstyle\sum_{i=1}^{r} x^i \left(\sum_{j=1}^{i} u_j\right) \mid x^i \geqslant 0,\ \sum_{i=1}^{r} x^i \leqslant 1\right\}, \\ &= \left\{p_0 + \textstyle\sum_{i=1}^{r} \left(\sum_{j=i}^{r} x^i\right) u_i \mid x^i \geqslant 0,\ \sum_{i=1}^{r} x^i \leqslant 1\right\}, \end{aligned} \tag{3.8}$$

one can write

$$s = \left\{p_0 + \textstyle\sum_{i=1}^{r} y^i u_i \mid 0 \leqslant y^i \leqslant 1,\ y^{i+1} \leqslant y^i\right\}. \tag{3.9}$$

For the simplex s as above, we will use the notation

$$s = [p_0, \ldots, p_r]. \tag{3.10}$$

In case we are considering a fixed point p_0 or ignoring it, we will also write (abusing the notation)

$$s = [v_1, \ldots, v_r] = [u_1, \ldots, u_r] \tag{3.11}$$

for the two alternative specifications of the simplex.

3.3 Cubes, Prisms, and Simplices

Consider the collection of points $p_0, \ldots, p_r \in \mathcal{A}$, such that the vectors $u_i = p_i - p_0 \in \mathbf{V}$, $i = 1, \ldots, r$, are linearly independent. The r-dimensional *cube*[1] generated by the points $p_0, \ldots, p_r \in \mathcal{A}$, contains all the points of the form

$$p_0 + \textstyle\sum_{i=1}^{r} x^i u_i, \quad 0 \leqslant x^i \leqslant 1. \tag{3.12}$$

In analogy, one may consider the open cube for which $0 < x^i < 1$. (See an illustration in Fig. 3.2.)

The points in the cube may be subdivided into prisms. Given an ordered pair (i, j) of indices $0 \leqslant i, j \leqslant r$, the prism P_{ij} is the collection of points of the cube such that $x^i \geqslant x^j$. (See an illustration in Fig. 3.3.)

The points of the prism may be subdivided further by ordering the rest of the components. Thus, for a sequence of indices $i_1, \ldots, i_r$ such that $i_j = 1, \ldots, r$ and $i_j \neq i_k$ for $j \neq k$, one may consider the set

$$s_{i_1 \ldots i_r} = \left\{ p_0 + \textstyle\sum_{j=1}^{r} x^{i_j} u_{i_j} \mid 1 \geqslant x^{i_j} \geqslant 0,\ x^{i_1} \geqslant x^{i_2} \geqslant \cdots \geqslant x^{i_r} \right\}. \tag{3.13}$$

(See an illustration in Fig. 3.4.)

We will use the notation Π_r for the collection of permutations of r ordered symbols (taken as the set $\{1, \ldots, r\}$). Thus, every permutation $\pi \in \Pi_r$ is a bijection of the form

$$\pi : \{1, \ldots, r\} \longrightarrow \{1, \ldots, r\}. \tag{3.14}$$

We will often view the mapping π as a mapping of sequences. We will use the notation π_i for $\pi(i)$, $i = 1, \ldots, r$ and write

$$\pi : (1, \ldots, r) \longmapsto (\pi_1, \ldots, \pi_r) = (\pi(1), \ldots, \pi(r)). \tag{3.15}$$

The number i' occupies the i-th position in the n-tuple $(\pi(1), \ldots, \pi(r))$ if and only if $\pi(i) = i'$. In other words, the number i' occupies the i-th position in this n-tuple if and only if $i = \pi^{-1}(i')$. Thus, $i = \pi^{-1}(i')$ is the "new" position of $i' \in \{1, \ldots, r\}$.

A sequence of r distinct vectors chosen from $u_1, \ldots, u_r$ is associated with a permutation π by

$$u_{\pi(1)}, \ldots, u_{\pi(i)}, \ldots, u_{\pi(r)} \tag{3.16}$$

so that $\pi(i)$ indicates the vector that occupies the i-th position in the sequence. Conversely, $\pi^{-1}(j)$ is the position of u_j in the sequence.

[1] The term *parallelepiped* would describe it better as lengths and angles are meaningless in this setting.

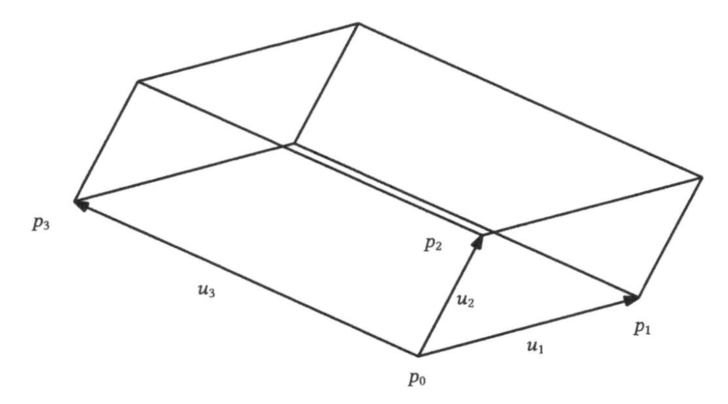

Fig. 3.2 Illustrating a cube

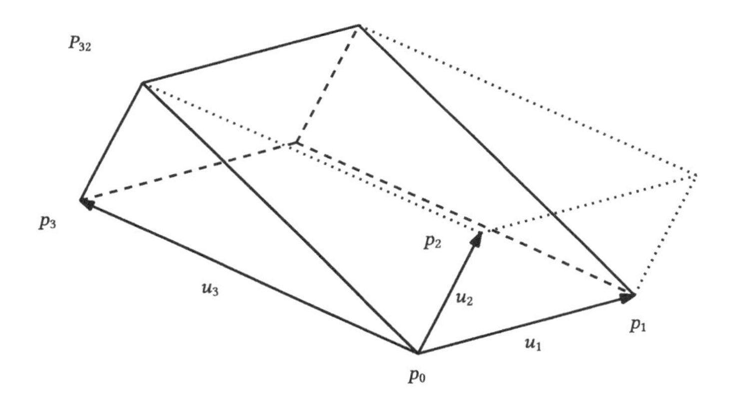

Fig. 3.3 Illustrating a prism

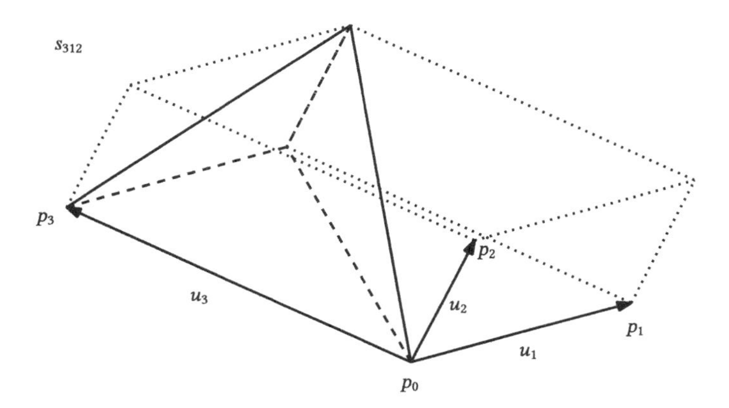

Fig. 3.4 Illustrating a typical simplex in a cube

Formally, one could define the mapping $\widehat{u} : \{1, \dots, r\} \to \mathbf{V}$ whereby $\widehat{u}(i) = u_i$. One may pullback $\widehat{u}$ by the mapping π to obtain $\pi^*(\widehat{u}) : \{1, \dots, r\} \to \mathbf{V}$, defined as $\pi^*(\widehat{u}) := \widehat{u} \circ \pi$. Thus,

$$\pi^*(\widehat{u})(i) = \widehat{u}(\pi(i)) = u_{\pi(i)}. \tag{3.17}$$

We can write (3.13) alternatively as

$$s_\pi = \left\{ p_0 + \textstyle\sum_{i=1}^r x^{\pi_i} u_{\pi_i} \mid 1 \geqslant x^{\pi_i} \geqslant 0,\ x^{\pi_1} \geqslant x^{\pi_2} \geqslant \cdots \geqslant x^{\pi_r} \right\}, \tag{3.18}$$

where $\pi_j = \pi(j) = i_j$.

Consider the set s_π. If one starts at p_0, and since x^{π_1} is not bounded by the other components, the path from p_0 to $p_{\pi_1} := p_0 + u_{\pi_1}$ is contained in the boundary of s_π. We will also use the notation $(s_\pi)_0 := p_0$ and $(s_\pi)_1 := p_{\pi_1}$, as these are the "origin" and first vertex of s_π. Now that $x^{\pi_1} = 1$, one can vary x^{π_2} from 0 to 1 to obtain the path from p_{π_1} to $(s_\pi)_2 = p_0 + u_{\pi_1} + u_{\pi_2} = (s_\pi)_1 + u_{\pi_2}$. Continuing this procedure r times, one obtains r points $(s_\pi)_i$, $i = 1, \dots, r$ such that

$$(s_\pi)_i = p_0 + \textstyle\sum_{j=1}^i u_{\pi_j}, \qquad u_{\pi_i} = (s_\pi)_i - (s_\pi)_{i-1}. \tag{3.19}$$

Clearly, each of these points is located at one of the vertices of the cube.

We will use the notation $v_{\pi_i} = (s_\pi)_i - (s_\pi)_0$ so that

$$(s_\pi)_i = p_0 + v_{\pi_i}, \qquad u_{\pi_i} = v_{\pi_i} - v_{\pi_{i-1}}. \tag{3.20}$$

The set s_π is the simplex determined by the points $p_0, (s_\pi)_1, (s_\pi)_2, \dots, (s_\pi)_r$. To realize that s_π is indeed a simplex, let

$$p = p_0 + \textstyle\sum_{i=1}^r x^{\pi_i} u_{\pi_i}, \tag{3.21}$$

so that $p \in s_\pi$ if an only if $1 \geqslant x^{\pi_i} \geqslant 0$, $x^{\pi_1} \geqslant x^{\pi_2} \geqslant \cdots \geqslant x^{\pi_r}$. Then, setting $v_0 = 0$ formally,

$$\begin{aligned}
p &= p_0 + \textstyle\sum_{i=1}^r x^{\pi_i}(v_{\pi_i} - v_{\pi_{i-1}}),\\
&= p_0 + \textstyle\sum_{i=1}^r x^{\pi_i} v_{\pi_i} - \sum_{i=1}^r x^{\pi_i} v_{\pi_{i-1}},\\
&= p_0 + \textstyle\sum_{i=1}^r x^{\pi_i} v_{\pi_i} - \sum_{i=0}^{r-1} x^{\pi_{i+1}} v_{\pi_i},\\
&= p_0 + \textstyle\sum_{i=1}^{r-1} (x^{\pi_i} - x^{\pi_{i+1}}) v_{\pi_i} + x^{\pi_r} v_{\pi_r},\\
&= p_0 + \textstyle\sum_{i=1}^r y^{\pi_i} v_{\pi_i}, \qquad y^{\pi_i} \geqslant 0, \qquad \sum_{i=1}^r y^{\pi_i} \leqslant 1.
\end{aligned} \tag{3.22}$$

Here, the numbers y^{π_i} are related to the x^{π_j} by $y^{\pi_i} = x^{\pi_i} - x^{\pi_{i+1}}$, $i = 1, \dots, r-1$, $y^{\pi_r} = x^{\pi_r}$, so that $y^{\pi_i} \geqslant 0$, $\sum_{i=1}^r y^{\pi_i} = x^{\pi_1} \leqslant 1$. Conversely, given $y^{\pi_i} \geqslant 0$ with

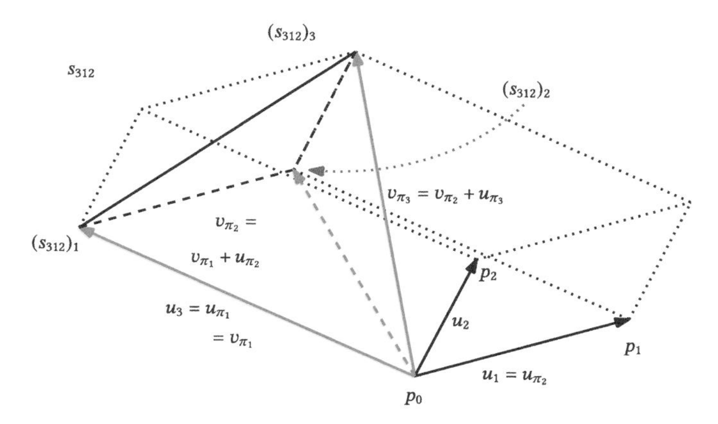

Fig. 3.5 Enumerating the vertices and vectors of a typical simplex in a cube

$\sum_{i=1}^{r} y^{\pi_i} \leqslant 1$, then, $x^{\pi_i} = \sum_{j=i}^{r} y^{\pi_j} \leqslant 1$ with $x^{\pi_i} \geqslant x^{\pi_{i+1}} \geqslant 0$. We conclude that s_π is indeed a simplex parametrized by the barycentric coordinates y^{π_i}. (See an illustration of the enumeration of the vertices and vectors in Fig. 3.5.)

The simplices s_π, for the various permutations $\pi \in \Pi_r$, cover the cube. They intersect one another on their boundaries where equality signs hold in (3.18), so that their interiors are disjoint. Since there are $r!$ permutations of r symbols, there are $r!$ simplices of this form in the cube.

Consider the prism P_{ij}, containing points for which $x^i \geqslant x^j$. In view of Eq. (3.18), a simplex s_π is contained in P_{ij} if the position of i precedes the position of j in the sequence $(\pi_1, \ldots, \pi_r)$, that is, if $\pi^{-1}(i) < \pi^{-1}(j)$. Evidently, there are $r!/2$ such simplices that will fit in P_{ij}—those corresponding to permutations π such that $\pi^{-1}(i) < \pi^{-1}(j)$.

3.4 Orientation

By an r-dimensional *hyperplane* in the affine space $\mathcal{A}$, we mean an r-dimensional affine subspace or $\mathcal{A}$.[2] An orientation in an r-dimensional hyperplane is determined by an ordered collection of r linearly independent vectors $v_1, \ldots, v_r$ in the translation space of the hyperplane. As these vectors may be used as a basis, any other collection of r linearly independent vectors, e.g., $u_1, \ldots, u_r$, may be represented by a matrix A such that $u_i = \sum_{j=1}^{r} A_i^j v_j$. If the determinant satisfies $\det(A_i^j) > 0$, one says that the collection of vectors $u_1, \ldots, u_r$ *determines the same orientation as* $v_1, \ldots, v_r$, or alternatively, one says that the basis $u_1, \ldots, u_r$

[2] Note that the term linear variety is also used for an affine subspace and that, often, the term hyperplane is used only for the case $r = n - 1$.

has positive orientation relative to the basis $v_1, \ldots, v_r$. Thus, any other basis of the translation space of the hyperplane may be either of a positive orientation or a negative orientation relative to $v_1, \ldots, v_r$. Using the elementary properties of determinants, it follows that switching the position of two vectors in a list reverses the orientation of the collection.

Given a permutation π, we use the notation $|\pi|$ for the number of swaps, or transpositions, required to transform $1, \ldots, r$ into $\pi_1, \ldots, \pi_r$. Clearly, $|\pi^{-1}| = |\pi|$. We will use the notation

$$(-1)^{|\pi|} = \operatorname{sign} \pi = \varepsilon^{\pi_1 \ldots \pi_r}_{1 \ldots\ldots r}, \tag{3.23}$$

where $\varepsilon^{\pi_1 \ldots \pi_r}_{1 \ldots\ldots r}$ is the (generalized) Levi-Civita permutation symbol. As expected, one has

$$(-1)^{|\pi' \circ \pi|} = (-1)^{|\pi'| + |\pi|} = (-1)^{|\pi'|}(-1)^{|\pi|}, \tag{3.24}$$

for the composition of the permutation π' with the permutation π.

An ordered collection of vectors $v_1, \ldots, v_r$, $v_i = p_i - p_0$, defining a simplex, induces an orientation on the r-dimensional hyperplane containing it. It is easy to see that the collection of vectors $u_1, \ldots, u_r$, $u_i = p_i - p_{i-1} = v_i - v_{i-1}$, where we use the convention that $v_0 = 0$, has the same orientation. The *orientation of the simplex* is the orientation determined by any one of these collections of vectors. Henceforth, we will refer to a simplex together with a given orientation on it as an *oriented simplex*. In order to simplify the terminology, we will often omit the adjective "oriented."

Transposing any two points p_i and p_j, $i, j > 0$, in the simplex $s = [p_0, p_1, \ldots, p_r]$ is equivalent to transposing the v_i and v_j. Hence, a transposition of the points will reverse the orientation. Similarly, it can be easily verified that a transposition of any point p_i, $i > 0$ with p_0 reverses the orientation. Thus, any transposition of a pair of points defining a simplex will reverse its orientation.

Given an oriented simplex s, the oriented simplex containing the same set of points but having the opposite orientation is denoted by $-s$. Two simplices are defined to be equal if they have the same vertices and the same orientation. Clearly, two equal simplices contain the same set of points. We may write, therefore,

$$[v_{\pi_1}, \ldots, v_{\pi_r}] = (-1)^{|\pi|}[v_1, \ldots, v_r], \quad \pi : \{1, \ldots, r\} \to \{1, \ldots, r\}, \tag{3.25}$$

$$[p_{\pi_0}, \ldots, p_{\pi_r}] = (-1)^{|\pi|}[p_0, \ldots, p_r], \quad \pi : \{0, \ldots, r\} \to \{0, \ldots, r\}. \tag{3.26}$$

It is noted that Eq. (3.11) still holds as the vectors v_i and the vectors u_i induce the same orientation on a simplex.

3.5 Simplices on the Boundaries and Their Orientations

Given an oriented simplex $s = [p_0, p_1, \dots, p_r]$ of dimension r, a collection of $r+1$ oriented simplices of dimension $r-1$ is obtained by omitting, each time, one of the vertices. Thus, denoting by a superimposed "hat" an omitted item, for each $i = 0, \dots, r$, the oriented $(r-1)$-dimensional simplex τ_i is defined by omitting the i-th vertex as

$$\tau_i = (-1)^i [p_0, \dots, \widehat{p_i}, \dots, p_r]. \tag{3.27}$$

The oriented simplex τ_i is referred to as the i-th *face* of the simplex s. (See an illustration in Fig. 3.6.) In general, faces, or edges, of lower dimensions may be constructed by omitting additional vertices.

In terms of the vectors $v_i = p_i - p_0$, we observe that τ_0 is the simplex starting at p_1 and determined by $v_i - v_1$. Using the notation of Eq. (3.11),

$$\tau_0 = [v_2 - v_1, \dots, v_r - v_1], \quad \text{and} \quad \tau_i = (-1)^i [v_1, \dots, \widehat{v_i}, \dots, v_r], \quad i = 1, \dots, r. \tag{3.28}$$

Using the vectors $u_i = p_i - p_{i-1}$, one has

$$\begin{aligned} \tau_0 &= [u_2, \dots, u_r], \\ \tau_i &= (-1)^i [u_1, \dots, u_{i-1}, u_i + u_{i+1}, u_{i+2}, \dots, u_r], \quad i = 1, \dots, r-1, \\ \tau_r &= (-1)^r [u_1, \dots, u_{r-1}]. \end{aligned} \tag{3.29}$$

The collection of simplices $\tau_0, \dots, \tau_r$ is referred to as the *boundary of the simplex* and it will be denoted by ∂s.

Note that $[p_{\pi_0}, \dots, p_{\pi_r}] = (-1)^{|\pi|} [p_0, \dots, p_r]$ implies that

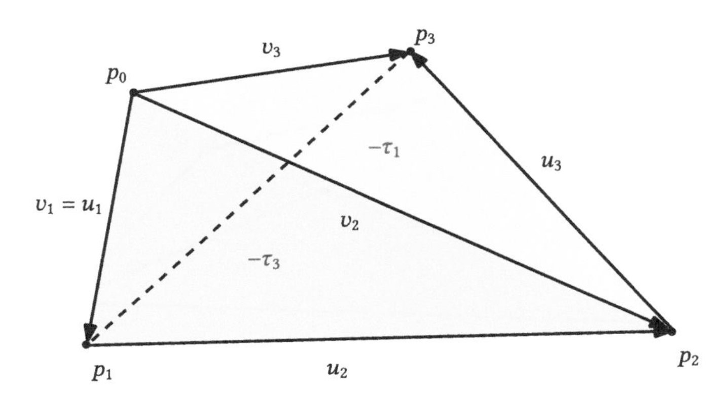

Fig. 3.6 Illustrating faces of a simplex

$$s = [p_0, \ldots, p_r] = (-1)^i [p_i, p_0, \ldots, \widehat{p_i}, \ldots, p_r]. \tag{3.30}$$

With the definition of τ_i in (3.27), the last representation provides some motivation for the sign of τ_i.

3.6 Subdivisions

Simplices may be subdivided into simplices. A subdivision of an r-simplex s is a collection of r-simplices $s_1, \ldots, s_A$ having the following properties (see [Kur72, p. 261]):

- $s = \bigcup_{a=1}^{A} s_a$.
- $s_1, \ldots, s_A$ and s induce the same orientation on the subspace containing them.
- The interiors of $s_1, \ldots, s_A$ are disjoint, and any two simplices intersect on faces.

If $s_1, \ldots, s_A$ is a subdivision of s, we will write $s = s_1 + \cdots + s_A$.

The following example will be used below and is described therefore in some detail. Consider the simplex $s = [p_0, \ldots, p_r]$ and let

$$q = \tfrac{1}{2}(p_0 + p_1) = p_0 + \tfrac{1}{2}(p_1 - p_0) \tag{3.31}$$

be the midpoint between p_0 and p_1. One may define the simplices $s_1 = [p_0, q, p_2, \ldots, p_r]$ and $s_2 = [q, p_1, \ldots, p_r]$. (See an illustration in Fig. 3.7.) Each point y in s_1 is of the form

$$y = y^0 p_0 + \tfrac{1}{2} y^1 (p_0 + p_1) + y^2 p_2 + \cdots + y^r p_r, \quad y^i \geqslant 0,\ \textstyle\sum_{i=1}^{r} y^i = 1. \tag{3.32}$$

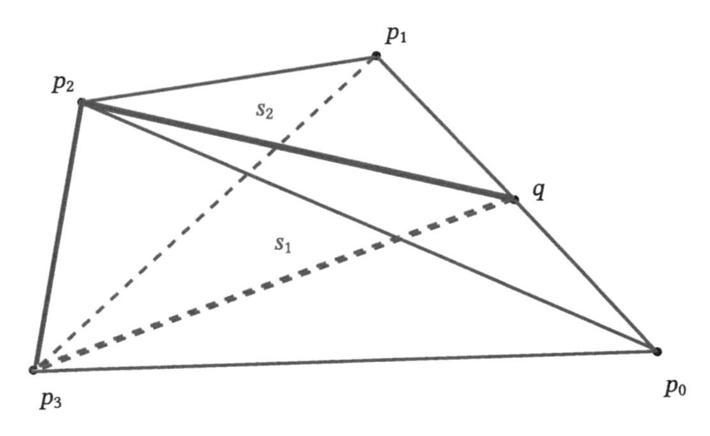

Fig. 3.7 Illustrating a subdivision of a simplex

Rearranging the terms, we have

$$y = (y^0 + \tfrac{1}{2}y^1)p_0 + \tfrac{1}{2}y^1 p_1 + y^2 p_2 + \cdots + y^r p_r. \tag{3.33}$$

Setting $x^0 = y^0 + \frac{1}{2}y^1$, $x^1 = \frac{1}{2}y^1$, $x^2 = y^2, \ldots, x^r = y^r$, we observe that $\sum x^i = 1$, $x^i \geqslant 0$, $x^0 \geqslant x^1$ and $y = \sum_{i=0}^r x^i p_i$. Thus, each point in s_1 belongs also to s. Similarly, any $z \in s_2$ is of the form

$$z = \tfrac{1}{2}z^0(p_0 + p_1) + z^1 p_1 + z^2 p_2 + \cdots + z^r p_r, \quad z^i \geqslant 0,\ \textstyle\sum_{i=1}^r z^i = 1. \tag{3.34}$$

Rearranging the terms we have

$$z = \tfrac{1}{2}z^0 p_0 + (\tfrac{1}{2}z^0 + z^1)p_1 + z^2 p_2 + \cdots + z^r p_r, \tag{3.35}$$

and setting $x^0 = \frac{1}{2}z^0$, $x^1 = \frac{1}{2}z^0 + z^1$, $x^2 = z^2, \ldots, x^r = z^r$, we observe that $\sum x^i = 1$, $x^i \geqslant 0$, $x^0 \leqslant x^1$ and $z = \sum_{i=0}^r x^i p_i$. Thus, each point in s_2 belongs to s, too.

In addition, any point $x = \sum x^i p_i$ in s belongs to s_1 if $x^0 > x^1$, it belongs to s_2 if $x^0 < x^1$, and it belongs to both simplices if $x^0 = x^1$. If $x^0 = x^1$, then, $y^0 = 0$, $z^1 = 0$, and so x belongs to the face $[q, p_2, \ldots, p_r]$ in ∂s_1 and to the face $(-1)^1[q, p_2, \ldots, p_r]$ of ∂s_2.

It is also noted that if $u_j = p_j - p_{j-1}$, $j = 1, \ldots, r$, are the vectors determining s, then,

$$s_1 = [\tfrac{1}{2}u_1, \tfrac{1}{2}u_1 + u_2, u_3, \ldots, u_r], \quad \text{and} \quad s_2 = [\tfrac{1}{2}u_1, u_2, \ldots, u_r]. \tag{3.36}$$

Clearly, the orientations of s_1 and s_2 are positive relative to s. We conclude that $s = s_1 + s_2$.

Using Eq. (3.29) and denoting the faces of s_1 and s_2 by τ_i^1 and τ_i^2, respectively, we have

$$\begin{aligned}
\tau_0^1 &= [\tfrac{1}{2}u_1 + u_2, u_3, \ldots, u_r],\\
\tau_1^1 &= (-1)^1[u_1 + u_2, u_3, \ldots, u_r],\\
\tau_2^1 &= (-1)^2[\tfrac{1}{2}u_1, \tfrac{1}{2}u_1 + u_2 + u_3, u_4, \ldots, u_r],\\
\tau_i^1 &= (-1)^i[\tfrac{1}{2}u_1, \tfrac{1}{2}u_1 + u_2, u_3, \ldots, u_i + u_{i+1}, u_{i+2}, \ldots, u_r],\ 2 < i < r,\\
\tau_r^1 &= (-1)^r[\tfrac{1}{2}u_1, \tfrac{1}{2}u_1 + u_2, u_3, \ldots, u_{r-1}],
\end{aligned} \tag{3.37}$$

and

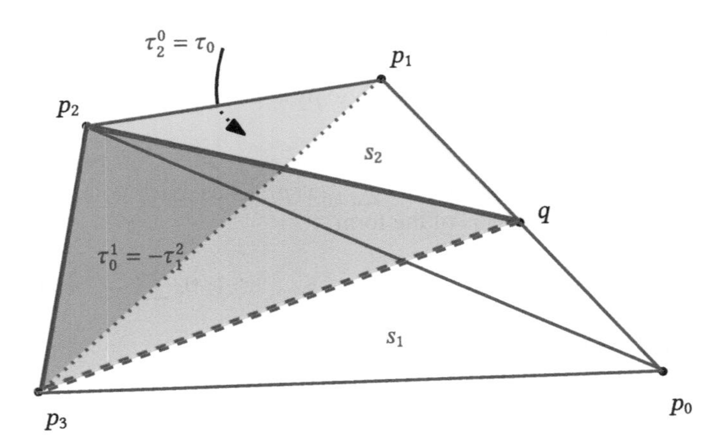

Fig. 3.8 The faces of the subdivided simplex

$$
\begin{aligned}
\tau_0^2 &= [u_2, u_3, \ldots, u_r], \\
\tau_1^2 &= (-1)^1[\tfrac{1}{2}u_1 + u_2, u_3, \ldots, u_r], \\
\tau_2^2 &= (-1)^2[\tfrac{1}{2}u_1, u_2 + u_3, u_4, \ldots, u_r], \\
\tau_i^2 &= (-1)^i[\tfrac{1}{2}u_1, u_2, u_3, \ldots, u_i + u_{i+1}, u_{i+2}, \ldots, u_r],\ \ 2 < i < r, \\
\tau_r^2 &= (-1)^r[\tfrac{1}{2}u_1, u_2, u_3, \ldots, u_{r-1}].
\end{aligned}
\tag{3.38}
$$

Using Eq. (3.29) again, we finally arrive at

$$
\begin{aligned}
\tau_0^1 &= -\tau_1^2, \\
\tau_1^1 &= \tau_1, \\
\tau_0^2 &= \tau_0, \\
\tau_2^1 + \tau_2^2 &= \tau_2, \\
\tau_i^1 + \tau_i^2 &= \tau_i,\ \ i > 2.
\end{aligned}
\tag{3.39}
$$

(See an illustration in Fig. 3.8.)

Chapter 4
Uniform Fluxes in Affine Spaces

Flux vector fields, or flow vector fields, are of fundamental significance in standard formulations of the foundations of continuum mechanics. Let P be a scalar extensive property, such as mass, electric charge, internal energy, etc., over some three-dimensional region $\mathcal{R}$. The property, P, is usually assumed to be an additive set function defined on a certain class of subregions of $\mathcal{R}$. Flux theory relates the time rate of change of the property $P(\mathcal{R}')$, for a generic subregion $\mathcal{R}'$, to the flux density $\varphi_{\mathcal{R}'}$—a scalar field defined on the boundary, $\partial\mathcal{R}'$, of $\mathcal{R}'$. Integration of the flux density over a subset S of the boundary yields the total flux, ω_S, which is understood as the rate of transfer of the property across S. Thus, the total flux may be viewed as a set function defined on two-dimensional oriented surfaces. The basic theorem of flux theory in continuum mechanics, the Cauchy flux theorem, gives conditions for the flux density function to be given in terms of a flux vector field J, defined globally on $\mathcal{R}$. It implies that the flux density at a point on the boundary is given by the Cauchy formula, $\varphi_{\mathcal{R}'} = J \cdot \mathbf{n}$, where $\mathbf{n}$ is the unit normal to the boundary of $\mathcal{R}'$ at that point.[1] In addition to the relevance of flux theory to scalar extensive properties, it provides a basis for the proof Cauchy theorem for the existence of the stress tensor.

A number of aspects of flux theory are generalized in this book. In the smooth case, the theory is formulated on n-dimensional differentiable manifolds devoid of a Riemannian metric. Subsequently, for the non-smooth case, irregular regions and flux distributions are considered. In this chapter, we start the study of Cauchy fluxes on manifolds, by preparing the geometric-algebraic tool needed for the smooth case—the algebraic version of the Cauchy theorem or the version for uniform fluxes. These are analogous to the case where the flux vector field is uniform and the class of regions includes polyhedral chains. Thus, we propose a generalization of the tetrahedron argument for the case of higher dimension general affine spaces. Making several assumptions as to the properties of the total flux set function, it is shown that

[1] To simplify the notation, we use here the same symbols for functions and their values.

R. Segev, *Foundations of Geometric Continuum Mechanics*, Advances in Continuum Mechanics 49, https://doi.org/10.1007/978-3-031-35655-1_4

the total flux for a simplex is multilinear and completely skew-symmetric in the vectors defining the simplex.

4.1 Basic Assumptions

By a *uniform flux* or an *algebraic* flux, we mean a real-valued function defined on the collection of all $(r-1)$-oriented simplices in the n-dimensional affine space $\mathcal{A}$, satisfying some postulates. For a uniform flux ω and an oriented $(r-1)$-simplex τ, the value $\omega(\tau)$ may be interpreted as the total flux of a certain property that flows through the simplex τ relative to the orientation specified. Note also that the simplices may be of any dimension $r \leqslant n-1$. For example, one may consider the flux through lines in a three-dimensional space, or two-dimensional hyperplanes in a five-dimensional space.

Temporarily, we suspend the convention exhibited by Eq. (3.11) whereby a simplex $s=[p_0,\dots,p_r]$ may be written as $s=[v_1,\dots,v_r]$ or $s=[u_1,\dots,u_r]$ and for the flux through the simplex s we will write $\omega([p_0,\dots,p_r])$.

Assumption 1 (Translation Invariance) Consider $\tau = [p_0,\dots,p_{r-1}]$; then, $\omega(\tau)$ is invariant under translation, i.e.,

$$\omega([p_0,\dots,p_{r-1}]) = \omega([p_0+w,\dots,p_{r-1}+w]) \tag{4.1}$$

for any vector $w \in \mathbf{V}$.

It is observed that $\tau=[p_0,\dots,p_{r-1}]$ and $\tau+w=[p_0+w,\dots,p_{r-1}+w]$ share the same vectors $u_i = p_i - p_{i-1}$ and $v_i = p_i - p_0$. Thus, given a function $\hat{\omega}$ of $r-1$ vectors, one may define $\omega([p_0,\dots,p_{r-1}]) = \hat{\omega}(p_1-p_0, p_2-p_1,\dots,p_{r-1}-p_{r-2})$ which is clearly invariant under translation. Conversely, we may write

$$\begin{aligned}\omega([p_0,\dots,p_{r-1}]) &= \omega\left(\left[p_0, p_0+u_1,\dots,p_0+\textstyle\sum u_i\right]\right),\\ &= \omega\left(\left[p_0+w, p_0+w+u_1,\dots,p_0+w+\textstyle\sum u_i\right]\right).\end{aligned} \tag{4.2}$$

Since the last expression is independent of w, $\omega([p_0,\dots,p_{r-1}])$ depends only on $u_1,\dots,u_{r-1}$ and may be represented by a function $\hat{\omega}$ as above. In the sequel, we will omit the "hat" and will use $\omega(u_1,\dots,u_{r-1})$ to denote the corresponding value of the flux mapping.

Assumption 2 (Homogeneity) We recall (see Sect. 3.3) that one can fit $(r-1)!$ simplices into an $(r-1)$-cube. Thus, if we multiply one of the vectors u_i that determine a simplex by a positive number a, one side of the cube will be multiplied by the same number. It is natural therefore to assume that the flux will be multiplied by a also. In case $a=0$, then, $au_i=0$, and one does not have, any longer, a nontrivial $(r-1)$-simplex. Thus, formally, it is assumed that for all $a \geqslant 0$ and $i=1,\dots,r-1$,

$$\omega(u_1, \dots, u_{i-1}, au_i, u_{i+1}, \dots, u_{r-1}) = a\omega(u_1, \dots, u_{i-1}, u_i, u_{i+1}, \dots, u_{r-1}). \tag{4.3}$$

(The case $a = 0$ also follows from Eq. (4.9).) Evidently, it is implied that $\omega(u_1, \dots, u_{r-1}) = 0$ if any one of the vectors vanishes.

Let s be an r-simplex in $\mathcal{A}$ the boundary of which consists of the $(r-1)$-simplices τ_i, $i = 0, \dots, r$. Basic properties of uniform fluxes are concerned with the total flux out of the various faces on the boundary of s, i.e., with

$$\omega(\partial s) := \textstyle\sum_{i=0}^{r} \omega(\tau_i). \tag{4.4}$$

Assumption 3 (Additivity Under Subdivision) Let τ be an oriented $(r-1)$-simplex and let $\tau_1, \dots, \tau_A$ be a subdivision of τ. We assume that

$$\omega(\tau) = \textstyle\sum_{a=1}^{A} \omega(\tau_a). \tag{4.5}$$

Let s be an oriented r-simplex and let $s_1, \dots, s_B$ be a subdivision of s. It is assumed that

$$\omega(\partial s) = \textstyle\sum_{b=1}^{B} \omega(\partial s_b). \tag{4.6}$$

This assumption, associated with the locality of the flux operator, states that the total flux out of s is the sum of the fluxes out of its various parts.

We now apply the additivity property for r-simplices, (4.6), to the subdivision, s_1, s_2, of a simplex s as in the example presented in Sect. 3.6. Using Eq. (3.39), we obtain

$$\begin{aligned} \omega(\tau_0) + \omega(\tau_1) + \textstyle\sum_{i=2}^{r} \omega(\tau_r) &= \omega(\tau_0^1) + \omega(\tau_1^1) + \textstyle\sum_{i=2}^{r} \omega(\tau_r^1) \\ &\quad + \omega(\tau_0^2) + \omega(\tau_1^2) + \textstyle\sum_{i=2}^{r} \omega(\tau_r^2), \\ &= \omega(-\tau_1^2) + \omega(\tau_1) + \omega(\tau_0) + \omega(\tau_1^2) + \textstyle\sum_{i=2}^{r} \omega(\tau_r), \end{aligned} \tag{4.7}$$

where we used the additivity property (4.5) to obtain the last line above. It follows immediately that

$$\omega(\tau_1^2) = -\omega(-\tau_1^2). \tag{4.8}$$

Since for any oriented $(r-1)$-simplex τ one may construct a simplex $\tilde{s}$ with faces $\tilde{\tau}_i$ and subdivision $\tilde{s} = \tilde{s}_1 + \tilde{s}_2$ such that $\tilde{\tau}_1^2 = \tau$, we conclude that for any oriented $(r-1)$-simplex τ,

$$\omega(\tau) = -\omega(-\tau). \tag{4.9}$$

Next, we make use of the additivity assumption of Eq. (4.5) for $(r-1)$-simplices and apply it to a subdivision $\tau = \tau_1 + \tau_2$, $\tau = [p_0, \dots, p_{r-1}]$,

$\tau_1 = [p_0, q, p_2, \ldots, p_{r-1}]$, $\tau_2 = [q, p_1, \ldots, p_{r-1}]$, as in Sect. 3.6 but pertaining to $(r-1)$-simplices. Keeping the same notation, the simplices are determined by the vectors u_i, $i = 1, \ldots, r-1$. Thus, $\omega(\tau) = \omega(\tau_1) + \omega(\tau_2)$ implies that

$$\begin{aligned}\omega(u_1, \ldots, u_{r-1}) &= 2\omega(\tfrac{1}{2}u_1, u_2, \ldots, u_{r-1}), \\ &= \omega(\tfrac{1}{2}u_1, \tfrac{1}{2}u_1 + u_2, u_3, \ldots, u_{r-1}) + \omega(\tfrac{1}{2}u_1, u_2, \ldots, u_{r-1}),\end{aligned} \tag{4.10}$$

where in the first line we used the homogeneity property and in the second line we used the subdivision (3.36) and the additivity property. Since u_1 is arbitrary, we obtain

$$\omega(u_1, u_1 + u_2, u_3, \ldots, u_{r-1}) = \omega(u_1, u_2, u_3, \ldots, u_{r-1}). \tag{4.11}$$

We can re-enumerate the points on the simplex, or equivalently, we can make a subdivision along a different vector. Hence,

$$\omega(u_1, \ldots, u_{i-1}, u_{i-1}+u_i, u_{i+1}, \ldots, u_{r-1}) = \omega(u_1, \ldots, u_{i-1}, u_i, u_{i+1}, \ldots, u_{r-1}). \tag{4.12}$$

In particular, since the flux vanishes if any of the vectors determining the simplex vanishes,

$$\omega(u_1, \ldots, u_i, u_i, u_{i+2}, \ldots, u_{r-1}) = 0. \tag{4.13}$$

Using homogeneity, we also have

$$\begin{aligned}&\omega(u_1, \ldots, u_{i-1}, au_{i-1} + u_i, u_{i+1}, \ldots, u_{r-1}) \\ &\quad= \frac{1}{a}\omega(u_1, \ldots, au_{i-1}, au_{i-1} + u_i, u_{i+1}, \ldots, u_{r-1}), \\ &\quad= \frac{1}{a}\omega(u_1, \ldots, au_{i-1}, u_i, u_{i+1}, \ldots, u_{r-1}),\end{aligned} \tag{4.14}$$

and therefore,

$$\begin{aligned}&\omega(u_1, \ldots, u_{i-1}, au_{i-1} + u_i, u_{i+1}, \ldots, u_{r-1}) \\ &\quad= \omega(u_1, \ldots, u_{i-1}, u_i, u_{i+1}, \ldots, u_{r-1}).\end{aligned} \tag{4.15}$$

Next, consider the vectors $v_i = p_i - p_0 = \sum_{j=1}^{i} u_j$, $i = 1, \ldots, r-1$. Using Eq. (4.12) repeatedly,

$$\omega(v_1, \ldots, v_{r-1}) = \omega\left(u_1, u_1 + u_2, \ldots, \sum_{j=1}^{r-2} u_j, \sum_{j=1}^{r-1} u_j\right),$$
$$= \omega\left(u_1, u_1 + u_2, \ldots, \sum_{j=1}^{r-3} u_j, \sum_{j=1}^{r-2} u_j, u_{r-1}\right), \tag{4.16}$$
$$= \omega\left(u_1, u_1 + u_2, \ldots, \sum_{j=1}^{r-3} u_j, u_{r-2}, u_{r-1}\right)$$

etc. It is implied that

$$\omega(v_1, \ldots, v_{r-1}) = \omega(u_1, \ldots, u_{r-1}). \tag{4.17}$$

Since $\tau_\pi = [p_0, p_0 + v_{\pi_1}, \ldots, p_0 + v_{\pi_{r-1}}] = (-1)^{|\pi|}[p_0, p_0 + v_1, \ldots, p_0 + v_{r-1}]$, Eqs. (4.9) and (4.17) give

$$\omega(v_{\pi_1}, \ldots, v_{\pi_{r-1}}) = (-1)^{|\pi|}\omega(v_1, \ldots, v_{r-1}),$$
$$\omega(u_{\pi_1}, \ldots, u_{\pi_{r-1}}) = (-1)^{|\pi|}\omega(u_1, \ldots, u_{r-1}), \tag{4.18}$$

which makes the flux mapping an *alternating* function of its arguments. It is noted that Eqs. (4.13) and (4.18) imply that the total flux vanishes if some of the vectors are recurring (in which case the simplex collapses).

Remark 4.1 For the case where $r - 1 = n$, the assumptions above are sufficient in order to prove that the action of the flux mapping is additive. Consider, for example, the $n + 1$ vectors $v'_1, v_1, \ldots, v_n$. Assume that $v_1, \ldots, v_n$ are linearly independent so that we may write $v'_1 = \sum_i a^i v_i$. Using Eq. (4.15) repeatedly, one has

$$\omega(v'_1, v_2, \ldots, v_n) = \omega\left(\sum_i a^i v_i, v_2, \ldots, v_n\right),$$
$$= \omega(a^1 v_1, v_2, \ldots, v_n), \tag{4.19}$$
$$= a^1 \omega(v_1, v_2, \ldots, v_n),$$

$$\omega(v_1 + v'_1, v_2, \ldots, v_n) = \omega\left(v_1 + \sum_i a^i v_i, v_2, \ldots, v_n\right),$$
$$= \omega\left(v_1 + a^1 v_1, v_2, \ldots, v_n\right), \tag{4.20}$$
$$= (1 + a^1)\omega(v_1, v_2, \ldots, v_n),$$

and so

$$\omega(v_1 + v'_1, v_2, \ldots, v_n) = \omega(v_1, v_2, \ldots, v_n) + \omega(v'_1, v_2, \ldots, v_n). \tag{4.21}$$

Assumption 4 (Balance) Let s be an r-simplex in $\mathscr{A}$, the boundary of which consists of the $(r - 1)$-simplices τ_i, $i = 0, \ldots, r$. It is assumed that

$$\omega(\partial s) := \textstyle\sum_{i=0}^{r}\omega(\tau_i) = 0. \tag{4.22}$$

Thus, we assume that the sum of all fluxes through the various faces of a simplex vanishes.

Remark 4.2 It is noted that the assumption of balance as in Eq. (4.22) implies the additivity property under subdivision of Eq. (4.6). Thus, once balance is assumed, the additivity under subdivision assumption of Eq. (4.6) is redundant.

4.2 Balance and Linearity

In this section, we prove the second basic property of uniform fluxes: the assumption of balance implies multi-linearity. The notation Π' will be used for the collection of all permutations π of r-symbols such that $\pi^{-1}(1) < \pi^{-1}(2)$, i.e., u_1 precedes u_2 in the list of vectors $u_{\pi(1)}, u_{\pi(2)}, \dots, u_{\pi(r)}$. For each $\pi \in \Pi'$, let $s_\pi = (-1)^{|\pi|}[u_{\pi_1}, \dots, u_{\pi_r}]$ so that all the simplices s_π are of the same orientation, and their union is the prism $x^1 \geqslant x^2$ as in Sect. 3.3. (See an illustration in Fig. 4.1.) The assumption of balance, together with Eqs. (3.29) and (4.9), implies that

$$\begin{aligned} 0 = \omega(\partial s_\pi) = (-1)^{|\pi|}\Bigg[&\omega(u_{\pi_2}, \dots, u_{\pi_r}) \\ &+ \sum_{i=1}^{r-1}(-1)^i\omega(u_{\pi_1}, \dots, u_{\pi_{i-1}}, u_{\pi_i} + u_{\pi_{i+1}}, u_{\pi_{i+2}}, \dots, u_{\pi_r}) \\ &+ (-1)^r\omega(u_{\pi_1}, \dots, u_{\pi_{r-1}})\Bigg]. \end{aligned} \tag{4.23}$$

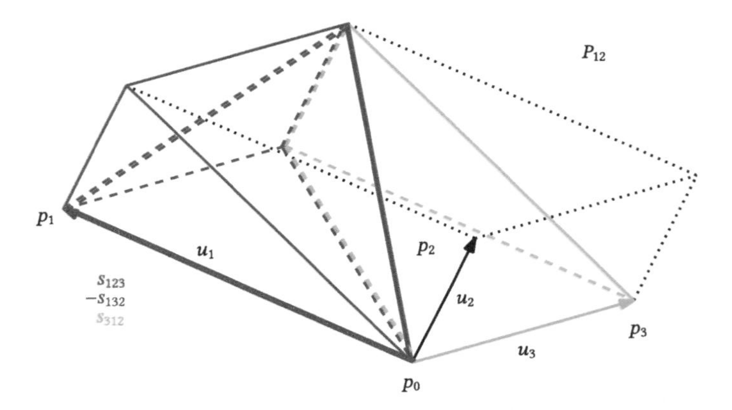

Fig. 4.1 Subdividing the prism P_{12}

The sum of the variants of the equation above over all $\pi \in \Pi'$ gives

$$0 = \sum_{\pi \in \Pi'} (-1)^{|\pi|} \omega(u_{\pi_2}, \ldots, u_{\pi_r}) + \sum_{\pi \in \Pi'} (-1)^{|\pi|} (-1)^r \omega(u_{\pi_1}, \ldots, u_{\pi_{r-1}})$$
$$+ \sum_{\pi \in \Pi'} \sum_{i=1}^{r-1} (-1)^{|\pi|} (-1)^i \omega(u_{\pi_1}, \ldots, u_{\pi_{i-1}}, u_{\pi_i} + u_{\pi_{i+1}}, u_{\pi_{i+2}}, \ldots, u_{\pi_r}). \tag{4.24}$$

Denoting the three terms on the right-hand side of the last equation by t_a, t_b, and t_c, respectively, it may be written in the form

$$0 = t_a + t_b + t_c. \tag{4.25}$$

Consider first the terms in the sum of t_a for which $\pi_1 = 1$. Using the fact that ω is alternating, i.e., Eq. (4.18), one has

$$\begin{aligned} \sum_{\pi \in \Pi', \pi_1 = 1} (-1)^{|\pi|} \omega(u_{\pi_2}, \ldots, u_{\pi_r}) &= \sum_{\pi \in \Pi', \pi_1 = 1} (-1)^{2|\pi|} \omega(u_2, \ldots, u_r), \\ &= (r-1)! \omega(u_2, \ldots, u_r), \end{aligned} \tag{4.26}$$

as the positions of $u_2, \ldots, u_r$ are not constrained by the requirement that $\pi^{-1}(1) < \pi^{-1}(2)$.

Similarly, consider the terms in t_b such that $\pi_r = 2$. Note that any permutation π for which $\pi(r) = 2$ may be factored in the form $\pi = \pi^{(2)} \circ \pi^{(1)}$, where $\pi^{(1)} : (1, 2, \ldots, r) \mapsto (1, 3, \ldots, r, 2)$, so that $\left|\pi^{(1)}\right| = r - 2$, and $\pi^{(2)} : \{1, 2, \ldots, r-1, r\} \mapsto \{\pi_1, \pi_2, \ldots, \pi_{r-1}, r\}$, keeping the last term invariant. Hence, $|\pi| = \left|\pi^{(2)}\right| + \left|\pi^{(1)}\right| = \left|\pi^{(2)}\right| + r - 2$. One has

$$\begin{aligned} &\sum_{\pi \in \Pi', \pi_r = 2} (-1)^{|\pi|} (-1)^r \omega(u_{\pi_1}, \ldots, u_{\pi_{r-1}}) \\ &\quad = \textstyle\sum_{\pi \in \Pi', \pi_r = 2} (-1)^{|\pi| + r} (-1)^{\left|\pi^{(2)}\right|} \omega(u_1, u_3, \ldots, u_r), \\ &\quad = \textstyle\sum_{\pi \in \Pi', \pi_r = 2} (-1)^{2\left|\pi^{(2)}\right| + 2r - 2} \omega(u_1, u_3, \ldots, u_r), \\ &\quad = (r-1)! \omega(u_1, u_3, \ldots, u_r). \end{aligned} \tag{4.27}$$

Again, the fact that u_2 is in the last position implies that there is no restriction on the positions of the other vectors.

Let $\Pi_1 \subset \Pi'$ be the collection of permutations not included in (4.26), i.e., permutations π for which $\pi^{-1}(1) < \pi^{-1}(2)$ and $\pi_1 \neq 1$. Similarly, let $\Pi_2 \subset \Pi'$ be the collection of permutations not included in (4.27), i.e., permutations π for which $\pi^{-1}(1) < \pi^{-1}(2)$ and $\pi_r \neq 2$. We now consider the sum of terms

$$t_a' + t_b' := \sum_{\pi \in \Pi_1} (-1)^{|\pi|} \omega(u_{\pi_2}, \ldots, u_{\pi_r}) + \sum_{\pi' \in \Pi_2} (-1)^{|\pi'|+r} \omega(u_{\pi_1'}, \ldots, u_{\pi_{r-1}'}) \tag{4.28}$$

in t_a and t_b not included above. Since $\pi_1 \neq 1$, both u_1 and u_2 must appear in each element of Π_1 and as $\pi_r \neq 2$ the two vectors must appear in each element of Π_2. Consider the permutation $\rho : (1, \ldots, r) \mapsto (2, \ldots, r, 1)$. For each permutation $\pi = \{\pi_1, \ldots, \pi_r\}$, set $\pi' = \rho \circ \pi : (1, \ldots, r) \mapsto (\pi_2, \ldots, \pi_r, \pi_1)$, such that $|\pi'| = |\rho \circ \pi| = |\pi| + r - 1$. It is noted that $\pi' = \rho \circ \pi$ satisfies the condition $\pi'^{-1}(1) < \pi'^{-1}(2)$ and so composition with ρ on the left, $\pi \mapsto \rho \circ \pi$, is a bijection of Π_1 onto Π_2. Hence,

$$\begin{aligned} t_a' + t_b' &= \sum_{\pi \in \Pi_1} \left[(-1)^{|\pi|} \omega(u_{\pi_2}, \ldots, u_{\pi_r}) + (-1)^{|\rho\circ\pi|+r} \omega(u_{(\rho\circ\pi)_1}, \ldots, u_{(\rho\circ\pi)_{r-1}}) \right], \\ &= \sum_{\pi \in \Pi_1} \left[(-1)^{|\pi|} \omega(u_{\pi_2}, \ldots, u_{\pi_r}) + (-1)^{|\pi|+2r-1} \omega(u_{\pi_2}, \ldots, u_{\pi_r}) \right], \\ &= 0. \end{aligned} \tag{4.29}$$

As a result

$$t_a + t_b = (r-1)! \left[\omega(u_2, \ldots, u_r) + \omega(u_1, u_3, \ldots, u_r) \right]. \tag{4.30}$$

Next, we compute

$$\begin{aligned} t_c &= \sum_{\pi \in \Pi'} \sum_{i=1}^{r-1} (-1)^{|\pi|} (-1)^i \omega(u_{\pi_1}, \ldots, u_{\pi_{i-1}}, u_{\pi_i} + u_{\pi_{i+1}}, u_{\pi_{i+2}}, \ldots, u_{\pi_r}), \\ &= \sum_{i=1}^{r-1} (-1)^i t_{ci}, \end{aligned} \tag{4.31}$$

where

$$t_{ci} := \sum_{\pi \in \Pi'} (-1)^{|\pi|} \omega(u_{\pi_1}, \ldots, u_{\pi_{i-1}}, u_{\pi_i} + u_{\pi_{i+1}}, u_{\pi_{i+2}}, \ldots, u_{\pi_r}). \tag{4.32}$$

For a fixed value of i, let

$$\rho_i : (1, \ldots, i-1, i, i+1, i+2, \ldots, r) \longmapsto (1, \ldots, i-1, i+1, i, i+2, \ldots, r) \tag{4.33}$$

be the permutation that swaps the positions of i and $i+1$. The permutation ρ_i acts on Π' by composition $\pi \mapsto \rho_i \circ \pi$ with $(-1)^{|\rho_i \circ \pi|} = (-1)^{|\pi|+1}$. In addition, ρ_i acts as an automorphism on the subset of permutations $\Pi_{3i} \subset \Pi'$ containing permutations for which we exclude the case where $\pi_i = 1$ and $\pi_{i+1} = 2$. That is, let $\Pi_{4i} = \{\pi \in \Pi' \mid \pi_i = 1,\ \pi_{i+1} = 2\}$; then, $\Pi_{3i} = \Pi' \setminus \Pi_{4i}$. Setting

$$t'_{ci} = \sum_{\pi \in \Pi_{3i}} (-1)^{|\pi|} \omega(u_{\pi_1}, \ldots, u_{\pi_{i-1}}, u_{\pi_i} + u_{\pi_{i+1}}, u_{\pi_{i+2}}, \ldots, u_{\pi_r}), \tag{4.34}$$

one obtains

$$\begin{aligned} t'_{ci} &= \frac{1}{2}\left[\sum_{\pi \in \Pi_{3i}} (-1)^{|\pi|} \omega(u_{\pi_1}, \ldots, u_{\pi_i} + u_{\pi_{i+1}}, \ldots, u_{\pi_r}) \right. \\ &\qquad \left. + \sum_{\pi \in \Pi_{3i}} (-1)^{|\rho_i \circ \pi|} \omega(u_{(\rho_i \circ \pi)_1}, \ldots, u_{(\rho_i \circ \pi)_i} + u_{(\rho_i \circ \pi)_{i+1}}, \ldots, u_{(\rho_i \circ \pi)_r})\right], \\ &= \frac{1}{2}\left[\sum_{\pi \in \Pi_{3i}} (-1)^{|\pi|} \omega(u_{\pi_1}, \ldots, u_{\pi_i} + u_{\pi_{i+1}}, \ldots, u_{\pi_r}) \right. \\ &\qquad \left. + \sum_{\pi \in \Pi_{3i}} (-1)^{|\pi|+1} \omega(u_{\pi_1}, \ldots, u_{\pi_i} + u_{\pi_{i+1}}, \ldots, u_{\pi_r})\right], \\ &= 0. \end{aligned} \tag{4.35}$$

Thus,

$$\begin{aligned} t_{ci} &= \sum_{\pi \in \Pi_{4i}} (-1)^{|\pi|} \omega(u_{\pi_1}, \ldots, u_{\pi_{i-1}}, u_{\pi_i} + u_{\pi_{i+1}}, u_{\pi_{i+2}}, \ldots, u_{\pi_r}), \\ &= \sum_{\pi \in \Pi_{4i}} (-1)^{|\pi|+i-1} \omega(u_1 + u_2, u_{\pi_1}, \ldots, u_{\pi_{i-1}}, u_{\pi_{i+2}}, \ldots, u_{\pi_r}), \\ &= \sum_{\pi \in \Pi_{4i}} (-1)^{|\pi|+i-1+|\rho|} \omega(u_1 + u_2, u_3, \ldots, u_r), \end{aligned} \tag{4.36}$$

where ρ is a permutation of the form

$$\rho : (1, 2, 3, \ldots, r) \longmapsto (1, 2, \rho_3, \ldots, \rho_r).$$

Let η_i be the permutation

$$\eta_i : (1, 2, 3, \ldots, r) \longmapsto (3, 4, \ldots, i-1, 1, 2, i+2, \ldots, r), \tag{4.37}$$

so that for each $\pi \in \Pi_{4i}$ there is a unique ρ with $\pi = \eta_i \circ \rho$. Since $|\eta_i|$ is even, $(-1)^{|\rho|} = (-1)^{|\pi|}$. We conclude that

$$t_{ci} = \sum_{\pi \in \Pi_{4i}} (-1)^{2|\pi|+i-1} \omega(u_1 + u_2, u_3, \ldots, u_r). \tag{4.38}$$

Therefore,

$$\begin{aligned} t_c &= \sum_{i=1}^{r-1} (-1)^i t_{ci}, \\ &= \sum_{i=1}^{r-1} (-1)^i \left[\sum_{\pi \in \Pi_{4i}} (-1)^{2|\pi|+i-1} \omega(u_1 + u_2, u_3, \ldots, u_r) \right], \\ &= -\sum_{i=1}^{r-1} \left[\sum_{\pi \in \Pi_{4i}} \omega(u_1 + u_2, u_3, \ldots, u_r) \right]. \end{aligned} \tag{4.39}$$

Using the fact that for each of the $r-1$ values of i there are $(r-2)!$ permutations in Π_{4i}, we obtain

$$t_c = -(r-1)!\omega(u_1 + u_2, u_3, \ldots, u_r). \tag{4.40}$$

Combining Equations (4.30) and (4.40) into the balance (4.25), one arrives at

$$\omega(u_1 + u_2, u_3, \ldots, u_r) = \omega(u_1, u_3, \ldots, u_r) + \omega(u_2, u_3, \ldots, u_r). \tag{4.41}$$

Since ω is alternating by Eq. (4.18), for each $i = 1, \ldots, r$, and pair of vectors u_i, $u_{i'}$,

$$\begin{aligned} \omega(u_1, \ldots, u_i + u_i', \ldots, u_{r-1}) &= \omega(u_1, \ldots, u_i, \ldots, u_{r-1}) \\ &\quad + \omega(u_1, \ldots, u_i', \ldots, u_{r-1}). \end{aligned} \tag{4.42}$$

It is noted that originally, each of the terms in the equations above was multiplied by $(r-1)!$ the reason being the fact that we computed the total flux for the cubes on the faces of the prism and there are $(r-1)!$ simplices in each such $(r-1)$-cube. (See an illustration in Fig. 4.2.)

4.3 Immediate Implications of Skew Symmetry and Multi-Linearity

Equations (4.42) and (4.3) imply immediately

$$\begin{aligned} &\omega(u_1, \ldots, u_i, \ldots, u_{r-1}) + \omega(u_1, \ldots, -u_i, \ldots, u_{r-1}) \\ &\quad = \omega(u_1, \ldots, 0, \ldots, u_{r-1}) = 0 \end{aligned} \tag{4.43}$$

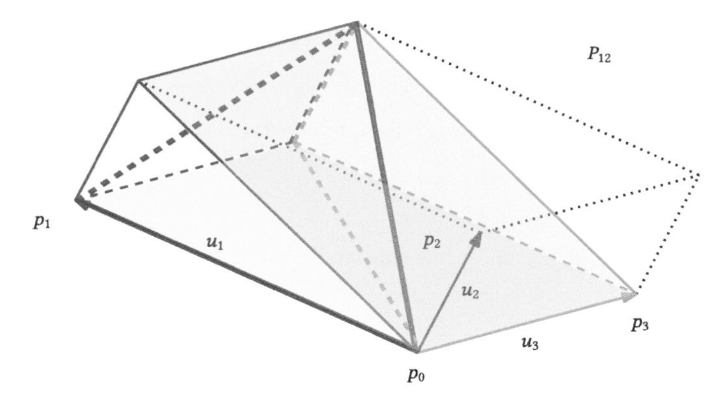

Fig. 4.2 Illustrating the faces pertaining to the linearity relation

and so we obtain

$$\omega(u_1, \ldots, -u_i, \ldots, u_{r-1}) = -\omega(u_1, \ldots, u_i, \ldots, u_{r-1}). \tag{4.44}$$

It follows that the homogeneity property (4.3) applies also to the case $a < 0$. We have obtained that the flux operator is both additive and homogeneous in each of its arguments, and we conclude that

$$\omega : \mathbf{V}^{r-1} \longrightarrow \mathbb{R} \tag{4.45}$$

is an alternating multilinear mapping.

Next, we show that the properties we obtained above imply that the flux vanishes in the case where the vectors $u_1, \ldots, u_{r-1}$ are not linearly independent. This is, of course, expected as linear dependence implies that the simplex collapses to a hyperplane of a lower dimension. If the vectors are not linearly independent, one of the vectors, say u_k, may be represented as a linear combination of the others in the form $u_k = \sum_j a_j u_j$ where $j = 1, \ldots, \widehat{k}, \ldots, r-1$. Thus,

$$\begin{aligned} \omega(u_1, \ldots, u_{r-1}) &= \omega(u_1, \ldots, u_{k-1}, \textstyle\sum_j a_j u_j, u_{k+1}, \ldots, u_{r-1}), \\ &= \sum_j a_j \omega(u_1, \ldots, u_{k-1}, u_j, u_{k+1}, \ldots, u_{r-1}), \\ &= 0 \end{aligned} \tag{4.46}$$

because each term in the sum contains u_j twice (see Eq. (4.13)). Conversely, a multilinear map that vanishes for recurring vectors is necessarily alternating;

$$
\begin{aligned}
0 &= \omega(u_1, \dots, u_i + u_i', \dots, u_i + u_i', \dots u_{r-1}), \\
&= \omega(u_1, \dots, u_i, \dots, u_i, \dots u_{r-1}) + \omega(u_1, \dots, u_i', \dots, u_i', \dots u_{r-1}) \\
&\quad + \omega(u_1, \dots, u_i, \dots, u_i', \dots u_{r-1}) + \omega(u_1, \dots, u_i', \dots, u_i, \dots u_{r-1}),
\end{aligned} \tag{4.47}
$$

implies

$$
\omega(u_1, \dots, u_i', \dots, u_i, \dots u_{r-1}) = -\omega(u_1, \dots, u_i, \dots, u_i', \dots u_{r-1}). \tag{4.48}
$$

4.4 The Algebraic Cauchy Theorem

The results of the foregoing sections may be summarized as follows: Let ω be a function of $(r-1)$-simplices satisfying Assumptions 1–4 of Sect. 4.1; then, $\omega(\tau)$ depends on the vectors $(v_1, \dots, v_{r-1})$, determining τ, multi-linearly and skew-symmetrically. Thus, there is an alternating $(r-1)$-tensor $J \in \otimes^{r-1}(\mathbf{V}^*)$, where $\mathbf{V}$ is the translation space of the affine space $\mathcal{A}$, such that

$$
\omega(\tau) = J(v_1, \dots, v_{r-1}). \tag{4.49}
$$

We will refer to this conclusion as the *algebraic Cauchy theorem*. The theorem may be thought of as a representation theorem asserting that the function ω is represented by the alternating tensor J. In the rest of the text, we will often identify ω and J.

The classical algebraic Cauchy theorem applies in the case where the space $\mathcal{A}$ is a three-dimensional Euclidean space so that the vector space $\mathbf{V}$ has an inner product structure. It states that if τ is determined by the vectors v and u, then,

$$
\omega(\tau) = (\mathbf{J} \cdot \mathbf{n})A, \tag{4.50}
$$

where $\mathbf{J}$ is a vector, $\mathbf{n}$ is the unit normal to the simplex pointing in a direction so that $(\mathbf{n}, v, u)$ are positively oriented (i.e., generate a right-handed triad), and A is the area of the simplex. Thus, one has $\mathbf{n}A = \frac{1}{2}v \times u$, where $\times$ denotes the vector product of the two vectors in the three-dimensional Euclidean space. Thus,

$$
\omega(\tau) = \tfrac{1}{2}\mathbf{J} \cdot (v \times u), \tag{4.51}
$$

which is clearly bilinear and alternating in the arguments v and u. For more details on the comparison of the classical Cauchy formula and the metric-independent version described above, see Example 5.3.

Chapter 5
From Uniform Fluxes to Exterior Algebra

In the previous chapter, we considered fluxes ω defined on $(r-1)$-dimensional simplices. A generic $(r-1)$-simplex was denoted by τ. Uniform fluxes were shown to be represented by $(r-1)$-alternating tensors. To simplify the notation, we slightly modify it and consider below fluxes, ω, through a generic r-simplex s in $\mathscr{A}$. Accordingly, a flux, ω, will be represented by an r-alternating tensor. Specifically, for a simplex $s = [v_1, \ldots, v_r] = [u_1, \ldots, u_r]$, the action of the flux $\omega(s)$ is computed by applying the alternating multilinear mappings to the collection of vectors $v_1, \ldots, v_r$, or $u_1, \ldots, u_r$, corresponding to s.

In this chapter, we first extend the action of fluxes to a wider class of geometric objects—the vector space of r-polyhedral chains. A polyhedral chain is defined as a formal linear combination of simplices and may be thought of as a simple function defined on the union of the simplices. A natural example of a polyhedral chain is the boundary of a simplex, which may be written as the sum of the faces.

The collection of vectors $v_1, \ldots, v_r$ that corresponds to the simplex, and determines the action of a flux on it, contains more information than necessary for the determination of the total flux. The actions of all fluxes on two distinct simplices may be identical. For example, for each flux ω,

$$\omega(v_1, v_2, \ldots, v_r) = \omega(v_1 + v_2, v_2, \ldots, v_r). \tag{5.1}$$

The property of a simplex, or a polyhedral chain, that determines the action of fluxes on it will be referred to as the *extent* of the simplex, or the chain, respectively. The study of the extents of simplices and chains will lead us to exterior algebra. In particular, the extents of polyhedral r-chains are represented mathematically by r-multivectors. Thus, exterior algebra is reviewed below in the context of algebraic fluxes.

R. Segev, *Foundations of Geometric Continuum Mechanics*, Advances in Continuum Mechanics 49, https://doi.org/10.1007/978-3-031-35655-1_5

5.1 Polyhedral Chains and Cochains

A formal finite linear combination of r-simplices of the form

$$B = \sum_{l=1}^{q} a_l s_l, \tag{5.2}$$

where q is some finite integer, determines an r-polyhedral chain. (See [Whi57, pp. 152–153] and [Mun84]). It is assumed that if some of the simplices in the linear combination are contained in one affine subspace of $\mathscr{A}$, then, their interiors are disjoint. This can always be achieved by subdivision of these simplices. So far, there is a formal distinction between a chain of the form (5.2) and a chain in which the simplices s_l are subdivided further.

In order to avoid this distinction and actually identify the two chains, we associate with a chain B a real-valued function $\widehat{B}$ defined on $\mathscr{A}$ as follows: One chooses orientations for each of the affine r-subspaces that contain simplices in the linear combination. For example, one may choose the orientation determined by one of the simplices contained in the subspace. Then, for $x \in s_l^{\circ}$, where s_l° is the interior of s_l, set $\widehat{B}(x) := a_l$ if s_l is positively oriented relative to the orientation chosen on the subspace containing it. If s_l is negatively oriented, set $\widehat{B}(x) := -a_l$. Otherwise, if there is no l such that $x \in s_l^{\circ}$, set $\widehat{B}(x) := 0$.

Now, if B and B' are two linear combinations of simplices with corresponding functions $\widehat{B}$ and $\widehat{B}'$, respectively, we identify the two chains if $\widehat{B}(x) = \widehat{B}'(x)$, for all points $x \in \mathscr{A}$ except for points on a finite number of simplices of lower dimensions. The representation of an r-chain by a function, defined for all $x \in \mathscr{A}$ except for a finite number of simplices of lower dimensional, is unique. For r-chains A, B, B' and real numbers α, α', one set

$$A = \alpha B + \alpha' B', \qquad \text{if} \qquad \widehat{A} = \alpha \widehat{B} + \alpha' \widehat{B}', \tag{5.3}$$

where equality between the functions is in the sense described above. In particular, for a subdivision $s_1, \dots, s_q$ of a simplex s,

$$s = \sum_{l=1}^{q} s_l. \tag{5.4}$$

If $B = \sum_{l=1}^{q} a_l s_l$ and $B' = \sum_{l'=1}^{q'} b_{l'} \tau_{l'}$, using appropriate subdivisions one may represent B and B' in the forms

$$B = \sum_{l''=1}^{q''} a'_{l''} s'_{l''}, \qquad B' = \sum_{l''=1}^{q''} b'_{l''} s'_{l''}, \tag{5.5}$$

for one collection of simplices $s'_1, \dots, s'_{q''}$. Then, $\alpha B + \alpha' B'$ may be represented in the form

$$\alpha B + \alpha' B' = \sum_{l''=1}^{q''} (\alpha a'_{l''} + \alpha' b'_{l''}) s'_{l''}. \tag{5.6}$$

In Sects. 4.1 and 4.2, we used examples of chains in the proofs. In particular, the boundary of an r-simplex may be regarded as a chain

$$\partial s = \sum_{l=0}^{r} \tau_l. \tag{5.7}$$

With the definitions of the algebraic operations above, the collection of polyhedral r-chains, $\mathbf{C}_r(\mathscr{A})$, is provided with a structure of a vector space.

The boundary of a chain $B = \sum_{l=1}^{q} a_l s_l$ is defined as

$$\partial B = \sum_{l=1}^{q} a_l \partial s_l. \tag{5.8}$$

The boundary mapping

$$\partial : \mathbf{C}_r(\mathscr{A}) \longrightarrow \mathbf{C}_{r-1}(\mathscr{A}) \tag{5.9}$$

is a linear mapping. The fact that for any simplex s, $\partial(\partial s) = 0$ (see, e.g., [Kur72, pp. 309-310]) implies that for polyhedral chains,

$$\partial^2 := \partial \circ \partial = 0. \tag{5.10}$$

Once chains have been defined as linear combinations of simplices, one may extend by linearity the action of fluxes on simplices to action on polyhedral chains. Thus, for the chain $B = \sum_{i=1}^{q} a_l s_l$, one sets

$$\omega(B) = \omega(\textstyle\sum_{l=1}^{q} a_l s_l) := \displaystyle\sum_{l=1}^{q} a_l \omega(s_l). \tag{5.11}$$

The action of fluxes on chains is linear. Given chains B and B', we can represent them using a common subdivision, as in (5.5), so that

$$\begin{aligned}\omega(\alpha B + \alpha' B') &= \omega(\textstyle\sum_{l''=1}^{q''}(\alpha a'_{l''} + \alpha' b'_{l''}) s'_{l''}, \\ &= \alpha \sum_{l''=1}^{q''} a'_{l''} \omega(s'_{l''}) + \alpha' \sum_{l''=1}^{q''} b'_{l''} \omega(s'_{l''}), \\ &= \alpha \omega(B) + \alpha' \omega(B').\end{aligned} \tag{5.12}$$

Let $\mathbf{C}^*_r(\mathscr{A})$ be the algebraic dual space to $\mathbf{C}_r(\mathscr{A})$, where it is noted that $\mathbf{C}_r(\mathscr{A})$ is infinite-dimensional. We will refer to an element of $\mathbf{C}^*_r(\mathscr{A})$ as an *algebraic cochain*. Thus, every uniform flux ω represents an algebraic cochain of a particularly simple form.

5.2 Component Representation of Fluxes

The Cauchy flux theorem brings to light the significance of alternating tensors in continuum mechanics. In this section and some of the following, we introduce the notation and review the basic facts pertaining to alternating multilinear mappings and multivectors.

Let $\{e_i\}_{i=1}^n$ be a basis of $\mathbf{V}$, and let $\{e^i\}_{i=1}^n$ be the corresponding dual basis in the dual space, $\mathbf{V}^*$. The elements of the dual basis satisfy the condition $e^i(v) = e^i\left(\sum_j v^j e_j\right) = v^i$ and so $e^i(e_j) = \delta^i_j$, where δ^i_j is the Kronecker symbol. Given a covector $\varphi \in \mathbf{V}^*$, the components of φ relative to $\{e^i\}$ are given by $\varphi_i = \varphi(e_i)$, and so

$$\varphi = \sum_i \varphi_i e^i = \sum_i \varphi(e_i) e^i.$$

The array of a multilinear function $\omega : \mathbf{V}^r \to \mathbb{R}$ is given by

$$\omega_{i_1 \dots i_r} = \omega(e_{i_1}, \dots, e_{i_r}). \tag{5.13}$$

With

$$e^{i_1} \otimes \cdots \otimes e^{i_r}(v_1, \dots, v_r) = v_1^{i_1} \cdots v_r^{i_r}, \tag{5.14}$$

one has

$$\omega = \sum_{i_1,\dots,i_r} \omega_{i_1 \dots i_r} e^{i_1} \otimes \cdots \otimes e^{i_r}, \quad \text{and} \quad \omega(v_1, \dots, v_r) = \sum_{i_1,\dots,i_r} \omega_{i_1 \dots i_r} v_1^{i_1} \cdots v_r^{i_r}. \tag{5.15}$$

In order to simplify the notation, one can write the equations above using multi-indices. An r-multi-index, I, is viewed as a function

$$I : \{1, \dots, r\} \longrightarrow \{1, \dots, n\}, \tag{5.16}$$

and we write (with some abuse of notation)

$$\begin{aligned} &i_p := I(p), \qquad |I| := r, \\ &I : (1, \dots, r) \longmapsto (i_1, \dots, i_r), \qquad \text{and} \\ &I = (i_1, \dots, i_r). \end{aligned} \tag{5.17}$$

As an example, a permutation may be viewed as a multi-index. For the components of a multilinear function ω, set $\omega_I = \omega_{i_1 \dots i_r}$, so that $\omega = \sum_I \omega_I e^{i_1} \otimes \dots \otimes e^{i_r}$. (We do not write $e^I = e^{i_1} \otimes \dots \otimes e^{i_r}$ because we reserve the notation e^I for another object.) Multi-indices may be concatenated naturally such that $|IJ| = |I| + |J|$.

Next, we consider the consequences of the fact that a flux is an alternating multilinear function, as in (4.18). Since ω vanishes on a collection of linearly dependent vectors, in case an index j appears twice,

$$\omega_{i_1 \dots j \dots j \dots i_r} = \omega(e_{i_1}, \dots, e_j, \dots, e_j, \dots e_{i_r}) = 0. \tag{5.18}$$

In addition, for any permutation $\pi : \{1, \dots, r\} \to \{1, \dots, r\}$,

$$\omega_{I \circ \pi} = (-1)^{|\pi|} \omega_I. \tag{5.19}$$

It follows that the array of ω is completely determined by the components corresponding to increasing multi-indices.

Consider a multi-index, $\gamma : \{1, \dots, r\} \to \{1, \dots, n\}$, which is a monotonically increasing injection, i.e., $\gamma_p = \gamma(p) > \gamma_q = \gamma(q)$ if $p > q$. When we want to indicate that a multi-index γ is increasing, we will write (γ). Given ω, for any I for which $i_p = I(p) \neq i_p = I(q)$, $p \neq q$, there is a unique increasing γ containing the same elements. Let the generalized Levi-Civita symbol ε^I_J be defined by

$$\varepsilon^I_J = \begin{cases} 0, & \text{if there is no permutation } \pi \text{ such that } J = I \circ \pi, \\ 0, & \text{if } I \text{ and } J \text{ contain the same index twice or more,} \\ (-1)^{|\pi|}, & \text{if } J = I \circ \pi, \pi \in \Pi, \text{ and no index is repeated.} \end{cases} \tag{5.20}$$

We will also set

$$\varepsilon_J := \varepsilon^{1,\dots,r}_J, \qquad \varepsilon^I := \varepsilon^I_{1,\dots,r}. \tag{5.21}$$

Thus,

$$\omega_I = \sum_{(\gamma)} \varepsilon_I^{\gamma} \omega_{\gamma}, \tag{5.22}$$

where the sum contains only one nonvanishing term. It follows that the multilinear mapping can now be expressed by

$$\begin{aligned} \omega &= \sum_I \omega_I e^{i_1} \otimes \cdots \otimes e^{i_r}, \\ &= \sum_{I,(\gamma)} \varepsilon_I^{\gamma} \omega_{\gamma} e^{i_1} \otimes \cdots \otimes e^{i_r}, \end{aligned} \tag{5.23}$$

and we obtain

$$\omega = \sum_{(\gamma)} \omega_{\gamma} e^{\gamma}, \qquad \text{where} \qquad e^{\gamma} = \sum_I \varepsilon_I^{\gamma} e^{i_1} \otimes \cdots \otimes e^{i_r}. \tag{5.24}$$

It is observed that

$$\begin{aligned} e^{\gamma}(e_{i_1}, \ldots, e_{i_r}) &= \sum_J \varepsilon_J^{\gamma} e^{j_1} \otimes \cdots \otimes e^{j_r}(e_{i_1}, \ldots, e_{i_r}), \\ &= \varepsilon_I^{\gamma}. \end{aligned} \tag{5.25}$$

Hence, e^{γ} is an alternating tensor. Moreover, we introduce formally the notation

$$e^{\gamma_1} \wedge \cdots \wedge e^{\gamma_r} := e^{\gamma}. \tag{5.26}$$

Note that the definition of e^{γ} in (5.24) applies also for the case where the multi-index γ is not increasing, and it implies that

$$e^{\gamma \circ \pi} = (-1)^{|\pi|} e^{\gamma}, \quad e^{(\gamma \circ \pi)_1} \wedge \cdots \wedge e^{(\gamma \circ \pi)_r} = (-1)^{|\pi|} e^{\gamma_1} \wedge \cdots \wedge e^{\gamma_r}. \tag{5.27}$$

In fact, we may also consider e^J where the multi-index J need not even be an injection. We just set formally $e^J = 0$ if J contains two or more identical indices.

In view of the skew symmetry property, as exhibited in Eq. (5.27), and observing that an injective multi-index I may be written in the form $I = \gamma \circ \pi$ for a unique increasing multi-index $\gamma : \{1, \ldots, r\} \to \{1, \ldots, n\}$ and a unique permutation π of r symbols, we have $e^I = \sum_{(\gamma)} \varepsilon_{\gamma_1 \ldots \gamma_r}^{i_1 \ldots i_r} e^{\gamma}$, where the sum contains only one nonvanishing term. Thus,

$$\begin{aligned} \sum_I \omega_I e^I &= \sum_{I,(\gamma)} \omega_I \varepsilon_{\gamma_1 \ldots \gamma_r}^{i_1 \ldots i_r} e^{\gamma}, \\ &= \sum_{(\gamma)} r! \omega_{\gamma} e^{\gamma}, \end{aligned} \tag{5.28}$$

where in the last line we used the fact that both the Levi-Civita symbol and the array ω_I are skew-symmetric (as in (5.19)). Thus, Eq. (5.24) may be rewritten in the form

$$\omega = \frac{1}{r!}\sum_I \omega_I e^{i_1} \wedge \cdots \wedge e^{i_r}. \tag{5.29}$$

To show that the collection of mappings $\{e^\gamma\}$, for increasing multi-indices γ, forms a basis for the space of alternating r-multilinear mappings, it remains to prove that they are linearly independent. If $\sum_{(\gamma)} \omega_\gamma e^\gamma = 0$, then for every multi-index I,

$$\begin{aligned} 0 &= \sum_{(\gamma)} \omega_\gamma e^\gamma (e_{i_1}, \dots, e_{i_r}), \\ &= \sum_{(\gamma)} \omega_\gamma \varepsilon_I^\gamma . \end{aligned} \tag{5.30}$$

Since I is arbitrary in the equation above, for each increasing γ_0, we may choose $I = \gamma_0$ so that $\varepsilon_I^{\gamma_0} = 1$ and $\varepsilon_I^\gamma = 0$ for $\gamma \neq \gamma_0$. It follows that $\omega_{\gamma_0} = 0$.

With the representation of alternating tensors as in (5.24), the action of a flux may be represented as

$$\begin{aligned} \omega(v_1, \dots, v_r) &= \sum_{(\gamma)} \omega_\gamma e^{\gamma_1} \wedge \cdots \wedge e^{\gamma_r} \left(\textstyle\sum_{i_1} v_1^{i_1} e_{i_1}, \dots, \sum_{i_r} v_r^{i_r} e_{i_r}\right), \\ &= \sum_{(\gamma),I} \omega_\gamma v_1^{i_1} \cdots v_r^{i_r} e^{\gamma_1} \wedge \cdots \wedge e^{\gamma_r} \left(e_{i_1}, \dots, e_{i_r}\right), \end{aligned} \tag{5.31}$$

and so, as expected,

$$\omega(v_1, \dots, v_r) = \sum_{(\gamma),I} \omega_\gamma v_1^{i_1} \cdots v_r^{i_r} \varepsilon_I^\gamma = \sum_I \omega_I v_1^{i_1} \cdots v_r^{i_r}. \tag{5.32}$$

5.3 Multivectors

We will roughly refer to the collection of properties an of r-polyhedral chain that determines the total flux through it, for an arbitrary flux, as the *extent* of the chain. Mathematically, the extent of an r-chain will be represented by an r-multivector as described below. Evidently, two distinct chains may have the same extent. For example, by the balance assumption (4.22), the boundary of a simplex and the zero chain have the same extent.

Let $\mathbf{O} \subset \mathbf{C}_r(\mathscr{A})$ be the collection of chains such that for $B \in \mathbf{O}$, $\omega(B) = 0$ for every uniform flux ω. By the linearity of the action of fluxes on polyhedral r-chains, it follows that $\mathbf{O}$ is a vector subspace of $\mathbf{C}_r(\mathscr{A})$. We consider now the quotient space

$$\bigwedge^r \mathbf{V} := \mathbf{C}_r(\mathscr{A})/\mathbf{O}. \tag{5.33}$$

Two polyhedral r-chains B and B' belong to the same equivalence class if $\omega(B) = \omega(B')$ for every uniform flux ω. Denoting the equivalence class of B by $\{B\}$, we have

$$\mathfrak{v} = \{B\} = \{B'\} \in \bigwedge^r \mathbf{V}. \tag{5.34}$$

An element $\mathfrak{v} \in \bigwedge^r \mathbf{V}$ will be referred to as a *multivector*. The space of r-multivectors $\bigwedge^r \mathbf{V}$ has a natural structure of a vector space.

Since the action of any flux on a chain B is uniquely determined by the associated multivector $\{B\}$, a flux ω determines a linear mapping $\widehat{\omega} \in \left(\bigwedge^r \mathbf{V}\right)^*$ by

$$\widehat{\omega}(\{B\}) = \omega(B). \tag{5.35}$$

It is noted that the notation, $\bigwedge^r \mathbf{V}$, for the space of multivectors makes reference only to $\mathbf{V}$ and not to $\mathscr{A}$ because of the fact that fluxes are uniform and their action is invariant under translations of simplices and chains.

As mentioned above, by the balance assumption (4.22),

$$\widehat{\omega}(\{\partial s\}) = \sum_{i=0}^{r} \widehat{\omega}\{\tau_i\} = 0 \tag{5.36}$$

for any flux ω. Hence, as an r-multivector,

$$\{\partial s\} = 0. \tag{5.37}$$

Note that the chain ∂s is not the zero chain, but the multivector, $\{\partial s\}$, associated with it vanishes. This is the metric free, n-dimensional, analog of the identity $\sum_{l=0}^{3} A_l \mathbf{n}_l = 0$ for a tetrahedron in the three-dimensional Euclidean space, where A_l is the area of the l-th face and $\mathbf{n}_l$ is the unit normal to the l-th face.

We conclude that ω, which is an alternating multilinear function of vectors, may be represented as a linear function of multivectors. The algebraic Cauchy theorem as presented above may be viewed as an analog to Whitney's theorem on an algebraic criterion for a multivector (see [Whi57, pp. 165–167]). Here, we used only elementary notions and arguments in order to motivate the use of alternating tensors.

A multivector $\mathfrak{v}$ is referred to as a *simple*, or a *decomposable*, *multivector* if there is a simplex s such that $\mathfrak{v} = \{s\}$. If the simplex s determines the vectors $v_1, \ldots, v_r$, we write

$$\mathfrak{v} = \{s\} = v_1 \wedge \cdots \wedge v_r. \tag{5.38}$$

We mention again that the representation $v_1 \wedge \cdots \wedge v_r$ of $\mathfrak{v}$ is not unique. For example, using the notation of Chap. 3, in view of (4.17),

$$v_1 \wedge \cdots \wedge v_r = u_1 \wedge \cdots \wedge u_r. \tag{5.39}$$

Due to the linearity of the projection $\mathrm{pr} : \mathbf{C}_r(\mathscr{A}) \to \bigwedge^r \mathbf{V}$, every multivector can be represented as a linear combination of decomposable multivectors. Specifically, if $B = \sum_{l=1}^{q} a_l s_l$ and each simplex s_l is associated with the vectors $v_1^{(l)}, \ldots, v_r^{(l)}$, then,

$$\mathfrak{v} = \{B\} = \sum_{l=1}^{q} a_l v_1^{(l)} \wedge \cdots \wedge v_r^{(l)}. \tag{5.40}$$

Henceforth, we will omit the "hat" for the action of fluxes on multivectors so that the notation will be identical to that pertaining to the action on chains. Thus, by linearity,

$$\omega(\mathfrak{v}) = \omega(\{B\}) = \sum_{l=1}^{q} a_l \omega(v_1^{(l)} \wedge \cdots \wedge v_r^{(l)}). \tag{5.41}$$

Some properties of multivectors follow immediately from the properties of fluxes. For example,

$$a(u_1 \wedge \cdots \wedge u_r) = u_1 \wedge \cdots \wedge a u_i \wedge \cdots \wedge u_j \wedge \cdots \wedge u_r = u_1 \wedge \cdots \wedge u_i \wedge \cdots \wedge a u_j \wedge \cdots \wedge u_r, \tag{5.42}$$

and clearly,

$$a(u_1 \wedge \cdots \wedge u_r) + b(u_1 \wedge \cdots \wedge u_r) = (a+b)(u_1 \wedge \cdots \wedge u_r). \tag{5.43}$$

In addition, the additivity property (4.42) implies that

$$u_1 \wedge \cdots \wedge (u_i + u_i') \wedge \cdots \wedge u_{r-1} = u_1 \wedge \cdots \wedge u_i \wedge \cdots \wedge u_{r-1} + u_1 \wedge \cdots \wedge u_i' \wedge \cdots \wedge u_{r-1}. \tag{5.44}$$

Furthermore, it follows from the fact that the flux mapping is alternating, i.e., Eq. (4.18), that

$$u_{\pi_1} \wedge \cdots \wedge u_{\pi_r} = (-1)^{|\pi|} u_1 \wedge \cdots \wedge u_r = \varepsilon_{\pi_1 \ldots \pi_r} u_1 \wedge \cdots \wedge u_r. \tag{5.45}$$

It is concluded that the mapping

$$\bigwedge : \mathbf{V}^r \longrightarrow \bigwedge^r \mathbf{V}, \qquad (v_1, \ldots, v_r) \longmapsto v_1 \wedge \cdots \wedge v_r, \tag{5.46}$$

is a multilinear alternating mapping.

From (4.46) it follows that for r nonvanishing vectors $v_1, \dots, v_r$, the condition $v_1 \wedge \cdots \wedge v_r \neq 0$ implies that the vectors are linearly independent. Conversely, if the vectors are linearly independent, they may be completed to form a basis. Thus, by (5.25), a flux ω may be constructed such that $\omega(v_1 \wedge \cdots \wedge v_r) \neq 0$. Thus,

$$v_1 \wedge \cdots \wedge v_r \neq 0 \tag{5.47}$$

if and only if the vectors are linearly independent.

Remark 5.1 It is sometimes convenient to associate a decomposable multivector with a cube rather than a simplex so that $\omega(v_1 \wedge \cdots \wedge v_r)$ is the total flux through the parallelepiped (cube) determined by the vectors $v_1, \dots, v_r$. With such an interpretation of a decomposable multivector, the total flux is multiplied by $r!$. This interpretation results in the difference in notation between various authors on these subjects, e.g., [War83, p. 60], [Whi57], and [AMR88].

5.4 Component Representation of Multivectors

Since the mapping $\bigwedge$ defined above is multilinear, a decomposable multivector $\mathfrak{v} = v_1 \wedge \cdots \wedge v_r$ may be represented in the form

$$\begin{aligned} \mathfrak{v} &= (\textstyle\sum_{i_1=1}^{n} v_1^{i_1} e_{i_1}) \wedge \cdots \wedge (\textstyle\sum_{i_r=1}^{n} v_r^{i_r} e_{i_r}), \\ &= \sum_I v_1^{i_1} \cdots v_r^{i_r} e_{i_1} \wedge \cdots \wedge e_{i_r}, \\ &= \sum_{I,(\gamma)} v_1^{i_1} \cdots v_r^{i_r} \varepsilon_I^{\gamma} e_{\gamma_1} \wedge \cdots \wedge e_{\gamma_r}. \end{aligned} \tag{5.48}$$

We set

$$\mathfrak{v}^{\gamma} := (v_1 \wedge \cdots \wedge v_r)^{\gamma} := \sum_I v_1^{i_1} \cdots v_r^{i_r} \varepsilon_I^{\gamma}, \tag{5.49}$$

and

$$e_I := e_{i_1} \wedge \cdots \wedge e_{i_r}, \qquad \text{in particular,} \quad e_{\gamma} := e_{\gamma_1} \wedge \cdots \wedge e_{\gamma_r}, \tag{5.50}$$

for an increasing multi-index, γ. With these definitions, the decomposable multivector is represented in the form

$$\mathfrak{v} = v_1 \wedge \cdots \wedge v_r = \sum_{(\gamma)} \mathfrak{v}^{\gamma} e_{\gamma}. \tag{5.51}$$

Thus, decomposable multivectors may be represented as linear combinations of the multivectors generated by basis vectors with increasing indices.

Moreover, since any multivector can be represented as a linear combination of decomposable multivectors as in (5.40), in view of the last representation (5.51) of decomposable multivectors, we conclude that every multivector, whether decomposable or not, may be represented in the form

$$\mathfrak{v} = \{B\} = \sum_{(\gamma)} \mathfrak{v}^{\gamma} e_{\gamma}, \quad \text{for some} \quad \mathfrak{v}^{\gamma} \in \mathbb{R}. \tag{5.52}$$

In order to show that the vectors $\{e_\gamma\}$, for increasing multi-indices γ, may serve as a basis for $\bigwedge^r \mathbf{V}$, we still have to prove that they are linearly independent. First, it is observed that Eq. (5.25) may now be rewritten as

$$e^{\gamma}(e_I) := e^{\gamma}(e_{i_1} \wedge \cdots \wedge e_{i_r}) := e^{\gamma}(e_{i_1}, \ldots, e_{i_r}) = \varepsilon_I^{\gamma}, \tag{5.53}$$

and in particular,

$$e^{\gamma}(e_\beta) = \varepsilon_\beta^{\gamma}. \tag{5.54}$$

Since both multi-indices are increasing, this expression equals either 1 or 0.

Hence, if $\sum_{(\gamma)} \mathfrak{v}^{\gamma} e_{\gamma} = 0$, applying e^{β}, for an arbitrary increasing multi-index β, to both sides yields $\mathfrak{v}^{\beta} = 0$. It is concluded that the collection $\{e_\gamma\}$, for all increasing multi-indices γ, may serve as a basis for the space of multivectors. In addition, Eq. (5.54) implies that $\{e^{\gamma}\}$ and $\{e_\gamma\}$ are dual bases for the space of alternating multilinear mappings and the space of multivectors, respectively. In particular

$$\mathfrak{v}^{\gamma} = e^{\gamma}(\mathfrak{v}). \tag{5.55}$$

For this reason, the space of alternating multilinear mappings will be denoted by $(\bigwedge^r \mathbf{V})^*$.

As $\{e_\gamma\}$ and $\{e^{\gamma}\}$, for increasing multi-indices γ, may serve as bases for $\bigwedge^r \mathbf{V}$ and $(\bigwedge^r \mathbf{V})^*$, respectively, and as each of the values $\gamma_1, \ldots, \gamma_r$ may assume the values $1, \ldots, n$, one concludes that

$$\dim \bigwedge^r \mathbf{V} = \dim (\bigwedge^r \mathbf{V})^* = \frac{n!}{(n-r)!r!}. \tag{5.56}$$

In particular, in view of the algebraic Cauchy theorem in Sect. 4.4, the flux alternating tensor J is uniquely determined by the function ω, or, equivalently, by its action on all multivectors.

5.5 Alternation

Alternation is the action of extracting an alternating r-tensor from an arbitrary r-tensor ω. It is desirable that the alternation action will be a projection, i.e., when applied to an alternating tensor, it will not change it. The alternation operator is defined by

$$\mathfrak{A}(\omega)(v_1, \ldots, v_r) = \frac{1}{r!} \sum_{\pi} (-1)^{|\pi|} \omega(v_{\pi_1}, \ldots, v_{\pi_r}). \tag{5.57}$$

To show that $\mathfrak{A}(\omega)$ is indeed alternating, we observe that (5.57) implies that for a permutation ρ,

$$\begin{aligned}
\mathfrak{A}(\omega)(v_{\rho_1}, \ldots, v_{\rho_r}) &= \frac{1}{r!} \sum_{\pi} (-1)^{|\pi|} \omega(v_{(\pi\circ\rho)_1}, \ldots, v_{(\pi\circ\rho)_r}), \\
&= \frac{1}{r!} \sum_{\mu} (-1)^{|\mu|-|\rho|} \omega(v_{\mu_1}, \ldots, v_{\mu_r}), \\
&= \frac{1}{r!} (-1)^{|\rho|} \sum_{\mu} (-1)^{|\mu|} \omega(v_{\mu_1}, \ldots, v_{\mu_r}), \\
&= (-1)^{|\rho|} \mathfrak{A}(\omega)(v_1, \ldots, v_r),
\end{aligned} \tag{5.58}$$

where in the second line we used $\mu = \pi \circ \rho$, $|\mu| = |\pi| + |\rho|$.

In case ω is alternating,

$$\begin{aligned}
\mathfrak{A}(\omega)(v_1, \ldots, v_r) &= \frac{1}{r!} \sum_{\pi} (-1)^{|\pi|} \omega(v_{\pi_1}, \ldots, v_{\pi_r}), \\
&= \frac{1}{r!} \sum_{\pi} (-1)^{|\pi|} (-1)^{|\pi|} \omega(v_1, \ldots, v_r), \\
&= \omega(v_1, \ldots, v_r),
\end{aligned} \tag{5.59}$$

so $\mathfrak{A}$ leaves alternating tensors invariant.

For an r-tensor $\omega = \sum_I \omega_{i_1 \ldots i_r} e^{i_1} \otimes \cdots \otimes e^{i_r}$, one has

$$
\begin{aligned}
\mathfrak{A}(\omega)(v_1,\ldots,v_r) &= \frac{1}{r!}\sum_{\pi,I}(-1)^{|\pi|}\omega_{i_1\ldots i_r}e^{i_1}\otimes\cdots\otimes e^{i_r}(v_{\pi_1},\ldots,v_{\pi_r}),\\
&= \frac{1}{r!}\sum_{\pi,I}(-1)^{|\pi|}\omega_{i_1\ldots i_r}e^{i_1}(v_{\pi_1})\cdots e^{i_r}(v_{\pi_r}),\\
&= \frac{1}{r!}\sum_{\pi,I}(-1)^{|\pi|}\omega_{i_1\ldots i_r}e^{(I\circ\pi^{-1})_1}(v_1)\cdots e^{(I\circ\pi^{-1})_r}(v_r),\\
&= \frac{1}{r!}\sum_{\pi,I}(-1)^{|\pi|}\omega_{i_1\ldots i_r}e^{(I\circ\pi^{-1})_1}\otimes\cdots\otimes e^{(I\circ\pi^{-1})_r}(v_1,\ldots,v_r),
\end{aligned}
\tag{5.60}
$$

whereby

$$
\mathfrak{A}(\omega) = \frac{1}{r!}\sum_{\pi,I}(-1)^{|\pi|}\omega_{i_1\ldots i_r}e^{(I\circ\pi^{-1})_1}\otimes\cdots\otimes e^{(I\circ\pi^{-1})_r}. \tag{5.61}
$$

In particular, using δ_i^j for the Kronecker symbol and setting $\omega_{i_1\ldots i_r} = \delta_{i_1}^{j_1}\cdots\delta_{i_r}^{j_r}$, for some multi-index J,

$$
\begin{aligned}
\mathfrak{A}(e^{j_1}\otimes\cdots\otimes e^{j_r}) &= \frac{1}{r!}\sum_{\pi}(-1)^{|\pi|}e^{(J\circ\pi^{-1})_1}\otimes\cdots\otimes e^{(J\circ\pi^{-1})_r},\\
&= \frac{1}{r!}\sum_{I}\varepsilon^{j_1\ldots j_r}_{i_1\ldots i_r}e^{i_1}\otimes\cdots\otimes e^{i_r},
\end{aligned}
\tag{5.62}
$$

where in the last equality we set $I = J\circ\pi^{-1}$ and used $|I| = |J| + |\pi|$, $(-1)^{|\pi|} = (-1)^{|I|-|J|} = \varepsilon^{j_1\ldots j_r}_{i_1\ldots i_r}$ for each nonvanishing term. As expected, it follows that $\mathfrak{A}(e^{j_1}\otimes\cdots\otimes e^{j_r}) = 0$ if two or more of the indices are identical. In addition, using the definitions of $e^I = e^{i_1}\wedge\cdots\wedge e^{i_r}$ in (5.24) and (5.27), we record that

$$
e^J = e^{j_1}\wedge\cdots\wedge e^{j_r} = r!\mathfrak{A}(e^{j_1}\otimes\cdots\otimes e^{j_r}) = \sum_I \varepsilon^{j_1\ldots j_r}_{i_1\ldots i_r}e^{i_1}\otimes\cdots\otimes e^{i_r}. \tag{5.63}
$$

The previous equations are not limited to elements of the dual basis. If $\varphi^1,\ldots,\varphi^r \in \mathbf{V}^*$ are r covectors, then, for $j_1,\ldots,j_r \in \{1,\ldots,r\}$

$$
\begin{aligned}
\mathfrak{A}(\varphi^{j_1}\otimes\cdots\otimes\varphi^{j_r}) &= \frac{1}{r!}\sum_{\pi}(-1)^{|\pi|}\varphi^{(J\circ\pi^{-1})_1}\otimes\cdots\otimes\varphi^{(J\circ\pi^{-1})_r},\\
&= \frac{1}{r!}\sum_{I}\varepsilon^{j_1\ldots j_r}_{i_1\ldots i_r}\varphi^{i_1}\otimes\cdots\otimes\varphi^{i_r}.
\end{aligned}
\tag{5.64}
$$

The last equation implies that for any permutation π,

$$\mathfrak{A}(\varphi^{\pi_1} \otimes \cdots \otimes \varphi^{\pi_r}) = (-1)^{|\pi|}\mathfrak{A}(\varphi^1 \otimes \cdots \otimes \varphi^r). \tag{5.65}$$

It is observed that by (5.61) the alternation operator is linear and, in particular,

$$\mathfrak{A}(\omega) = \sum_I \omega_{i_1 \ldots i_r} \mathfrak{A}\left(e^{i_1} \otimes \cdots \otimes e^{i_r}\right). \tag{5.66}$$

We conclude that $\mathfrak{A}$ is indeed a linear projection of the vector space of r-tensors on the space of r-alternating tensors.

5.6 Exterior Products

The notation $e^{\gamma_1} \wedge \cdots \wedge e^{\gamma_r} := e^\gamma$ was introduced formally by (5.24), (5.26) without an explanation or motivation. In this section, we review the exterior product of alternating tensors, denoted by $\wedge$, for which the expression for e^γ above is a special case.

We observe first that Eqs. (5.24) and (5.26) may be extended to any collection $\varphi^1, \ldots, \varphi^r \in \mathbf{V}^*$ of r covector. Thus, one may consider *simple,* or *decomposable, multi-covectors* of the form

$$\varphi^1 \wedge \cdots \wedge \varphi^r := r!\mathfrak{A}(\varphi^1 \otimes \cdots \otimes \varphi^r) = \sum_\pi (-1)^{|\pi|} \varphi^{\pi_1} \otimes \cdots \otimes \varphi^{\pi_r}. \tag{5.67}$$

From Eq. (5.65), it follows that

$$\begin{aligned} \varphi^{\pi_1} \wedge \cdots \wedge \varphi^{\pi_r} &= (-1)^{|\pi|} \varphi^1 \wedge \cdots \wedge \varphi^r, \\ \varphi^{(\pi \circ \rho)_1} \wedge \cdots \wedge \varphi^{(\pi \circ \rho)_r} &= (-1)^{|\pi|} \varphi^{\rho_1} \wedge \cdots \wedge \varphi^{\rho_r}. \end{aligned} \tag{5.68}$$

For a collection of r linearly independent vectors, $v_1, \ldots, v_r$, one has

$$\begin{aligned} \varphi^1 \wedge \cdots \wedge \varphi^r(v_1, \ldots, v_r) &= \sum_\pi (-1)^{|\pi|} \varphi^{\pi_1} \otimes \cdots \otimes \varphi^{\pi_r}(v_1, \ldots, v_r), \\ &= \sum_\pi (-1)^{|\pi|} \varphi^{\pi_1}(v_1) \cdots \varphi^{\pi_r}(v_r), \\ &= \frac{1}{r!} \sum_{I,J} \varepsilon^I_J \varphi^{j_1}(v_{i_1}) \cdots \varphi^{j_r}(v_{i_r}), \\ &= \det\left[\varphi^i(v_j)\right]. \end{aligned} \tag{5.69}$$

In particular, the tensors e^I satisfy

$$e^{i_1} \wedge \cdots \wedge e^{i_r}(v_1, \ldots, v_r) = \det\left[(v_j)^{i_k}\right], \quad k = 1, \ldots, r. \tag{5.70}$$

The action of the alternation operator may be represented now using (5.66) and (5.63) as

$$\mathfrak{A}(\omega) = \frac{1}{r!}\sum_I \omega_I e^{i_1} \wedge \cdots \wedge e^{i_r} = \frac{1}{r!}\sum_I \omega_I e^I = \sum_{(\gamma)} \mathfrak{A}(\omega)_\gamma e^\gamma, \tag{5.71}$$

where

$$\begin{aligned} \mathfrak{A}(\omega)_\gamma &= \mathfrak{A}(\omega)(e_{\gamma_1}, \ldots, e_{\gamma_r}), \\ &= \frac{1}{r!}\sum_I \omega_I e^I(e_{\gamma_1}, \ldots, e_{\gamma_r}), \\ &= \frac{1}{r!}\sum_I \omega_I \varepsilon^I_\gamma. \end{aligned} \tag{5.72}$$

The "wedge" operation for a collection of covectors is extended now to the exterior product between covariant tensors. First, for the covectors $\varphi^1, \ldots, \varphi^{r+p}$, let $\omega = \varphi^1 \wedge \cdots \wedge \varphi^r$ and let $\psi = \varphi^{r+1} \wedge \cdots \wedge \varphi^{r+p}$. We set (see [Whi57, p. 41])

$$\omega \wedge \psi = \varphi^1 \wedge \cdots \wedge \varphi^r \wedge \varphi^{r+1} \wedge \cdots \wedge \varphi^{r+p}. \tag{5.73}$$

It is noted that this definition is consistent with our notation $e^{\gamma_1} \wedge \cdots \wedge e^{\gamma_r} := e^\gamma$ because

$$e^{\gamma_1} \wedge \cdots \wedge e^{\gamma_r} = (\cdots(e^{\gamma_1} \wedge e^{\gamma_2}) \wedge e^{\gamma_3} \wedge \cdots) \wedge e^{\gamma_r}. \tag{5.74}$$

Consider the relation between $\omega \wedge \psi$ and $\mathfrak{A}(\omega \otimes \psi)$. One has

$$\begin{aligned} \omega \otimes \psi &= \left(\sum_\beta (-1)^{|\beta|}\varphi^{\beta_1} \otimes \cdots \otimes \varphi^{\beta_r}\right) \otimes \left(\sum_\gamma (-1)^{|\gamma|}\varphi^{\gamma_{r+1}} \otimes \cdots \otimes \varphi^{\gamma_{r+p}}\right), \\ &= \sum_{\beta,\gamma} (-1)^{|\beta|}(-1)^{|\gamma|}\varphi^{\beta_1} \otimes \cdots \otimes \varphi^{\beta_r} \otimes \varphi^{\gamma_{r+1}} \otimes \cdots \otimes \varphi^{\gamma_{r+p}}, \end{aligned} \tag{5.75}$$

where β is a permutation of $\{1, \ldots, r\}$ and γ is a permutation of $\{r+1, \ldots, r+p\}$. Thus, letting π denote a permutation of $\{1, \ldots, r+p\}$, (5.64) yields

$$\begin{aligned} &(r+p)!\mathfrak{A}(\omega \otimes \psi) \\ &= \sum_{\pi,\beta,\gamma} (-1)^{|\pi|}(-1)^{|\beta|}(-1)^{|\gamma|}\varphi^{(\pi\circ\beta)_1} \otimes \cdots \otimes \varphi^{(\pi\circ\beta)_r} \otimes \varphi^{(\pi\circ\gamma)_{r+1}} \otimes \cdots \end{aligned}$$

$$
\begin{aligned}
&\otimes \varphi^{(\pi\circ\gamma)_{r+p}},\\
&=\sum_{\beta,\gamma}(-1)^{|\beta|}(-1)^{|\gamma|}\left(\sum_{\pi}(-1)^{|\pi|}\varphi^{(\pi\circ\beta)_1}\otimes\cdots\otimes\varphi^{(\pi\circ\beta)_r}\otimes\varphi^{(\pi\circ\gamma)_{r+1}}\otimes\cdots\right.\\
&\left.\otimes\varphi^{(\pi\circ\gamma)_{r+p}}\right),\\
&=\sum_{\beta,\gamma}(-1)^{|\beta|}(-1)^{|\gamma|}\varphi^{\beta_1}\wedge\cdots\wedge\varphi^{\beta_r}\wedge\varphi^{\gamma_{r+1}}\wedge\cdots\wedge\varphi^{\gamma_{r+p}},\\
&=\sum_{\beta,\gamma}(-1)^{|\beta|}(-1)^{|\gamma|}(-1)^{|\beta|}(-1)^{|\gamma|}\varphi^{1}\wedge\cdots\wedge\varphi^{r}\wedge\varphi^{r+1}\wedge\cdots\wedge\varphi^{r+p},\\
&=r!p!\varphi^{1}\wedge\cdots\wedge\varphi^{r}\wedge\varphi^{r+1}\wedge\cdots\wedge\varphi^{r+p},
\end{aligned}
\tag{5.76}
$$

where we used (5.68) in the fifth line. We conclude that

$$
\omega\wedge\psi=\frac{(r+p)!}{r!p!}\mathfrak{A}(\omega\otimes\psi). \tag{5.77}
$$

The exterior product of decomposable alternating tensors, as defined above, may be extended to the exterior product of any two tensors, particularly, two alternating tensors. For any two tensors ω and ψ, one may define the exterior product $\omega\wedge\psi$ by using Eq. (5.77). Since the alternation operator is linear, it follows that this definition is compatible with the definition of the exterior product of two decomposable alternating tensors and with the representations of alternating tensors in terms of the decomposable multi-covectors $\{e^{\gamma_1}\wedge\cdots\wedge e^{\gamma_r}\}$. Specifically, for an r-alternating tensor, ω, and a p-alternating tensor ψ, one has

$$
\begin{aligned}
\omega\wedge\psi&=\frac{(r+p)!}{r!p!}\mathfrak{A}(\omega\otimes\psi)\\
&=\frac{(r+p)!}{r!p!}\mathfrak{A}\left[\left(\frac{1}{r!}\sum_{I}\omega_I e^{i_1}\wedge\cdots\wedge e^{i_r}\right)\right.\\
&\quad\left.\otimes\left(\frac{1}{p!}\sum_{J}\psi_J e^{j_{r+1}}\wedge\cdots\wedge e^{j_{r+p}}\right)\right],\\
&=\frac{(r+p)!}{r!p!}\sum_{I,J}\frac{1}{r!p!}\omega_I\psi_J\mathfrak{A}\left[\left(e^{i_1}\wedge\cdots\wedge e^{i_r}\right)\otimes\left(e^{j_{r+1}}\wedge\cdots\wedge e^{j_{r+p}}\right)\right],\\
&=\frac{(r+p)!}{r!p!}\sum_{I,J}\frac{1}{(r+p)!}\omega_I\psi_J e^{i_1}\wedge\cdots\wedge e^{i_r}\wedge e^{j_{r+1}}\wedge\cdots\wedge e^{j_{r+p}},\\
&=\frac{1}{r!p!}\sum_{I,J}\omega_I\psi_J e^{i_1}\wedge\cdots\wedge e^{i_r}\wedge e^{j_{r+1}}\wedge\cdots\wedge e^{j_{r+p}},
\end{aligned}
\tag{5.78}
$$

where $I : \{1, \dots, r\} \to \{1, \dots, n\}$, $J : \{r+1, \dots, r+p\} \to \{1, \dots, n\}$. (It is noted that one could write $j_1, \dots, j_p$ for $j_{r+1}, \dots, j_{r+p}$ so that $J : \{1, \dots, p\} \to \{1, \dots, n\}$.) Each nonvanishing term above may be represented by a multi-index $K : \{1, \dots, r+p\} \to \{1, \dots, n\}$ such that $K|_{\{1,\dots,r\}} = I$, $K|_{\{r+1,\dots,r+p\}} = J$, and

$$\omega \wedge \psi = \frac{1}{r!p!} \sum_K \omega_{k_1 \dots k_r} \psi_{k_{r+1} \dots k_{r+p}} e^{k_1} \wedge \dots \wedge e^{k_{r+p}}. \tag{5.79}$$

Returning to the last line of (5.78), we may write $I = \alpha \circ \pi'$, $J = \beta \circ \pi''$, where α and β are increasing and π', π'' are permutations of $\{1, \dots, r\}$ and $\{r+1, \dots, r+p\}$, respectively. The decomposition is obviously unique. Thus,

$$\begin{aligned}
\omega \wedge \psi &= \frac{1}{r!p!} \sum_{(\alpha),(\beta)} \left(\sum_{\pi',\pi''} \omega_{\alpha\circ\pi'} \psi_{\beta\circ\pi''} e^{(\alpha\circ\pi')_1} \wedge \cdots \right. \\
&\qquad \left. \wedge\, e^{(\alpha\circ\pi')_r} \wedge e^{(\beta\circ\pi'')_{r+1}} \wedge \cdots \wedge e^{(\beta\circ\pi'')_{r+p}} \right), \\
&= \frac{1}{r!p!} \sum_{(\alpha),(\beta)} \left(\sum_{\pi',\pi''} (-1)^{2(|\pi'|+|\pi''|)} \omega_\alpha \psi_\beta e^{\alpha_1} \wedge \cdots \right. \\
&\qquad \left. \wedge\, e^{\alpha_r} \wedge e^{\beta_{r+1}} \wedge \cdots \wedge e^{\beta_{r+p}} \right), \\
&= \sum_{(\alpha),(\beta)} \omega_\alpha \psi_\beta e^{\alpha_1} \wedge \cdots \wedge e^{\alpha_r} \wedge e^{\beta_{r+1}} \wedge \cdots \wedge e^{\beta_{r+p}}, \\
&= \sum_\mu \omega_{\mu_1 \dots \mu_r} \psi_{\mu_{r+1} \dots \mu_{r+p}} e^{\mu_1} \wedge \cdots \wedge e^{\mu_{r+p}},
\end{aligned} \tag{5.80}$$

where the multi-indices μ are injections $\{1, \dots, r+p\} \to \{1, \dots, n\}$ such that the restrictions $\mu|_{\{1,\dots,r\}}$ and $\mu|_{\{r+1,\dots,r+p\}}$ are increasing. Each such multi-index μ may be represented uniquely in the form $\mu = \gamma \circ \rho$, where γ is increasing and ρ is a permutation of $\{1, \dots, r+p\}$ such that $\rho|_{\{1,\dots,r\}}$ and $\rho|_{\{r+1,\dots,r+p\}}$ are increasing. (See [dR84a, p. 18] and [Whi57, p. 41].) A permutation such as ρ is referred to as an (r, p)-*shuffle* (e.g., [AMR88, p. 394]). Using this representation in (5.80), we have

$$\begin{aligned}
\omega \wedge \psi &= \sum_{(\gamma)} \left(\sum_\rho \omega_{(\gamma\circ\rho)_1 \dots (\gamma\circ\rho)_r} \psi_{(\gamma\circ\rho)_{r+1} \dots (\gamma\circ\rho)_{r+p}} e^{(\gamma\circ\rho)_1} \wedge \cdots \wedge e^{(\gamma\circ\rho)_{r+p}} \right), \\
&= \sum_{(\gamma)} \left(\sum_\rho (-1)^{|\rho|} \omega_{(\gamma\circ\rho)_1 \dots (\gamma\circ\rho)_r} \psi_{(\gamma\circ\rho)_{r+1} \dots (\gamma\circ\rho)_{r+p}} e^{\gamma_1} \wedge \cdots \wedge e^{\gamma_{r+p}} \right),
\end{aligned} \tag{5.81}$$

and so

$$
\begin{aligned}
(\omega \wedge \psi)_\gamma &= \sum_\rho (-1)^{|\rho|} \omega_{(\gamma\circ\rho)_1\dots(\gamma\circ\rho)_r} \psi_{(\gamma\circ\rho)_{r+1}\dots(\gamma\circ\rho)_{r+p}}, \\
&= \sum_\mu \varepsilon^{\mu_1\dots\mu_{r+p}}_{\gamma_1\dots,\gamma_{r+p}} \omega_{\mu_1\dots\mu_r} \psi_{\mu_{r+1}\dots\mu_{r+p}},
\end{aligned} \tag{5.82}
$$

where the sums are taken over shuffles ρ and $\mu = \gamma \circ \rho$. Clearly, one could obtain the last equation by starting with the representations of ω and ψ in terms of the multivectors $e^{\gamma_1} \wedge \cdots \wedge e^{\gamma_r}$ and $e^{\gamma_{r+1}} \wedge \cdots \wedge e^{\gamma_{r+p}}$ as in Eq. (5.24).

Using Eqs. (5.57), (5.77), and (5.78), the action of the exterior product may be written as

$$
\omega \wedge \psi(v_1, \dots, v_{r+p}) = \frac{1}{r!p!} \sum_\pi (-1)^{|\pi|} \omega(v_{\pi_1}, \dots, v_{\pi_r}) \psi(v_{\pi_{r+1}}, \dots, v_{\pi_{r+p}}) \tag{5.83}
$$

Here, π is not a shuffle, so that, for example, $\pi_1, \dots, \pi_r$ need not be increasing. However, we can write each such permutation as follows: For $l = 1, \dots, r$, set $\pi_l = (\alpha \circ \mu)_l$, where μ is an (r, p)-shuffle and α is a permutation of $\{1, \dots, r\}$. For $l = r + 1, \dots, r + p$, set $\pi_l = (\beta \circ \mu)_l$, where β is a permutation of p symbols. Hence, with $|\pi| = |\mu| + |\alpha| + |\beta|$,

$$
\begin{aligned}
&\omega \wedge \psi(v_1, \dots, v_{r+p}) \\
&= \frac{1}{r!p!} \sum_{\mu,\alpha,\beta} (-1)^{|\mu|+|\alpha|+|\beta|} \omega(v_{(\alpha\circ\mu)_1}, \dots, v_{(\alpha\circ\mu)_r}) \psi(v_{(\beta\circ\mu)_{r+1}}, \dots, v_{(\beta\circ\mu)_{r+p}}), \\
&= \frac{1}{r!p!} \sum_{\mu,\alpha,\beta} (-1)^{|\mu|+|\alpha|+|\beta|} (-1)^{|\alpha|} (-1)^{|\beta|} \omega(v_{\mu_1}, \dots, v_{\mu_r}) \psi(v_{\mu_{r+1}}, \dots, v_{\mu_{r+p}}) \\
&= \frac{1}{r!p!} \sum_{\mu,\alpha,\beta} (-1)^{|\mu|} \omega(v_{\mu_1}, \dots, v_{\mu_r}) \psi(v_{\mu_{r+1}}, \dots, v_{\mu_{r+p}}), ,
\end{aligned} \tag{5.84}
$$

so taking account of the number of permutations α and β, we have

$$
\omega \wedge \psi(v_1, \dots, v_{r+p}) = \sum_\mu (-1)^{|\mu|} \omega(v_{\mu_1}, \dots, v_{\mu_r}) \psi(v_{\mu_{r+1}}, \dots, v_{\mu_{r+p}}), \tag{5.85}
$$

where the sum is taken over all (r, p)-shuffles μ.

Remark 5.2 There are other conventions for the definition of the exterior product. See, for example, [War83, p. 60], [Whi57], and [AMR88]. From the point of view of fluxes, as presented here, one may think of the other conventions as pertaining to the interpretation of $v_1 \wedge \cdots \wedge v_r$ as the extent of the parallelepiped generated by the vectors rather than the extent of the simplex.

Example 5.1 Consider the case where ω above is a 1-alternating tensor, i.e., an element of the dual space $\mathbf{V}^*$. Thus, if we have the component representations $\omega = \sum_i \omega_i e^i$, $\psi = \sum_{(\beta)} \psi_\beta e^{\beta_1} \wedge \cdots \wedge e^{\beta_p}$, Eq. (5.73) and linearity imply that

$$\omega \wedge \psi = \sum_{(\beta),i} \omega_i \psi_\beta e^i \wedge e^{\beta_1} \wedge \cdots \wedge e^{\beta_p}. \tag{5.86}$$

Alternatively, we may use Eq. (5.82) noting that for the case $r = 1$ considered, any $(1, p)$ shuffle ρ is of the form $(1, \dots, l, \dots, p+1) \mapsto (l, 1, \dots, \widehat{l}, \dots, p+1)$, so that $|\rho| = l - 1$. It follows that

$$(\omega \wedge \psi)_\gamma = \sum_l (-1)^{l-1} \omega_{\gamma_l} \psi_{\gamma_1 \dots \widehat{\gamma_l} \dots \gamma_{p+1}}, \tag{5.87}$$

and

$$\begin{aligned} \omega \wedge \psi &= \sum_{(\gamma),l} (-1)^{l-1} \omega_{\gamma_l} \psi_{\gamma_1 \dots \widehat{\gamma_l} \dots \gamma_{p+1}} e^{\gamma_1} \wedge \cdots \wedge e^{\gamma_{p+1}}, \\ &= \sum_{(\gamma),l} \omega_{\gamma_l} \psi_{\gamma_1 \dots \widehat{\gamma_l} \dots \gamma_{p+1}} e^{\gamma_l} \wedge e^{\gamma_1} \wedge \cdots \wedge \widehat{e^{\gamma_l}} \wedge \cdots \wedge e^{\gamma_{p+1}}. \end{aligned} \tag{5.88}$$

Example 5.2 Consider the relation between $\psi \wedge \omega$ and $\omega \wedge \psi$ for an r-alternating tensor ω and a p-alternating tensor ψ. From Eq. (5.78) and counting the number of permutations needed in order to arrive from $e^{i_1} \wedge \cdots \wedge e^{i_r} \wedge e^{j_1} \wedge \cdots \wedge e^{j_p}$ to $e^{j_1} \wedge \cdots \wedge e^{j_p} \wedge e^{i_1} \wedge \cdots \wedge e^{i_r}$, it follows that

$$\psi \wedge \omega = (-1)^{rp} \omega \wedge \psi. \tag{5.89}$$

In particular, if either r or p is even, $\psi \wedge \omega = \omega \wedge \psi$.

In this section, we have considered exterior products of alternating covariant tensors—multi-covectors. An analogous treatment applies to exterior products of contravariant tensors—multivectors. In particular, the vector space, $\mathbf{V}$, may be identified naturally with the dual space to $\mathbf{V}^*$, the double dual. Thus, we will use below exterior products of multivectors.

5.7 The Spaces of Multivectors and Multi-Covectors

Now that the base vectors of $(\bigwedge^r \mathbf{V})^*$ have been represented by exterior products, $e^{\gamma_1} \wedge \cdots \wedge e^{\gamma_r}$, of basis elements in $\mathbf{V}^*$ for all increasing multi-indices γ, we may write $\bigwedge^r \mathbf{V}^* := \bigwedge^r (\mathbf{V}^*) = (\bigwedge^r \mathbf{V})^*$. Recalling Equation (5.56),

$$\dim \bigwedge^r \mathbf{V}^* = \frac{n!}{r!(n-r)!}. \tag{5.90}$$

In particular, $\dim \bigwedge^0 \mathbf{V}^* = \dim \bigwedge^n \mathbf{V}^* = 1$, and $\dim \bigwedge^1 \mathbf{V}^* = \dim \bigwedge^{n-1} \mathbf{V}^* = n$. The basis of $\bigwedge^0 \mathbf{V}^*$ contains no elements of the basis of $\mathbf{V}^*$ and so $\bigwedge^0 \mathbf{V}^*$ may be identified with the real numbers. For $r > n$, the vector space $\bigwedge^r \mathbf{V}^*$ is trivial.

Example 5.3 (Algebraic Cauchy Formula with Components) Consider the case where $r = n$ and let $\{e_1, \ldots, e_n\}$ be a basis for $\mathbf{V}$. We want to compute the total flux through the $(n-1)$-simplex τ determined by the vectors $v_1, \ldots, v_{n-1}$ for a flux ω represented as in Eq. (4.49) by an alternating tensor $J = \sum_{(\alpha)} J_{\alpha_1 \ldots \alpha_{n-1}} e^{\alpha_1} \wedge \cdots \wedge e^{\alpha_{n-1}}$. Since α is increasing and contains $n-1$ out of the n numbers $\{1, \ldots, n\}$, there is a unique $k \in \{1, \ldots, n\}$ such that $(\alpha_1, \ldots, \alpha_{n-1}) = (1, \ldots, \widehat{k}, \ldots, n\}$. It follows that any $(n-1)$-alternating tensor J is of the form

$$J = \sum_k J_{1 \ldots \widehat{k} \ldots n} e^1 \wedge \cdots \wedge \widehat{e^k} \wedge \cdots \wedge e^n, \tag{5.91}$$

for components $J_{1 \ldots \widehat{k} \ldots n}$, $k = 1, \ldots, n$. Thus,

$$\begin{aligned}
\omega(\tau) &= J(v_1, \ldots, v_{n-1}), \\
&= \sum_k J_{1 \ldots \widehat{k} \ldots n} e^1 \wedge \cdots \wedge \widehat{e^k} \wedge \cdots \wedge e^n \left(\sum_{i_1} v_1^{i_1} e_{i_1}, \ldots, \sum_{i_{n-1}} v_{n-1}^{i_{n-1}} e_{i_{n-1}} \right), \\
&= \sum_{k,I} J_{1 \ldots \widehat{k} \ldots m} v_1^{i_1} \cdots v_{n-1}^{i_{n-1}} e^1 \wedge \cdots \wedge \widehat{e^k} \wedge \cdots \wedge e^n \left(e_{i_1}, \ldots, e_{i_{n-1}} \right), \\
&= \sum_{k,I} J_{1 \ldots \widehat{k} \ldots n} v_1^{i_1} \cdots v_{n-1}^{i_{n-1}} \varepsilon^{1 \ldots \widehat{k} \ldots n}_{i_1 \ldots \ldots i_{n-1}}, \qquad i_p = 1, \ldots, n, \ p = 1, \ldots, n-1.
\end{aligned} \tag{5.92}$$

Since

$$\varepsilon^{1 \ldots \widehat{k} \ldots n}_{i_1 \ldots \ldots i_{n-1}} = \varepsilon^{k 1 \ldots \widehat{k} \ldots n}_{k\, i_1 \ldots \ldots i_{n-1}} = (-1)^{k-1} \varepsilon^{1 \ldots \ldots n}_{k\, i_1 \ldots i_{n-1}}, \tag{5.93}$$

we have

$$\begin{aligned}
\omega(\tau) &= \sum_{k,I} \left[(-1)^{k-1} J_{1 \ldots \widehat{k} \ldots n} \right] v_1^{i_1} \cdots v_{n-1}^{i_{n-1}} \varepsilon^{1 \ldots \ldots n}_{k\, i_1 \ldots i_{n-1}}, \\
&= \det \left(\left[(-1)^{k-1} J_{1 \ldots \widehat{k} \ldots n} \right] ; \ v_1^{i_1} ; \ \cdots \ ; \ v_{n-1}^{i_{n-1}} \right),
\end{aligned} \tag{5.94}$$

where the determinant is evaluated for the column vectors indicated by the indices $k, i_1, \ldots, i_{n-1} = 1, \ldots, n$.

Consider the expression for the flux in a three-dimensional Euclidean space using the traditional Cauchy formula (4.51) relative to some orthonormal basis,

$$\omega(\tau) = \tfrac{1}{2}\mathbf{J}\cdot(v\times u) = \tfrac{1}{2}\det\begin{bmatrix} \mathbf{J}_1 & v_1 & u_1 \\ \mathbf{J}_2 & v_2 & u_2 \\ \mathbf{J}_3 & v_3 & u_3 \end{bmatrix} = \tfrac{1}{2}\sum_{i,j,k}\varepsilon^{ijk}\mathbf{J}_i v_j u_k. \tag{5.95}$$

Comparing the last equality with Eq. (5.94), we see that the two equations are equivalent if we make the identification $\mathbf{J}_k = 2(-1)^{k-1}J_{1\ldots\widehat{k}\ldots r}$. (The $\frac{1}{2}$ factor is due to the fact that the vector product evaluates the area of the parallelogram, while the alternating tensors compute the fluxes through simplices.) It is also noted that $v\times u$ is the analog of the 2-vector $v\wedge u$.

5.8 Contraction

The *contraction* $v \lrcorner \omega$ of an alternating r-tensor ω and a vector v is an $(r-1)$-alternating tensor defined by

$$v \lrcorner \omega(v_2,\ldots,v_r) = \omega(v, v_2,\ldots,v_r). \tag{5.96}$$

Clearly, the contraction operation $v \lrcorner \omega$ is linear in both v and ω.

Using (5.32) and (5.70), one has

$$\begin{aligned} v \lrcorner \omega(v_2,\ldots,v_r) &= \sum_{(\gamma),I}\omega_\gamma v^{i_1}v_2^{i_2}\cdots v_r^{i_r}\varepsilon_I^\gamma \\ &= \sum_{(\gamma),I,J}\omega_\gamma v^{i_1}\frac{1}{(r-1)!}\varepsilon^{\gamma_1\ldots\gamma_r}_{i_1 j_1\ldots j_{r-1}}\varepsilon^{j_1\ldots j_{r-1}}_{i_2\ldots,i_r}v_2^{i_2}\cdots v_r^{i_r}, \\ &= \sum_{(\gamma),I,J}\omega_\gamma v^{i_1}\frac{1}{(r-1)!}\varepsilon^{\gamma_1\ldots\gamma_r}_{i_1 j_1\ldots j_{r-1}}e^{j_1}\wedge\cdots\wedge e^{j_{r-1}}(v_2,\ldots,v_r), \\ &= \sum_{(\gamma),i,(\beta)}\omega_\gamma v^i\varepsilon^{\gamma_1\ldots\gamma_r}_{i\beta_1\ldots\beta_{r-1}}e^{\beta_2}\wedge\cdots\wedge e^{\beta_{r-1}}(v_2,\ldots,v_r). \end{aligned} \tag{5.97}$$

Here, I is an r-multi-index, J is an $(r-1)$-multi-index, and β is an $(r-1)$-increasing multi-index. It follows that

$$v \lrcorner \omega = \sum_{(\gamma),i,(\beta)}\omega_\gamma v^i\varepsilon^{\gamma_1\ldots\gamma_r}_{i\beta_1\ldots\beta_{r-1}}e^{\beta_2}\wedge\cdots\wedge e^{\beta_{r-1}}, \tag{5.98}$$

and

$$(v \lrcorner \omega)_{\beta_1\dots\beta_{r-1}} = \sum_{(\gamma),i} \omega_\gamma v^i \varepsilon^{\gamma_1\dots\gamma_r}_{i\beta_1\dots\beta_{r-1}}. \tag{5.99}$$

It is noted that in (5.99), for a given value of β and any increasing multi-index γ, there is a unique value $i = \gamma \setminus \beta$ for which the Levi-Civita symbol does not vanish. Thus, the sum over i in that equation is superfluous. Its location in the sequence $\gamma_1 \dots, \gamma_r$ is $\gamma^{-1}\{\gamma \setminus \beta\}$, and as $(\beta_1, \dots, \beta_{r-1}) = (\gamma_1, \dots, \widehat{\gamma \setminus \beta}, \dots, \gamma_r)$, one has

$$\begin{aligned}
(v \lrcorner \omega)_{\beta_1\dots\beta_{r-1}} &= \sum_{(\gamma)} \omega_\gamma v^i \varepsilon^{\gamma_1\dots\gamma_r}_{i\beta_1\dots\beta_{r-1}}, \\
&= \sum_{(\gamma)} \omega_{\gamma_1\dots\gamma\setminus\beta\dots\gamma_r} v^{\gamma\setminus\beta} \varepsilon^{\gamma_1\dots\gamma\setminus\beta\dots\gamma_r}_{\gamma\setminus\beta\,\beta_1\dots\beta_{r-1}}, \\
&= \sum_{(\gamma)} \omega_{\gamma_1\dots\gamma\setminus\beta\dots\gamma_r} v^{\gamma\setminus\beta} (-1)^{\gamma^{-1}\{\gamma\setminus\beta\}-1}.
\end{aligned} \tag{5.100}$$

We conclude that

$$(v \lrcorner \omega)_{\beta_1\dots\beta_{r-1}} = \sum_{(\gamma)} \omega_{\gamma\setminus\beta\,\gamma_1\dots\widehat{\gamma\setminus\beta}\dots\gamma_r} v^{\gamma\setminus\beta}. \tag{5.101}$$

Example 5.4 In the special case where $r = n$, an alternating tensor θ is represented by a single component $\theta_{1\dots n}$ as $\theta = \theta_{1\dots n} e^1 \wedge \dots \wedge e^n$. As in Example 5.3, any $(n-1)$-alternating tensor ψ is of the form $\psi = \sum_k \psi_{1\dots\widehat{k}\dots n} e^1 \wedge \dots \wedge \widehat{e^k} \wedge \dots e^n$. For an alternating n-tensor θ and a vector v, we consider $v \lrcorner \theta$. Using the third line of Eq. (5.100) with $\beta = \{1, \dots, \widehat{k}, \dots, n\}$, $\gamma \setminus \beta = k$, $\gamma^{-1}(\gamma \setminus \beta) = k$,

$$(v \lrcorner \theta)_{1\dots\widehat{k}\dots n} = \sum_k (-1)^{k-1} \theta_{1\dots n} v^k, \quad v \lrcorner \theta = \sum_k (-1)^{k-1} \theta_{1\dots n} v^k e^1 \wedge \dots \wedge \widehat{e^k} \wedge \dots e^n. \tag{5.102}$$

We conclude that any nonzero n-alternating tensor, θ, induces an isomorphism

$$\mathcal{J}_\theta : \bigwedge\nolimits^{n-1} \mathbf{V}^* \longrightarrow \mathbf{V} \tag{5.103}$$

defined by the condition that

$$\mathcal{J}_\theta(\omega) \lrcorner \theta = \omega. \tag{5.104}$$

Equation (5.102) implies that in terms of components

$$\mathcal{J}_\theta(\omega)^k = (-1)^{k-1} \frac{\omega_{1\dots\widehat{k}\dots n}}{\theta_{1\dots n}}. \tag{5.105}$$

Example 5.5 Consider Equation (5.83). One may represent any permutation π of $r+p$ symbols in the form $\pi = \rho_i \circ \mu$, where μ is a permutation of $\{1, \ldots, r+p\}$ with $\mu_1 = 1$ and ρ_i is the permutation of $r+p$ symbols given by

$$(1, 2, \ldots, i, \ldots, r+p) \longmapsto (2, \ldots, i, 1, i+1, \ldots, r+p).$$

Clearly, $|\pi| = |\rho_i| + |\mu| = i - 1 + |\mu|$, so that Eq. (5.83) may be rewritten as

$$\begin{aligned}
&\omega \wedge \psi(v_1, \ldots, v_{r+p}) \\
&= \frac{1}{r!p!} \sum_{\mu,\rho_i} (-1)^{|\mu|}(-1)^{i-1} \omega(v_{(\rho_i \circ \mu)_1}, \ldots, v_{(\rho_i \circ \mu)_r}) \psi(v_{(\rho_i \circ \mu)_{r+1}}, \ldots, v_{(\rho_i \circ \mu)_{r+p}}), \\
&= \frac{1}{r!p!} \sum_{\mu,\rho_i,i\leqslant r} (-1)^{|\mu|}(-1)^{i-1} \omega(v_{(\rho_i \circ \mu)_1}, \ldots, v_{(\rho_i \circ \mu)_r}) \\
&\quad \times \psi(v_{(\rho_i \circ \mu)_{r+1}}, \ldots, v_{(\rho_i \circ \mu)_{r+p}}) \\
&\quad + \frac{1}{r!p!} \sum_{\mu,\rho_i,i>r} (-1)^{|\mu|}(-1)^{i-1} \omega(v_{(\rho_i \circ \mu)_1}, \ldots, v_{(\rho_i \circ \mu)_r}) \\
&\quad \times \psi(v_{(\rho_i \circ \mu)_{r+1}}, \ldots, v_{(\rho_i \circ \mu)_{r+p}}), \\
&= \frac{1}{r!p!} \sum_{\mu,\rho_i,i\leqslant r} (-1)^{|\mu|}(-1)^{2(i-1)} \omega(v_1, v_{\mu_2} \ldots, v_{\mu_r}) \psi(v_{\mu_{r+1}}, \ldots, v_{\mu_{r+p}}) \\
&\quad + \frac{1}{r!p!} \sum_{\mu,\rho_i,i>r} (-1)^{|\mu|}(-1)^{i-1} \omega(v_{\mu_2}, \ldots, v_{\mu_{r+1}})(-1)^{i-r-1} \\
&\quad \times \psi(v_1, v_{\mu_{r+2}}, \ldots, v_{\mu_{r+p}}), \\
&= \frac{1}{r!p!} \sum_{\mu,\rho_i,i\leqslant r} (-1)^{|\mu|}(v_1 \lrcorner \omega)(v_{\mu_2} \ldots, v_{\mu_r}) \psi(v_{\mu_{r+1}}, \ldots, v_{\mu_{r+p}}) \\
&\quad + \frac{(-1)^r}{r!p!} \sum_{\mu,\rho_i,i>r} (-1)^{|\mu|} \omega(v_{\mu_2}, \ldots, v_{\mu_{r+1}})(v_1 \lrcorner \psi)(v_{\mu_{r+2}}, \ldots, v_{\mu_{r+p}}), \\
&= \frac{r}{r!p!} \sum_{\mu} (-1)^{|\mu|}(v_1 \lrcorner \omega)(v_{\mu_2} \ldots, v_{\mu_r}) \psi(v_{\mu_{r+1}}, \ldots, v_{\mu_{r+p}}) \\
&\quad + \frac{(-1)^r p}{r!p!} \sum_{\mu} (-1)^{|\mu|} \omega(v_{\mu_2}, \ldots, v_{\mu_{r+1}})(v_1 \lrcorner \psi)(v_{\mu_{r+2}}, \ldots, v_{\mu_{r+p}}), \\
&= [(v_1 \lrcorner \omega) \wedge \psi](v_2, \ldots, v_{r+p}) + (-1)^r [\omega \wedge (v_1 \lrcorner \psi)](v_2, \ldots, v_{r+p}).
\end{aligned} \tag{5.106}$$

It is concluded therefore that

$$v \lrcorner (\omega \wedge \psi) = (v \lrcorner \omega) \wedge \psi + (-1)^r \omega \wedge (v \lrcorner \psi). \tag{5.107}$$

Example 5.6 Consider the expression $e_{i_p} \lrcorner (e^{i_p} \wedge e^{i_1} \wedge \cdots \wedge \widehat{e^{i_p}} \wedge \cdots \wedge e^{i_r})$ where the indices $i_1, \ldots, i_r$ are all distinct so that the exterior product does not vanish. Then, using the last equation

$$\begin{aligned} e_{i_p} &\lrcorner (e^{i_p} \wedge e^{i_1} \wedge \cdots \wedge \widehat{e^{i_p}} \wedge \cdots \wedge e^{i_r}) \\ &= e_{i_p} \lrcorner (e^{i_p} \wedge (e^{i_1} \wedge \cdots \wedge \widehat{e^{i_p}} \wedge \cdots \wedge e^{i_r})), \\ &= (e_{i_p} \lrcorner e^{i_p}) \wedge (e^{i_1} \wedge \cdots \wedge \widehat{e^{i_p}} \wedge \cdots \wedge e^{i_r}) \\ &\quad - e^{i_p} \wedge (e_{i_p} \lrcorner (e^{i_1} \wedge \cdots \wedge \widehat{e^{i_p}} \wedge \cdots \wedge e^{i_r})), \\ &= e^{i_1} \wedge \cdots \wedge \widehat{e^{i_p}} \wedge \cdots \wedge e^{i_r}. \end{aligned} \tag{5.108}$$

Using the skew symmetry of the exterior product, we conclude that

$$e_{i_p} \lrcorner (e^{i_1} \wedge \cdots \wedge e^{i_p} \wedge \cdots \wedge e^{i_r}) = (-1)^{p-1} e^{i_1} \wedge \cdots \wedge \widehat{e^{i_p}} \wedge \cdots \wedge e^{i_r}, \tag{5.109}$$

and

$$e_j \lrcorner (e^{i_1} \wedge \cdots \wedge e^{i_r}) = \sum_p \delta_j^{i_p} (-1)^{p-1} e^{i_1} \wedge \cdots \wedge \widehat{e^{i_p}} \wedge \cdots \wedge e^{i_r}. \tag{5.110}$$

Clearly, there can be no more than one nonvanishing term in the last sum.

Example 5.7 Let $v = \sum_i v^i e_i$; then, by linearity

$$\begin{aligned} v \lrcorner (e^{i_1} \wedge \cdots \wedge e^{i_r}) &= \sum_{j,p} v^j \delta_j^{i_p} (-1)^{p-1} e^{i_1} \wedge \cdots \wedge \widehat{e^{i_p}} \wedge \cdots \wedge e^{i_r}, \\ &= \sum_p (-1)^{p-1} v^{i_p} e^{i_1} \wedge \cdots \wedge \widehat{e^{i_p}} \wedge \cdots \wedge e^{i_r}. \end{aligned} \tag{5.111}$$

The following examples will be used in Chap. 11 concerning metric-free properties of electromagnetism.

Example 5.8 Consider the contraction $v \lrcorner \omega$ for an $(n-1)$-form $\omega = \sum_k \omega_{1\ldots\widehat{k}\ldots n} e^1 \wedge \cdots \wedge \widehat{e^k} \wedge \ldots e^n$. Writing $v = \sum_i v^i e_i$ one has

$$
\begin{aligned}
v \lrcorner \omega &= \Big(\sum_i v^i e_i\Big) \lrcorner \Big(\sum_k \omega_{1\ldots\widehat{k}\ldots n} e^1 \wedge \cdots \wedge \widehat{e^k} \wedge \ldots e^n\Big), \\
&= \sum_{i,k} v^i \omega_{1\ldots\widehat{k}\ldots n} e_i \lrcorner \big(e^1 \wedge \cdots \wedge \widehat{e^k} \wedge \ldots e^n\big).
\end{aligned}
\tag{5.112}
$$

Note that for $i < k$,

$$
e_i \lrcorner \big(e^1 \wedge \cdots \wedge \widehat{e^k} \wedge \ldots e^n\big) = (-1)^{i-1} e^1 \wedge \cdots \wedge \widehat{e^i} \wedge \cdots \wedge \widehat{e^k} \wedge \cdots \wedge e^n, \tag{5.113}
$$

and for $i > k$

$$
e_i \lrcorner \big(e^1 \wedge \cdots \wedge \widehat{e^k} \wedge \ldots e^n\big) = (-1)^{i-2} e^1 \wedge \cdots \wedge \widehat{e^k} \wedge \cdots \wedge \widehat{e^i} \wedge \cdots \wedge e^n. \tag{5.114}
$$

Thus, one has

$$
\begin{aligned}
v \lrcorner \omega = &\sum_{i<k} v^i \omega_{1\ldots\widehat{k}\ldots n} (-1)^{i-1} e^1 \wedge \cdots \wedge \widehat{e^i} \wedge \cdots \wedge \widehat{e^k} \wedge \cdots \wedge e^n \\
&- \sum_{i>k} v^i \omega_{1\ldots\widehat{k}\ldots n} (-1)^{i-1} e^1 \wedge \cdots \wedge \widehat{e^k} \wedge \cdots \wedge \widehat{e^i} \wedge \cdots \wedge e^n,
\end{aligned}
\tag{5.115}
$$

Switching the names of the indices in the preceding line, we arrive at

$$
v \lrcorner \omega = \sum_{i<k} \Big((-1)^{i-1} v^i \omega_{1\ldots\widehat{k}\ldots n} - (-1)^{k-1} v^k \omega_{1\ldots\widehat{i}\ldots n}\Big) e^1 \wedge \cdots \wedge \widehat{e^i} \wedge \cdots \wedge \widehat{e^k} \wedge \cdots \wedge e^n. \tag{5.116}
$$

Example 5.9 Using the notation of the preceding example, let $\phi = \sum_{i<j} \phi_{ij} e^i \wedge e^i$ be a 2-form. Using Eq. (5.107), and noting that $\phi \wedge \omega$ vanishes as an $(n+1)$-form in an n-dimensional space,

$$
0 = v \lrcorner (\phi \wedge \omega) = (v \lrcorner \phi) \wedge \omega + (-1)^2 \phi \wedge (v \lrcorner \omega), \tag{5.117}
$$

and so

$$
(v \lrcorner \phi) \wedge \omega = -\phi \wedge (v \lrcorner \omega). \tag{5.118}
$$

We will also demonstrate this equation by components computation. We have computed the components of $v \lrcorner \omega$ above, and we may write

$$
\begin{aligned}
&\phi \wedge (v \lrcorner \omega) \\
&= \Big(\sum_{i<j} \phi_{ij} e^i \wedge e^j\Big) \wedge \Big[\sum_{p<q} \Big(\big((-1)^{p-1} v^p \omega_{1\ldots\widehat{q}\ldots n} - (-1)^{q-1} v^q \omega_{1\ldots\widehat{p}\ldots n}\big) \\
&\qquad\qquad \cdot e^1 \wedge \cdots \wedge \widehat{e^p} \wedge \cdots \wedge \widehat{e^q} \wedge \cdots \wedge e^n\Big)\Big], \\
&= \sum_{i<j,\, p<q} \phi_{ij} \big((-1)^{p-1} v^p \omega_{1\ldots\widehat{q}\ldots n} - (-1)^{q-1} v^q \omega_{1\ldots\widehat{p}\ldots n}\big) \\
&\qquad\qquad \cdot (-1)^{j-2} \delta^{jq} (-1)^{i-1} \delta^{ip} e^1 \wedge \cdots \wedge e^n, \\
&= -\sum_{i<j} \phi_{ij} (-1)^{j-1} v^i \omega_{1\ldots\widehat{j}\ldots n} e^1 \wedge \cdots \wedge e^n \\
&\qquad\qquad + \sum_{i<j} \phi_{ij} (-1)^{i-1} v^j \omega_{1\ldots\widehat{i}\ldots n} e^1 \wedge \cdots \wedge e^n.
\end{aligned} \tag{5.119}
$$

Now, switching the names of the indices in the first sum of the last equality above, one has

$$
\begin{aligned}
\sum_{i<j} \phi_{ij} (-1)^{j-1} v^i \omega_{1\ldots\widehat{j}\ldots n} &= \sum_{j<i} \phi_{ji} (-1)^{i-1} v^j \omega_{1\ldots\widehat{i}\ldots n}, \\
&= -\sum_{j<i} \phi_{ij} (-1)^{i-1} v^j \omega_{1\ldots\widehat{i}\ldots n},
\end{aligned} \tag{5.120}
$$

where for $j < i$ we set $\phi_{ij} = -\phi_{ji}$. (Recall that ϕ_{ij} was defined originally only for $i < j$.) It follows that

$$
\phi \wedge (v \lrcorner \omega) = \Big(\sum_{j<i} \phi_{ij} (-1)^{i-1} v^j \omega_{1\ldots\widehat{i}\ldots n} + \sum_{i<j} \phi_{ij} (-1)^{i-1} v^j \omega_{1\ldots\widehat{i}\ldots n}\Big) e^1 \wedge \cdots \wedge e^n, \tag{5.121}
$$

and we conclude that

$$
\phi \wedge (v \lrcorner \omega) = \Big(\sum_{i,j} \phi_{ij} (-1)^{i-1} v^j \omega_{1\ldots\widehat{i}\ldots n}\Big) e^1 \wedge \cdots \wedge e^n. \tag{5.122}
$$

We now compute the components of $(v \lrcorner \phi) \wedge \omega$. Firstly,

$$
\begin{aligned}
v \lrcorner \phi &= \Big(\sum_i v^i e_i\Big) \lrcorner \Big(\sum_{p<q} \phi_{pq} e^p \wedge e^q\Big), \\
&= \sum_{i,p<q} \phi_{pq} v^i e_i \lrcorner \Big(e^p \wedge e^q\Big), \\
&= \sum_{i,p<q} \phi_{pq} v^i (\delta_i^p e^q - \delta_i^q e^p), \\
&= \sum_{p<q} \phi_{pq} v^p e^q - \sum_{p<q} \phi_{pq} v^q e^p, \\
&= \sum_{p<q} \phi_{pq} v^p e^q + \sum_{q<p} \phi_{pq} v^p e^q,
\end{aligned}
\tag{5.123}
$$

where in the second sum of the last line we switched the names of the indices and for $p > q$, we set $\phi_{pq} = -\phi_{qp}$. Thus,

$$
v \lrcorner \phi = \sum_{p,q} \phi_{pq} v^p e^q = -\sum_{p,q} \phi_{qp} v^p e^q. \tag{5.124}
$$

Continuing, we obtain

$$
\begin{aligned}
(v \lrcorner \phi) \wedge \omega &= \Big(\sum_{p,q} \phi_{pq} v^p e^q\Big) \wedge \Big(\sum_k \omega_{1\ldots\widehat{k}\ldots n} e^1 \wedge \cdots \wedge \widehat{e^k} \wedge \cdots \wedge e^n\Big), \\
&= \sum_{p,q,k} \phi_{pq} v^p \omega_{1\ldots\widehat{k}\ldots n} e^q \wedge e^1 \wedge \cdots \wedge \widehat{e^k} \wedge \cdots \wedge e^n, \\
&= \sum_{p,q} \phi_{pq} v^p \omega_{1\ldots\widehat{q}\ldots n} (-1)^{q-1} e^1 \wedge \cdots \wedge e^n,
\end{aligned}
\tag{5.125}
$$

which confirms Eq. (5.118).

Example 5.10 Keeping the notation scheme of the previous examples, let $\theta = \theta_{1\ldots n} e^1 \wedge \cdots \wedge e^n$ be an n-form, and consider the n-form $(\mathcal{J}_\theta(\omega) \lrcorner \phi) \wedge \mathcal{J}_\theta^{-1}(v)$ where $\mathcal{J}_\theta$ is the isomorphism defined in Example 5.4. Using Eqs. (5.105) and (5.125),

$$
\begin{aligned}
&(\mathcal{J}_\theta(\omega) \lrcorner \phi) \wedge \mathcal{J}_\theta^{-1}(v) \\
&\quad = \sum_{p,q} \phi_{pq} \mathcal{J}_\theta(\omega)^p \mathcal{J}_\theta^{-1}(v)_{1\ldots\widehat{q}\ldots n} (-1)^{q-1} e^1 \wedge \cdots \wedge e^n, \\
&\quad = \sum_{p,q} \phi_{pq} (-1)^{p-1} \omega_{1\ldots\widehat{p}\ldots n} (-1)^{q-1} v^q (-1)^{q-1} e^1 \wedge \cdots \wedge e^n, \\
&\quad = \sum_{p,q} \phi_{pq} (-1)^{p-1} \omega_{1\ldots\widehat{p}\ldots n} v^q e^1 \wedge \cdots \wedge e^n.
\end{aligned}
\tag{5.126}
$$

As $\phi_{pq} = -\phi_{qp}$, comparing the last expression with (5.125), we conclude that

$$
(\mathcal{J}_\theta(\omega) \lrcorner \phi) \wedge \mathcal{J}_\theta^{-1}(v) = -(v \lrcorner \phi) \wedge \omega = \phi \wedge (v \lrcorner \omega) \tag{5.127}
$$

which is evidently independent of the n-form θ.

In Sect. 6.2.3 we consider an extension of the notion of contraction and some additional properties to the material presented above.

5.9 Pullback of Alternating Tensors

Let $A : \mathbf{V} \to \mathbf{U}$ be a linear mapping between vector spaces and $\omega \in \bigwedge^r(\mathbf{U}^*)$ be an alternating r-tensor defined on $\mathbf{U}$. Then, ω induces an alternating r-tensor $A^*(\omega) \in \bigwedge^r \mathbf{V}^*$ by

$$
A^*(\omega)(w_1, \ldots, w_r) = \omega(A(w_1), \ldots, A(w_r)). \tag{5.128}
$$

Let $\{e_i\}$ and $\{f_p)$ be bases in $\mathbf{V}$ and $\mathbf{U}$, respectively, so that $A(e_i) = \sum_p A_i^p f_p$, where A_i^p are the elements of the matrix of A. Then,

$$
\begin{aligned}
A^*(\omega)_\gamma &= A^*(\omega)(e_{\gamma_1}, \ldots, e_{\gamma_r}), \\
&= \omega\left(A(e_{\gamma_1}), \ldots, A(e_{\gamma_r})\right), \\
&= \omega\Bigl(\sum_{i_1} A_{\gamma_1}^{i_1} f_{i_1}, \ldots, \sum_{i_r} A_{\gamma_r}^{i_r} f_{i_r}\Bigr), \\
&= \sum_I A_{\gamma_1}^{i_1} \cdots A_{\gamma_r}^{i_r} \omega\bigl(f_{i_1}, \ldots, f_{i_r}\bigr), \\
&= \sum_I A_{\gamma_1}^{i_1} \cdots A_{\gamma_r}^{i_r} \omega_{i_1 \ldots i_r}, \\
&= \sum_{I,(\beta)} \varepsilon_{i_1 \ldots i_r}^{\beta_1 \ldots \beta_r} A_{\gamma_1}^{i_1} \cdots A_{\gamma_r}^{i_r} \omega_{\beta_1 \ldots \beta_r}.
\end{aligned}
\tag{5.129}
$$

Using Eq. (5.83), for an r-alternating tensor ω and a p-alternating tensor ψ,

$$\begin{aligned}
&A^*(\omega \wedge \psi)(w_1, \dots, w_{r+p}) \\
&\quad = \omega \wedge \psi(A(w_1), \dots, A(w_{r+p})), \\
&\quad = \tfrac{1}{r!p!} \textstyle\sum_\pi (-1)^{|\pi|} \omega(A(w_{\pi_1}), \dots, A(w_{\pi_r}))\psi(A(w_{\pi_{r+1}}), \dots, A(w_{\pi_{r+p}})), \\
&\quad = \tfrac{1}{r!p!} \textstyle\sum_\pi (-1)^{|\pi|} A^*(\omega)(w_{\pi_1}, \dots, w_{\pi_r}) A^*(\psi)(w_{\pi_{r+1}}, \dots, w_{\pi_{r+p}}), \\
&\quad = A^*(\omega) \wedge A^*(\psi)(w_1, \dots, w_{r+p}).
\end{aligned} \tag{5.130}$$

One obtains

$$A^*(\omega \wedge \psi) = A^*(\omega) \wedge A^*(\psi). \tag{5.131}$$

5.10 Abstract Algebraic Cauchy Formula

Considering the classical Cauchy formula as in Sect. 4.4, Eq. (4.50) may be written in the standard form

$$\varphi := \frac{\omega(\tau)}{A} = \mathbf{J} \cdot \mathbf{n}, \tag{5.132}$$

where φ is the flux density induced by the flux vector $\mathbf{J}$ on the plane $\mathbf{H}$ determined by the normal unit vector $\mathbf{n}$. The metric-independent version of (5.132) may be written as follows:

Let $\mathbf{H}$ be an r-dimensional affine subspace of the affine space $\mathscr{A}$ whose translation space is $\mathbf{U}$—an r-dimensional subspace of $\mathbf{V}$. Let $\mathcal{J}_\mathbf{U} : \mathbf{U} \to \mathbf{V}$ be the natural inclusion. Then, the pullback $\mathcal{J}_\mathbf{U}^* : \bigwedge^r \mathbf{V}^* \to \bigwedge^r \mathbf{U}^*$, given by

$$\mathcal{J}_\mathbf{U}^*(\omega)(v_1, \dots, v_r) = \omega(\mathcal{J}_\mathbf{U}(v_1), \dots, \mathcal{J}_\mathbf{U}(v_r)) = \omega(v_1, \dots, v_r), \tag{5.133}$$

is simply the restriction of the tensor ω to vectors in $\mathbf{U}$. We will use the notation $\rho_\mathbf{U} = \mathcal{J}_\mathbf{U}^*$ for the restriction. Consider the case where ω is an $(n-1)$-alternating tensor and $\mathbf{H}$ is an $(n-1)$-dimensional hyperplane. Then, ω induces an $(n-1)$-alternating tensor ψ on $\mathbf{U}$ given by

$$\psi = \rho_\mathbf{U}(\omega) = \mathcal{J}_\mathbf{U}^*(\omega), \tag{5.134}$$

the *abstract Cauchy formula*.

Part II
Smooth Theory

This part of the book presents a formulation of various notions of continuum physics on differentiable manifolds devoid of a metric structure. In particular, we concentrate our attention on flux fields and stress fields. The smooth setting of the classical theory is kept as well as the reliance on balance principles. An effort was made to make the notes self-contained and so they contain a short review of some elements of the theory of differentiable manifolds and integration theory of differential forms.

In a metric independent formulation of continuum mechanics, various fields such as flux fields and stress fields may be represented by objects independent of configuration. For example, the transformations of vector fields associated with electromagnetic theory from the Eulerian description to the Lagrangian description in [DO05, DO09] should be viewed merely as transformation rules for the components of tensors under coordinate transformations. A flux field, such as the heat flux field, should be regarded as a 2-form in the three-dimensional body rather than as a vector field. As a 2-form, a heat flux field may be considered independently of the configuration of the body in space.

Chapter 6
Smooth Analysis on Manifolds: A Short Review

In this chapter we introduce the notation and summarize basic elements pertaining to smooth analysis on manifolds that will be used in subsequent chapters. This summary is by no means comprehensive. For general introductions to differential geometry, consult [AMR88, AM77, BC70, Fra04, Lee02, War83].

6.1 Manifolds and Bundles

Various manifolds will be considered in this text. For example, we already considered the configuration space $\mathbb{Q}$ of a mechanical system having a finite number of degrees of freedom. In the context of continuum mechanics, one considers the body manifold $\mathcal{B}$ and the physical space manifold $\mathcal{S}$. In addition, electrodynamics is formulated in the event space, or spacetime $\mathcal{E}$. Finally, when considering theories of microstructure, one may consider a fiber bundle $\mathcal{Y}$ with a typical fiber $\mathcal{M}$ whose points represent possible values of micro-configurations. In the generic case, $\mathcal{X}$, $\mathcal{Y}$, and $\mathcal{Z}$ will be used to denote manifolds with $\mathcal{X}$ used mainly for the base manifold of a fiber bundle or the domain of definition of a mapping. The notation for coordinates on the various manifolds will vary accordingly.

6.1.1 Manifolds

Manifolds, Atlases, and Charts For an n-dimensional manifold $\mathcal{X}$, an *atlas* consists of a collection $\{(U_a, \phi_a)\}_{a\in A}$ of *charts* or *coordinate neighborhoods*. Each U_a is an open subset of $\mathcal{X}$ and

$$\phi_a : U_a \longrightarrow \mathbb{R}^n, \qquad x \longmapsto (x^1, \ldots, x^n) \tag{6.1}$$

R. Segev, *Foundations of Geometric Continuum Mechanics*, Advances in Continuum Mechanics 49, https://doi.org/10.1007/978-3-031-35655-1_6

is a homeomorphism onto an open subset of $\mathbb{R}^n$. In case $U_a \cap U_b \neq \varnothing$, then, $\phi_b \circ \phi_a^{-1}$ defined on $\phi_a\{U_a \cap U_b\}$ is the transformation of coordinates for the intersection $U_a \cap U_b$. If $x \in U_a \cap U_b$, let $(x^1, \dots, x^n) = \phi_a(x)$ and $(x^{1'}, \dots, x^{n'}) = \phi_b(x)$ be the coordinates under the two charts. Then, the transformation of coordinates is of the form $x^{i'} = x^{i'}(x^i)$. All such transformations and their inverses are smooth. In particular, the Jacobian matrices $[\partial x^{i'}/\partial x^j]$ should be invertible at all points.

Submanifold A subset $\mathfrak{Z} \subset \mathfrak{X}$ is an m-dimensional *submanifold* of $\mathfrak{X}$, $m \leqslant n$, if for every $z \in \mathfrak{Z}$ there is a chart $\phi : U \to \mathbb{R}^n$ on $\mathfrak{X}$ such that $z \in U$ and for each $x \in U \cap \mathfrak{Z}$, the last $n - m$ coordinates of x vanish, i.e., $\phi(x) = (x^1, \dots, x^m, 0, \dots, 0)$. The collection of charts satisfying this condition induces an atlas and an m-dimensional manifold structure on $\mathfrak{Z}$.

Manifold with Boundary A manifold with boundary is a manifold where the charts assume values in the closed half-space $\overline{\mathbb{R}}^{n-} = \{x \in \mathbb{R}^n \mid x^n \leqslant 0\}$. In the construction of the manifold structure, it is understood that a function defined on a closed half-space is said to be differentiable if it is the restriction to the half-space of a differentiable function defined on $\mathbb{R}^n$. For a manifold with boundary, $\mathfrak{X}$, a point $y \in \mathfrak{X}$ is a *boundary point* if there is a chart ϕ in a neighborhood U of y such that the n-th coordinate of y vanishes, i.e., $\phi(y)^n = 0$. The collection of boundary points is the *boundary* $\partial\mathfrak{X}$ of the manifold. By the definition of a submanifold, $\partial\mathfrak{X}$ is a submanifold of $\mathfrak{X}$ (see [Hir76, p. 29]).

Mappings Let $f : \mathfrak{X} \to \mathcal{Y}$ be a mapping between two manifolds. For a chart $\psi : V \to \mathbb{R}^{\dim \mathcal{Y}}$ on $\mathcal{Y}$, one can intersect the open set $f^{-1}\{V\}$ with the domains of charts in $\mathfrak{X}$ and obtain a chart $\phi : U \to \mathbb{R}^{\dim \mathfrak{X}}$ in $\mathfrak{X}$ and a mapping

$$\psi \circ f \circ \phi^{-1} : \phi(U) \longrightarrow \psi(V) \tag{6.2}$$

between an open subset of $\mathbb{R}^{\dim \mathfrak{X}}$ and $\mathbb{R}^{\dim \mathcal{Y}}$. The mapping $\psi \circ f \circ \phi^{-1}$ is a *local representative* of f relative to the charts under consideration. If x^i and y^j are the local coordinates under ϕ and ψ, respectively, then the local representative is of the form $y^j = y^j(x^i)$. A mapping $f : \mathfrak{X} \to \mathcal{Y}$ is said to be of class C^r if its local representatives are of class C^r. As a result of the compatibility condition, the differentiability properties of f are well-defined.

A C^1-mapping between two manifolds is an *immersion* if for each local representative $\psi \circ f \circ \phi^{-1}$, the derivative $D(\psi \circ f \circ \phi^{-1})$ is an injective linear mapping at each point in the domain. Thus, in terms of local coordinates, the matrix $[\partial y^j/\partial x^i]$ is of rank $\dim \mathfrak{X}$. The mapping is a *submersion* if the derivatives of local representatives are surjective so that $[\partial y^j/\partial x^i]$ is of rank $\dim \mathcal{Y}$.

An *embedding* is an immersion whose image is homeomorphic to $\mathfrak{X}$. The image of an embedding is a submanifold of $\mathcal{Y}$. A *diffeomorphism* is an invertible mapping $f : \mathfrak{X} \to \mathcal{Y}$ which is C^1 and whose inverse $f^{-1} : \mathcal{Y} \to \mathfrak{X}$ is also C^1. In case there is a diffeomorphism between $\mathfrak{X}$ and $\mathcal{Y}$, one says that the two manifolds are *diffeomorphic*. If $g : \mathfrak{X} \to \mathcal{Y}$ is an embedding, $\mathfrak{X}$ is diffeomorphic with its image.

Example 6.1 (Configurations of Bodies) In a general geometric setting of continuum mechanics, both a body $\mathcal{B}$ and the physical space $\mathcal{S}$ are represented by differentiable manifolds. Traditionally, one assumes in continuum mechanics that the principle of material impenetrability holds. This implies that the deformation gradient is injective at each point and that the configuration mapping is injective. In the framework of mechanics on differentiable manifolds, the impenetrability principle together with the requirement for differentiability implies that a configuration,

$$\kappa : \mathcal{B} \longrightarrow \mathcal{S}, \tag{6.3}$$

should be an embedding of the body manifold in the space manifold.

Example 6.2 In the continuum mechanics of generalized media,[1] in addition to the variable specifying the location in space of the various material points, the configuration of a body is described by additional degrees of freedom (sometimes referred also as order parameters or internal degrees of freedom) describing the micro-configuration of a material point (see [Cap89, CM02]). These variables usually assume values on manifolds different from the space manifold so that even in the case where the physical space is modeled as a Euclidean space, the micro-configuration of a point is valued on a manifold, $\mathcal{M}$, which may be devoid of a natural metric structure. As a simple example, the void fraction of a porous material may be described by a real number so that the space of micro-configuration $\mathcal{M}$ may be modeled by the segment $[0, 1] \in \mathbb{R}$. Clearly, this parametrization is arbitrarily chosen and other charts on $\mathcal{M}$ may be given. Thus, in general, the micro-configuration of a body $\mathcal{B}$ is given by a mapping $m : \mathcal{B} \to \mathcal{M}$.

6.1.2 Tangent Vectors and the Tangent Bundle

A *curve* in the smooth manifold $\mathcal{X}$ is a smooth mapping $c : (-\varepsilon, \varepsilon) \to \mathcal{X}$, where $(-\varepsilon, \varepsilon)$ is some interval in $\mathbb{R}$ containing zero. For the interpretation of a manifold as a configuration space of a mechanical system, a curve represents a motion of the system where the variable in $\mathbb{R}$ represents time. We say that c is a *curve at* $y \in \mathcal{X}$ if $c(0) = y$. The local representative of a curve is of the form $x^i = x^i(t)$.

Let c_1 and c_2 be two curves at $x \in \mathcal{X}$. The two curves are defined to be *tangent* at x if their local representatives $x_1^i(t)$ and $x_2^i(t)$, under some chart $\phi : U \to \mathbb{R}^n$ defined in a coordinate neighborhood of x, satisfy

$$\frac{\mathrm{d}}{\mathrm{d}t} x_1^i(t)|_{t=0} = \frac{\mathrm{d}}{\mathrm{d}t} x_2^i(t)|_{t=0}, \quad i = 1, \ldots, n. \tag{6.4}$$

---

[1] This includes media referred to by various authors as bodies with microstructure, bodies with substructure, Cosserat media, multipolar media, etc.

A *tangent vector* at x is defined as an equivalence class of curves under the equivalence relation of tangency of curves at x. Using a superimposed dot to denote the derivative of a curve in $\mathbb{R}^n$, the tangency condition may be written in the form $\dot{x}_1^i(0) = \dot{x}_2^i(0)$. For a given tangent vector v at x and a chart ϕ in a neighborhood of x, $(\dot{x}^1(0), \ldots, \dot{x}^n(0)) \in \mathbb{R}^n$ represents v.

The *tangent space* $T_x\mathcal{X}$ *to the manifold* $\mathcal{X}$ *at* x is the collection of the various tangent vectors at x. Given a chart ϕ in a neighborhood U of $x \in \mathcal{X}$, consider the mapping

$$T_x\phi : T_x\mathcal{X} \longrightarrow \mathbb{R}^n, \qquad T_x\phi(v) = \frac{\mathrm{d}}{\mathrm{d}t}c(t)|_{t=0} = (\dot{x}^1(0), \ldots, \dot{x}^n(0)), \tag{6.5}$$

where c is any curve that represents the tangent vector v. This mapping is a bijection, and it induces an n-dimensional vector space structure on $T_x\mathcal{X}$ whereby $T_x\phi$ becomes a vector space isomorphism. Let $\mathbf{e}_i$ be the i-th natural base vector of $\mathbb{R}^n$. Then, the isomorphism $T_x\phi$ induces a chart-dependent basis $\{e_i\}_{i=1}^n$ for $T_x\mathcal{X}$ by $e_i = (T_x\phi)^{-1}(\mathbf{e}_i)$. It is also customary to use the notation

$$\partial_i := \frac{\partial}{\partial x^i} := e_i. \tag{6.6}$$

Thus,

$$v = \sum_i v^i e_i = \sum_i v^i \partial_i, \tag{6.7}$$

where v^i are the components of $T_x\phi(v)$.

In what follows, we will use a superimposed dot both for the derivative with respect to the parameter of a curve and for a possible value of such a derivative. For example, we may write $\dot{x} = v = \sum_i \dot{x}^i \partial_i$ instead of the equation above.

If ϕ and ϕ' are charts in a neighborhood of $x \in \mathcal{X}$ and the local representatives of a curve c relative to the two charts are given in the form $(x^1(t), \ldots, x^n(t))$, and $(x^{1'}(t), \ldots, x^{n'}(t))$, then, the transformation rule for the components of tangent vectors is

$$T_x\phi' \circ (T_x\phi)^{-1} : \mathbb{R}^n \longrightarrow \mathbb{R}^n, \quad \text{given by} \quad \dot{x}^{i'} = \sum_i \frac{\partial x^{i'}}{\partial x^i}\dot{x}^i = \sum_i x^{i'}_{,i}\dot{x}^i. \tag{6.8}$$

In the last expression and in what follows, partial differentiation by x^i is indicated by a subscript comma followed by i. Thus, denoting the derivative of the transformation of coordinates by $D(\phi' \circ \phi^{-1})$, one has

$$T_x\phi' \circ (T_x\phi)^{-1} = D(\phi' \circ \phi^{-1})(\phi(x)). \tag{6.9}$$

The *tangent bundle* $T\mathcal{X}$ is the disjoint union of the tangent spaces at the various points of the manifold, i.e.,

$$T\mathcal{X} = \bigsqcup_{x\in\mathcal{X}} T_x\mathcal{X}. \tag{6.10}$$

The tangent bundle projection mapping is

$$\tau : T\mathcal{X} \longrightarrow \mathcal{X}, \qquad \tau(v) = x, \text{ if } v \in T_x\mathcal{X}. \tag{6.11}$$

Clearly, τ is a surjection and $\tau^{-1}\{x\} = T_x\mathcal{X}$.

An atlas $\{(U_a\phi_a)\}_{a\in A}$ on $\mathcal{X}$ induces naturally a $2n$-dimensional atlas $\{(\tau^{-1}\{U_a\}, \Phi_a\}_{a\in A}$ on $T\mathcal{X}$, where

$$\Phi_a : \tau^{-1}\{U_a\} \longrightarrow \mathbb{R}^{2n} \quad \text{is given by} \quad \Phi_a(v) = (\phi_a(\tau(v)), T_{\tau(v)}\phi_a(v)). \tag{6.12}$$

For the two charts (ϕ_a, U_a) and (ϕ_b, U_b), $U_a \cap U_b \neq \varnothing$, the transformation of coordinates $\Phi_b \circ \Phi_a^{-1} : \mathbb{R}^{2n} \to \mathbb{R}^{2n}$ is given by

$$\Phi_b \circ \Phi_a^{-1}(x^1, \dots, x^n, v^1, \dots, v^n) = (x^{1'}, \dots, x^{n'}, v^{1'}, \dots, v^{n'}), \tag{6.13}$$

where

$$(x^{1'}, \dots, x^{n'}) = (\phi_b \circ \phi_a^{-1})(x^1, \dots, x^n), \qquad v^{i'} = \sum_i x^{i'}_{,i} v^i. \tag{6.14}$$

Let $c : (-\varepsilon, \varepsilon) \to \mathcal{X}$ be a curve and let $s \in (-\varepsilon, \varepsilon)$. Then, there is an interval $(-\varepsilon', \varepsilon') \subset (-\varepsilon, \varepsilon)$ on which one can consider the curve $c' : (-\varepsilon', \varepsilon') \to \mathcal{X}$ given by $c'(t) = c(t + s)$. Thus, the curve c' induces a tangent vector v_s at $c'(0) = c(s)$. If we let s vary, we get a curve,

$$\dot{c} : (-\varepsilon, \varepsilon) \longrightarrow T\mathcal{X}, \qquad s \longmapsto v_s. \tag{6.15}$$

Evidently, $\tau \circ \dot{c} = c$. The curve $\dot{c}$ is the *lift* of the curve c in $\mathcal{X}$ to $T\mathcal{X}$.

6.1.3 Fiber Bundles

A *fiber bundle* structure on a manifold $\mathcal{Y}$ having a *base manifold* $\mathcal{X}$ consists of a projection mapping $\pi : \mathcal{Y} \to \mathcal{X}$ and a fiber bundle atlas that have the following properties. For all $x \in \mathcal{X}$, $\pi^{-1}\{x\}$ is diffeomorphic to a m-dimensional manifold $\mathfrak{T}$, called the *typical fiber* of the fiber bundle. The fiber bundle atlas consists of a collection $\{(U_a, \phi_a, \Phi_a)\}_{a\in A}$ such that $\{(U_a, \phi_a)\}$ is an atlas of $\mathcal{X}$ and each $\Phi_a : \pi^{-1}\{U_a\} \to U_a \times \mathfrak{T}$, called a *local trivialization*, is a diffeomorphism such that

$$\mathrm{pr}_1 \circ \Phi_a(y) = \pi(y), \quad \text{for every } y \in \pi^{-1}\{U_a\}. \tag{6.16}$$

Here and in the sequel, pr_1 and pr_2 denote the natural projections on the factors of a Cartesian product. We refer to $\mathcal{Y}$ as the *total space* of the fiber bundle and will denote the fiber bundle by $\pi : \mathcal{Y} \to \mathcal{X}$. Occasionally, we will also refer to either $\mathcal{Y}$ or π as the fiber bundle, for short. The manifold $\mathcal{Y}_x := \pi^{-1}\{x\}$ is referred to as the *fiber* of the bundle *at* x.

For a nonempty intersection, $U_a \cap U_b$, one has the transformation

$$\Phi_b \circ \Phi_a^{-1} : (U_a \cap U_b) \times \mathcal{Z} \longrightarrow (U_a \cap U_b) \times \mathcal{Z}, \tag{6.17}$$

where $\Phi_b \circ \Phi_a^{-1}(x, z) = (x, z')$. It follows that for any $x \in U_a \cap U_b$ there is a diffeomorphism $\Phi_{ab}(x) : \mathcal{Z} \to \mathcal{Z}$ such that $z' = \Phi_{ab}(x)(z)$. Thus, $\Phi_b \circ \Phi_a^{-1}(x, z) = (x, \Phi_{ab}(x)(z))$.

Let $\{(V_l, \psi_l)\}_{l \in L}$ be an atlas on $\mathcal{Z}$. Then, given a point $y \in \mathcal{Y}$, there is a chart $\phi_a : U_a \to \mathbb{R}^n$ such that $\mathrm{pr}_1(\Phi(y)) = \pi(y) \in U_a$ and a chart $\psi_l : V_l \to \mathbb{R}^m$ such that $\mathrm{pr}_2(\Phi(y)) \in V_l$. Let $\overline{U}_a = \mathrm{Image}\,\phi_a$, $\overline{V}_l = \mathrm{Image}\,\psi_l$. Then on $R_{al} = \Phi_a^{-1}\{\phi_a^{-1}\{\overline{U}_a\} \times \psi^{-1}\{\overline{V}_l\}\} \subset \mathcal{Y}$, an open neighborhood of y, one has the chart

$$\Psi_{al} := (\phi_a \circ \mathrm{pr}_1, \psi_l \circ \mathrm{pr}_2) \circ \Phi_a : R_{al} \longrightarrow \mathbb{R}^n \times \mathbb{R}^m. \tag{6.18}$$

Let $(z^1, \dots, z^m)$ be the local coordinates in $\mathbb{R}^m$, then, under the chart (Ψ_{al}, R_{al}), elements in the neighborhood of y will be represented in the form

$$(x^1, \dots, x^n, z^1, \dots, z^m),$$

and π is represented locally by

$$(x^1, \dots, x^n, z^1, \dots, z^m) \mapsto (x^1, \dots, x^n).$$

These will often be written in a compact form as (z^α), (x^i, z^α), and $(x^i, z^\alpha) \mapsto (x^i)$, respectively.

A fiber bundle is *trivial* if it has a global structure of a Cartesian product, i.e., $\mathcal{Y} = \mathcal{X} \times \mathcal{Z}$.

Sections of Fiber Bundles and Fiber Bundle Morphisms A C^r*-section* of a fiber bundle $\pi : \mathcal{Y} \to \mathcal{X}$ is a C^r-mapping $\kappa : \mathcal{X} \to \mathcal{Y}$ satisfying $\pi(\kappa(x)) = x$. A section κ of the fiber bundle is represented locally in the form

$$(x^i) \longmapsto (x^i, z^\alpha) = (x^i, \kappa^\alpha(x^j)), \qquad \text{or} \qquad \boldsymbol{x} \longmapsto (\boldsymbol{x}, \boldsymbol{z} = \boldsymbol{\kappa}(\boldsymbol{x})), \tag{6.19}$$

where $\kappa^1, \dots, \kappa^m$ are m real-valued functions defined on $\mathbb{R}^n$ representing the principal components of the mapping κ. The collection of all C^r-sections of the fiber bundle $\pi : \mathcal{Y} \to \mathcal{X}$, $r = 1, 2, \dots$, will be denoted by $C^r(\pi)$, or by $C^r(\mathcal{Y})$

when no confusion may arise. For $r = \infty$, the collection of sections will be denoted by $\Omega(\pi)$, or $\Omega(\mathscr{Y})$.

A C^r-fiber bundle morphism is a C^r-mapping between two fiber bundles that preserves the bundle structure. Specifically, a *C^r-fiber bundle morphism* (f, f_0) between the fiber bundles $\pi_1 : \mathscr{Y}_1 \to \mathscr{X}_1$ and $\pi_2 : \mathscr{Y}_2 \to \mathscr{X}_2$ consists of a C^r-mapping $f : \mathscr{Y}_1 \to \mathscr{Y}_2$ and a C^r-mapping $f_0 : \mathscr{X}_1 \to \mathscr{X}_2$, called the *base mapping*, such that $\pi_2 \circ f = f_0 \circ \pi_1$. In other words, the restriction, f_x, of f to the fiber $\mathscr{Y}_{1x}$ maps $\mathscr{Y}_{1x}$ into the fiber $\mathscr{Y}_{2 f_0(x)}$. Thus, the following diagram is commutative:

$$\begin{array}{ccc} \mathscr{Y}_1 & \xrightarrow{f} & \mathscr{Y}_2 \\ {\scriptstyle \pi_1}\downarrow & & \downarrow{\scriptstyle \pi_2} \\ \mathscr{X}_1 & \xrightarrow[f_0]{} & \mathscr{X}_2 \end{array} \tag{6.20}$$

Example 6.3 In Galilean mechanics, time is an absolute property of an event, while the location in space depends on the frame, or the observer, wherein the event is observed. Thus, while any two copies of space at distinct time instances are diffeomorphic, there is no natural way to specify a fixed location in space at distinct time instances. The geometrical structure suitable for the description of Galilean spacetime is that of a fiber bundle $\pi : \mathscr{E} \to \mathscr{T}$. The base manifold, $\mathscr{T}$, is the time manifold, and the fiber bundle projection assigns to each event $e \in \mathscr{E}$ a unique time. The typical fiber is some manifold $\mathscr{S}$ and at each time $t \in \mathscr{T}$, $\pi^{-1}(t)$ is diffeomorphic to $\mathscr{S}$. A physical frame is a (global) trivialization $\Phi : \mathscr{E} \to \mathscr{T} \times \mathscr{S}$. (See Chap. 9 and Sect. 13.2 for further discussions on the structure of classical spacetime.)

6.1.4 Vector Bundles

A *vector bundle* $\pi : W \to \mathscr{X}$ is a fiber bundle for which the typical fiber is a m-dimensional vector space $\mathbf{V}$. Consider the nonempty intersection $U_a \cap U_b$ of the domains of charts and the transformation $\Phi_{ab}(x)$ defined above by $(x, \Phi_{ab}(x)(v)) = \Phi_b \circ \Phi_a^{-1}(x, v)$ for $v \in \mathbf{V}$. It is required that for every $x \in U_a \cap U_b$, $\Phi_{ab}(x)$ be a linear transformation of $\mathbf{V}$. As a result of the definition, the mapping

$$\Phi_{ax} := \Phi_a|_{W_x} : W_x \longrightarrow \mathbf{V}. \tag{6.21}$$

is an isomorphism of the fiber $\pi^{-1}\{x\} = W_x$ with $\mathbf{V}$. Thus, Φ_{ax} induces a vector space structure on W_x. The requirement that the various Φ_{ab} are linear implies that this vector space structure is independent of the choice of a chart. It is noted that no natural isomorphism between the two spaces is available in general.

A basis $\{\mathbf{g}_1, \ldots, \mathbf{g}_m\}$ of $\mathbf{V}$ induces at each $x \in U_a$ a basis $\{g_1, \ldots, g_m\}$ of W_x by $g_\alpha = \Phi_{ax}^{-1}(\mathbf{g}_\alpha)$. Thus, an element w of the vector bundle may be represented in the forms $(x_1, \ldots, x_n, w_1, \ldots, w_m)$, and $w = \sum_\alpha w^\alpha g_\alpha \in W_x$. A section of a vector bundle is represented in the form

$$(x^i) \longmapsto (x^i, w^\alpha(x^j)) \quad \text{or} \quad w(x) = \sum_\alpha w^\alpha(x^j) g_\alpha. \tag{6.22}$$

The only example of a vector bundle we have so far is the tangent bundle. Naturally, a section of a vector bundle, and, in particular, a section of the tangent bundle, is referred to as a *vector field*. Since a coordinate system induces locally the bases $e_i = \partial/\partial x^i$ in the various spaces $T_x\mathcal{X}$, a vector field w in $T\mathcal{X}$ is represented locally in the form

$$w(x) = \sum_i w^i(x^j)\partial_i. \tag{6.23}$$

A *vector bundle morphism* between the vector bundles $\pi_1 : W_1 \to \mathcal{X}_1$ and $\pi_2 : W_2 \to \mathcal{X}_2$ is a fiber bundle morphism (f, f_0) such that for each $x \in \mathcal{X}_1$ the restriction $f_x = f|_{W_{1x}}$ is a linear mapping of W_{1x} into $W_{2 f_0(x)}$.

Consider the sections of a vector bundle. Using the vector space structure of each fiber, linear combinations of sections of a vector bundle are defined by

$$(aw + a'w')(x) = aw(x) + a'w'(x). \tag{6.24}$$

This definition endows the collection of sections of a vector bundle with the structure of a vector space. The space of sections of a vector bundle has a natural structure of an infinite-dimensional vector space.

6.1.5 Tangent Mappings

Given a C^r mapping $f : \mathcal{X}_1 \to \mathcal{X}_2$, $\dim \mathcal{X}_1 = n$, $\dim \mathcal{X}_2 = n'$, the *tangent mapping* is a C^{r-1}-vector bundle morphism $Tf : T\mathcal{X}_1 \to T\mathcal{X}_2$, defined as follows: Let v be a tangent vector at $x \in \mathcal{X}_1$ and let the curve $c : \mathbb{R} \to \mathcal{X}_1$ be a curve representing v. Then, $f \circ c : \mathbb{R} \to \mathcal{X}_2$ is a curve at $f(x)$ and $Tf(v) \in T_{f(x)}\mathcal{X}_2$ is defined to be the tangent vector that $f \circ c$ induces. Consider local coordinates x^i in a neighborhood of x and local coordinates y^α in a neighborhood of $f(x)$. Then, the vector v is represented by $(v^1, \ldots, v^n) = (\dot{x}^1(0), \ldots, \dot{x}^n(0))$, where $x^i(t)$ represent the curve c. Let a local representative of f be given by the functions $y^\alpha = f^\alpha(x^i)$ so that $f \circ c$ is represented by $y^\alpha(t) = f^\alpha(x^i(t))$. It follows that $Tf(v)$ is represented by $(u^1, \ldots, u^{n'}) = (\dot{y}^1(0), \ldots, \dot{y}^{n'}(0))$, given as

$$u^\alpha = \sum_i f^\alpha_{,i} v^i, \tag{6.25}$$

where we omitted the indication of the points where the derivatives are evaluated. Thus, the restriction,

$$T_x f := Tf|_{T_x\mathcal{X}_1} : T_x\mathcal{X}_1 \longrightarrow T_{f(x)}\mathcal{X}_2,$$

of the tangent mapping to the tangent space $T_x\mathcal{X}_1$, is represented by the Jacobian matrix of the partial derivatives of the local representatives of f. We conclude that the tangent mapping is represented locally in the form

$$(x^1, \dots, x^n, v^1, \dots, v^n) \longmapsto \left(f^1(x^i), \dots, f^{n'}(x^i), \textstyle\sum_i f^1_{,i} v^i, \dots, \sum_i f^{n'}_{,i} v^i\right). \tag{6.26}$$

Let $f_1 : \mathcal{X}_1 \to \mathcal{X}_2$ and $f_2 : \mathcal{X}_2 \to \mathcal{X}_3$ be smooth mappings. From the local representation of a tangent mapping above, and using the chain rule of multivariable calculus, it follows that

$$T_x(f_2 \circ f_1) = T_{f_1(x)} f_2 \circ T_x f_1, \qquad T(f_2 \circ f_1) = Tf_2 \circ Tf_1. \tag{6.27}$$

Example 6.4 For a formulation of continuum mechanics in which both the body and space are modeled as general differentiable manifolds, as in Example 6.1, the traditional deformation gradient should be replaced by the tangent

$$T\kappa : T\mathcal{B} \longrightarrow T\mathcal{S}, \tag{6.28}$$

to the configuration mapping between the body and space manifolds. Thus, it can no longer be represented as a tensor. The fact that the deformation gradient is a "two-point tensor" is indicative of the fact that it is a vector bundle morphism between the tangent bundles of two different manifolds. Even in the case where a body is an abstract manifold and space is a Euclidean three-dimensional space, the derivative of a configuration should be represented by the tangent mapping rather than a tensor.

6.1.6 The Tangent Bundle of a Fiber Bundle and Its Vertical Subbundle

Let $\pi : \mathcal{Y} \to \mathcal{X}$ be a fiber bundle, where $\mathcal{X}$ is n-dimensional and the typical fiber is m-dimensional. An element $w \in T_y\mathcal{Y}$, for some $y \in \mathcal{Y}$, is represented locally by coordinates $(x^i, y^\alpha, \dot{x}^j, \dot{y}^\beta)$, where (x^i, y^α) represent y. Alternatively, if $w \in T_y\mathcal{Y}$, it may be represented in the form

$$\textstyle\sum_j \dot{x}^j \partial_j + \sum_\beta \dot{y}^\beta \partial_\beta. \tag{6.29}$$

The tangent bundle projection

$$\tau_{\mathcal{Y}} : T\mathcal{Y} \longrightarrow \mathcal{Y}, \tag{6.30}$$

is given locally by

$$(x^i, y^\alpha, \dot{x}^j, \dot{y}^\beta) \longmapsto (x^i, y^\alpha). \tag{6.31}$$

In addition, one may consider the tangent mapping to the fiber bundle projection, π,

$$T\pi : T\mathcal{Y} \longrightarrow T\mathcal{X}, \tag{6.32}$$

represented locally by

$$(x^i, y^\alpha, \dot{x}^j, \dot{y}^\beta) \longmapsto (x^i, \dot{x}^j) \tag{6.33}$$

(in distinction with (6.31)). The *vertical subbundle* $V\mathcal{Y}$ of $T\mathcal{Y}$ is the kernel of $T\pi$. An element $v \in V\mathcal{Y}$ is represented locally as $(x^i, y^\alpha, 0, \dot{y}^\beta)$. With some abuse of notation, we will write both $\tau : T\mathcal{Y} \to \mathcal{Y}$ and $\tau : V\mathcal{Y} \to \mathcal{Y}$. For $v \in V\mathcal{Y}$ with $\tau(v) = y$ and $\pi(y) = x$, we may view v as an element of $T_y(\mathcal{Y}_x) = T_y(\pi^{-1}(x))$. In other words, elements of the vertical subbundle are tangent vectors to $\mathcal{Y}$ that are tangent to the fibers,

$$(V\mathcal{Y})_y = T_y(\mathcal{Y}_{\pi(y)}). \tag{6.34}$$

Let $f : \mathcal{Y} \to \mathcal{Z}$ be a fiber bundle morphism for the fiber bundles $\pi : \mathcal{Y} \to \mathcal{X}$ and $\rho : \mathcal{Z} \to \mathcal{X}$. Then, one can easily verify that, by linearity, the tangent mapping $Tf : T\mathcal{Y} \to T\mathcal{Z}$ maps the fibers of $V\mathcal{Y}$ into the fibers of $V\mathcal{Z}$—vectors tangent to the fibers are mapped into vectors tangent to the corresponding fibers. For short, we will keep the notation Tf for the restriction to the vertical bundle. (Cf. Palais [Pal68, p. 43], where a different notation is used for the restriction of the tangent mapping to the vertical subbundle.) Thus we have

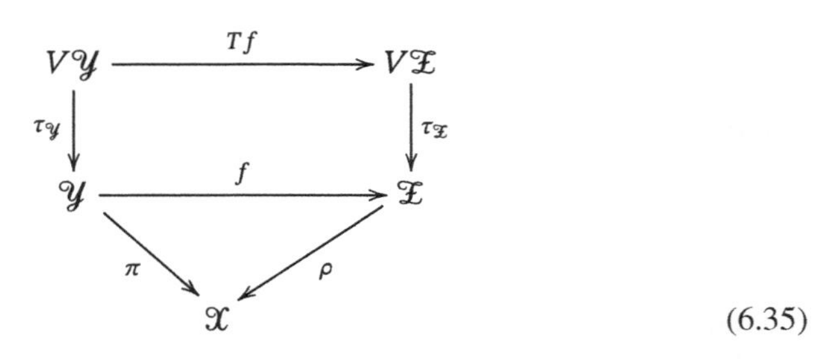

(6.35)

6.1.7 Jet Bundles

Let $I = i_1 \cdots i_k, i_r = 1, \ldots, n$, be a $|I| = k$-multi-index and let $u : \mathbb{R}^n \to \mathbb{R}$ be a k-times differentiable function. Then, I induces a partial differential operator ∂_I such that

$$\partial_I u := u_{,I} := \frac{\partial^{|I|} u}{\partial x^{i_1} \cdots \partial x^{i_{|I|}}}. \tag{6.36}$$

In case an r-array A, with components A_I, $|I| = r$, is completely symmetric, such as the array $\partial_I u$ above, the independent components of the array may be listed as $A_{i_1 \cdots i_k}$with $i_l \leq i_{l+1}$. An *ordered*, nondecreasing multi-index, that is, a multi-index that satisfies the condition $i_l \leq i_{l+1}$, will be denoted by boldface, uppercase Roman characters so that a symmetric array A is represented uniquely by the components $(A_{\boldsymbol{I}})$, $|\boldsymbol{I}| = k$. Evidently, any multi-index I may be ordered to produce the corresponding ordered multi-index $\boldsymbol{I}$. If an array A is completely symmetric, $A_{\boldsymbol{I}} = A_I$ for any multi-index I that will produce $\boldsymbol{I}$ under reordering.

In addition, a multi-index I induces a sequence $\tilde{I} = (I_1, \ldots, I_n)$ in which I_p is the number of times the index p appears in the sequence $i_1 \cdots i_k$. Thus, $|I| = \sum_p I_p$. The ordered multi-index $\boldsymbol{I}$ is determined uniquely by the sequence $\tilde{I} = (I_1, \ldots, I_n)$. In particular, for a differentiable function u as above, a particular partial derivative of order $|I|$ is written in the form

$$\partial_I u = \partial_{\boldsymbol{I}} u = u_{,\boldsymbol{I}} = \frac{\partial^{|\boldsymbol{I}|} u}{(\partial x^1)^{I_1} \cdots (\partial x^n)^{I_n}}. \tag{6.37}$$

Let $\pi : \mathcal{Y} \to \mathcal{X}$ be a fiber bundle and let $x \in \mathcal{X}$. Consider two sections, $\kappa : \mathcal{X} \to \mathcal{Y}$ and $\kappa' : \mathcal{X} \to \mathcal{Y}$, of the fiber bundle represented in a neighborhood of x in the forms

$$(x^i) \longmapsto (x^i, \kappa^\alpha(x^j)), \quad \text{and} \quad (x^i) \longmapsto (x^i, \kappa'^\alpha(x^j)), \tag{6.38}$$

respectively. These two sections are said to *have the same r-jet* at $x \in \mathcal{X}$, if for all I with $|I| \leqslant r$, and $\alpha = 1, \ldots, m$,

$$\partial_I \kappa^\alpha(x^i) = \partial_I \kappa'^\alpha(x^i). \tag{6.39}$$

By the rules for the transformations of partial derivatives, the definition is independent of the charts chosen. The relation of having the same r-jet at x is an equivalence relation, and an equivalence class is referred to as a *jet* at x. The equivalence class of a section κ is denoted by $j^r_x \kappa$. Under charts as above $j^r_x \kappa$ is represented by the collection of coordinates $A^\alpha_{\boldsymbol{I}} = \partial_{\boldsymbol{I}} \kappa^\alpha(x^i) = \kappa^\alpha_{,\boldsymbol{I}}(x^i)$, for all $\boldsymbol{I}$, such that $|\boldsymbol{I}| \leqslant r$ and all $\alpha = 1, \ldots, m$. The r-th jet space at x is the collection, $J^r(\mathcal{X}, \mathcal{Y})_x$, of r-jets at x. The disjoint union of jet spaces for the various points $x \in \mathcal{X}$ is the jet bundle

$J^r(\mathcal{X}, \mathcal{Y})$. Thus,

$$J^r(\mathcal{X}, \mathcal{Y}) = \bigsqcup_{x \in \mathcal{X}} J^r(\mathcal{X}, \mathcal{Y})_x. \tag{6.40}$$

Notation Equivalently, as in [Sau89], we will use the notation $J^r\pi$ for $J^r(\mathcal{X}, \mathcal{Y})$. In addition, to simplify the notation, when no ambiguity arises, we will write $J^r\mathcal{Y}$ for $J^r(\mathcal{X}, \mathcal{Y})$.

The jet bundle has several natural fiber bundle structures. The obvious one is

$$\pi^r : J^r\pi = J^r(\mathcal{X}, \mathcal{Y}) \to \mathcal{X}, \qquad j^r_x\kappa \longmapsto x. \tag{6.41}$$

Since $(\pi^r)^{-1}\{x\} = J^r(\mathcal{X}, \mathcal{Y})_x$, an element $A \in J^r(\mathcal{X}, \mathcal{Y})$ is represented using coordinates in the form

$$(x^i, A^\alpha_{\boldsymbol{I}}), \qquad 0 \le |\boldsymbol{I}| \le r, \quad \alpha = 1, \ldots, m, \quad i = 1, \ldots, n. \tag{6.42}$$

An additional fiber bundle structure is

$$\pi^r_0 : J^r(\mathcal{X}, \mathcal{Y}) \to J^0(\mathcal{X}, \mathcal{Y}) := \mathcal{Y}, \qquad j^r_x\kappa \longmapsto j^0_x\kappa = \kappa(x). \tag{6.43}$$

Similarly, one has the natural projections

$$\pi^r_l : J^r(\mathcal{X}, \mathcal{Y}) \to J^l(\mathcal{X}, \mathcal{Y}), \quad l < r, \qquad j^r_x\kappa \longmapsto j^l_x\kappa. \tag{6.44}$$

A section $\kappa : \mathcal{X} \to \mathcal{Y}$ induces a section $j^r\kappa : \mathcal{X} \to J^r(\mathcal{X}, \mathcal{Y})$ of the r-jet bundle by $j^r\kappa(x) := j^r_x\kappa$. The section $j^r\kappa$ is the *r -jet extension of* κ. Locally, the jet extension is represented in the form

$$(x^i) \longmapsto (x^i, \partial_{\boldsymbol{I}}\kappa^\alpha(x^j)), \qquad 0 \le |\boldsymbol{I}| \le r, \quad \alpha = 1, \ldots, m, \quad i = 1, \ldots, n. \tag{6.45}$$

Let $\pi_1 : \mathcal{Y}_1 \to \mathcal{X}_1$ and $\pi_2 : \mathcal{Y}_2 \to \mathcal{X}_2$ be two fiber bundles, and let $f : \mathcal{Y}_1 \to \mathcal{Y}_2$ be a fiber bundle morphism such that the base mapping $f_0 : \mathcal{X}_1 \to \mathcal{X}_2$ is a diffeomorphism. Then, the r-jet of f

$$J^r f : J^r\mathcal{Y}_1 \longrightarrow J^r\mathcal{Y}_2 \tag{6.46}$$

is defined as follows: Let $\chi \in (J^r\mathcal{Y}_1)_x$ and let $\kappa : \mathcal{X}_1 \to \mathcal{Y}_1$ be a section of π_1 such that $\chi = j^r_x\kappa$. Then, we have a section $f \circ \kappa \circ f_0^{-1} : \mathcal{X}_2 \to \mathcal{Y}_2$, and one defines

$$J^r f(\chi) := j^r_x(f \circ \kappa \circ f_0^{-1}) \in J^r\mathcal{Y}_2, \tag{6.47}$$

as illustrated in the following diagram:

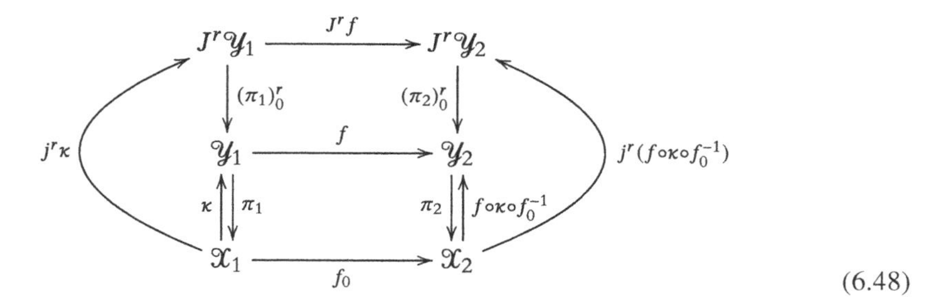

(6.48)

If $\mathcal{Z}$ is a manifold, we will also write $J^r(\mathcal{X}, \mathcal{Z})$ for the jet bundle associated with the trivial fiber bundle

$$\pi = \mathrm{pr}_1 : Y := \mathcal{X} \times \mathcal{Z} \longrightarrow \mathcal{X}. \tag{6.49}$$

For further details on jet bundles, see [Sau89, KSM93].

6.1.8 The First Jet of a Vector Bundle

Consider the particular case of a vector bundle $\pi : W \to \mathcal{X}$, with typical fiber $\mathbf{V}$. Under a local chart and a local collection of induced bases, $\{g_\alpha\}_{\alpha=1}^m$, for the various fibers, a section, $w : \mathcal{X} \to W$, of π is represented in the form $(x^i) \longmapsto (x^i, \sum_\alpha w^\alpha(x^j)g_\alpha)$. Therefore, the first jet $j^1_x w$ is represented by $(x^i, w^\alpha(x^j), w^\beta_{,k}(x^j))$. Using tensor notation, the 1-jet extension of w is represented by

$$(x^i) \longmapsto (x^i, \textstyle\sum_\alpha w^\alpha(x^j)g_\alpha + \sum_{\alpha,k} w^\alpha_{,k}(x^j)\mathrm{d}x^k \otimes g_\alpha). \tag{6.50}$$

6.1.9 The Pullback of a Fiber Bundle

Let $\pi : \mathcal{Y} \to \mathcal{X}$ be a fiber bundle whose typical fiber is $\mathcal{Z}$, and for a manifold $\mathcal{M}$, let $f : \mathcal{M} \to \mathcal{X}$ be a smooth mapping. The mapping f determines a fiber bundle

$$f^*\pi : f^*\mathcal{Y} \longrightarrow \mathcal{M} \tag{6.51}$$

over $\mathcal{M}$ called the *pullback* of the bundle π by f. For each $m \in \mathcal{M}$, we make the identification

$$(f^*\pi)^{-1}\{m\} = (f^*\mathcal{Y})_m = \pi^{-1}\{f(m)\} = \mathcal{Y}_{f(m)}, \tag{6.52}$$

i.e., the fiber over $m \in \mathcal{M}$ of the pullback fiber bundle is identified with the fiber at $f(m)$. Thus, the typical fiber of $f^*(\pi)$ is also $\mathcal{Z}$. In addition, one has a natural fiber bundle morphism $(\pi^* f, f)$, where the mapping $\pi^* f : \pi^*\mathcal{Y} \to \mathcal{Y}$ is defined by the requirement that for $\tilde{y} \in (f^*\mathcal{Y})_m$, i.e., $m = (f^*\pi)(\tilde{y})$, $\pi^*(f)(\tilde{y})$ is the unique element in $\mathcal{Y}_{f(y)}$ which is identified with $\tilde{y}$ under the construction of the pullback bundle.

If $\kappa : \mathcal{X} \to \mathcal{Y}$ is a section of the bundle π, one has an induced section $f_\pi^*\kappa : \mathcal{M} \to f^*\mathcal{Y}$ (the notation follows [Pal68, p. 5]) of the pullback bundle—*the pullback of the section* κ—defined by

$$(f_\pi^*\kappa)(m) = \kappa(f(m)). \tag{6.53}$$

When no ambiguity may ensue, we may omit the indication of the projection π and write simply $f^*\kappa$ for $f_\pi^*\kappa$, as in the following commutative diagram:

$$\begin{array}{ccc} f^*\mathcal{Y} & \xrightarrow{\pi^* f} & \mathcal{Y} \\ {\scriptstyle f^*w := f_\pi^* w}\ \uparrow\!\!\downarrow {\scriptstyle f^*\pi} & & {\scriptstyle \pi}\ \downarrow\!\!\uparrow\ {\scriptstyle w} \\ \mathcal{M} & \xrightarrow[f]{} & \mathcal{X}. \end{array} \tag{6.54}$$

The pullback has the universal property described in the following diagram. If $\phi : U \to \mathcal{Y}$ and $\psi : U \to \mathcal{M}$ are mappings such that $f \circ \psi = \pi \circ \phi$, then, there is a unique mapping $u : U \to f^*\mathcal{Y}$ such that $\pi^* f \circ u = \phi$ and $f^*\pi \circ u = \psi$.

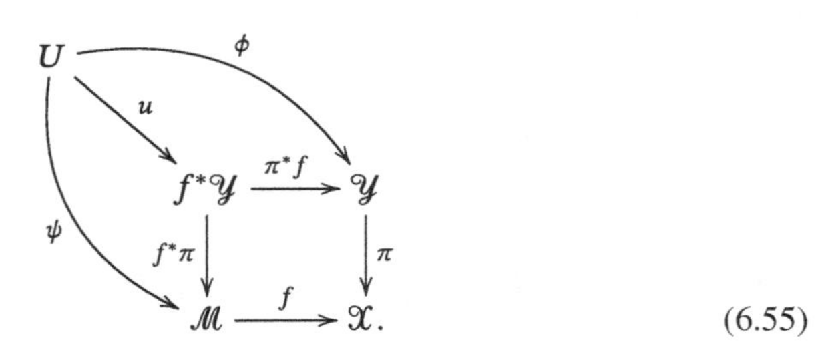

(6.55)

In fact, the induced section $f_\pi^*\kappa$ of the section κ, as described above, may be defined in the context of the universal property as shown in the diagram:

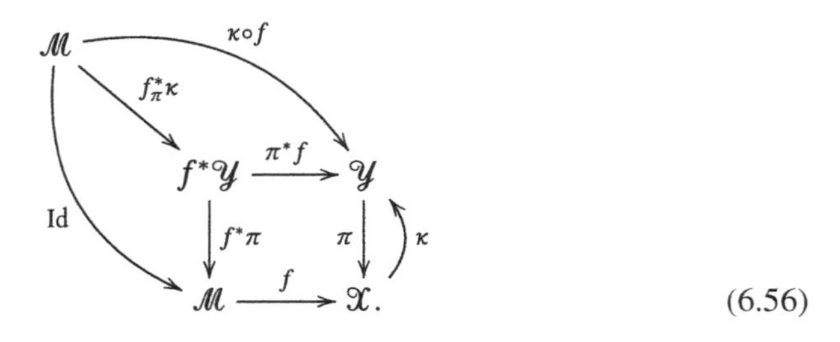

(6.56)

As a special case, the pullback of a vector bundle $\pi : W \to \mathcal{X}$ by the mapping $f : \mathcal{M} \to \mathcal{X}$ is a vector bundle

$$f^*\pi : f^*W \longrightarrow \mathcal{M}. \tag{6.57}$$

For example, let $\kappa : \mathcal{X} \to \mathcal{Y}$ be a section of a fiber bundle $\pi : \mathcal{Y} \to \mathcal{X}$, and let

$$\kappa^*\tau : \kappa^*V\mathcal{Y} \longrightarrow \mathcal{X} \tag{6.58}$$

be the pullback of the vertical subbundle. Then, we may identify $\kappa^*V\mathcal{Y}$ with the restriction of the vertical bundle to Image κ.

Example 6.5 (Virtual Velocity Fields) Let the embedding $\kappa : \mathcal{B} \to \mathcal{S}$ be a configuration of a body manifold, $\mathcal{B}$, into the space manifold, $\mathcal{S}$. A *virtual velocity* field (or a *virtual displacement field*) w assigns to each material point $X \in \mathcal{B}$ a tangent vector at $\kappa(X)$. That is, using the notation $\tau : T\mathcal{S} \to \mathcal{S}$ for the natural tangent bundle projection, a virtual velocity field is a mapping $w : \mathcal{B} \to T\mathcal{S}$ such that $\tau \circ w = \kappa$. Since κ is an embedding, its inverse κ^{-1} : Image $\kappa \to \mathcal{B}$ is well-defined and the virtual displacement field induces a section $w \circ \kappa^{-1}$: Image $\kappa \to T\mathcal{S}$. Clearly, $\tau \circ w \circ \kappa^{-1}(x) = x$ for all $x \in$ Image κ. The section $w \circ \kappa^{-1}$ is traditionally referred to as the Eulerian version of the velocity field.

Alternatively, consider the pullback of the tangent bundle, $\kappa^*\tau : \kappa^*T\mathcal{S} \to \mathcal{B}$, onto the body. The velocity field w induces a section, $u : \mathcal{B} \to \kappa^*T\mathcal{S}$, by setting $u(X)$ to be the unique element of $(\kappa^*T\mathcal{S})_X$ which is equal to $w(X) \in T_{\kappa(X)}\mathcal{S}$ by the construction of the pullback bundle. It is noted that while the section u is defined whether κ is invertible or not, the Eulerian vector field cannot be defined if κ is not invertible. Thus, it is convenient to regard a virtual velocity field as a section of the pullback $\kappa^*T\mathcal{S}$.

In the case of bodies with microstructure described in Example 6.2, a virtual velocity field at the micro-configuration $m : \mathcal{B} \to \mathcal{M}$ will be a section w of $m^*T\mathcal{M}$ or, equivalently, a mapping $w : \mathcal{B} \to T\mathcal{M}$ with $\tau_{\mathcal{M}} \circ w = m$. It is noted that as a micro-configuration is not expected to be an embedding, there is no Eulerian version of the micro-virtual velocity field. Thus, for all the situations described above, a virtual velocity field at a given configuration of a body may be modeled as a section of some vector bundle $\pi : W \to \mathcal{X}$.

6.1.10 Dual Vector Bundles and the Cotangent Bundle

Given a vector bundle $\pi_W : W \to \mathcal{X}$ with typical fiber $\mathbf{V}$, a vector bundle $\pi_{W^*} : W^* \to \mathcal{X}$, the typical fiber of which is the dual space $\mathbf{V}^*$, is defined naturally. The mapping

$$(\Phi^*_{ax})^{-1} : W^*_x \longrightarrow \mathbf{V}^* \tag{6.59}$$

(see Eq. (6.21)) determines the required isomorphism for each $x \in U_a$. As the basis $\{\mathbf{g}_1, \ldots, \mathbf{g}_m\}$ of $\mathbf{V}$ induces a dual basis $\{\mathbf{g}^1, \ldots, \mathbf{g}^m\}$ of $\mathbf{V}^*$, a dual basis $\{g^1, \ldots, g^m\}$, $g^\alpha = \Phi^*_{ax}(\mathbf{g}^\alpha)$ will be defined in W^*_x, for a given chart. An element $\phi \in W^*_x$ will be represented in the form $(x^1, \ldots, x^n, \phi_1, \ldots, \phi_m)$ or $\phi = \sum_\alpha \phi_\alpha g^\alpha$.

Let $\pi_1 : W_1 \to \mathcal{X}_1$ and $\pi_2 : W_2 \to \mathcal{X}_2$ be two vector bundles and let (f, f_0) be a vector bundle morphism so that $f : W_1 \to W_2$ and $f_0 : \mathcal{X}_1 \to \mathcal{X}_2$. Then, for each $x \in \mathcal{X}_1$, one has the dual linear mapping $f^*_x : W^*_{2f_0(x)} \to W^*_{1x}$ given by $f^*_x(\phi_2)(w_1) = \phi_2(f_x(w_1))$, $\phi_2 \in W^*_{2f_0(x)}$, $w_1 \in W_{1x}$. If $\phi : \mathcal{X}_2 \to W^*_2$ is a section, then a section[2] $f^*(\phi) : \mathcal{X}_1 \to W^*_1$, written also as $f^*\phi$, is induced by setting

$$f^*(\phi)(x) = f^*_x(\phi(f_0(x))). \tag{6.60}$$

Note the distinction between the last definition and that of the pullback of a section as in (6.53), which may also be applied to sections of dual bundles.

The most common example for a dual bundle is the *cotangent bundle*

$$\tau^*_{\mathcal{X}} : T^*\mathcal{X} \to \mathcal{X} \tag{6.61}$$

—the dual to the tangent bundle, which contains covectors. The fiber of $T^*\mathcal{X}$ at a point $x \in \mathcal{X}$ will be denoted by $T^*_x\mathcal{X} := (T_x\mathcal{X})^* = (T^*\mathcal{X})_x$. The local dual basis is denoted by $e^i =: \mathrm{d}x^i$, so that by the definition of the dual basis

$$\mathrm{d}x^i(\partial_j) = \delta^i_j. \tag{6.62}$$

The action of the covector $\phi = \sum_i \phi_i \mathrm{d}x^i$ on the tangent vector $v = \sum_i v^i \partial_i$ is computed by $\phi(v) = \sum_i \phi_i v^i$. Using Eq. (6.14), the transformation rules for the base vectors and for the components of elements of the cotangent bundle are, respectively,

$$\mathrm{d}x^{i'} = \sum_i x^{i'}_{,i} \mathrm{d}x^i, \qquad \phi_{i'} = \sum x^i_{,i'} \phi_i, \tag{6.63}$$

i.e., the components of a covector transform using the transpose of the Jacobian. The equation on the left motivates the notation $\mathrm{d}x^i$ for the dual base vectors.

Let $f : \mathcal{X}_1 \to \mathcal{X}_2$ be a differentiable mapping and let $\phi \in T^*_{f(x)}\mathcal{X}_2$ for a point $x \in \mathcal{X}_1$. Then, ϕ induces the element

$$T^*_x f(\phi) := (T_x f)^*(\phi) \in T^*_x\mathcal{X}_1. \tag{6.64}$$

A section of the cotangent bundle is referred to as a *differential* 1*-form*. As a special case of the definition in Eq. (6.60), if $f : \mathcal{X}_1 \to \mathcal{X}_2$ is a C^r-mapping and $\phi : \mathcal{X}_2 \to$

[2] Note that we use the same notation for a section of a vector bundle and for elements in it. The relevant interpretation should be inferred from the context.

$T^*\mathcal{X}_2$ is a C^r 1-form, then, $(Tf)^*(\phi) : \mathcal{X}_1 \to T^*\mathcal{X}_1$ is the C^{r-1}, 1-form over $\mathcal{X}_1$ induced by the dual of the tangent mapping $Tf : T\mathcal{X}_1 \to T\mathcal{X}_2$. That is, for $x \in \mathcal{X}_1$, $(Tf)^*(\phi)(x) := T_x^* f(\phi(x))$, so that specifically,

$$((Tf)^*(\phi)(x))(v) = \phi(f(x))(T_x f(v)), \quad x \in \mathcal{X}_1,\ v \in T_x\mathcal{X}_1. \tag{6.65}$$

The matrix representing $(T_x f)^*$ is evidently the transpose of the Jacobian matrix corresponding to the local representative of f. In case $\mathcal{X}_1$ is a submanifold of $\mathcal{X}_2$ and $\mathcal{J} : \mathcal{X}_1 \to \mathcal{X}_2$ is the inclusion embedding with $T\mathcal{J}(v) = v$, then, for a 1-form ϕ over $\mathcal{X}_2$, $T\mathcal{J}^*(\phi)(v) = \phi(T\mathcal{J}(v)) = \phi(v)$. Hence, $T\mathcal{J}^*(\phi)$ is simply the restriction of the form to vectors tangent to $\mathcal{X}_1$. We will refer to $Tf^*(\phi)$ as the *pullback of the form* ϕ by f.

In Chap. 2 we have demonstrated the relevance of these notions to statics of systems having finite numbers of degrees of freedom.

6.2 Tensor Bundles and Differential Forms

6.2.1 Tensor Bundles and Their Sections

Let $\pi_1 : W_1 \to \mathcal{X}$ and $\pi_2 : W_2 \to \mathcal{X}$ be two vector bundles over the same manifold $\mathcal{X}$ having typical fibers $\mathbf{V}_1$ and $\mathbf{V}_2$, respectively. Clearly, local trivializations of these vector bundles allow one to consider local trivializations that are common to both. These will induce local trivializations for vector bundles over $\mathcal{X}$ with typical fibers such as $\mathbf{V}_1 \times \mathbf{V}_2$, $\mathbf{V}_1 \otimes \mathbf{V}_2$, $L(\mathbf{V}_1, \mathbf{V}_2)$—the vector space of linear transformations $\mathbf{V}_1 \to \mathbf{V}_2$, etc. The corresponding vector bundles will be denoted respectively by $\pi_1 \times \pi_2 : W_1 \times W_2 \to \mathcal{X}$, $\pi_1 \otimes \pi_2 : W_1 \otimes W_2 \to \mathcal{X}$, $L(\pi_1, \pi_2) : L(W_1, W_2) \to \mathcal{X}$, etc. In what follows, we make the identification

$$\mathbf{V}_1^* \otimes \mathbf{V}_2 \cong L(\mathbf{V}_1, \mathbf{V}_2), \qquad \mathbf{V}_1^* \otimes \mathbf{V}_2^* \cong (\mathbf{V}_1 \otimes \mathbf{V}_2)^*. \tag{6.66}$$

Hence,

$$L(W_1, W_2) \cong W_1^* \otimes W_2. \tag{6.67}$$

In view of the natural isomorphism, in what follows, we will often identify the two vector bundles and use the equality symbol.

Various algebraic operations between vector spaces carry over to operations between elements of the corresponding bundles and to sections of these tensor bundles. For example, let S be a section of $L(W_1, W_2) = W_1^* \otimes W_2$ and let χ be a section of W_1. The notation $S \cdot \chi$ is used for the section of W_2 given by

$$(S \cdot \chi)(x) = S(x)(\chi(x)). \tag{6.68}$$

In particular, such constructions may be applied to the tangent and cotangent bundles of the manifold $\mathcal{X}$. The constructions of Sect. 5 will produce

$$\tau^p_{\mathcal{X}} : \bigwedge^p T\mathcal{X} \to \mathcal{X} \tag{6.69}$$

—the bundle whose fiber at $x \in \mathcal{X}$ is $\bigwedge^p T_x\mathcal{X}$, the space of p-multivectors—and the corresponding dual bundle

$$\tau^{*p}_{\mathcal{X}} : \bigwedge^p T^*\mathcal{X} \to \mathcal{X} \tag{6.70}$$

the bundle whose fiber at $x \in \mathcal{X}$ is $\bigwedge^p T^*_x\mathcal{X}$, the space of p-alternating tensors.

6.2.2 Differential Forms

A differentiable section of $\bigwedge^p T^*\mathcal{X} \to \mathcal{X}$ is referred to as a *differential p-form.*

Using the properties of alternating tensors we studied in Sect. 5, an r-form ω may be represented locally as

$$\omega = \sum_{(\gamma)} \omega_\gamma(x^i)\mathrm{d}x^{\gamma_1} \wedge \cdots \wedge \mathrm{d}x^{\gamma_p} = \sum_I \frac{1}{r!}\omega_I \mathrm{d}x^{i_1} \wedge \cdots \wedge \mathrm{d}x^{i_p}. \tag{6.71}$$

In accordance with the notation introduced in Sect. 5.2, for a multi-index I and an increasing multi-index γ, set

$$\mathrm{d}x^I := \mathrm{d}x^{i_1} \wedge \cdots \wedge \mathrm{d}x^{i_p}, \qquad \mathrm{d}x^\gamma := \mathrm{d}x^{\gamma_1} \wedge \cdots \wedge \mathrm{d}x^{\gamma_p}. \tag{6.72}$$

We will also use the notation

$$\mathrm{d}x = \mathrm{d}x^1 \wedge \cdots \wedge \mathrm{d}x^n. \tag{6.73}$$

Thus, the local expression of a p-form ω assumes the forms

$$\omega = \sum_{(\gamma)} \omega_\gamma(x^i)\mathrm{d}x^\gamma = \sum_I \frac{1}{r!}\omega_I \mathrm{d}x^I. \tag{6.74}$$

The operations available for alternating tensors, e.g., exterior products and contractions, apply by pointwise computations to differential forms. Thus, for example, for two forms ω and θ, one has the form $\omega \wedge \theta$ given by $(\omega \wedge \theta)(x) = \omega(x) \wedge \theta(x)$. For the local expressions, the base vectors $\mathrm{d}x^1, \ldots, \mathrm{d}x^n$ should be substituted for $e^1, \ldots, e^n$ in the expressions obtained in Chap. 5.

The transformation rules for elements of the cotangent bundle in (6.63) give the transformation rules for elements of $\bigwedge^p(T^*\mathcal{X})$. One has

$$
\begin{aligned}
\mathrm{d}x^{\gamma'} = \mathrm{d}x^{\gamma'_1} \wedge \cdots \wedge \mathrm{d}x^{\gamma'_p} &= \sum_I \left(x^{\gamma'_1}_{,i_1}\mathrm{d}x^{i_1}\right) \wedge \cdots \wedge \left(x^{\gamma'_p}_{,i_p}\mathrm{d}x^{i_p}\right), \\
&= \sum_I x^{\gamma'_1}_{,i_1} \cdots x^{\gamma'_p}_{,i_p}\mathrm{d}x^{i_1} \wedge \cdots \wedge \mathrm{d}x^{i_p} \\
&= \sum_{I,(\gamma)} x^{\gamma'_1}_{,i_1} \cdots x^{\gamma'_p}_{,i_p}\varepsilon^{i_1\ldots i_p}_{\gamma_1\ldots\gamma_p}\mathrm{d}x^{\gamma_1} \wedge \cdots \wedge \mathrm{d}x^{\gamma_p},
\end{aligned}
\tag{6.75}
$$

and using $\omega = \sum_{(\gamma')} \omega_{\gamma'}\mathrm{d}x^{\gamma'_1} \wedge \cdots \wedge \mathrm{d}x^{\gamma'_p} = \sum_{(\gamma)} \omega_\gamma \mathrm{d}x^{\gamma_1} \wedge \cdots \wedge \mathrm{d}x^{\gamma_p}$,

$$
\omega_{\gamma'_1\ldots\gamma'_p} = \sum_{I',(\gamma)} \varepsilon^{i'_1\ldots i'_p}_{\gamma'_1\ldots\gamma'_p} x^{\gamma_1}_{,i'_1} \cdots x^{\gamma_p}_{,i'_p}\omega_{\gamma_1\ldots\gamma_p}, \quad \text{or} \quad \omega_{\gamma'} = \sum_{I',(\gamma)} \varepsilon^{I'}_{\gamma'} x^{\gamma_1}_{,i'_1} \cdots x^{\gamma_p}_{,i'_p}\omega_\gamma .
\tag{6.76}
$$

A particularly simple expression is obtained for the transformation rule in $\bigwedge^n(T^*\mathcal{X})$. Since the only increasing multi-index is $(\gamma_1, \ldots, \gamma_n) = (1, \ldots, n)$,

$$
\omega_{1'\ldots n'} = \sum_I \varepsilon^{i'_1\ldots i'_r}_{1'\ldots n'} x^1_{,i'_1} \cdots x^n_{,i'_p}\omega_{1\ldots n} = \det(x^i_{,i'})\omega_{1\ldots n}.
\tag{6.77}
$$

As 0-alternating tensors may be regarded as real numbers (see Sect. 5.7), 0-differential forms are real-valued functions on the manifold. In particular, for any coordinate neighborhood $U_a \subset \mathcal{X}$, the coordinate functions $x^i(x)$ induced by the chart ϕ_a may be regarded as local 0-forms. A 1-tensor may be identified with an element of the dual vector space so that a 1-form may be identified with a section of the cotangent bundle. We will use the notation $\Omega^p(\mathcal{X})$ for the vector space of p-forms on $\mathcal{X}$, that is, sections of $\bigwedge^p T^*\mathcal{X}$.

The definition of the pullback of a 1-form can be extended naturally to all differential forms. Let $f : \mathcal{X}_1 \to \mathcal{X}_2$ be a smooth mapping. For a 0-form, a real-valued function $\phi : \mathcal{X}_2 \to \mathbb{R}$, set $((Tf)^*(\phi))(x) = \phi(f(x))$. For $0 < p \leqslant n$,

$$
((Tf)^*(\phi))(x)(v_1, \ldots, v_p) = \phi(f(x))(T_x f(v_1), \ldots, T_x f(v_p)),
\tag{6.78}
$$

for all $x \in \mathcal{X}_1,\ v_1, \ldots, v_p \in T_x\mathcal{X}_1$.

In analogy with the bundle of multicovectors—alternating multilinear mappings—the bundle of p-multivectors will be denoted by $\bigwedge^p T\mathcal{X}$. A section of $\bigwedge^p T\mathcal{X}$ is a multivector field. In analogy with the notation of (6.72) and (6.73), we set

$$
\partial_I := \partial_{I_1} \wedge \cdots \wedge \partial_{I_r}, \quad \partial_\gamma := \partial_{\gamma_1} \wedge \cdots \wedge \partial_{\gamma_r}, \quad \partial_x := \partial_1 \wedge \cdots \wedge \partial_n .
\tag{6.79}
$$

6.2.3 Contraction and Related Mappings

We expand here the review of the contraction operation and related mappings. The notions described below, which are sometimes missing from standard texts, will be used in Chap. 18.

For an element $\xi \in \bigwedge^r T\mathcal{X}$, consider the mapping

$$\wedge^{\xi}_{\llcorner} : \bigwedge^p T\mathcal{X} \longrightarrow \bigwedge^{p+r} T\mathcal{X}, \qquad \wedge^{\xi}_{\llcorner}(\eta) = \eta \wedge \xi, \quad \eta \in \bigwedge^p T\mathcal{X}. \tag{6.80}$$

The dual mapping,

$$(\wedge^{\xi}_{\llcorner})^* : \bigwedge^{p+r} T^*\mathcal{X} \longrightarrow \bigwedge^p T^*\mathcal{X}, \qquad (\wedge^{\xi}_{\llcorner})^*(\omega) = \omega \circ \wedge^{\xi}_{\llcorner}, \tag{6.81}$$

satisfies

$$((\wedge^{\xi}_{\llcorner})^*(\omega))(\eta) = (\omega \circ \wedge^{\xi}_{\llcorner})(\eta) = \omega(\eta \wedge \xi). \tag{6.82}$$

We use the notation

$$\omega \llcorner \xi := (\wedge^{\xi}_{\llcorner})^*(\omega), \qquad \xi \in \bigwedge^r T\mathcal{X}, \quad \omega \in \bigwedge^{r+p} T^*\mathcal{X}, \tag{6.83}$$

so that

$$(\omega \llcorner \xi)(\eta) = \omega(\eta \wedge \xi), \quad \text{for all} \quad \eta \in \bigwedge^p T\mathcal{X}. \tag{6.84}$$

We refer to this operation as *right contraction* or *interior product on the right.*

Analogous definitions, leading to the operation of a *left contraction*, apply to

$$\wedge^{\xi}_{\lrcorner} : \bigwedge^p T\mathcal{X} \longrightarrow \bigwedge^{p+r} T\mathcal{X}, \qquad \wedge^{\xi}_{\lrcorner}(\eta) = \xi \wedge \eta. \tag{6.85}$$

One sets,

$$\xi \lrcorner \omega := (\wedge^{\xi}_{\lrcorner})^*(\omega), \qquad \text{so that} \qquad \xi \lrcorner \omega(\eta) = \omega(\xi \wedge \eta). \tag{6.86}$$

The simple computation

$$\begin{aligned} ((\eta \wedge \xi) \lrcorner \omega)(\zeta) &= \omega(\eta \wedge \xi \wedge \zeta), \\ &= (\eta \lrcorner \omega)(\xi \wedge \zeta), \\ &= (\xi \lrcorner (\eta \lrcorner \omega))(\zeta), \end{aligned} \tag{6.87}$$

implies that

$$(\eta \wedge \xi) \lrcorner \omega = \xi \lrcorner (\eta \lrcorner \omega), \qquad \text{and similarly,} \qquad \omega \llcorner (\eta \wedge \xi) = (\omega \llcorner \xi) \llcorner \eta. \tag{6.88}$$

By standard properties of the exterior product, the left and right contractions differ only by a factor of $(-1)^{rp}$.

Let γ and β be increasing multi-indices and let $\gamma \setminus \beta$ be the increasing multi-index containing the indices in γ not included in β. Then (see [Ste83, p. 21]),

$$\partial_\beta \lrcorner \, \mathrm{d}x^\gamma = \varepsilon^{\gamma}_{\beta,\gamma\setminus\beta} \mathrm{d}x^{\gamma\setminus\beta}. \tag{6.89}$$

In particular, let γ be an increasing multi-index and let $\hat{\gamma}$ be the increasing multi-index that completes γ to $\{1, \dots, n\}$, implying that $\mathrm{d}x = \varepsilon_{\hat{\gamma}\gamma} \mathrm{d}x^{\hat{\gamma}} \wedge \mathrm{d}x^\gamma = \varepsilon_{\gamma\hat{\gamma}} \mathrm{d}x^\gamma \wedge \mathrm{d}x^{\hat{\gamma}}$. Then,

$$\partial_{\hat{\gamma}} \lrcorner \, (\mathrm{d}x^{\hat{\gamma}} \wedge \mathrm{d}x^\gamma) = \mathrm{d}x^\gamma = \varepsilon^{\hat{\gamma}\gamma} \partial_{\hat{\gamma}} \lrcorner \, \mathrm{d}x = \varepsilon^{\gamma\hat{\gamma}} \mathrm{d}x \llcorner \partial_{\hat{\gamma}}, \tag{6.90}$$

so that

$$\partial_{\hat{\gamma}} \lrcorner \, \mathrm{d}x = \varepsilon_{\hat{\gamma}\gamma} \mathrm{d}x^\gamma, \qquad \mathrm{d}x \llcorner \partial_{\hat{\gamma}} = \varepsilon_{\gamma\hat{\gamma}} \mathrm{d}x^\gamma. \tag{6.91}$$

When the summation convention is used, we view λ and $\hat{\lambda}$ as two distinct indices, so summation is not implied in a term such as $\mathrm{d}x^\lambda \wedge \mathrm{d}x^{\hat{\lambda}}$.

If $\theta \in \bigwedge^n T_x^*\mathcal{X}$, $\xi \in \bigwedge^p T_x\mathcal{X}$, and $\psi \in \bigwedge^{n-p} T_x^*\mathcal{X}$ are represented, respectively, as

$$\theta = \theta_0 \mathrm{d}x, \quad \xi = \sum_{(\beta)} \xi^\beta \partial_\beta, \quad \psi = \sum_{(\gamma)} \psi_\gamma \mathrm{d}x^\gamma, \tag{6.92}$$

then,

$$\begin{aligned}
(\theta \llcorner \xi) \wedge \psi &= [\theta_0 \mathrm{d}x \llcorner (\textstyle\sum_{(\beta)} \xi^\beta \partial_\beta)] \wedge (\textstyle\sum_{(\gamma)} \psi_\gamma \mathrm{d}x^\gamma), \\
&= \theta_0 \sum_{(\beta),(\gamma)} \xi^\beta \psi_\gamma (\mathrm{d}x \llcorner \partial_\beta) \wedge \mathrm{d}x^\gamma, \\
&= \theta_0 \sum_{(\beta),(\gamma),(\beta')} \xi^\beta \psi_\gamma \varepsilon_{\hat{\beta}\beta} \mathrm{d}x^{\hat{\beta}} \wedge \mathrm{d}x^\gamma, \\
&= \theta_0 \sum_{(\beta),(\gamma),(\beta')} \xi^\beta \psi_\gamma \varepsilon_{\hat{\beta}\beta} \varepsilon^{\hat{\beta}\gamma} \mathrm{d}x, \\
&= \theta_0 \sum_{(\beta),(\gamma)} \xi^\beta \psi_\gamma \varepsilon^\gamma_\beta \mathrm{d}x, \\
&= \psi(\xi)\theta.
\end{aligned} \tag{6.93}$$

For a multicovector $\omega \in \bigwedge^r T^*\mathcal{X}$, we may also consider the linear mapping

$$\wedge_{\llcorner}^{\omega} : \bigwedge^p T^*\mathcal{X} \longrightarrow \bigwedge^{p+r} T^*\mathcal{X}, \qquad \wedge_{\llcorner}^{\omega}(\psi) = \psi \wedge \omega. \tag{6.94}$$

Since $\wedge_{\llcorner}^{\omega} \in L(\bigwedge^p T^*\mathcal{X}, \bigwedge^{p+r} T^*\mathcal{X}) \cong \bigwedge^p T\mathcal{X} \otimes \bigwedge^{p+r} T^*\mathcal{X}$ is linear in ω, we have a linear mapping

$$\wedge_{\llcorner} : \bigwedge^r T^*\mathcal{X} \longrightarrow L(\bigwedge^p T^*\mathcal{X}, \bigwedge^{p+r} T^*\mathcal{X}), \qquad \omega \longmapsto \wedge_{\llcorner}^{\omega}, \quad \wedge_{\llcorner}(\omega)(\psi) = \psi \wedge \omega. \tag{6.95}$$

Identifying a vector space with its double dual, the dual mapping

$$(\wedge_{\llcorner}^{\omega})^* : \bigwedge^{p+r} T\mathcal{X} \longrightarrow \bigwedge^p T\mathcal{X}, \qquad (\wedge_{\llcorner}^{\omega})^*(\xi) = \xi \circ \wedge_{\llcorner}^{\omega}, \tag{6.96}$$

satisfies

$$(\wedge_{\llcorner}^{\omega})^*(\xi)(\psi) = \xi \circ \wedge_{\llcorner}^{\omega}(\psi) = \xi(\psi \wedge \omega) = (\psi \wedge \omega)(\xi). \tag{6.97}$$

Thus, one sets

$$\xi \llcorner \omega := (\wedge_{\llcorner}^{\omega})^*(\xi), \qquad \text{so that} \qquad \xi \llcorner \omega(\psi) = \xi(\psi \wedge \omega) = (\psi \wedge \omega)(\xi), \tag{6.98}$$

where $\omega \in \bigwedge^r T^*\mathcal{X}$, $\xi \in \bigwedge^{r+p} T\mathcal{X}$, and $\psi \in \bigwedge^p T^*\mathcal{X}$. Similarly, we have

$$\wedge_{\lrcorner}^{\omega} : \bigwedge^p T^*\mathcal{X} \longrightarrow \bigwedge^{p+r} T^*\mathcal{X}, \qquad \wedge_{\lrcorner}^{\omega}(\psi) = \omega \wedge \psi, \tag{6.99}$$

together with the corresponding linear mapping,

$$\wedge_{\lrcorner} : \bigwedge^r T^*\mathcal{X} \longrightarrow L(\bigwedge^p T^*\mathcal{X}, \bigwedge^{p+r} T^*\mathcal{X}), \quad \omega \longmapsto \wedge_{\lrcorner}^{\omega}, \quad \wedge_{\lrcorner}(\omega)(\psi) = \omega \wedge \psi, \tag{6.100}$$

and the contraction

$$\omega \lrcorner \xi := (\wedge_{\lrcorner}^{\omega})^*(\xi), \qquad \text{so that} \qquad (\omega \lrcorner \xi)(\psi) = \xi(\omega \wedge \psi) = (\omega \wedge \psi)(\xi). \tag{6.101}$$

Being bilinear, the contraction operators may be represented as linear mappings defined on the tensor products. Thus, we set

$$\mathsf{C}_{\llcorner} : \bigwedge^p T\mathcal{X} \otimes \bigwedge^{p+r} T^*\mathcal{X} \cong L(\bigwedge^p T^*\mathcal{X}, \bigwedge^{p+r} T^*\mathcal{X}) \longrightarrow \bigwedge^r T^*\mathcal{X}, \tag{6.102}$$

and

$$\mathsf{C}_{\lrcorner} : \bigwedge^p T\mathcal{X} \otimes \bigwedge^{p+r} T^*\mathcal{X} \cong L(\bigwedge^p T^*\mathcal{X}, \bigwedge^{p+r} T^*\mathcal{X}) \longrightarrow \bigwedge^r T^*\mathcal{X}, \tag{6.103}$$

by

$$\mathsf{C}_{\llcorner}(\xi \otimes \theta) = \theta \llcorner \xi, \qquad \text{and} \qquad \mathsf{C}_{\lrcorner}(\xi \otimes \theta) = \xi \lrcorner \theta, \tag{6.104}$$

respectively. Again, identifying double dual spaces with the corresponding primal spaces, analogous mappings may be defined on $\bigwedge^p T^*\mathcal{X} \otimes \bigwedge^{p+r} T\mathcal{X}$. As noted above, the left and right contractions differ by a factor of $(-1)^{rp}$.

We will be particularly interested in the case where $r + p = n = \dim \mathcal{X}$. In this case, since $\bigwedge^n T^*\mathcal{X}$ is one-dimensional, $\dim L(\bigwedge^p T^*\mathcal{X}, \bigwedge^n T^*\mathcal{X}) = \dim \bigwedge^{n-p} T^*\mathcal{X}$. Moreover, the contraction mappings are isomorphisms

$$L(\textstyle\bigwedge^p T^*\mathcal{X}, \bigwedge^n T^*\mathcal{X}) \longrightarrow \bigwedge^{n-p} T^*\mathcal{X}, \tag{6.105}$$

with

$$\wedge_\lrcorner = \mathsf{C}_\llcorner^{-1}, \qquad \text{and} \qquad \wedge_\llcorner = \mathsf{C}_\lrcorner^{-1}. \tag{6.106}$$

For example,

$$\begin{aligned} \wedge_\lrcorner(\mathsf{C}_\llcorner(\xi \otimes \theta))(\psi) &= \mathsf{C}_\llcorner(\xi \otimes \theta) \wedge \psi, \\ &= (\theta \llcorner \xi) \wedge \psi, \\ &= \psi(\xi)\theta, \\ &= (\xi \otimes \theta)(\psi), \end{aligned} \tag{6.107}$$

where in the third line above we have used the identity (6.93). We record, therefore, that

$$\mathsf{C}_\llcorner^{-1} = \wedge_\lrcorner : \textstyle\bigwedge^{n-p} T^*\mathcal{X} \longrightarrow L(\bigwedge^p T^*\mathcal{X}, \bigwedge^n T^*\mathcal{X}), \qquad \wedge_\lrcorner(\omega)(\psi) = \omega \wedge \psi, \tag{6.108}$$

is an isomorphism.

6.2.4 Vector-Valued Forms

Let $\pi : W \to \mathcal{X}$ be a vector bundle over $\mathcal{X}$. At each $x \in \mathcal{X}$ one may consider the vector space

$$\textstyle\bigwedge^p (T_x^*\mathcal{X}, W_x) := L_{\mathfrak{A}}^p(T_x\mathcal{X}; W_x)$$

of alternating p-linear mappings of the tangent space $T_x\mathcal{X}$ into W_x. Clearly, $\bigwedge^p (T_x^*\mathcal{X}, W_x) = \bigwedge^p (T_x^*\mathcal{X}) \otimes W_x$. This induces the vector bundle

$$\textstyle\bigwedge^p (T^*\mathcal{X}, W) = (\bigwedge^p T^*\mathcal{X}) \otimes W = L_{\mathfrak{A}}^p(T\mathcal{X}; W) \longrightarrow \mathcal{X}. \tag{6.109}$$

The transposition operation for tensor product of vector spaces, $\mathbf{V} \otimes \mathbf{U} \cong \mathbf{U} \otimes \mathbf{V}$, induces the following natural isomorphisms of vector bundles

$$\begin{aligned}\bigwedge^p(T^*\mathcal{X}, W) = (\bigwedge^p T^*\mathcal{X}) \otimes W &= L^p_{\mathfrak{A}}(T\mathcal{X}; W) \\ &\cong W \otimes \bigwedge^p T^*\mathcal{X} = L(W^*, \bigwedge^p T^*\mathcal{X}) \longrightarrow \mathcal{X}.\end{aligned} \tag{6.110}$$

Specifically, we will write

$$\mathrm{tr} : \bigwedge^p(T^*\mathcal{X}, W) = (\bigwedge^p T^*\mathcal{X}) \otimes W \longrightarrow \bigwedge^p(T^*\mathcal{X}, W) = (\bigwedge^p T^*\mathcal{X}) \otimes W \tag{6.111}$$

for the transposition isomorphism and for an element $\omega \in W \otimes \bigwedge^p T^*\mathcal{X}$, we will use the notation $\omega^{\mathsf{T}} := \mathrm{tr}(\omega)$. Evidently, $\mathrm{tr}^{-1} = \mathrm{tr}$. The same notation will apply to sections of these tensor bundles.

Replacing W in the last equation by its dual, we have

$$\begin{aligned}\bigwedge^p(T^*\mathcal{X}, W^*) = (\bigwedge^p T^*\mathcal{X}) \otimes W^* &= L^p_{\mathfrak{A}}(T\mathcal{X}; W^*) \cong W^* \otimes \bigwedge^p T^*\mathcal{X} \\ &= L(W, \bigwedge^p T^*\mathcal{X}).\end{aligned} \tag{6.112}$$

We will refer to sections of $\bigwedge^p(T^*\mathcal{X}, W)$, or the other variants, as *W-valued p-forms.*

Various smooth fields of continuum mechanics, e.g., the body force field and the surface force field, are modeled below by vector-valued forms.

If $\{\mathbf{g}_1, \ldots, \mathbf{g}_m\}$ is a basis of the typical fiber $\mathbf{V}$ of the vector bundle W, and $\{g_1, \ldots, g_m\}$ are the induced local base vectors for a chart in W, then, a W-valued r-form is represented locally by

$$\omega(x) = \sum_{(\gamma),\alpha} \omega^\alpha_\gamma(x^i)\mathrm{d}x^\gamma \otimes g_\alpha = \sum_{I,\alpha} \frac{1}{r!}\omega^\alpha_I(x^i)\mathrm{d}x^I \otimes g_\alpha. \tag{6.113}$$

In analogy with the space of scalar-valued differential forms, $\Omega^p(T^*\mathcal{X}, W)$ will denote the space of W-valued differential forms, i.e., sections of $\bigwedge^p(T^*\mathcal{X}, W)$. Analogous notation applies to the other variants.

Given an element, $\sigma \in (W^* \otimes \bigwedge^p T^*\mathcal{X})_x$, and an element $w \in W_x$, one has the natural contraction

$$\sigma \cdot w \in \bigwedge^p T^*_x\mathcal{X}, \quad \text{defined by the condition,} \quad (f \otimes \omega) \cdot w := f(w)\omega. \tag{6.114}$$

Given, $\sigma \in W^* \otimes \bigwedge^q T^*\mathcal{X}$, one can define the bilinear action $g \dot{\wedge} \sigma$ by setting

$$(f \otimes \omega)\dot{\wedge}(w \otimes \psi) := f(w)\omega \wedge \psi, \tag{6.115}$$

for elements f, w, ω, ψ of $W^*, W, \bigwedge^q T^*\mathcal{X}, \bigwedge^p T^*\mathcal{X}$, at some $x \in \mathcal{X}$, respectively. Because of the natural identification of W^{**} with W, we may also write $\chi \dot{\wedge} g$ for

$$(w \otimes \psi)\dot{\wedge}(f \otimes \omega) := f(w)\psi \wedge \omega. \tag{6.116}$$

Evidently,

$$\sigma \dot{\wedge} (w \otimes \psi) = (\sigma \cdot w) \wedge \psi, \qquad (w \otimes \psi) \dot{\wedge} \sigma = \psi \wedge (\sigma \cdot w). \tag{6.117}$$

The elements $g \dot{\wedge} \chi$ and $\chi \dot{\wedge} g$ evidently belong to $\bigwedge^{p+q} T^* \mathcal{X}$.

The binary operation, $\dot{\wedge}$, induces linear isomorphism $\dot{\wedge}_{\lrcorner}$ and $\dot{\wedge}_{\llcorner}$,

$$\dot{\wedge}_{\lrcorner}, \dot{\wedge}_{\llcorner} : W^* \otimes \bigwedge^{n-p} T^* \mathcal{X} \longrightarrow L(W \otimes \bigwedge^p T^* \mathcal{X}, \bigwedge^n T^* \mathcal{X}), \tag{6.118}$$

whereby

$$\begin{aligned} \dot{\wedge}_{\lrcorner}(g \otimes \omega)(w \otimes \psi) &= (g \otimes \omega) \dot{\wedge} (w \otimes \psi) = g(w) \omega \wedge \psi, \\ \dot{\wedge}_{\llcorner}(g \otimes \omega)(w \otimes \psi) &= (w \otimes \psi) \dot{\wedge} (g \otimes \omega) = g(w) \psi \wedge \omega. \end{aligned} \tag{6.119}$$

These operations may be extended to sections of the corresponding bundles by applying them pointwise. In what follows, we will use the same notation for the actions on sections—vector-valued forms.

6.2.5 Density-Dual Spaces

Consider, in analogy with the dual of a vector bundle, the vector bundle of linear mappings into another one-dimensional vector bundle, that of n-alternating tensors. Thus, for a given vector bundle, W, we use the notation (see Atiyah and Bott [AB67])

$$W' = L\big(W, \bigwedge^n T^* \mathcal{X}\big) \cong W^* \otimes \bigwedge^n T^* \mathcal{X}. \tag{6.120}$$

Thus, transposition gives an isomorphism

$$W' \cong \bigwedge^n (T^* \mathcal{X}, W^*) \tag{6.121}$$

under which a section of W' may be represented as a covector-valued form.

Let $A : W_1 \to W_2$ be a vector bundle morphism over $\mathcal{X}$. Then, in analogy with the dual mapping, one may consider

$$A' : {W_2}' \longrightarrow {W_1}', \qquad \text{given by} \qquad f \longmapsto f \circ A. \tag{6.122}$$

It is also noted that we have

$$\begin{aligned}(W')' &= (W^* \otimes \textstyle\bigwedge^n T^*\mathcal{X})' \\ &= (W^* \otimes \textstyle\bigwedge^n T^*\mathcal{X})^* \otimes \textstyle\bigwedge^n T^*\mathcal{X}, \\ &= W \otimes \textstyle\bigwedge^n T\mathcal{X} \otimes \textstyle\bigwedge^n T^*\mathcal{X},\end{aligned} \tag{6.123}$$

and as $\bigwedge^n T\mathcal{X} \otimes \bigwedge^n T^*\mathcal{X}$ is naturally isomorphic with $\mathbb{R}$, one has a canonical isomorphism

$$(W')' \cong W. \tag{6.124}$$

For the vector bundles W, U,

$$(W \otimes U)' \cong W^* \otimes U^* \otimes \textstyle\bigwedge^n T^*\mathcal{X} \cong W^* \otimes U'. \tag{6.125}$$

We will refer to W' as the *density-dual bundle* and to A' as the *density-dual mapping*.

As an example, for the case $W = \bigwedge^p T^*\mathcal{X}$, it is implied by Eq. (6.108) that we have an isomorphism

$$\wedge_\lrcorner : \textstyle\bigwedge^{n-p} T^*\mathcal{X} \longrightarrow (\textstyle\bigwedge^p T^*\mathcal{X})', \tag{6.126}$$

given by

$$\wedge_\lrcorner (\omega)(\psi) = \omega \wedge \psi. \tag{6.127}$$

For the case of vector-valued forms, the mappings $\dot{\wedge}_\lrcorner$ defined in (6.118) is, in fact,

$$\dot{\wedge}_\lrcorner : W^* \otimes \textstyle\bigwedge^{n-p} T^*\mathcal{X} \longrightarrow \left(W \otimes \textstyle\bigwedge^p T^*\mathcal{X}\right)', \qquad \dot{\wedge}_\lrcorner(f)(\chi) = f \dot{\wedge} \chi. \tag{6.128}$$

Evidently, analogous observations apply to $\wedge_\llcorner$ and $\dot{\wedge}_\llcorner$.

6.3 Differentiation and Integration

6.3.1 Integral Curves and the Flow of a Vector Field

A vector field $w : \mathcal{X} \to T\mathcal{X}$ on an n-dimensional manifold is equivalent to a system of n ordinary differential equations. Any curve $c : (-\varepsilon, \varepsilon) \to \mathcal{X}$ induces a curve $\dot{c} : (-\varepsilon, \varepsilon) \to T\mathcal{X}$, by requiring that for each $t_0 \in (-\varepsilon, \varepsilon)$, $\dot{c}(t_0)$ is the tangent vector to the curve c at the point $c(t_0)$. Clearly, $\dot{c}$ is a *lift* of c in the sense that $c = \tau \circ \dot{c}$, where τ is the tangent bundle projection. A curve c in $\mathcal{X}$ is an *integral line* of the vector field $w : \mathcal{X} \to T\mathcal{X}$ if for each $t_0 \in (-\varepsilon, \varepsilon)$, $\dot{c}(t_0) = w(c(t_0))$. If the vector field, w, is represented locally in the form $\sum_i w^i \partial_i$ and the integral line is represented locally by $x^k(t)$, then the tangency condition is represented locally by

$$\frac{\mathrm{d}x^i}{\mathrm{d}t}(t) = w^i(x^k(t)), \tag{6.129}$$

a system of n, first order, ordinary differential equations. An initial condition for the problem is a point $x_0 \in \mathcal{X}$, and thus the *integral curve at* x_0 is the integral curve c such that $c(0) = x_0$.

It is noted that for a given smooth vector field and an initial condition x_0, the integral curve at x_0 need not necessarily be defined for all $t \in \mathbb{R}$. However, the existence theorem for ordinary differential equations implies that there is an open interval $a < t < b$ containing zero for which the solution exists and on which it is unique. The interval (a, b) depends on the initial condition and may tend to zero for some sequence of initial conditions. If the integral lines for all points in $\mathcal{X}$ may be defined for all values $t \in \mathbb{R}$, the vector field is said to be *complete*. Even if the vector field is not complete, for each $x_0 \in \mathcal{X}$ there is a neighborhood U_{x_0} containing x_0 and an interval (a, b) such that the integral curves for all initial conditions $x \in U_{x_0}$ are defined for all $t \in (a, b)$. Let $D \subset \mathcal{X} \times \mathbb{R}$ be the collection of points (x_0, t_0) such that the integral line starting at x_0 is defined for $t = t_0$. One can define the *flow* of the vector field by

$$\Phi : D \longrightarrow \mathcal{X}, \qquad \Phi(x, t) = c(t), \tag{6.130}$$

where c is the integral curve with $c(0) = x$. We observe that the flow is a smooth mapping because the solution of a differential equation, for a smooth vector field, depends smoothly on the initial condition. Clearly, for a complete vector field, $D = \mathcal{X} \times \mathbb{R}$. A sufficient condition for a vector field w to be complete is that the support of w is compact (see, e.g., [AMR88, p. 249]). In particular, every vector field on a compact manifold without boundary is complete. Furthermore, it can be shown (e.g., [BC70, pp. 139–142], [War83, pp. 40–41], [AMR88, pp. 247–248]) that if $w(x_0) \neq 0$, then, there is a chart in an open neighborhood of x_0 such that, locally, w is represented by ∂_1.

6.3.2 Exterior Derivatives

The exterior derivative, $\mathrm{d}\omega$, of the zero form, ω—a real-valued function on $\mathcal{X}$—is defined to be the 1-form, a covector field, given by

$$\mathrm{d}\omega(x)(v) = \frac{\mathrm{d}}{\mathrm{d}t}\omega(c(t))|_{t=0}, \tag{6.131}$$

where c is any curve at x representing the tangent vector v. The linear dependence of the result on v and its independence of the particular choice of curve $c \in v$ follow immediately from the local expression

$$\begin{aligned} \mathrm{d}\omega(x)(v) &= \frac{\mathrm{d}}{\mathrm{d}t}\tilde{\omega}(x^i(t))|_{t=0}, \\ &= \sum_i \frac{\partial \tilde{\omega}}{\partial x^i}\dot{x}^i(0), \end{aligned} \tag{6.132}$$

where $\tilde{\omega}$ is the local representative of the real function ω. It follows from the expression above that $\mathrm{d}\omega$ is represented locally by

$$\sum_i \tilde{\omega}_{,i}\mathrm{d}x^i. \tag{6.133}$$

If $\tilde{\omega} = x^k$ locally, then, $\mathrm{d}\omega$ is represented by $\mathrm{d}x^k$.

The exterior derivative of higher-order forms may be defined as follows: Since, at least locally, an r-form may be represented as a sum of expressions such as

$$\omega = \theta \mathrm{d}\theta^1 \wedge \cdots \wedge \mathrm{d}\theta^r, \tag{6.134}$$

where $\theta, \theta^1, \ldots, \theta^r$ are real-valued functions, we need to define the exterior derivative only for expressions of this pattern. Thus, one sets

$$\mathrm{d}(\theta \mathrm{d}\theta^1 \wedge \cdots \wedge \mathrm{d}\theta^r) = \mathrm{d}\theta \wedge \mathrm{d}\theta^1 \wedge \cdots \wedge \mathrm{d}\theta^r, \tag{6.135}$$

an $(r+1)$-form. If locally $\omega = \sum_{(\gamma)} \omega_\gamma \mathrm{d}x^{\gamma_1} \wedge \cdots \wedge \mathrm{d}x^{\gamma_r}$, the local expression for the exterior derivative is

$$\mathrm{d}\omega = \sum_{(\gamma),i} \omega_{\gamma,i}\mathrm{d}x^i \wedge \mathrm{d}x^{\gamma_1} \wedge \cdots \wedge \mathrm{d}x^{\gamma_r}. \tag{6.136}$$

Comparing this expression with Example 5.1, we note that the exterior derivative may be regarded as an exterior product on the left with a differential operator $\sum_i \partial/\partial x^i\, \mathrm{d}x^i$. Similar to that example, the local expression for the exterior derivative may be written as

$$\begin{aligned} \mathrm{d}\omega &= \sum_{(\gamma),i} \omega_{\gamma_1 \ldots \widehat{\gamma_i} \ldots \gamma_{r+1}, \gamma_i} \mathrm{d}x^{\gamma_i} \wedge \mathrm{d}x^{\gamma_1} \wedge \cdots \wedge \widehat{\mathrm{d}x^{\gamma_i}} \wedge \cdots \wedge \mathrm{d}x^{\gamma_{r+1}}, \\ &= \sum_{(\gamma),i} (-1)^{i-1} \omega_{\gamma_1 \ldots \widehat{\gamma_i} \ldots \gamma_{r+1}, \gamma_i} \mathrm{d}x^{\gamma_1} \wedge \cdots \wedge \mathrm{d}x^{\gamma_{r+1}}. \end{aligned} \tag{6.137}$$

Equation (6.136) implies that the exterior derivative is linear, i.e., $\mathrm{d}(a\omega + a'\omega') = a\mathrm{d}\omega + a'\mathrm{d}\omega'$. In addition,

$$\mathrm{d}^2\omega := \mathrm{d}(\mathrm{d}\omega) = \sum_{(\gamma),i} \omega_{\gamma,ik}\mathrm{d}x^k \wedge \mathrm{d}x^i \wedge \mathrm{d}x^{\gamma_1} \wedge \cdots \wedge \mathrm{d}x^{\gamma_r} = 0 \tag{6.138}$$

since the second derivative is symmetric in the i, k indices, while the exterior product is antisymmetric.

A differential form ω is *closed* if $\mathrm{d}\omega = 0$. A differential r-form ω is *exact* if there is an $(r-1)$-form α such that $\omega = \mathrm{d}\alpha$. The identity $\mathrm{d}^2\omega = 0$ implies immediately that every exact form is closed. The Poincaré lemma implies that locally every closed form is exact. That is, in a neighborhood of each point x, $\omega = \mathrm{d}\alpha$, for an $(r-1)$-form α (e.g., [AMR88, pp. 435–437] or [Lee02, pp. 400–401]). The problem of a global representation of a closed form as the exterior derivative is a topological problem that is the subject of de Rham cohomology theory. However, for a simply connected manifold, every closed form is exact (loc. cit.).

By (6.136),

$$\mathrm{d}\omega(v_1, \ldots, v_{r+1}) = \sum_{(\gamma),i,K} \omega_{\gamma_1\ldots\gamma_r,i}\varepsilon^{i\gamma_1\ldots\gamma_r}_{k_1\ldots k_{r+1}} v_1^{k_1} \cdots v_{r+1}^{k_{r+1}}. \tag{6.139}$$

Writing in the previous equation

$$\varepsilon^{i\gamma_1\ldots\gamma_r}_{k_1\ldots k_{r+1}} = \sum_{l=1}^{r+1} \delta^i_{k_l}(-1)^{l-1}\varepsilon^{\gamma_1\ldots\ \ldots\gamma_r}_{k_1\ldots\widehat{k_l}\ldots k_{r+1}}, \tag{6.140}$$

one has

$$\mathrm{d}\omega(v_1, \ldots, v_{r+1}) = \sum_{(\gamma),i,K,l} \omega_{\gamma_1\ldots\gamma_r,i}\delta^i_{k_l}(-1)^{l-1}\varepsilon^{\gamma_1\ldots\ \ldots\gamma_r}_{k_1\ldots\widehat{k_l}\ldots k_{r+1}} v_1^{k_1} \cdots v_{r+1}^{k_{r+1}}, \tag{6.141}$$

which implies

$$\begin{aligned}
\mathrm{d}\omega(v_1, \ldots, v_{r+1}) &= \sum_{l,i,K}(-1)^{l-1}\omega_{k_1\ldots\widehat{k_l}\ldots k_{r+1},i} v_l^i v_1^{k_1} \cdots \widehat{v_l^{k_l}} \cdots v_{r+1}^{k_{r+1}}, \\
&= \sum_{l,i}(-1)^{l-1} D\tilde{\omega}(\tilde{v}_l)(\tilde{v}_1, \ldots, \widehat{\tilde{v}_l}, \ldots, \tilde{v}_{r+1}),
\end{aligned} \tag{6.142}$$

where in the last line the superimposed tildes indicate local representatives, $D\tilde{\omega}$ is the derivative of the tensor field, and so the directional derivative $D\tilde{\omega}(\tilde{v}_l)$ is regarded as an alternating r-tensor.

Example 6.6 (The Generalization of the Divergence Operator) Let ω be an $(n-1)$-form. Locally, ω is of the form

$$\sum_{(\beta)} \omega_{\beta_1\ldots\beta_{n-1}}(x^i)\mathrm{d}x^{\beta_1} \wedge \cdots \wedge \mathrm{d}x^{\beta_{n-1}}. \tag{6.143}$$

As in Example 5.4, ω may be represented alternatively in the form

$$\sum_k \omega_{1\ldots\widehat{k}\ldots n}(x^i)\mathrm{d}x^1 \wedge \cdots \wedge \widehat{\mathrm{d}x^k} \wedge \cdots \wedge \mathrm{d}x^n = \sum_k \omega_{1\ldots\widehat{k}\ldots n}(x^i)(-1)^{k-1}\partial_k \lrcorner\, \mathrm{d}x. \tag{6.144}$$

Evidently, ω has n components, and given a Riemannian metric, ω defines a unique vector. In the general non-Riemannian case, many physical objects that are modeled in the Riemannian case by vectors are modeled as $(n-1)$-forms, as elaborated in Chap. 8. This holds in particular for fluxes, e.g., fluxes of mass, heat, and electric charge—current density. When a flux is modeled as an $(n-1)$-form, the divergence of the flux is not a scalar as in the Riemannian situation but an n-form determined by a single component. The divergence operator in this setting is the exterior derivative of ω represented locally as

$$\sum_k \omega_{1\ldots\widehat{k}\ldots n,k}(x^i)\mathrm{d}x^k \wedge \mathrm{d}x^1 \wedge \cdots \wedge \widehat{\mathrm{d}x^k} \wedge \cdots \wedge \mathrm{d}x^n$$

$$= \sum_k (-1)^{k-1}\omega_{1\ldots\widehat{k}\ldots n,k}(x^i)\mathrm{d}x, \tag{6.145}$$

where we used the fact that in Eq. (6.137), for each k, there is only one term that does not vanish—the term that does not contain $\mathrm{d}x^k$. We conclude that the single component of $\mathrm{d}\omega$ is given by $\sum_k (-1)^{k-1}\omega_{1\ldots\widehat{k}\ldots n,k}$.

Note that setting

$$\omega_{\hat{k}} = (-1)^{k-1}\omega_{1\ldots\widehat{k}\ldots n}, \tag{6.146}$$

we have the local representation

$$\omega = \sum_k \omega_{\hat{k}}\partial_k \lrcorner\, \mathrm{d}x, \qquad \mathrm{d}\omega = \sum_k \omega_{\hat{k},k}\mathrm{d}x, \tag{6.147}$$

which is even more reminiscent of the traditional expression for the divergence.

Let ω be an r-form, and let τ be a p-form. Using (5.80) in (6.136), one has

$$
\begin{aligned}
&\mathrm{d}(\omega \wedge \tau) \\
&= \mathrm{d}\left(\sum_{(\alpha),(\beta)} \omega_{\alpha_1\ldots\alpha_r} \tau_{\beta_1\ldots\beta_p} \mathrm{d}x^{\alpha_1} \wedge \cdots \wedge \mathrm{d}x^{\alpha_r} \wedge \mathrm{d}x^{\beta_1} \wedge \cdots \wedge \mathrm{d}x^{\beta_p} \right), \\
&= \sum_{(\alpha),(\beta),i} \omega_{\alpha_1\ldots\alpha_r,i} \tau_{\beta_1\ldots\beta_p} \mathrm{d}x^{i} \wedge \mathrm{d}x^{\alpha_1} \wedge \cdots \wedge \mathrm{d}x^{\alpha_r} \wedge \mathrm{d}x^{\beta_1} \wedge \cdots \wedge \mathrm{d}x^{\beta_p} \\
&\quad + \sum_{(\alpha),(\beta),i} \omega_{\alpha_1\ldots\alpha_r} \tau_{\beta_1\ldots\beta_p,i} \mathrm{d}x^{i} \wedge \mathrm{d}x^{\alpha_1} \wedge \cdots \wedge \mathrm{d}x^{\alpha_r} \wedge \mathrm{d}x^{\beta_1} \wedge \cdots \wedge \mathrm{d}x^{\beta_p}, \\
&= \sum_{(\alpha),(\beta),i} \omega_{\alpha_1\ldots\alpha_r,i} \tau_{\beta_1\ldots\beta_p} \mathrm{d}x^{i} \wedge \mathrm{d}x^{\alpha_1} \wedge \cdots \wedge \mathrm{d}x^{\alpha_r} \wedge \mathrm{d}x^{\beta_1} \wedge \cdots \wedge \mathrm{d}x^{\beta_p} \\
&\quad + (-1)^r \sum_{(\alpha),(\beta),i} \omega_{\alpha_1\ldots\alpha_r} \tau_{\beta_1\ldots\beta_p,i} \mathrm{d}x^{\alpha_1} \wedge \cdots \wedge \mathrm{d}x^{\alpha_r} \wedge \mathrm{d}x^{i} \wedge \mathrm{d}x^{\beta_1} \wedge \cdots \wedge \mathrm{d}x^{\beta_p}.
\end{aligned}
\tag{6.148}
$$

Hence,

$$
\mathrm{d}(\omega \wedge \tau) = \mathrm{d}\omega \wedge \tau + (-1)^r \omega \wedge \mathrm{d}\tau. \tag{6.149}
$$

Finally, one can show that for a differentiable mapping between manifolds $f : \mathcal{X} \to \mathcal{X}'$ and a differential form ω on $\mathcal{X}'$,

$$
\mathrm{d}((Tf)^*(\omega)) = (Tf)^*(\mathrm{d}\omega), \tag{6.150}
$$

that is, the pullback of forms commutes with exterior differentiation.

6.3.3 Partitions of Unity

A partition of unity is a technical tool that enables one to relate local and global fields.

Let $\{(U_a, \phi_a)\}_{a \in A}$ be an atlas on a manifold $\mathcal{X}$. *A partition of unity on* $\mathcal{X}$ *subordinate to* $\{(U_a, \phi_a)\}$ is a collection $\{(V_b, u_b)\}_{b \in B}$ of real-valued functions $u_b : \mathcal{X} \to \mathbb{R}$, and a corresponding open covering $\{V_b\}$ of $\mathcal{X}$, satisfying the following conditions for all b (B need not be a finite set): (a) The support, $\operatorname{supp} u_b$, of u_b, is a subset of V_b, and V_b is a subset of a coordinate neighborhood U_a for some $a \in A$; (b) $u_b(x) \geqslant 0$ for all $x \in \mathcal{X}$; (c) $\sum_b u_b(x) = 1$ for all $x \in \mathcal{X}$; (d) each $x \in \mathcal{X}$ belongs to a finite number of sets V_b.

It is noted that $\{(V_b, \phi_a|_{V_b})\}_{b \in B}$ may serve as an atlas that may contain an infinite number of charts, but each point is covered by a finite number of coordinate neighborhoods. Thus, in the sequel we will write $\{(V_b, \phi_b, u_b)\}$ for the partition of unity subordinate to this atlas.

The existence of partitions of unity is a standard result of differential geometry (see [dR84a, pp. 4–7], [AMR88, pp. 377–388], [Lee02, pp. 54–55]). Partitions of unity have various applications. For example, if $\omega \in \Omega^r(\mathcal{X})$ is a differential r-form, then for an atlas and a subordinate partition of unity $\{(V_b, \phi_b, u_b)\}$, one has $\omega = \sum_b \omega_b$, where $\omega_b = u_b\omega$, so that each of the forms ω_b may be represented locally using ϕ_b. Conversely, let $\{\omega_b : V_b \to \bigwedge^r V_b,\ b \in B\}$ be a collection of local forms. A local ω_b may be defined using the chart (ϕ_b, V_b) and a form $\tilde{\omega}_b$ in $\phi_b\{V_b\}$ so that $\omega_b(x) = (T\phi_b)^*(\tilde{\omega}_b)(x)$ for $x \in V_b$ and $\omega_b(x) = 0$ outside V_b. Then, $\sum_b u_b\omega_b$ is a smooth r-form on $\mathcal{X}$ determined by the local forms ω_b and the partition of unity. If the local forms ω_b agree on the intersections of charts, the resulting ω does not depend on the partition of unity.

6.3.4 Orientation on Manifolds

The n-dimensional manifold $\mathcal{X}$ is *orientable* if there is an n-form θ on $\mathcal{X}$ such that $\theta(x) \neq 0$ for all $x \in \mathcal{X}$. If $\mathcal{X}$ is orientable, such an n-form θ that vanishes nowhere is referred to as a *volume element*. From the point of view of continuum mechanics, a volume element may represent a positive (or negative) density of a positive (negative, respectively) extensive property such as mass density, energy density, etc. A local representative of a volume element is of the form $\theta_{1\dots n}(x^i)\mathrm{d}x^1 \wedge \dots \wedge \mathrm{d}x^n$ for which the smooth function $\theta_{1\dots n}(x^i)$ does not vanish anywhere. We can assume that the coordinate neighborhoods that make up an atlas are connected sets, for otherwise, we can restrict the charts to connected components of coordinate neighborhoods. It follows that for such an atlas, the local representatives $\theta_{1\dots n}(x^i)$ are either positive everywhere or negative everywhere. Since by reordering the coordinates we can change the sign of $\mathrm{d}x^1 \wedge \dots \wedge \mathrm{d}x^n$, we can assume that $\theta_{1\dots n}(x^i)$ are positive everywhere for every chart in a specific atlas. It follows from Eq. (6.77) that the Jacobian determinants of the coordinate transformations between all pairs of charts having this property are positive, $\det(\partial x^i/\partial x^{i'}) > 0$. In fact, if the Jacobian determinants are positive, one can use a partition of unity to construct a volume element. Hence, orientability is equivalent to the existence of an atlas for which the Jacobian determinants of coordinate transformations are positive. Given a volume element θ on an orientable manifold $\mathcal{X}$, any n-form ρ may be represented as $\rho = u\theta$ for a real-valued function u defined on $\mathcal{X}$. A typical local representative $\tilde{u}$ of u is given in terms of the local representatives of ρ and θ by $\tilde{u}(x^i) = \rho_{1\dots n}(x^i)/\theta_{1\dots n}(x^i)$, where we use the fact that $\theta_{1\dots n}$ does not vanish.

A volume element induces an orientation on an orientable manifold $\mathcal{X}$. Two volume elements θ and θ' induce the *same orientation* on $\mathcal{X}$ if $\theta' = u\theta$ for a positive function u. Evidently, there may be only two orientations on a connected manifold, the one induced by θ and the one induced by $-\theta$. An orientable manifold $\mathcal{X}$ together with a choice of orientations is referred to as an *oriented manifold*. Let θ be a volume element that represents the given orientation. An atlas for which all the local representatives of θ are positive will be referred to as an *oriented atlas*.

An orientation of the manifold $\mathcal{X}$ induces an orientation on each of its tangent spaces. An oriented collection of n linearly independent tangent vectors $(v_1, \dots, v_n)$, $v_i \in T_x\mathcal{X}$, is positively oriented if $\theta(x)(v_1, \dots, v_n) > 0$. For a positively oriented chart, the collection of base vectors $(\partial/\partial x^1, \dots, \partial/\partial x^n)$ is positively oriented.

An orientation on a manifold with a boundary induces an orientation on the boundary. In order to do that, one has to examine further the tangent spaces to the manifold at points on the boundary. Let $\mathcal{X}$ be a manifold with a boundary and $x \in \partial\mathcal{X}$. A curve c at x need not be defined on an interval $(-\varepsilon, \varepsilon)$ because such a curve is represented locally by a curve $\tilde{c}$ valued in $\mathbb{R}^{n-}$ with $\tilde{c}(0) = (x^1, \dots, x^{n-1}, 0)$ and the manifold does not contain points with $x^n > 0$. Thus, for defining the tangent space at a boundary point x, we admit curves represented by the restrictions to $(-\varepsilon, 0]$ or to $[0, \varepsilon)$ of curves $\tilde{c} : (-\varepsilon, \varepsilon) \to \mathbb{R}^n$, for some $\varepsilon > 0$, with $\tilde{c}(0) = \phi(x) = (x^1, \dots, x^{n-1}, 0)$. Tangents to such curves are well-defined since they are induced by the derivatives of the representing curves $\tilde{c}$ at zero. This makes the tangent space to $\mathcal{X}$ at x isomorphic to $\mathbb{R}^n$ as expected.

A tangent vector $v \in T_x\mathcal{X}, x \in \partial\mathcal{X}$ is said to be *inward-pointing* if for an oriented chart in a neighborhood of x (which is compatible with the manifold with boundary structure), it is represented in the form $(v^1, \dots, v^n)$ with $v^n < 0$; v is *outward-pointing* if $v^n > 0$. These properties do not depend on the chart chosen. Let $\mathcal{J} : \partial\mathcal{X} \to \mathcal{X}$ be the natural embedding of the boundary and denote by $\tau_{\mathcal{X}} : T\mathcal{X} \to \mathcal{X}$ and $\tau_{\partial\mathcal{X}} : T\partial\mathcal{X} \to \partial\mathcal{X}$ the tangent bundle projections for the manifold and its boundary, respectively. We also have the vector bundle $\mathcal{J}^*\tau_{\mathcal{X}} : \mathcal{J}^*T\mathcal{X} \to \partial\mathcal{X}$ whose fiber at any point $y \in \partial\mathcal{X}$ is $T_y\mathcal{X}$ so that $\mathcal{J}^*T\mathcal{X}$ is simply the restriction of $T\mathcal{X}$ to the submanifold $\partial\mathcal{X}$. Assuming that $\mathcal{X}$ is oriented, using a partition of unity subordinate to an oriented atlas, one can construct a nowhere vanishing field of outward-pointing vectors. One has to patch together local representatives of vector fields of the form $(0, \dots, 0, 1)$. The nowhere vanishing field of outward-pointing vectors is a section $w : \partial\mathcal{X} \to \mathcal{J}^*T\mathcal{X}$. Given a volume element θ on $\mathcal{X}$, a collection of linearly independent vectors $(v_2, \dots, v_n)$, $v_i \in T_x\partial\mathcal{B}$ and an outward-pointing vector v, the vectors $(v, v_2, \dots, v_n)$ are linearly independent and so $\theta(x)(v, v_2, \dots, v_n) \neq 0$. Thus, for any nowhere vanishing outward-pointing vector field w, $w \lrcorner \theta$ is nowhere vanishing on $\partial\mathcal{X}$. It follows that $w \lrcorner \theta$ may serve as a volume element on $\partial\mathcal{X}$. The orientation on $\mathcal{X}$ induced by θ determines an orientation on $\partial\mathcal{X}$ for which the vectors $(v_2, \dots, v_n)$ are positively oriented if $(v, v_2, \dots, v_n)$ are positively oriented in $\mathcal{X}$.

Remark 6.1 There is a slight complication that stems from the definition given above for the positive orientation of the boundary. Given a coordinate neighborhood at the boundary belonging to a positively oriented atlas in $\mathcal{X}$, the form $\mathrm{d}x^1 \wedge \cdots \wedge \mathrm{d}x^n$ is a positive volume element. The vector $\partial/\partial x^n$ is outward-pointing, and so the induced orientation is determined by the form represented locally as $\partial/\partial x^n \lrcorner \mathrm{d}x^1 \wedge \cdots \wedge \mathrm{d}x^n$ and using (5.102), one has

$$\frac{\partial}{\partial x^n} \lrcorner \, dx^1 \wedge \cdots \wedge dx^n = (-1)^{n-1} dx^1 \wedge \cdots \wedge dx^{n-1}. \tag{6.151}$$

This implies that the form $dx^1 \wedge \cdots \wedge dx^{n-1}$ and the chart with coordinates $x^1, \ldots, x^{n-1}$ are negatively oriented if the dimension is even.

6.3.5 Integration on Oriented Manifolds

The natural objects to be integrated on an n-dimensional oriented manifold $\mathcal{X}$ are n-forms. An intuitive motivation, for the role of differential forms as integrands, is given below in Sect. 6.3.7 on integration over chains in manifolds.

Let ρ be a compactly supported n-form defined on the oriented manifold $\mathcal{X}$ and let $\{(V_b, \phi_b, u_b)\}$ be a partition of unity subordinate to an oriented atlas. Expecting that the integral will be linear in the integrands, we expect that

$$\int_{\mathcal{X}} \rho = \int_{\mathcal{X}} \Big(\sum_b u_b\Big)\rho = \sum_b \int_{\mathcal{X}} u_b \rho. \tag{6.152}$$

Thus, it will be sufficient to define the integral for a form that is compactly supported on the domain of a chart. Let α be an n-form that is compactly supported in the domain of the oriented chart (V_b, ϕ_b), and let the local representation of α be $\alpha_{1\ldots n}(x^i)dx = \alpha_{1\ldots n}(x^i)dx^1 \wedge \cdots \wedge dx^n$. Then, the integral of α over $\mathcal{X}$ is defined by

$$\int_{\mathcal{X}} \alpha := \int_{\phi_b\{V_b\}} \alpha_{1\ldots n}(x^i)dx^1 \cdots dx^n. \tag{6.153}$$

It is noted that locally,

$$\alpha_{1\ldots n} = (T\phi_b)^*(\alpha)(\mathbf{e}_1, \ldots, \mathbf{e}_n) = \alpha(\partial_1, \ldots, \partial_n) \tag{6.154}$$

where $\mathbf{e}_1, \ldots, \mathbf{e}_n$ is a positively oriented basis of $\mathbb{R}^n$.

The integral of a form on a manifold is computed now using Eq. (6.152). One can show, e.g., [Lee02, pp. 353–355], that the definition is independent of the choice of oriented atlas and partition of unity. A basic element in proving the independence of the integral on the choice of an atlas is the transformation rule (6.77) for n-forms and its relation to the transformation rule for integrands of functions defined on $\mathbb{R}^n$.

Evidently, if one reverses the choice of orientation on the manifold $\mathcal{X}$, the opposite value will be obtained for integral. The following is a particularly useful property of the integration of forms. Let $f : \mathcal{X}_1 \to \mathcal{X}_2$ be an orientation preserving diffeomorphism between oriented manifolds of dimension n, and let ω be an n-form on $\mathcal{X}_2$. Then,

$$\int_{\mathcal{X}_2} \omega = \int_{\mathcal{X}_1} (Tf)^*\omega, \tag{6.155}$$

where $(Tf)^*\omega$ is the pullback of the differential form ω as in (6.78).

6.3.6 *Stokes's Theorem*

Let $\mathcal{X}$ be a compact manifold with a boundary and let $\mathcal{J} : \partial\mathcal{X} \to \mathcal{X}$, $\mathcal{J}(x) = x$, be the inclusion of the boundary in $\mathcal{X}$. The inclusion induces the tangent mapping $T\mathcal{J} : T\partial\mathcal{X} \to T\mathcal{X}$, $\mathcal{J}(v) = v$, the inclusion of vectors tangent to the boundary in the tangent spaces to the manifold, and $(T\mathcal{J})^* : \Omega^{n-1}(\mathcal{X}) \to \Omega^{n-1}(\partial\mathcal{X})$ that restricts $(n-1)$-forms defined on the manifold to vectors tangent to its boundary (cf. the analogous Sects. 5.10, 6.1.10, and 6.3.4).

Consider the local representative $\sum_k \omega_{1\ldots\widehat{k}\ldots n}(x^i)\mathrm{d}x^1 \wedge \cdots \wedge \widehat{\mathrm{d}x^k} \wedge \cdots \wedge \mathrm{d}x^n$ of an $(n-1)$-form ω in a chart for a manifold with a boundary as in Sect. 6.1.1. As an $(n-1)$-form on an $(n-1)$-manifold, $(T\mathcal{J})^*(\omega)$ has but one component $(T\mathcal{J})^*(\omega)_{1\ldots n-1}$ given by

$$\begin{aligned} (T\mathcal{J})^*(\omega)_{1\ldots n-1} &= (T\mathcal{J})^*(\omega)\,(\partial_1, \ldots, \partial_{n-1}), \\ &= \omega\,(T\mathcal{J}(\partial_1), \ldots, T\mathcal{J}(\partial_{n-1})), \\ &= \omega\,(\partial_1, \ldots, \partial_{n-1}), \\ &= \omega_{1\ldots n-1}. \end{aligned} \tag{6.156}$$

(See Sect. 8.1.3 for a more detailed representation of $(T\mathcal{J})^*$.)

Stokes's theorem in differential geometry is a generalization of the Gauss, Green, and Stokes's theorems of multivariable calculus. It does not use a Riemannian metric, and it has a very compact form using the integration of forms on oriented manifolds. It states that for a compactly supported $(n-1)$-form ω on $\mathcal{X}$,

$$\int_{\mathcal{X}} \mathrm{d}\omega = \int_{\partial\mathcal{X}} (T\mathcal{J})^*(\omega), \quad \text{or in short,} \quad \int_{\mathcal{X}} \mathrm{d}\omega = \int_{\partial\mathcal{X}} \omega. \tag{6.157}$$

Using a partition of unity to localize the forms to coordinate neighborhoods, it is sufficient to prove the theorem for a compactly supported form ω defined on an open subset U of $\mathbb{R}^{n-}$, where $\partial U = U \cap \partial\mathbb{R}^{n-}$ and $\partial\mathbb{R}^{n-} = \{x \in \mathbb{R}^{n-} \mid x^n = 0\}$. Thus, we consider the integral over ∂U of the $(n-1)$-form $(T\mathcal{J})^*(\omega)$, where ω is supported in a chart defined in a neighborhood U of the boundary with oriented coordinates $x^1, \ldots, x^n$. By the definition of the integral above, we have to evaluate the integral using a positively oriented chart on ∂U. However, by Remark 6.1, $(-1)^{n-1}\mathrm{d}x^1 \wedge \cdots \wedge \mathrm{d}x^{n-1}$ is positively oriented rather than $\mathrm{d}x^1 \wedge \cdots \wedge \mathrm{d}x^{n-1}$.

Thus,

$$\begin{aligned}\int_{\partial U}(T\mathcal{I})^*(\omega) &= \int_{\partial U}(-1)^{n-1}\omega_{1\ldots n-1}\mathrm{d}x^1\wedge\cdots\wedge\mathrm{d}x^{n-1},\\ &= \int_{\partial U}(-1)^{n-1}\omega_{1\ldots n-1}\mathrm{d}x^1\cdots\mathrm{d}x^{n-1}.\end{aligned} \tag{6.158}$$

Using Example 6.6 and Eq. (6.158), the local version of Eq. (6.157) becomes

$$\int_U\sum_k(-1)^{k-1}\omega_{1\ldots\widehat{k}\ldots n,k}(x^i)\mathrm{d}x^1\cdots\mathrm{d}x^n = \int_{\partial U}(-1)^{n-1}\omega_{1\ldots n-1}(x^i)\mathrm{d}x^1\cdots\mathrm{d}x^{n-1}. \tag{6.159}$$

For the integrals on the left with $k < n$, we have

$$\begin{aligned}\int_U\omega_{1\ldots\widehat{k}\ldots n,k}\mathrm{d}x^1\cdots\mathrm{d}x^n &= \int_{\mathbb{R}^{n-}}\omega_{1\ldots\widehat{k}\ldots n,k}\mathrm{d}x^1\cdots\mathrm{d}x^n,\\ &= \int_{\mathbb{R}^{(n-1)-}}\left(\int_{\mathbb{R}}\omega_{1\ldots\widehat{k}\ldots n,k}\mathrm{d}x^k\right)\mathrm{d}x^1\cdots\widehat{\mathrm{d}x^k}\ldots\mathrm{d}x^n,\\ &= 0,\end{aligned} \tag{6.160}$$

as the integral in the parenthesis vanishes due to the compact support of ω in U. If U does not intersect $\partial\mathbb{R}^{n-}$, the same holds for $k = n$, and in this case, the right-hand side of (6.159) vanishes too. If $\partial U \neq \varnothing$, then,

$$\begin{aligned}\int_U\omega_{1\ldots n-1,n}\mathrm{d}x^1\cdots\mathrm{d}x^n &= \int_{\mathbb{R}^{n-}}\omega_{1\ldots n-1,n}\mathrm{d}x^1\cdots\mathrm{d}x^n,\\ &= \int_{\mathbb{R}^{n-1}}\left(\int_{\mathbb{R}^-}\omega_{1\ldots n-1,n}\mathrm{d}x^n\right)\mathrm{d}x^1\ldots\mathrm{d}x^{n-1},\\ &= \int_{\mathbb{R}^{n-1}}\omega_{1\ldots n-1}(x)|_{x^n\to-\infty}^{x^n=0}\,\mathrm{d}x^1\ldots\mathrm{d}x^{n-1},\\ &= \int_{\mathbb{R}^{n-1}}\omega_{1\ldots n-1}(x)|_{x^n=0}\mathrm{d}x^1\ldots\mathrm{d}x^{n-1}.\end{aligned} \tag{6.161}$$

We conclude that for the left-hand side of (6.159),

$$\begin{aligned}&\int_U\sum_k(-1)^{k-1}\omega_{1\ldots\widehat{k}\ldots n,k}(x^i)\mathrm{d}x^1\cdots\mathrm{d}x^n\\ &\quad= \int_{\mathbb{R}^{n-1}}(-1)^{n-1}\omega_{1\ldots n-1}(x)|_{x^n=0}\mathrm{d}x^1\ldots\mathrm{d}x^{n-1},\end{aligned} \tag{6.162}$$

and as $\omega_{1\ldots n-1}(x)|_{x^n=0}$ vanishes outside U, (6.159) follows.

6.3.7 *Integration Over Chains on Manifolds*

In order that geometric objects such as simplices and chains, which do not have smooth boundaries, may serve as domains of integration, integration theory is extended as described below.

The *standard r-simplex* Δ^r is the simplex in $\mathbb{R}^r$ with the first vertex at the origin such that the vectors to the other vertices are $v_i = \mathbf{e}_i$, $i = 1, \dots, r$, where $\mathbf{e}_i$ is the i-th standard base vector in $\mathbb{R}^r$. The orientation of Δ^r is induced by the standard orientation of $\mathbb{R}^r$. A *singular r-simplex* in a manifold $\mathcal{X}$ is a smooth map $s : \Delta^r \to \mathcal{X}$. As the simplex is not an open set, smoothness means that there is an open set $U \in \mathbb{R}^r$ containing Δ^r such that s is the restriction of a smooth map $s_e : U \to \mathcal{X}$ to Δ^r.

The faces of singular r-simplices are represented as singular $(r-1)$-simplices. First, one introduces mappings $k_p{:}\Delta^{r-1} \to \mathbb{R}^r$, $p = 0, \dots, r$, and represents the faces of the standard r-simplex as singular $(r-1)$-simplices in $\mathbb{R}^r$. These mappings are defined as

$$k_p(y^1, \dots, y^{r-1}) = (y^1, \dots, y^{p-1}, 0, y^p, \dots, y^{r-1}), \quad \text{for} \quad p = 1, \dots, r, \tag{6.163}$$

and

$$k_0(y^1, \dots, y^{r-1}) = \left(1 - \textstyle\sum_{j=1}^{r-1} y^j, y^1, \dots, y^{r-1}\right). \tag{6.164}$$

The derivatives of these mappings are clearly uniform and are given in terms of the standard bases by the matrices

$$[K_p] = [Dk_p] = \begin{pmatrix} 1 & 0 & 0 & \cdots & 0 & 0 \\ 0 & 1 & 0 & \cdots & & 0 \\ & 0 & \ddots & 0 & & \\ 0 & \cdots & 0 & 0 & \cdots & 0 \\ \vdots & \vdots & & \ddots & & \\ 0 & & \cdots & & 1 & 0 \\ 0 & & \cdots & \cdots & 0 & 1 \end{pmatrix} \Leftarrow p\text{th row}, \quad p = 1, \dots, r \tag{6.165}$$

and

$$[K_0] = [Dk_0] = \begin{pmatrix} -1 & -1 & \cdots & -1 \\ 1 & 0 & \cdots & 0 \\ 0 & 1 & 0 & \\ \vdots & \vdots & \ddots & \vdots \\ 0 & \cdots & 0 & 1 \end{pmatrix}. \tag{6.166}$$

One can verify that when applied to the standard basis $\{\mathbf{e}_1, \dots \mathbf{e}_{r-1}\}$, one obtains for $\mathfrak{e}_{pl} := Dk_p(\mathbf{e}_l)$

$$\mathfrak{e}_{pl} := K_p(\mathbf{e}_l) = \begin{cases} \mathbf{e}_{l+1} - \mathbf{e}_1, & \text{for } p = 0, \\ \mathbf{e}_l, & \text{for } l < p, \\ \mathbf{e}_{l+1}, & \text{for } l \geqslant p. \end{cases} \tag{6.167}$$

in accordance with Eq. (3.28).

An *r-singular chain* on the manifold $\mathcal{X}$ is a formal linear combination $\sum_l a^l s_l$ of singular r-simplices. The chain $1s$ is identified with s and $(-1)s$ is identified with the simplex obtained from s by reversing the orientation of the standard simplex. Triangulation theory implies that differentiable manifolds may be represented as chains (see [Whi57, pp. 124–135]).

We now consider the boundary of a singular r-simplex s, a singular $(r-1)$-chain ∂s. In order that the orientations agree with Eq. (3.28), one defines $s_p = s \circ k_p$, and

$$\partial s = \sum_p (-1)^p s_p = \sum_p (-1)^p s \circ k_p. \tag{6.168}$$

The boundary operator is extended to chains by linearity. It is noted that in the last equation $(-1)s$ is regarded as a multiplication of the simple chain s by -1, thus reversing its orientation. It does not indicate the multiplication of the mapping s by -1 even in the case where such a multiplication makes sense, e.g., $(-1)^p k_p$.

Let s be a singular r-simplex on an n-dimensional manifold $\mathcal{X}$, and let ω be a smooth r-form defined in a neighborhood of $D = s(\Delta^r)$. We will refer to ω simply as a *smooth form over s*. The integral of ω over s (over D) is defined by

$$\int_s \omega := \int_D \omega = \int_{\Delta^r} (Ts)^*(\omega) = \int_{\Delta^r} \omega_{1\dots r} \mathrm{d}x^1 \cdots \mathrm{d}x^r, \tag{6.169}$$

where $\omega_{1\dots r}\mathrm{d}x^1 \wedge \cdots \wedge \mathrm{d}x^r = Ts^*(\omega)$ is the pullback of ω to $\Delta^r \subset \mathbb{R}^r$ using the simplex mapping s. The definition of the integral implies that if one reverses the choice of orientation on the simplex s, the opposite value will be obtained for the integral of a form ω.

The definition of the integral of a form over a simplex suggests an intuitive motivation for the role of differential forms as integrands. For a singular simplex s, and using the mean value theorem for integration in $\mathbb{R}^r$, there is a point $P \in \Delta^r$ such that

$$\begin{aligned}\int_s \omega &= \int_{\Delta^r} \omega_{1\dots r} \mathrm{d}x^1 \cdots \mathrm{d}x^r, \\ &= \omega_{1\dots r}(P) \int_{\Delta^r} \mathrm{d}x^1 \cdots \mathrm{d}x^r, \\ &= \frac{1}{r!}\omega_{1\dots r}(P).\end{aligned} \tag{6.170}$$

Letting $\{\mathbf{e}_1, \dots, \mathbf{e}_r\}$ be the standard basis of $\mathbb{R}^r$ and $v_i = Ts(P)(\mathbf{e}_i)$,

$$\begin{aligned}\omega_{1\dots r}(P) &= Ts^*(\omega)(P)(\mathbf{e}_1, \dots, \mathbf{e}_r), \\ &= \omega(s(P))(v_1, \dots, v_r).\end{aligned} \tag{6.171}$$

We conclude that

$$\int_s \omega = \frac{1}{r!}\omega(s(P))(v_1, \dots, v_r) \tag{6.172}$$

so that the integral is approximated by the evaluation of the form on the vectors tangent to the edges of D.

Remark 6.2 The last equation contradicts our interpretation of the action of an alternating tensor on the r-tuple of vectors $(v_1, \dots, v_r)$ as providing the total flux through the oriented simplex they determine. The division by $r!$ implies that for an alternating tensor ω, $\omega(v_1, \dots, v_r)$ should be interpreted as the flux corresponding to the r-cube that the vectors determine. This inconsistency follows from the definition of the integral as in Eqs. (6.153) and (6.169). The interpretation of the action as the flux through the simplex would be accurate if the integral would be defined alternatively as

$$r! \int_{\Delta^r} \omega_{1\dots r} \mathrm{d}x^1 \cdots \mathrm{d}x^r = \int_{\Delta^r} \sum_\pi (-1)^{|\pi|} \omega_{\pi_1 \dots \pi_r} \mathrm{d}x^{\pi_1} \cdots \mathrm{d}x^{\pi_r}. \tag{6.173}$$

Nevertheless, since we rely on the simplex constructions in Sect. 3, since we do not want to change the traditional definition of the integral of a form, and since for simplicity of notation we do not want to write the flux through the simplex as $\omega(v_1, \dots, v_r)/r!$, we ignore this inconsistency.

Let $\pi : \mathcal{Y} \to \mathcal{X}$ be a fiber bundle and $s : \Delta^r \to \mathcal{X}$ a singular simplex. We say that κ is a *smooth section of* $\mathcal{Y}$ *over* s if it is the restriction of a smooth section of $\mathcal{Y}$ to Image s. If $B = \sum_l a^l s_l$ is an r-chain on $\mathcal{X}$, a smooth section κ of $\mathcal{Y}$ over B is a collection of smooth sections over the various simplices that make up B. (Thus, κ is not necessarily extendable to a global smooth section.)

If $B = \sum_l a^l s_l$ is an r-chain, the integral of a smooth form ω over B is defined by linearity as

$$\int_B \omega = \sum_l a^l \int_{D_l} \omega_l = \sum_l a^l \int_{s_l} \omega_l, \qquad D_l = s_l\{\Delta^r\}, \tag{6.174}$$

where ω_l is the smooth form over s_l in the definition of a smooth section over a chain. In particular, Eq. (6.168) implies that for an $(r-1)$-form defined on a singular r-simplex s,

$$\begin{aligned}
\int_{\partial s} \omega &= \int_{\sum_p (-1)^p s_p} \omega_p, \\
&= \int_{\sum_p (-1)^p s_p} \omega_p, \\
&= \sum_p (-1)^p \int_{s_p(\Delta^{r-1})} \omega_p, \\
&= \sum_p (-1)^p \int_{s \circ k_p} \omega_p, \\
&= \sum_p (-1)^p \int_{\Delta^{r-1}} [(T(s \circ k_p)]^*(\omega_p) \mathrm{d}x^1 \cdots \mathrm{d}x^{r-1}, \\
&= \sum_p (-1)^p \int_{k_p(\Delta^{r-1})} [(Ts)^*(\omega_p)](\mathrm{e}_{p1}, \ldots, \mathrm{e}_{pr-1}) \mathrm{d}x^1 \cdots \mathrm{d}x^{r-1}.
\end{aligned} \tag{6.175}$$

To arrive at the last line, we used

$$\begin{aligned}
&\int_{\Delta^{r-1}} [(T(s \circ k_p)]^*(\omega_p) \\
&= \int_{\Delta^{r-1}} [(T(s \circ k_p)]^*(\omega_p)_{1 \ldots r-1} \mathrm{d}x^1 \cdots \mathrm{d}x^{r-1}, \\
&= \int_{\Delta^{r-1}} \{[(T(s \circ k_p)]^*(\omega_p)\}_{1 \ldots r-1} \mathrm{d}x^1 \cdots \mathrm{d}x^{r-1}, \\
&= \int_{\Delta^{r-1}} \{[(T(s \circ k_p)]^*(\omega_p)\}(\mathbf{e}_1, \ldots \mathbf{e}_{r-1}) \mathrm{d}x^1 \cdots \mathrm{d}x^{r-1}, \\
&= \int_{\Delta^{r-1}} \omega_p[T(s \circ k_p)(\mathbf{e}_1), \ldots T(s \circ k_p)(\mathbf{e}_{r-1})] \mathrm{d}x^1 \cdots \mathrm{d}x^{r-1}, \\
&= \int_{\Delta^{r-1}} \omega_p[Ts(Tk_p(\mathbf{e}_1)), \ldots Ts(Tk_p \mathbf{e}_{r-1}))] \mathrm{d}x^1 \cdots \mathrm{d}x^{r-1}.
\end{aligned} \tag{6.176}$$

By subdividing an r-chain into small simplices s_l and letting $Q_l \in D_l = s_l(\Delta^r)$, the integral of a form over a chain may be approximated by a sum in the form

$$\frac{1}{r!}\sum_{l} a^{l}\omega(Q_{l})(v_{1l},\ldots,v_{rl}),\tag{6.177}$$

where v_{pl} is the tangent to the p-th edge of D_l.

Using an analogous procedure to the proof of the Stokes's theorem for oriented manifolds, in particular, the fundamental theorem of calculus, Stokes's theorem for chains on manifolds may be proved in the form (see, e.g., [War83, AMR88, Lee02])

$$\int_{B} \mathrm{d}\omega = \int_{\partial B} \omega,\tag{6.178}$$

for any $(r-1)$-form ω and r-chain B.

Remark 6.3 It is noted that in addition to allowing non-smooth boundaries and enabling integration of forms on regions of dimension lower than $\dim \mathcal{X}$, integration theory over chains does not require that the manifold be orientable. Furthermore, the term "singular" is used in order to indicate that, in general, there is no requirement that s will be injective. However, integration on chains was presented above as a tool that allows one to consider the non-smooth boundaries of regular (rather than singular) simplices on manifolds. Additional approaches for integration on objects with non-smooth boundaries are available, e.g., the *standard manifolds* of [Whi57, pp. 108–110], or *manifolds with corners* as considered below. Integration theory on chains seems to us to have an appealing geometric flavor. In the sequel, for all chains on manifolds that we will consider, the mappings s_e will be non-singular so that their inverses may serve as charts.

6.4 Manifolds with Corners

An n-dimensional manifold with corners (e.g., [DH73, Mic80, Mel96, Lee02, MRD08, Mic20]) is a manifold the charts of which assume values in the n-quadrant of $\mathbb{R}^n$, that is, in

$$\overline{\mathbb{R}}^{n}_{-} := \{x \in \mathbb{R}^n \mid x^i \leqslant 0,\ i = 1,\ldots,n\}.\tag{6.179}$$

In the construction of the manifold structure, it is understood that a function defined on a quadrant is said to be differentiable if it is the restriction to the quadrant of a differentiable function defined on $\mathbb{R}^n$. If $\mathcal{X}$ is an n-dimensional manifold with corners, a subset $\mathcal{Z} \subset \mathcal{X}$ is defined to be a k-dimensional, $k \leq n$, submanifold with corners of $\mathcal{X}$ if for any $z \in \mathcal{Z}$ there is a chart (U, ϕ), $z \in U$, such that $\phi(\mathcal{Z} \cap U) \subset \{x \in \overline{\mathbb{R}}^{n-} \mid x^l = 0,\ k < l \leqslant n\}$.

With an appropriate natural definition of the integral of an $(n-1)$-form over the boundary of a manifold with corners, Stokes's theorem holds for manifolds with corners. (See [Lee02, pp. 363–370] and [Mic20, Section 3.5].)

In studying analysis over manifolds with corners, the following result is significant. (See [DH73, Mic80, Mel96] and [Mic20, Section 3.2].) Every n-dimensional manifold with corners, $\mathfrak{X}$, is a submanifold with corners of a manifold $\tilde{\mathfrak{X}}$ without boundary of the same dimension. In addition, if $\mathfrak{X}$ is compact, it can be embedded as a submanifold with corners in a compact manifold without boundary $\tilde{\mathfrak{X}}$ of the same dimension [Mel96, pp. I.24–26]. Furthermore, C^k-forms defined on $\mathfrak{X}$ may be extended continuously and linearly to forms defined on $\tilde{\mathfrak{X}}$. Such a manifold $\tilde{\mathfrak{X}}$ is referred to as an *extension* of $\mathfrak{X}$. Each smooth vector bundle, $\pi : W \longrightarrow \mathfrak{X}$ over $\mathfrak{X}$, extends to a smooth vector bundle, $\tilde{\pi} : \tilde{W} \longrightarrow \tilde{\mathfrak{X}}$, over $\tilde{\mathfrak{X}}$. Each immersion (embedding) of $\mathfrak{X}$ into a smooth manifold $\mathfrak{Z}$ without boundary is the restriction of an immersion (embedding) of $\tilde{\mathfrak{X}}$ into $\mathfrak{Z}$.

It is emphasized that manifolds with corners do not model some basic geometric shapes, such as a pyramid with a rectangular base or a cone. However, much of the material presented in this volume is valid for a class of more general objects, *Whitney manifold germs,* as presented in [Mic20].

Chapter 7
Interlude: Smooth Distributions of Defects

This short interlude presents an example of an application of notions reviewed in the foregoing chapter to the geometric description of smooth distributions of dislocations in a body. In accordance with the main theme, the macroscopic, continuum mechanics, point of view is adopted. The corresponding non-smooth theory will be presented in Chap. 19. These two formulations are based on [ES14, ES15, ES20].

7.1 Introduction

From the continuum mechanics point of view, material defects are frequently described by the relative deformation of neighboring points in the material (e.g., [KA75, LK06, Sah84]). Sometimes a global approach is adopted (e.g., [Cer99]) and defects are viewed as obstructions to the construction of a global inverse deformation. Another frequent approach (e.g., [Kon55, Nol67, Wan67, EE07]) views the existence of defects, or inhomogeneities, as an inherent consequence of the constitutive relation for a body.

Following [ES14], the present framework differs from the first point of view above in the sense that the analysis involves no kinematics of the body in space. No deformations are considered and only the material structure of the body manifold is studied. Our approach differs from the theory of inhomogeneities in the sense that rather than associating the defects with a particular constitutive relation, the material structure is given explicitly. (See somewhat similar approaches in Toupin [Tou68] and Eringen and Claus [EC70] who use oriented or micromorphic media.) Essentially, for the case of dislocations, it is assumed that a field of Bravais hyperplanes is given by prescribing explicitly a family of parallel hyperplanes at each point in a body. In other words, we specify a continuously distributed analog of the Miller indices for the crystallographic planes.

R. Segev, *Foundations of Geometric Continuum Mechanics*, Advances in Continuum Mechanics 49, https://doi.org/10.1007/978-3-031-35655-1_7

It is recalled that the Miller indices for a plane in $\mathbb{R}^3$ that intersects the Cartesian axes at the points $(a, 0, 0)$, $(0, b, 0)$, and $(0, 0, c)$ are the smallest three integers (including infinity) (A, B, C) that are related to one another as do $(1/a, 1/b, 1/c)$. Alternatively, it is observed that $(1/a, 1/b, 1/c)$ are the components, relative to the natural basis, of a covector $\phi : \mathbb{R}^3 \to \mathbb{R}$ satisfying $\phi(a, 0, 0) = \phi(0, b, 0) = \phi(0, 0, c) = 1$. In other words, the plane containing these three points is $\phi^{-1}(1)$. Thus, it is natural to view the Miller indices and the family of planes they represent as a covector.

We dispense with the three-dimensional Euclidean structure for a material body and model it as a manifold $\mathcal{X}$ of dimension n. The family of crystallographic planes in an "infinitesimal" neighborhood of a point $X \in \mathcal{X}$ is represented by the value $\phi(X)$ of a differential 1-form ϕ. The action $\phi(X)(v)$, for a vector $v \in T_X\mathcal{X}$, is interpreted as the amount of parallel crystallographic planes penetrated by the "infinitesimal" line represented by v, where the orientation is reflected in the sign of the action.

From the microscopic point of view of material science, a dislocation is characterized by the Burgers vector. To construct it, one maps a closed circuit in the perfect lattice to the dislocated lattice, and the Burgers vector is determined by the opening—the failure of the image of the circuit to close. In the deformation theory of dislocations, the Burgers vector is defined using the gap that opens up between the positions of neighboring points.

Here, on the other hand, we do not make any reference to a perfect body. Generalizing the notion of crystallographic hyperplanes, we examine *layers*—crystallographic hypersurfaces. Roughly, we consider the total amount of layers that are penetrated when a closed circuit is followed, taking the orientation into account.

It is observed that other descriptions of the geometry of defects based on explicit geometric specifications of material structures are available in the literature, e.g., [DP91].

7.2 Forms and Hypersurfaces

Consider a family $\mathcal{F}$ of $(n-r)$-dimensional oriented hypersurfaces, or *layers*, in the body manifold $\mathcal{X}$. We view the family $\mathcal{F}$ as a given material structure in the body. For example, a family of two-dimensional surfaces in a three-dimensional body may be thought of as a family of lattice layers or lattice surfaces. A family of one-dimensional surfaces in a three-dimensional body may describe a fibrous structure in a solid body or the state of a nematic liquid crystal. Let $\mathcal{S}$ be an $(r+1)$-dimensional submanifold with boundary in the body manifold. The "amount" of layers belonging to $\mathcal{F}$ that cross the boundary $\partial\mathcal{S}$, if different from zero, indicates the generation or annihilation of such layers in $\mathcal{S}$. We view such creation or annihilation of material hypersurfaces as an indication of the presence of defects inside $\mathcal{S}$.

We start with linear, infinitesimal, considerations and recall (see [Ste83, pp. 16–17]) that an r-dimensional subspace W of a vector space V is associated with a decomposable r-vector $\mathfrak{v}$ which is unique up to a scalar factor; $u \in W$ if and only if $\mathfrak{v} \wedge u = 0$. In the sequel, we will use this property for subspaces D_X^* of the various cotangent spaces $T_X^*\mathcal{X}$, $X \in \mathcal{X}$ of dimension $p = n - r$. Thus, a p-dimensional subspace $D_X^* \subset T_X^*\mathcal{X}$ is determined by a decomposable, nonvanishing p-covector ϕ, so that

$$D_X^* := \{\omega \in T_X^*\mathcal{X} \mid \phi \wedge \omega = 0\}. \tag{7.1}$$

It is observed that each D_X^* determines a unique r-dimensional subspace

$$D_X := (D_X^*)^\perp = \{v \in T_X\mathcal{X} \mid \omega(v) = 0, \text{ for all } \omega \in D_X^*\}.$$

We prove that $v \in D_X$ if and only if $v \lrcorner \phi = 0$.

Assume, first, that $v \lrcorner \phi = 0$. We have to prove that $v \in D_X$, in other words, that for each $\omega \in T_X^*\mathcal{X}$ such that $\phi \wedge \omega = 0$, $\omega(v) = 0$. Let $\omega \in T_X^*\mathcal{X}$ be a covector, such that $\phi \wedge \omega = 0$. Then, recalling the identity (5.107)

$$v \lrcorner (\phi \wedge \omega) = (v \lrcorner \phi) \wedge \omega + (-1)^p \phi \wedge (v \lrcorner \omega), \tag{7.2}$$

for any tangent vector v, one has

$$\phi \wedge (v \lrcorner \omega) = 0. \tag{7.3}$$

Since $v \lrcorner \omega$ is a scalar and ϕ is assumed to be nonvanishing, one has $\omega(v) = 0$.

Conversely, let $v \in D_X$, so that $\omega(v) = 0$ for all $\omega \in T_X^*\mathcal{X}$, satisfying $\phi \wedge \omega = 0$. We have to prove that $v \lrcorner \phi = 0$. Since ϕ is decomposable, there is a basis $\{\phi^1, \dots, \phi^n\}$ of $T_X\mathcal{X}$ such that ϕ can be expressed as $\phi = \phi^1 \wedge \cdots \wedge \phi^p$. Evidently, $\phi \wedge \phi^i = 0$ for all $i = 1, \dots, p$. From the assumption that $v \in D_X$, it follows that $\phi^i(v) = 0$ for all $i = 1, \dots, p$. Hence,

$$v \lrcorner (\phi^1 \wedge \cdots \wedge \phi^p) = 0, \tag{7.4}$$

and we conclude that

$$v \lrcorner \phi = 0. \tag{7.5}$$

Evidently, $\dim D_X = r = n - p$. Indeed, it is the intersection of the hyperplanes of codimension one determined by the equations $\phi^i(v) = 0$ for $i = 1, \dots, p$.

The decomposable p-form ϕ determines a family $\mathcal{F}_X$ of r-dimensional affine subspaces of $T_X\mathcal{X}$. The family $\mathcal{F}_X$ is the quotient space of $T_X\mathcal{X}$ relative to the equivalence relation $v \sim u$ if $v - u \in D_X$. Each equivalence class may be labeled by $(a^1, \dots, a^p) \in \mathbb{R}^p$ and is determined by the equations $\phi^i(v) = a^i$, $i = 1, \dots, p$. In other words, it is the intersection of the affine subspaces of codimension one

$\phi^i(v) = a^i$. If the two vectors v, u in $T_X\mathcal{X}$ satisfy $\phi^i(v) = \phi^i(u)$, then, $\phi^i(v-u) = 0$ for all $i = 1, \ldots, p$. It follows that $u - v \in D_X$.

Let $v_1, \ldots, v_p \in T_X\mathcal{X}$. The number $\phi(v_1, \ldots, v_p)$ is interpreted as the "amount" of affine subspaces in $\mathcal{F}_X$ that cross the infinitesimal oriented element (a p-dimensional parallelepiped or a simplex) generated by the vectors $v_1, \ldots, v_p$. In particular, if for some $i = 1, \ldots, p$, $v_i \in D_X$, so that $v \lrcorner \phi = 0$, this quantity will vanish. This happens as the affine hyperplanes and the subspace generated by $v_1, \ldots, v_p$ intersect on a subspace of dimension greater than zero. For a positive number a, $a\phi$ naturally induces a family of affine subspaces which are parallel to those represented by ϕ and whose density is a times larger. For example, when $p = 1$, $\mathcal{F}_X$ is a family of $(n-1)$-dimensional affine subspaces of $T_X\mathcal{X}$, and $\phi(v)$ is interpreted as the amount of affine subspace that v intersects, taking the orientation into account. This collection of affine subspaces is the analog of a family of crystallographic planes for the infinitesimal continuum formulation.

Globally, a smooth decomposable differential p-form ϕ will induce a distribution D on $\mathcal{X}$ of dimension $r = n - p$. Here, by a "distribution," we mean a subbundle of the tangent bundle rather than a Schwartz distribution. Conversely, a distribution D of dimension $r = n - p$ will induce a collection of forms such that if ϕ induces D, so would the form $a\phi$ for any positive, real-valued function a on $\mathcal{X}$. At each point, $X \in \mathcal{X}$, the form ϕ will induce a family, $\mathcal{F}_X$, of affine subspaces of $T_X\mathcal{X}$ as above. Multiplying the form ϕ by a differentiable function $a : \mathcal{X} \to \mathbb{R}^+$, the resulting form $a\phi$ induces a family of hyperplanes which are parallel to those represented by ϕ and whose density at $X \in \mathcal{X}$ is $a(X)$ times larger. Such a decomposable p-form ϕ representing a crystallographic structure will be referred to as a *layering form*.

Even before we consider the geometry of defects, one has to verify that subspaces D_X, induced by the layering form for the various points in the body, are tangent to some $r = n - p$ submanifolds, or crystallographic layers in our interpretation. A distribution need not represent necessarily tangent spaces to a family of hypersurfaces, as we wish to consider. It is recalled that an r-dimensional submanifold $\mathcal{S}$ is an integral manifold of the distribution if $T_X\mathcal{S} = D_X$ for all $X \in \mathcal{S}$. A distribution D is referred to as involutive if, at each $X \in \mathcal{X}$, D_X is the tangent space of an r-dimensional submanifold. The Frobenius theorem implies (e.g., [AMR88, pp. 441-442]) that the distribution D is involutive if and only if there is a 1-form β on $\mathcal{X}$ such that

$$\mathrm{d}\phi = \beta \wedge \phi. \tag{7.6}$$

For example, consider the case where ϕ has an integrating factor. That is, there is a positive differentiable function $a : \mathcal{X} \to \mathbb{R}^+$ such that $\mathrm{d}(a\phi) = 0$. Recalling the identity

$$\mathrm{d}(\mu \wedge \nu) = \mathrm{d}\mu \wedge \nu + (-1)^q \mu \wedge \mathrm{d}\nu, \tag{7.7}$$

for the q-form μ and a form ν over $\mathcal{X}$, one has

$$\mathrm{d}(a\phi) = \mathrm{d}a \wedge \phi + a\mathrm{d}\phi = 0. \tag{7.8}$$

Then, the 1-form $\beta = -\mathrm{d}a/a$ satisfies the condition of Eq. (7.6). Note that we can write $-\mathrm{d}a/a = -\mathrm{d}(\ln a)$ which implies that β is exact in this case.

Conversely, assume that condition (7.6) holds for ϕ with an exact form β so that for a positive function a on $\mathcal{X}$,

$$\mathrm{d}(a\phi) = (\mathrm{d}a + a\beta) \wedge \phi. \tag{7.9}$$

Then, if the function a is a solution of the equation $\mathrm{d}a = -a\beta$, Eq. (7.9) implies that $\mathrm{d}(a\phi) = 0$. Note that the equation for a may be written as $\mathrm{d}(\ln a) = -\beta = -\mathrm{d}\alpha$ for some function α, as β is assumed to be exact. Hence, a solution, a, always exists under our conditions.

We conclude therefore that the distribution induced by a form ϕ is involutive, with an exact form β, if and only if it has an integrating factor. Intuitively, a form ϕ will induce an involutive distribution when the density of the hyperplanes at each point may be readjusted so that the exterior derivative of the resulting form vanishes. In particular, if $\mathrm{d}\phi = 0$, the distribution induced by ϕ is involutive.

7.3 Layering Forms, Defect Forms

The families of affine subspaces $\mathcal{F}_X$ at the various points in the body, induced by a decomposable p-form, ϕ, represent the crystallographic structure of the body. We will refer to such decomposable forms as *layering-forms*. For the case $p = 1$, the $(n-1)$-subspaces are the continuum analogs of the Bravais planes. If the distribution induced by ϕ is involutive, the subspaces D_X are tangent to submanifolds of dimension $r = n - p$ at the various material points. These submanifolds represent the crystallographic hypersurfaces.

The material structure described by an involutive layering p-form may still contain defects. Such defects are due to the creation or loss of crystallographic hypersurfaces in some subsets of the body. Let $\mathcal{S}$ be a $(p+1)$-dimensional submanifold with boundary of $\mathcal{X}$. The creation or loss of crystallographic hypersurfaces inside $\mathcal{S}$ will be reflected by the integral

$$\Phi = \int_{\partial\mathcal{S}} \phi \tag{7.10}$$

of the layering form over the boundary, $\partial\mathcal{S}$. Note that integrals of a form over $\mathcal{S}$ and its boundary make sense even if the form is not involutive.

Stokes's theorem asserts that

$$\Phi := \int_{\partial\mathcal{S}} \phi = \int_{\mathcal{S}} \mathrm{d}\phi. \tag{7.11}$$

Thus, if the exterior derivative $d\phi$ of the layering form vanishes, the total creation or annihilation of material hypersurfaces within any $(p+1)$-submanifold $\mathcal{S}$, as reflected in the total amount of hypersurfaces that cross the boundary $\partial\mathcal{S}$, will vanish. In other words, for a decomposable form ϕ satisfying (7.6), which, by the Frobenius theorem, induces a family of crystallographic hypersurfaces, the stronger condition, $d\phi = 0$, i.e., ϕ is closed, implies that the family of hypersurfaces has no sources or sinks. By the Poincaré lemma, it follows that locally $\phi = d\alpha$ for some $(p-1)$-form α. This suggests that $d\phi$ is the measure of the sources of crystallographic hypersurfaces inside the body $\mathcal{X}$—the distribution of defects. We will refer to $d\phi$ as the *defect form* corresponding to ϕ.

7.4 Smooth Distributions of Dislocations

Dislocations are associated with the case where $p = 1$. By definition, any 1-forms are decomposable. The absence of dislocations implies that locally, $\phi = d\alpha$ where α is a real-valued function, a 0-form, on $\mathcal{X}$. The level sets of the function α represent the crystallographic hypersurfaces. In other words, the values of α may label the crystallographic hypersurfaces, locally. For the three-dimensional case and a two-dimensional manifold with boundary $\mathcal{S}$, the one-dimensional $\partial\mathcal{S}$ is a closed circuit analogous to the Burgers circuit.

7.5 Inclinations and Disclinations, the Smooth Case

Disclinations are viewed here as defects in the arrangements of one-dimensional subspaces, or directors. As in [Fra58] and [Cha77], the field under consideration may indicate the inclinations of the optical axes of liquid crystals. The interpretation of disclinations as defects in the orientations of the Bravais planes (e.g., [KA75]) may be viewed in some cases as defects in the arrangements of the normal vectors to the respective Bravais planes. Such cases can be described using the framework outlined below.

The distribution of inclinations of the directors is represented by an $(n-1)$-form ϕ, the *inclination form*, and the structure of the disclinations is given by the *disclination n-form* $d\phi$. It is noted that any $(n-1)$-form is decomposable. (See [Ste83, Section 1.V], and see Sect. 9.2 for a continuum mechanical application.) The induced distribution is necessarily involutive, and the one-dimensional integral submanifolds, to which the directors are tangent, may be easily constructed as follows:

At each point $X \in \mathcal{X}$ where $\phi(X) \neq 0$, $\phi(X)$ determines a unique one-dimensional subspace $\mathbf{V}_X$ of the tangent space $T_X\mathcal{X}$ by $v \lrcorner \phi(X) = 0$ for each $v \in \mathbf{V}_X$. The collection of subspaces $\mathbf{V}_X$ forms a one-dimensional distribution. The one-dimensional subspace $\mathbf{V}_X$ may be determined as follows: Let θ be a volume element on $\mathcal{X}$. Locally, θ may be represented in the form

$$\theta = \theta_0 \mathrm{d}X^1 \wedge \cdots \wedge \mathrm{d}X^n \tag{7.12}$$

for a positive real-valued function θ_0, and ϕ may be represented locally in the form

$$\phi = \sum_{i=1}^{n} \phi_{1\ldots\hat{\imath}\ldots n} \mathrm{d}X^1 \wedge \cdots \wedge \widehat{\mathrm{d}X^i} \wedge \cdots \wedge \mathrm{d}X^n, \tag{7.13}$$

where a "hat" indicates the omission of an element. Then, there is a unique tangent vector u such that $u \lrcorner \theta = \phi$. If a vector u is represented by $u = \sum_i u^i \partial_i$, then, $u \lrcorner \theta$ is represented by

$$u \lrcorner \theta = \sum_{i=1}^{n} (-1)^{i-1} \theta_0 u^i \mathrm{d}X^1 \wedge \cdots \wedge \widehat{\mathrm{d}X^i} \wedge \cdots \wedge \mathrm{d}X^n. \tag{7.14}$$

Thus, as $\theta_0 \neq 0$, there is always a vector field u satisfying $u \lrcorner \theta = \phi$ and its components are given locally by

$$u^i = (-1)^{i-1} \frac{\phi_{1\ldots\hat{\imath}\ldots n}}{\theta_0}. \tag{7.15}$$

If we select a different volume element, the only parameter that will change in the equation above will be the positive number θ_0, and so the resulting vector will be in the same one-dimensional subspace. Thus, the form ϕ determines a unique oriented one-dimensional subspace $\mathbf{U}_X$ at each X such that $\phi(X) \neq 0$. If no particular orientation is chosen on $\mathcal{X}$, no orientation will be induced on $\mathbf{U}_X$. The spaces $\mathbf{V}_X$ and $\mathbf{U}_X$ are isomorphic. Let θ be a volume element and u the vector such that $\phi = u \lrcorner \theta$. Then, any nonzero $v \in \mathbf{U}_X$ is of the form $v = au$, $a \neq 0$. Thus, $v \lrcorner (u \lrcorner \theta) = au \lrcorner (u \lrcorner \theta) = 0$, because $\theta(u, u, v_3, \ldots, v_n) = 0$ for any collection of vectors $v_3, \ldots, v_n$.

For an $(n-1)$-form ϕ, we interpret the distribution $\mathbf{V}$ of one-dimensional subspaces of the tangent space as describing the inclinations of the directors in the body. Multiplying the form ϕ by a positive number will affect the "density" of the directors.

Since we have assumed that the form ϕ is differentiable, it follows that for a choice of a smooth volume element θ, the representing vector field u is differentiable. Hence, the theorems on the existence and uniqueness of the solutions of ordinary differential equations imply the existence of integral lines to the vector field u, i.e., at each point $X \in \mathcal{X}$ there is a curve $c_X : (-\varepsilon, \varepsilon) \to \mathcal{X}$, $\varepsilon > 0$, such that $c_X(0) = X$ and the tangent vector to the curve satisfies

$$\left.\frac{\mathrm{d}}{\mathrm{d}t} c_X\right|_{t=0} = u(X). \tag{7.16}$$

An inclination form may be integrated over $(n-1)$-dimensional submanifolds of $\mathcal{X}$. Let $\mathcal{S}$ be an oriented $(n-1)$-dimensional submanifold of $\mathcal{X}$. Then,

$$\Phi_{\mathcal{S}} = \int_{\mathcal{S}} \phi \tag{7.17}$$

is interpreted as the total amount of directors penetrating the hypersurface $\mathcal{S}$. It should be noted that $\Phi_{\mathcal{S}}$ depends on the orientation of $\mathcal{S}$ and that the restriction of ϕ to a point in $\mathcal{S}$ may be of the same orientation as $\mathcal{S}$ or the inverse orientation. Thus, for a nonvanishing inclination form, the total, $\Phi_{\mathcal{S}}$, may vanish, which implies that the total production of directors inside $\mathcal{S}$ vanishes.

For the inclination $(n-1)$-form, ϕ, the induced distribution of smooth disclinations is the exterior derivative, the n-form, $\mathrm{d}\phi$. Thus, for an n-dimensional submanifold with boundary $\mathcal{R} \subset \mathcal{X}$, letting $\mathcal{S} = \partial\mathcal{R}$ in (7.17), $\Phi_{\partial\mathcal{R}}$ is interpreted as the total amount of directors that penetrate $\partial\mathcal{X}$.

Stokes's theorem implies immediately that

$$\Phi_{\partial\mathcal{R}} = \int_{\mathcal{R}} \mathrm{d}\phi, \tag{7.18}$$

so that $\Phi_{\partial\mathcal{R}}$ is the integral of the disclination field over $\mathcal{R}$. Figuratively speaking, the disclination field represents the source term for the directors.

It is observed that for any given vector field, one can label the integral lines by a submanifold of dimension $n-1$ of initial conditions (see [AMR88, pp. 246–247]). However, the vector fields induced by ϕ depend on the choice of volume element θ. Thus, such labeling is not unique and the presence of disclinations will be reflected by $\mathrm{d}\phi$.

7.6 Frank's Rules for Smooth Distributions of Defects

Let $\psi = \mathrm{d}\phi$ be the defect form associated with the layering form ϕ. It follows from the identity $\mathrm{d}^2\phi = 0$ that ψ must satisfy the condition

$$\mathrm{d}\psi = 0. \tag{7.19}$$

This compatibility condition is the analog of Frank's rules for smoothly distributed defects of crystallographic hypersurfaces having any dimension on general manifolds. For the case of dislocations, given in terms of a 2-form ψ, $\mathrm{d}\psi$ is a 3-form. In the standard three-dimensional Euclidean space, $\mathrm{d}\psi$ is therefore equivalent to a real-valued function. For disclinations, the disclination form, ψ, is an n-form, so that $\mathrm{d}\psi$ is an $(n+1)$-form. It follows that for disclinations, Frank's rules are satisfied identically.

Chapter 8
Smooth Fluxes

Uniform fluxes, or algebraic fluxes, considered in Part I, were assumed to be invariant under translation and are not suitable to model variable flux fields. Fixed alternating tensors are replaced, therefore, by fields of alternating tensors—differential forms. We will postulate below the analogs of the Cauchy postulates for the geometric settings of differentiable manifolds. The results pertaining to uniform fluxes will be used in the proof of the existence of flux differential forms.

The representation of flux fields without using a Riemannian metric enables one to consider fluxes of various extensive properties, e.g., electric charge and heat, independently or a particular configuration of a body in space.

Flux differential forms extend the fundamental tool provided by flux vector fields of the classical formulation; they make it possible to compute the total flux of an extensive property out of any sub-region and through any oriented hypersurface.

8.1 Balance Principles and Fluxes

One of the basic notions of continuum mechanics is that of an extensive property. The term extensive property is used to describe a property that may be assigned to subsets of a given universe. This includes, for example, the mass of the various parts of a material body, the electrical charge enclosed in regions in space, etc. Thus, an extensive property is a set function, P. Fluxes are associated with real-valued extensive properties. This chapter considers extensive properties that may be represented by smooth densities. For less regular extensive properties, see Chap. 23 in Part III, and, for example, [Sil85, GWZ86, NV88, DMM99, RS03a].

R. Segev, *Foundations of Geometric Continuum Mechanics*, Advances in Continuum Mechanics 49, https://doi.org/10.1007/978-3-031-35655-1_8

8.1.1 Densities of Extensive Properties

The setting of the theory of extensive properties presented here considers a fixed physical space modeled by an n-dimensional differentiable manifold $\mathcal{S}$. Alternatively, one may wish to interpret $\mathcal{S}$ as the material manifold so a point $y \in \mathcal{S}$ is a material point having an invariant meaning. Since we are going to use integration later on, we will assume that $\mathcal{S}$ is orientable and that a particular orientation has been chosen.

We will refer to "regions" for which we can define the total amount of the property P as *control regions* when we interpret $\mathcal{S}$ as the space manifold, and refer to them as *subbodies* when we interpret $\mathcal{S}$ as the material manifold. The term "region" will be used when the particular interpretation is immaterial. Admissible regions are compact n-dimensional submanifolds with boundary of $\mathcal{S}$ or chains in $\mathcal{S}$. In accordance with the smooth setting, it is assumed that there is an n-form ρ defined on $\mathcal{S}$ that represents the density of the property P. Using integration of forms, one can now calculate the total amount of the property

$$P(\mathcal{R}) = \int_{\mathcal{R}} \rho \tag{8.1}$$

in any region $\mathcal{R}$.

From a naive point of view, spacetime, or the event space, has the structure of a Cartesian product $\mathbb{R} \times \mathcal{S}$ so that any event may be assigned a specific location in $\mathcal{S}$ and a specific time $t \in \mathbb{R}$. Our ability to assign a particular pair of time and place to any event implies that we have a particular global *frame* on spacetime (cf. Example 6.3). More general structures of spacetime will be considered in the next chapter and Sect. 13.2. This means that in general, the density ρ of a property P should be time-dependent. Since the value $\rho(y, t) \in \bigwedge^n(T_y^*\mathcal{S})$ belongs to a fixed vector space, we may differentiate it with respect to the time variable and obtain the n-form $\beta = \partial\rho/\partial t$ on $\mathcal{S}$. Thus, for a fixed region $\mathcal{R}$,

$$\frac{dP(\mathcal{R})}{dt} = \int_{\mathcal{R}} \beta \tag{8.2}$$

represents the rate of change of the amount of the property P inside $\mathcal{R}$.

In the classical setting of continuum mechanics, it is assumed that the change of the amount of property within the region $\mathcal{R}$ is a result of two phenomena: the rate at which the property is produced inside $\mathcal{R}$, and the rate at which the property leaves $\mathcal{R}$ through its boundaries—the *total flux*, $\Phi_{\partial\mathcal{R}}$, of P.

The total flux of the property is assumed to be distributed smoothly on the boundary of $\mathcal{R}$. Hence, whether the admissible regions are compact n-dimensional submanifolds of $\mathcal{S}$ or chains, integration of $(n-1)$-forms on their boundaries is well-defined. It is assumed that for each region $\mathcal{R}$, there is an $(n-1)$-form $\tau_{\mathcal{R}}$ called the *flux density* such that the flux of P is given as

$$\Phi_{\partial\mathcal{R}} = \int_{\partial\mathcal{R}} \tau_{\mathcal{R}}. \tag{8.3}$$

In accordance with integration theory, the action $(\tau_{\mathcal{R}}(x))(v_1, \ldots, v_{n-1})$, for a positively oriented collection of vectors $v_1, \ldots, v_{n-1} \in T_x\partial\mathcal{R}$, is interpreted as the infinitesimal flux of P out of $\mathcal{R}$ through the oriented infinitesimal element determined by the vectors. Similarly to the integral, $(\tau_{\mathcal{R}}(x))(v_1, \ldots, v_{n-1})$ is meaningful at this point only for a positively oriented collection of vectors. In the sequel, when no confusion can occur, we will omit the $\mathcal{R}$ subscript and use only τ.

The production rate of the property inside $\mathcal{R}$ is assumed to be represented as

$$\int_{\mathcal{R}} \varsigma,$$

where ς is an n-form on $\mathcal{S}$, the *source density*, which is independent of $\mathcal{R}$. Thus, the classical balance law assumes the form

$$\int_{\mathcal{R}} \beta = \int_{\mathcal{R}} \varsigma - \int_{\partial\mathcal{R}} \tau_{\mathcal{R}} \tag{8.4}$$

for each admissible region $\mathcal{R}$.

Equation (8.4) implies that

$$\begin{aligned} \left|\int_{\partial\mathcal{R}} \tau_{\mathcal{R}}\right| &= \left|\int_{\mathcal{R}} (\varsigma - \beta),\right| \\ &\leqslant \int_{\mathcal{R}} |\varsigma - \beta|, \end{aligned} \tag{8.5}$$

where the absolute value of an n-form is defined as follows. Let θ be a volume element on $\mathcal{S}$ that is compatible with its orientation, that is, $\theta(x)(v_1, \ldots, v_n) > 0$ for all $x \in \mathcal{S}$ and an oriented collection of linearly independent vectors $v_1, \ldots, v_n$. Let ω be an n-form. Then, since the space $\bigwedge^n(T_x^*\mathcal{S})$ is one-dimensional for each $x \in \mathcal{S}$, $\omega(x) = a(x)\theta(x)$. The sign of $\omega(x)$ is identified with the sign of $a(x)$ and $|\omega|(x) = |a(x)|\,\theta(x)$. It is noted that an absolute value $|\omega|$ of an n-form, ω, may be defined alternatively without any reference to orientation by setting $|\omega|(x)(v_1, \ldots, v_n) := |\omega(v_1, \ldots, v_n)|$. With this definition, $|\omega|$ is not a differential form but a density, in the sense of [AMR88, p. 407] or [Lee02, pp. 375–380]).

For results we present later, in particular, the Cauchy theorem on the existence of flux forms, the balance principle (8.4) is regarded, in light of the inequality above, as a boundedness or regularity postulate for the fluxes corresponding to the various regions. The boundedness postulate for the fluxes states that there is a bounded non-negative n-form ς_0 on $\mathcal{S}$ such that for any region $\mathcal{R}$

$$\left|\int_{\partial\mathcal{R}} \tau_{\mathcal{R}}\right| \leqslant \int_{\mathcal{R}} \varsigma_0. \tag{8.6}$$

On the one hand, this form of the balance principle has the essential ingredients needed later for the proof of Cauchy's theorem. On the other hand, it does not require the specification of both β and ς. Thus, Eq. (8.6) is adopted as the balance principle.

8.1.2 Flux Forms and Cauchy's Formula

While the rate of change of the density and the production term are specified by fields defined on the manifold $\mathcal{S}$, the flux term is specified by means of a collection of flux fields—a set function $\mathcal{R} \mapsto \tau_{\mathcal{R}}$, the domain of which is the collection of all admissible regions. Thus, it is noted that for a point $y \in \mathcal{S}$ that is on the boundary of two distinct smooth submanifolds $\mathcal{R}$ and $\mathcal{R}'$, the values of the flux densities $\tau_{\mathcal{R}}(y) \in \bigwedge^{n-1}(T_y^*\mathcal{R})$ and $\tau_{\mathcal{R}'}(y) \in \bigwedge^{n-1}(T_y^*\mathcal{R}')$ belong to different spaces and cannot be compared.

Nevertheless, integration theory of forms provides a simple means for specifying the flux densities for admissible regions. Let J be an $(n-1)$-form on $\mathcal{S}$. For every region $\mathcal{R}$, the inclusion $\mathcal{J}_{\partial\mathcal{R}} : \partial\mathcal{R} \to \mathcal{S}$ induces the restriction

$$\rho_{\partial\mathcal{R}}(J) = (T\mathcal{J}_{\partial\mathcal{R}})^*(J), \qquad \rho_{\partial\mathcal{R}} = (T\mathcal{J}_{\partial\mathcal{R}})^* : \Omega^{n-1}(\mathcal{S}) \to \Omega^{n-1}(\partial\mathcal{R}), \tag{8.7}$$

as in Sects. 5.10 and 6.3.6. We will refer to such an $(n-1)$-form J as a *flux form*.

Thus, a flux form induces a collection of flux densities for the boundaries of the various regions by

$$\tau_{\mathcal{R}} = (T\mathcal{J}_{\partial\mathcal{R}})^*(J) = \rho_{\partial\mathcal{R}}(J). \tag{8.8}$$

The last equation will be referred to as the generalized *Cauchy formula*, and we will often omit the $\partial\mathcal{R}$-index if the particular region under consideration is clear from the context. The definition of the restriction of forms implies that for a point $y_0 \in \mathcal{S}$ and any region $\mathcal{R}$ such that $y_0 \in \partial\mathcal{R}$, we have for any collection $v_1, \dots, v_{n-1}$ of vectors in $T_{y_0}\partial\mathcal{R}$,

$$\begin{aligned} \tau_{\mathcal{R}}(v_1, \dots, v_{n-1}) &= J\big(T\mathcal{J}(v_1), \dots, T\mathcal{J}(v_{n-1})\big), \\ &= J(v_1, \dots, v_{n-1}), \\ &= J(v_1 \wedge \dots \wedge v_{n-1}), \end{aligned} \tag{8.9}$$

where in the last line we emphasized the point of view that the flux form is applied to the multivector representing the infinitesimal oriented simplex defined by the vectors. The relation between the current presentation of flux theory in terms of a flux form and the traditional formulation in terms of a flux vector field follows from the discussion in Example 5.3.

One of the main objectives of theoretical continuum mechanics is to find general conditions for the flux fields on the various regions to be induced by a field defined on $\mathcal{S}$. Such conditions are referred to as the *Cauchy postulates*, and the assertion that flux fields on regions satisfying the Cauchy postulates are induced by a global field is referred to as the *Cauchy theorem for fluxes*. Below, we present a version of Cauchy's formula and theorem for the general setting of an n-dimensional manifold devoid of a Riemannian metric.

8.1.3 Extensive Properties and Cauchy Formula—Local Representation

We now present the coordinate description of the objects and relations given above. Let $x^1, \dots, x^n$ be a coordinate system in a neighborhood of a point x_0. Then, omitting the indication of time dependence, the n-forms ρ and β are represented locally, using the scalar functions $\rho_{1\dots n}(x^i)$ and $\beta_{1\dots n}(x^i)$, as $\rho(y) = \rho_{1\dots n}(x^i)\,\mathrm{d}x^1 \wedge \dots \wedge \mathrm{d}x^n$ and $\beta(y) = \beta_{1\dots n}(x^i)\,\mathrm{d}x^1 \wedge \dots \wedge \mathrm{d}x^n$, respectively.

The flux density $\tau_{\mathcal{R}}$ should be represented using a coordinate system on the $(n-1)$-dimensional manifold $\partial\mathcal{R}$, say $y^1, \dots, y^{n-1}$. Thus, in such a coordinate system, $\tau_{\mathcal{R}}$ is represented in a neighborhood of the boundary point y_0 using the scalar function $\tau_{\mathcal{R}1\dots n-1}$ in the form

$$\tau_{\mathcal{R}}(y) = \tau_{\mathcal{R}1\dots n-1}(y^j)\,\mathrm{d}y^1 \wedge \dots \wedge \mathrm{d}y^{n-1}. \tag{8.10}$$

The local expression for the flux form $J \in \Omega^{n-1}(\mathcal{S})$, in a coordinate neighborhood $x^1, \dots, x^n$, is (cf. Example 5.3)

$$\begin{aligned} J(x) &= \sum_{(\alpha)} J_{\alpha_1\dots\alpha_{n-1}}(x^l)\,\mathrm{d}x^{\alpha_1} \wedge \dots \wedge \mathrm{d}x^{\alpha_{n-1}} \\ &= \sum_k J_{1\dots\widehat{k}\dots n}(x^l)\,\mathrm{d}x^1 \wedge \dots \wedge \widehat{\mathrm{d}x^k} \wedge \dots \wedge \mathrm{d}x^n \end{aligned} \tag{8.11}$$

and

$$J_{\alpha_1\dots\alpha_{n-1}} = J(\partial_{\alpha_1}, \dots, \partial_{\alpha_{n-1}}).$$

Noting that $\mathrm{d}x^1 \wedge \dots \wedge \widehat{\mathrm{d}x^k} \wedge \dots \wedge \mathrm{d}x^n = (-1)^{k-1}\partial_k \lrcorner\, \mathrm{d}x$, one may also write

$$J(x) = \sum_k (-1)^{k-1} J_{1\dots\widehat{k}\dots n}(x^l)\,\partial_k \lrcorner\, \mathrm{d}x. \tag{8.12}$$

Locally, the inclusion $\partial\mathcal{R} \to \mathcal{S}$ is represented by n functions $x^i = x^i(y^1, \dots, y^{n-1})$. Using the expanded notation for the base vectors,

$$T\mathcal{J}\left(\frac{\partial}{\partial y^p}\right) = \sum_i x^i_{,p}\frac{\partial}{\partial x^i}, \qquad p = 1, \dots, n-1,$$

and for a vector $v \in T_{y_0}\partial\mathcal{R}$ represented locally by $v = \sum_p v^p\, \partial/\partial y^p$, we have $T\mathcal{J}(v) = \sum_{p,i} x^i_{,p} v^p\, \partial/\partial x^i$, which we may write with some abuse of notation as $v^i = \sum_p x^i_{,p} v^p$.

The evaluation $\tau_{\mathcal{R}}(v_1, \dots, v_{n-1})$ is represented as

$$\begin{aligned}
&\tau_{\mathcal{R}}(v_1, \dots, v_{n-1}) \\
&= \tau_{\mathcal{R}1\dots n-1}\, \mathrm{d}y^1 \wedge \dots \wedge \mathrm{d}y^{n-1}\Big(\sum_{p_1} v_1^{p_1}\frac{\partial}{\partial y^{p_1}}, \dots, \sum_{p_{n-1}} v_{n-1}^{p_{n-1}}\frac{\partial}{\partial y^{p_{n-1}}}\Big), \\
&= \sum_P \tau_{\mathcal{R}1\dots n-1} v_1^{p_1} \cdots v_{n-1}^{p_{n-1}}\, \mathrm{d}y^1 \wedge \dots \wedge \mathrm{d}y^{n-1}\Big(\frac{\partial}{\partial y^{p_1}}, \dots, \frac{\partial}{\partial y^{p_{n-1}}}\Big), \\
&= \sum_P \tau_{\mathcal{R}1\dots n-1} v_1^{p_1} \cdots v_{n-1}^{p_{n-1}} \varepsilon_{p_1\dots p_{n-1}}, \\
&= \tau_{\mathcal{R}1\dots n-1} \det\left[(v_p)^q\right], \qquad\qquad p, q, p_l = 1, \dots, n-1.
\end{aligned} \tag{8.13}$$

Alternatively, using the Cauchy formula with components relative to the bases induced by the x^i-coordinates as in Eq. (5.94),

$$\begin{aligned}
\tau_{\mathcal{R}}(v_1, \dots, v_{n-1}) &= J(v_1, \dots, v_{n-1}) \\
&= \det\left(\left[(-1)^{k-1} J_{1\dots\widehat{k}\dots n}\right] ; v_1^{i_1} ; \cdots \cdots ; v_{n-1}^{i_{n-1}}\right),
\end{aligned} \tag{8.14}$$

where $i_p = 1, \dots, n$, $p = 1, \dots, n-1$. Hence,

$$\begin{aligned}
\tau_{\mathcal{R}1\dots n-1} &= \tau\left(\frac{\partial}{\partial y^1}, \dots, \frac{\partial}{\partial y^{n-1}}\right), \\
&= J\left(T\mathcal{J}\left(\frac{\partial}{\partial y^1}\right), \dots, T\mathcal{J}\left(\frac{\partial}{\partial y^{n-1}}\right)\right), \\
&= J\left(\sum_{i_1} x^{i_1}_{,1}\frac{\partial}{\partial x^{i_1}}, \dots, \sum_{i_{n-1}} x^{i_{n-1}}_{,n-1}\frac{\partial}{\partial x^{i_{n-1}}}\right),
\end{aligned} \tag{8.15}$$

and the resulting local representation of the Cauchy formula is

$$\tau_{\mathcal{R}1\dots n-1} = \det\left(\left[(-1)^{k-1} J_{1\dots\widehat{k}\dots n}\right] ; x^{i_1}_{,1} ; \cdots \cdots ; x^{i_{n-1}}_{,n-1}\right), \tag{8.16}$$

where $k, i_p = 1, \dots, n$, $p = 1, \dots, n-1$. Thus, the density of the flux through the boundary is obtained by the determinant of the matrix containing the components of the flux form in the first column and the vectors of the derivatives of $x^i = x^i(y^p)$, which represent the embedding of the boundary $\partial\mathcal{R}$ in $\mathcal{R}$.

8.1.4 The Cauchy Flux Theorem

We now present sufficient conditions for the existence of flux forms, i.e., conditions for Cauchy's formula to hold for some $(n-1)$-form J over $\mathcal{S}$. Clearly, if such a form exists, it is unique.

8.1.4.1 Assumptions

Assumption 1: Boundedness For any admissible region $\mathcal{R}$, there is a smooth $(n-1)$-form $\tau_{\mathcal{R}}$ such that the bound (8.6) holds for some non-negative bounded n-form ς (we omit the subscript 0 for simplicity) defined on $\mathcal{S}$.

The next assumption traditionally known as Cauchy's postulate is a locality assumption. Before making it, we note that for a point $x \in \mathcal{S}$, the evaluation $\tau_{\mathcal{R}}(x)(v_1, \dots, v_{n-1})$ is well-defined only if $v_1, \dots, v_{n-1} \in T_x\partial\mathcal{R}$. We interpret $\tau_{\mathcal{R}}(x)(v_1, \dots, v_{n-1})$ as the infinitesimal flux of the property out of the region $\mathcal{R}$ through the infinitesimal oriented simplex induced by the positively oriented collection $v_1, \dots, v_{n-1}$. In case the collection of vectors is negatively oriented, $\tau_{\mathcal{R}}(x)(v_1, \dots, v_{n-1})$ is interpreted as the infinitesimal flux into the region.

Assumption 2: Cauchy's Postulate of Locality Let $v_1, \dots, v_{n-1} \in T_x\mathcal{S}$, $x \in \mathcal{S}$, be a collection of $n-1$ vectors. Let $\mathcal{R}$ be any region such that $x \in \partial\mathcal{R}$, $v_1, \dots, v_{n-1} \in T_x\partial\mathcal{R}$, and the collection of vectors is positively oriented relative to the orientation of $\partial\mathcal{R}$. Then, $\tau_{\mathcal{R}}(x)(v_1, \dots, v_{n-1})$ is independent of $\mathcal{R}$.

Cauchy's postulate implies that for the given point $x \in \mathcal{S}$, $\tau_{\mathcal{R}}(x)$ depends on $\mathcal{R}$ only through the tangent space $T_x\partial\mathcal{R}$ and the orientation induced on it. In other words, let $\mathcal{I}_{\partial\mathcal{R}} : \partial\mathcal{R} \to \mathcal{S}$ be the natural embedding, and let $T\mathcal{I}_{\partial\mathcal{R}} : T(\partial\mathcal{R}) \to T\mathcal{S}$ be its tangent mapping. It follows that if for two regions $\mathcal{R}$ and $\mathcal{R}'$, $T\mathcal{I}_{\partial\mathcal{R}}\{T_x\partial\mathcal{R}\} = T\mathcal{I}_{\partial\mathcal{R}'}\{T_x\partial\mathcal{R}'\}$, and the orientations induced on $T\mathcal{I}_{\partial\mathcal{R}}\{T_x\partial\mathcal{R}\}$ and $T\mathcal{I}_{\partial\mathcal{R}'}\{T_x\partial\mathcal{R}'\}$ are identical, then $\tau_{\mathcal{R}}(x) = \tau_{\mathcal{R}'}(x)$.

Cauchy's postulate makes it possible to define a mapping

$$\mathrm{t}_x : (T_y\mathcal{S})^{n-1} \longrightarrow \mathbb{R} \tag{8.17}$$

such that for any collection $v_1, \dots, v_{n-1}$ of vectors

$$\mathrm{t}_x(v_1, \dots, v_{n-1}) = \tau_{\mathcal{R}}(x)(v_1, \dots, v_{n-1}) \tag{8.18}$$

for any region $\mathcal{R}$ such that the collection $v_1, \ldots, v_{n-1} \in T_x\partial\mathcal{R}$ is positively oriented. Thus, the order in which the collection is written specifies the orientation for which $\mathrm{t}_x(v_1, \ldots, v_{n-1})$ applies. We observe that the mapping t_x is analogous to the mapping ω for homogeneous fluxes as in Eq. (4.17). At this point, for an odd permutation π, there is no relation between $\mathrm{t}_x(v_1, \ldots, v_{n-1})$ and $\mathrm{t}_x(v_{\pi_1}, \ldots, v_{\pi_{n-1}})$. It is noted that because of the compatibility with $\tau_{\mathcal{R}}$, $\mathrm{t}_x(v_1, \ldots, v_{n-1})$ depends only on the multivector $\mathfrak{v} = v_1 \wedge \cdots \wedge v_{n-1}$. In other words, $\mathrm{t}_x(v_1, \ldots, v_{n-1})$ depends on the arguments multi-linearly as long as the arguments are restricted to some arbitrary $(n-1)$-dimensional hyperplane. Let $T^{n-1}\mathcal{S} \to \mathcal{S}$ be the vector bundle whose fiber at $x \in \mathcal{S}$ is $(T_x\mathcal{S})^{n-1}$. Then, globally, we have a mapping

$$\mathrm{t} : T^{n-1}\mathcal{S} \longrightarrow \mathbb{R}, \qquad \text{such that} \quad \mathrm{t}|_{(T_x\mathcal{S})^{n-1}} = \mathrm{t}_x. \tag{8.19}$$

We will refer to t as the *Cauchy–Whitney* map. Since $T^{n-1}\mathcal{S}$ is a vector bundle over $\mathcal{S}$—a differentiable manifold, the next assumption may be made.

Assumption 3: Regularity The mapping t is smooth.

We will refer to the collection $\{\tau_{\mathcal{R}}\}$ for the various regions $\mathcal{R}$ satisfying the foregoing assumptions as a *system of Cauchy fluxes*. To prove the existence of an n-form J, such that $\tau_{\mathcal{R}}(x) = (T_x\mathcal{I}_{\partial\mathcal{R}})^*(J(x))$, we show that the assumptions of Sect. 4.1 are satisfied so that the algebraic Cauchy theorem as in Sect. 4.4 may be used.

We first note that by its definition in terms of $\tau_{\mathcal{R}}(x)$, the mapping t_x satisfies the additivity assumption (4.5). The homogeneity assumption in Eq. (4.3) also follows from the homogeneity of $\tau_{\mathcal{R}}(x)$. It remains to show that the balance assumption of Sect. 4.1 holds. As mentioned in Remark 4.2, once balance holds, the additivity assumption of Eq. (4.6) is satisfied.

8.1.4.2 Notation

We first link the objects pertaining to uniform fluxes in affine spaces to the analogous ones in integration theory on manifolds. In order to make a clear distinction between simplices in affine spaces as in Chap. 3, and singular simplices used in the theory of integration on chains, we will refer to the former as algebraic simplices.

Let $v_1, \ldots, v_n \in T_y\mathcal{S}$ be an oriented collection of linearly independent vectors. We use S to denote the simplex $[v_1, \ldots, v_n]$. Thus, in accordance with the balance property of Eq. (4.22), we have to show that $\mathrm{t}_x(\partial S) = 0$. We rewrite the expression for ∂S in analogy with (3.28) as $\partial S = \sum_{p=0}^{n}(-1)^p S_p$, where

$$S_0 = [v_2 - v_1, \ldots, v_r - v_1], \quad \text{and} \quad S_p = [v_1, \ldots, \widehat{v}_p, \ldots, v_r], \quad p = 1, \ldots, n. \tag{8.20}$$

We observe that S_i does not have the right orientation for odd values of i. (This is done in order that the notation is analogous to the definition of the boundary of a singular simplex in integration theory.)

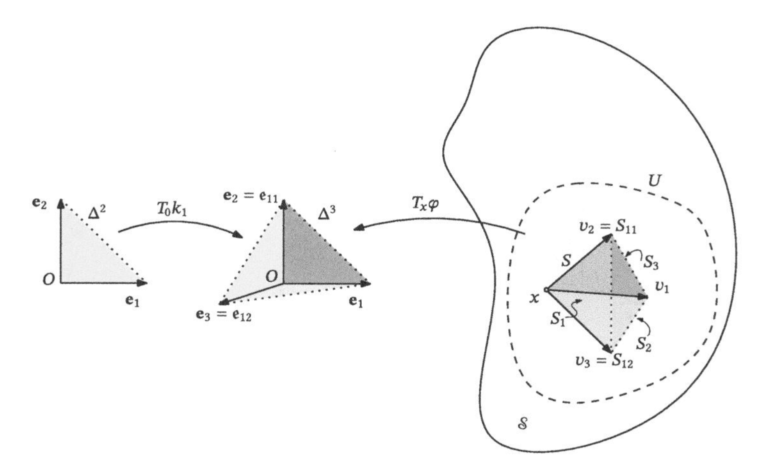

Fig. 8.1 Illustrating a simplex in $T_x\mathcal{S}$ and its image under $T_x\varphi$

We will represent each S_p, $p \geqslant 0$, in the form

$$S_p = [S_{p1}, \dots, S_{p(n-1)}], \tag{8.21}$$

where the vectors S_{pl} are given as in the previous equation. Thus, in view of (4.22) and (4.18), one has to show that for any given simplex $S = [v_1, \dots, v_n]$,

$$\mathrm{t}_x(\partial S) := \sum_{p=0}^{n} \mathrm{t}_x((-1)^p S_p) = \sum_{p=0}^{n} (-1)^p \mathrm{t}_x(S_p) = 0. \tag{8.22}$$

Consider a chart (φ, U) in a neighborhood of x. Since we can always adjust the chart mapping by a composition of a linear mapping and a translation, we may assume without loss of generality that $\varphi(x) = 0$ and $T\varphi(v_i) = \mathbf{e}_i$, where $\{\mathbf{e}_i\}_{i=1}^n$ is the standard basis of $\mathbb{R}^n$. It follows that the image of S under $T_x\varphi$ is the standard simplex Δ^n in $\mathbb{R}^n$ (see an illustration in Fig. 8.1).

Let $k_p : \Delta^{n-1} \to \mathbb{R}^n$ be the faces of the standard simplex as defined in (6.163), (6.164). Similarly to the notation S_{pl}, we will use $\mathfrak{e}_{pl}$ to denote the l-th vector in the p-th face of Δ^n (see Fig. 8.1). In view of the analogous enumeration of the faces and the vectors constructing them, as in (6.167), it follows that

$$\mathfrak{e}_{pl} = T_x\varphi(S_{pl}) = Dk_p(\mathbf{e}_l), \quad p = 0, \dots, n, \quad l = 1, \dots, n-1. \tag{8.23}$$

As $V = \varphi(U)$ is an open neighborhood of the origin in $\mathbb{R}^n$, there is a positive number a_0 such that the n-simplex Δ_0 in $\mathbb{R}^n$, generated by the origin and the points $a_0\mathbf{e}_i$, is contained in V. Thus, $\tilde{\Delta}_0 = \varphi^{-1}\{\Delta_0\}$ is an admissible region induced by the singular simplex $s_0 := \varphi^{-1} \circ (a_0\mathrm{Id}) : \Delta^n \to \mathcal{S}$, where $\mathrm{Id} : \Delta^n \to \mathbb{R}^n$ is the identity mapping. Similarly, for $a_m = 2^{-m}a_0$, let Δ_m be the n-simplex in $\mathbb{R}^n$

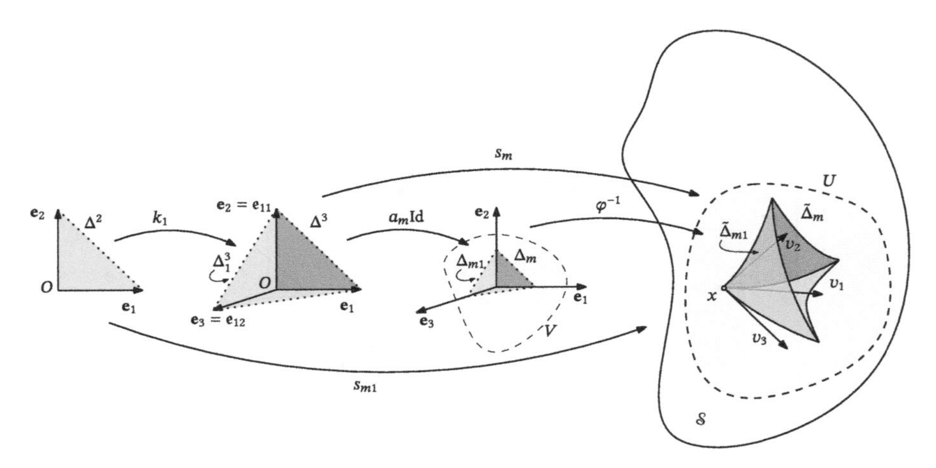

Fig. 8.2 Illustrating the singular simplex $\tilde{\Delta}_m$

generated by the origin and the points $a_m\mathbf{e}_i$. Set $\tilde{\Delta}_m := \varphi^{-1}\{\Delta_m\} \subset \mathcal{S}$. Let Δ_{mp} be the oriented simplices that make up $\partial\Delta_m$ and set $\tilde{\Delta}_{mp} := \varphi^{-1}\{\Delta_{mp}\}$ (see an illustration in Fig. 8.2). We observe that $\tilde{\Delta}_m$ is the image of the singular simplex $s_m = \varphi^{-1} \circ (a_m\mathrm{Id}) : \Delta^n \to \mathcal{S}$ and the faces $\tilde{\Delta}_{mp}$ are the images of the singular simplices $s_{mp} = \varphi^{-1} \circ (a_m\mathrm{Id}) \circ k_p : \Delta^{n-1} \to \mathcal{S}$, with $\partial s_m = \sum_p (-1)^p s_{mp}$.

8.1.4.3 Construction

Applying the boundedness assumption to $\tilde{\Delta}_m$, for $m = 0, 1, 2, \ldots,$ we have

$$\left|\sum_p (-1)^p \int_{\tilde{\Delta}_{mp}} \tau_{\tilde{\Delta}_{mp}}\right| = \left|\sum_p (-1)^p \int_{s_{mp}} \tau_{\tilde{\Delta}_{mp}}\right| \leqslant \int_{\tilde{\Delta}_m} \varsigma, \tag{8.24}$$

where $\tau_{\tilde{\Delta}_{mp}}$ is the restriction of $\tau_{\tilde{\Delta}_m}$ to the p-th face. Using integration on singular simplices,

$$\int_{\tilde{\Delta}_m} \varsigma = \int_{s_m} \varsigma = \int_{\Delta^n} [(Ts_m)^*(\varsigma)]_{1\ldots n} \mathrm{d}x^1 \cdots \mathrm{d}x^n. \tag{8.25}$$

By the mean value theorem for integration, there is a point $\boldsymbol{q}_m \in \Delta^n$ such that

$$\int_{\Delta^n} (Ts_m)^*(\varsigma)_{1,\ldots,n} \mathrm{d}x^1 \cdots \mathrm{d}x^n = \frac{1}{n!}[(Ts_m)^*(\varsigma)]_{1\ldots n}(\boldsymbol{q}_m), \tag{8.26}$$

where $1/n!$ is the measure of the n-simplex. Using $x_m = s_m(\boldsymbol{q}_m) = \varphi^{-1} \circ (a_m\mathrm{Id})(\boldsymbol{q}_m)$, we have

$$
\begin{aligned}
&[(Ts_m)^*(\varsigma)]_{1\dots,n}(\boldsymbol{q}_m)\\
&\quad= [(Ts_m)^*(\varsigma)](\boldsymbol{q}_m)(\mathbf{e}_1,\dots,\mathbf{e}_n),\\
&\quad= \varsigma(x_m)\left[T_{\boldsymbol{q}_m}(\varphi^{-1}\circ(a_m\mathrm{Id}))(\mathbf{e}_1),\dots,T_{\boldsymbol{q}_m}(\varphi^{-1}\circ(a_m\mathrm{Id}))(\mathbf{e}_n)\right],\\
&\quad= \varsigma(x_m)\left[T_{a_m\boldsymbol{q}_m}\varphi^{-1}(a_m\mathbf{e}_1),\dots,T_{a_m\boldsymbol{q}_m}\varphi^{-1}(a_m\mathbf{e}_n)\right],\\
&\quad= (a_m)^n\varsigma(x_m)\left[T_{a_m\boldsymbol{q}_m}\varphi^{-1}(\mathbf{e}_1),\dots,T_{a_m\boldsymbol{q}_m}\varphi^{-1}(\mathbf{e}_n)\right],\\
&\quad= (a_m)^n[(T\varphi^{-1})^*(\varsigma)(a_m\boldsymbol{q}_m)](\mathbf{e}_1,\dots,\mathbf{e}_n).
\end{aligned}
\tag{8.27}
$$

To arrive at the third line above, we used

$$
T_{\boldsymbol{q}_m}(\varphi^{-1}\circ(a_m\mathrm{Id})) = T_{a_m\mathrm{Id}(\boldsymbol{q}_m)}\varphi^{-1}\circ T_{\boldsymbol{q}_m}(a_m\mathrm{Id}) = T_{a_m\boldsymbol{q}_m}\varphi^{-1}\circ(a_m\mathrm{Id}), \tag{8.28}
$$

which follows from Eq. (6.27) and the linearity of $a_m\mathrm{Id}$. Thus,

$$
\int_{\tilde{\Delta}_m}\varsigma = \frac{(a_m)^n}{n!}[(T\varphi^{-1})^*(\varsigma)(a_m\boldsymbol{q}_m)](\mathbf{e}_1,\dots,\mathbf{e}_n). \tag{8.29}
$$

Similarly, using the mean value theorem for the integrals over the faces, $\tilde{\Delta}_{mp}$,

$$
\begin{aligned}
\int_{\tilde{\Delta}_{mp}}\tau_{\tilde{\Delta}_{mp}} &= \int_{\Delta^{n-1}}(Ts_{mp})^*(\tau_{\tilde{\Delta}_{mp}})_{1\dots,n-1}\mathrm{d}x^1\cdots\mathrm{d}x^{n-1},\\
&= \frac{1}{(n-1)!}(Ts_{mp})^*(\tau_{\tilde{\Delta}_{mp}})_{1\dots,n-1}(\boldsymbol{q}_{mp}),
\end{aligned}
\tag{8.30}
$$

for some point $\boldsymbol{q}_{mp}\in\Delta^{n-1}$. With $x_{mp}=s_{mp}(\boldsymbol{q}_{mp})=\varphi^{-1}\circ(a_m\mathrm{Id})\circ k_p(\boldsymbol{q}_{mp})$, one has

$$
\begin{aligned}
&[(Ts_{mp})^*(\tau_{\tilde{\Delta}_{mp}})_{1\dots,n-1}](\boldsymbol{q}_{mp})\\
&\quad= [(Ts_{mp})^*(\tau_{\tilde{\Delta}_{mp}})(\boldsymbol{q}_{mp})](\mathbf{e}_1,\dots,\mathbf{e}_{n-1}),\\
&\quad= \tau_{\tilde{\Delta}_{mp}}(x_{mp})\Big[[T_{\boldsymbol{q}_{mp}}(\varphi^{-1}\circ(a_m\mathrm{Id})\circ k_p)](\mathbf{e}_1),\dots\\
&\qquad\qquad\dots,[T_{\boldsymbol{q}_{mp}}(\varphi^{-1}\circ(a_m\mathrm{Id})\circ k_p)](\mathbf{e}_{n-1})\Big],\\
&\quad= a_m^{n-1}\tau_{\tilde{\Delta}_{mp}}(x_{mp})\Big[[T_{a_mk_p(\boldsymbol{q}_{mp})}\varphi^{-1}](\mathrm{e}_{p1}),\dots\\
&\qquad\qquad\dots,[T_{a_mk_p(\boldsymbol{q}_{mp})}\varphi^{-1}](\mathrm{e}_{pn-1})\Big].
\end{aligned}
\tag{8.31}
$$

To arrive at the third equality above, we used

$$
\begin{aligned}
[T_{\boldsymbol{q}_{mp}}(\varphi^{-1} \circ (a_m \mathrm{Id}) \circ k_p)](\mathbf{e}_l) &= [T_{a_m k_p(\boldsymbol{q}_{mp})}\varphi^{-1} \circ T_{k_p(\boldsymbol{q}_{mp})}(a_m \mathrm{Id}) \circ T_{\boldsymbol{q}_{mp}} k_p](\mathbf{e}_l), \\
&= a_m [T_{a_m k_p(\boldsymbol{q}_{mp})}\varphi^{-1} \circ T k_p](\mathbf{e}_l), \\
&= a_m [T_{a_m k_p(\boldsymbol{q}_{mp})}\varphi^{-1}](\mathrm{e}_{pl}),
\end{aligned}
\tag{8.32}
$$

noting that $T_{\boldsymbol{q}_m} k_p$ is independent of the point where it is evaluated, and clearly, $T_{k_p(\boldsymbol{q}_{mp})}(a_m \mathrm{Id}) = a_m \mathrm{Id}$.

The balance equation (8.24) for the simplices $\tilde{\Delta}_m$ assumes the form

$$
\begin{aligned}
&\left|\sum_p (-1)^p \frac{(a_m)^{n-1}}{(n-1)!} \tau_{\tilde{\Delta}_{mp}}(x_{mp}) \left[T_{a_m k_p(\boldsymbol{q}_{mp})}\varphi^{-1}(\mathrm{e}_{p1}), \dots, T_{a_m k_p(\boldsymbol{q}_{mp})}\varphi^{-1}(\mathrm{e}_{pn-1})\right]\right| \\
&\quad \leqslant a_m^n (T\varphi^{-1})^*(\varsigma)(a_m \boldsymbol{q}_m)(\mathbf{e}_1, \dots, \mathbf{e}_n).
\end{aligned}
\tag{8.33}
$$

Dividing both sides of the inequality above by $(a_m)^{n-1}$, we take the limit of the equation as $m \to \infty$. Noting that $\lim_{m\to\infty} a_m = 0$, $\lim_{m\to\infty} x_{mp} = x$, $\lim_{m\to\infty} a_m k_p(\boldsymbol{q}_{mp}) = 0$, it follows from $\tau_{\tilde{\Delta}_{mp}}(x_{mp}) = \mathrm{t}_{x_{mp}}$ and the regularity assumption that

$$
\sum_p (-1)^p \mathrm{t}_x \left[T_0\varphi^{-1}(\mathrm{e}_{p1}), \dots, T_0\varphi^{-1}(\mathrm{e}_{pn-1})\right] = 0.
\tag{8.34}
$$

As $S_{pl} = T_0(\varphi^{-1})(\mathrm{e}_{pl})$, see (8.23), we can finally write

$$
\sum_p (-1)^p \mathrm{t}_x(S_p) = \mathrm{t}_x(\partial S) = 0.
\tag{8.35}
$$

We conclude that all the assumptions made in Sect. 4.1 are satisfied. It follows that at each point $x \in \mathcal{S}$ there is an $(n-1)$-alternating tensor $J(x)$ such that $\mathrm{t}_x(v_1, \dots, v_{n-1}) = J(x)(v_1, \dots, v_{n-1})$. Equations (8.8) and (8.9) follow. Summarizing, we have

$$
\Phi_{\partial \mathcal{R}} = \int_{\partial \mathcal{R}} \tau_{\mathcal{R}} = \int_{\partial \mathcal{R}} J.
\tag{8.36}
$$

8.1.5 The Differential Balance Law

By the regularity assumption, the Cauchy mapping t is smooth. In addition, it follows from Cauchy's theorem that the flux $(n-1)$-form $J : \mathcal{S} \to \bigwedge^{n-1}(T^*\mathcal{S})$ satisfies $J(x) = \mathrm{t}_x$, and it follows that J is a smooth form. For a given region $\mathcal{R}$, using (8.8) in the balance equation (8.4), one obtains

$$\int_{\partial \mathcal{R}} (T\mathcal{I}_{\partial \mathcal{R}})^*(J) + \int_{\mathcal{R}} \beta = \int_{\mathcal{R}} \varsigma. \tag{8.37}$$

We can apply Stokes's theorem to the boundary integral above so that

$$\int_{\mathcal{R}} \mathrm{d}J + \int_{\mathcal{R}} \beta = \int_{\mathcal{R}} \varsigma. \tag{8.38}$$

Since the equation above holds for an arbitrary region, we conclude that the balance equation is equivalent to

$$\mathrm{d}J + \beta = \varsigma \tag{8.39}$$

—the *differential balance law*.

8.2 Properties of Fluxes

The following comments are concerned with issues pertaining to flux fields.

8.2.1 Flux Densities and Orientation

We collect here some notes regarding the orientation aspects of the foregoing analysis:

1. It is recalled that the expression for total outflux

$$\Phi_{\partial \mathcal{R}} = \int_{\partial \mathcal{R}} \tau_{\mathcal{R}} = \int_{\partial \mathcal{R}} (T\mathcal{I}_{\partial \mathcal{R}})^*(J) \tag{8.40}$$

   is meaningful for the particular orientation chosen for $\mathcal{S}$ and inherited by $\mathcal{R}$ and $\partial \mathcal{R}$. Similarly, $\tau_{\mathcal{R}}(x)(v_1, \dots, v_{n-1})$ is interpreted as the infinitesimal flux of the property out of the region $\mathcal{R}$ through the infinitesimal oriented simplex induced $v_1, \dots, v_{n-1}$ only when they are positively oriented. In case the collection of vectors is negatively oriented, $\tau_{\mathcal{R}}(x)(v_1, \dots, v_{n-1})$ is interpreted as the infinitesimal influx.
2. Let $\mathcal{V} \subset \mathcal{S}$ be an oriented $(n-1)$-dimensional submanifold. Then, the flux through $\mathcal{V}$ relative to the given orientation may be defined by setting

$$\tau_{\mathcal{V}} := (T\mathcal{I}_{\mathcal{V}})^*(J), \qquad \Phi_{\mathcal{V}} = \int_{\mathcal{V}} \tau_{\mathcal{V}} = \int_{\mathcal{V}} (T\mathcal{I}_{\mathcal{V}})^*(J). \tag{8.41}$$

3. Let $\mathcal{V} \subset \mathcal{S}$ be an orientable $(n-1)$-dimensional submanifold. Let $\Phi_{\mathcal{V}}$ be the flux through $\mathcal{V}$ when the orientation o is chosen for $\mathcal{V}$, and let $\Phi_{\mathcal{V}'}$ be the flux when the opposite orientation o' is chosen. By the definition of integration over oriented manifolds,

$$\Phi_{\mathcal{V}'} = -\Phi_{\mathcal{V}}. \tag{8.42}$$

4. Let $\tau_{\mathcal{V}}$ and $\tau_{\mathcal{V}'}$ be the corresponding flux densities for $\Phi_{\mathcal{V}}$ and $\Phi_{\mathcal{V}'}$, respectively; then, by (8.41),

$$\tau_{\mathcal{V}}(x)(v_1, \dots, v_{n-1}) = \tau_{\mathcal{V}'}(v_1, \dots, v_{n-1}) = J(v_1, \dots, v_{n-1}). \tag{8.43}$$

However, if $v_1, \dots, v_{n-1}$ are positively oriented with respect to o, they are negatively oriented relative to o'. Thus, $\tau_{\mathcal{V}'}(v_1, \dots, v_{n-1})$ is considered as the "influx" of the property.

5. The sign reversal in Eq. (8.42) is missing in (8.43). The following is the geometric analogy of a standard procedure in classical continuum mechanics as described below.

   Let o be an orientation of $\mathcal{V}$, and consider the symmetric, covariant $(n-1)$-tensor τ_o defined on $\mathcal{V}$ by

$$\tau_o(x)(v_1, \dots, v_{n-1}) = \tau_{\mathcal{V}}(x)(v_{\pi_1}, \dots, v_{\pi_{n-1}}), \tag{8.44}$$

   where π is any permutation such that $v_{\pi_1}, \dots, v_{\pi_{n-1}}$ are positively oriented relative to o. Note that there is no restriction here on the orientation determined by the vectors $v_1, \dots, v_{n-1}$. It follows that $\tau_o(x)(v_1, \dots, v_{n-1})$ is the well-defined infinitesimal outflux for the orientation o. For the opposite orientation, o', and a corresponding permutation π',

$$\begin{aligned} \tau_{o'}(x)(v_1, \dots, v_{n-1}) &:= \tau_{\mathcal{V}'}(x)(v_{\pi'_1}, \dots, v_{\pi'_{n-1}}), \\ &= \tau_{\mathcal{V}}(x)(v_{\pi'_1}, \dots, v_{\pi'_{n-1}}), \\ &= -\tau_{\mathcal{V}}(x)(v_{\pi_1}, \dots, v_{\pi_{n-1}}), \end{aligned} \tag{8.45}$$

   where in the last line we used the fact that $\pi \circ \pi'$ is an odd permutation. It is concluded that

$$\tau_{o'} = -\tau_o. \tag{8.46}$$

6. Let $A : T_x\mathcal{V} \to T_x\mathcal{V}$ be a linear isomorphism. Then,

$$\tau_o(x)(A(v_1), \dots, A(v_{n-1})) = \tau_{\mathcal{V}}(x)(A(v)_{\mu_1} \dots, A(v)_{\mu_{n-1}}), \tag{8.47}$$

where μ is a permutation that takes the sequence $A(v_1), \dots, A(v_{n-1})$ into the sequence $A(v)_{\mu_1}, \dots, A(v)_{\mu_{n-1}}$ that has the orientation o. If the determinant $\det A$ is positive, then A preserves orientations. Thus, if for a permutation π, $v_{\pi_1}, \dots, v_{\pi_{n-1}}$ are positively oriented, then the sequence $A(v_{\pi_1}), \dots, A(v_{\pi_{n-1}})$ will also be positively oriented, and we may set

$$A(v)_{\mu_1} = A(v_{\pi_1}), \dots, A(v)_{\mu_{n-1}} = A(v_{\pi_{n-1}}).$$

Thus,

$$\begin{aligned} \tau_o(x)(A(v_1), \dots, A(v_{n-1})) &= \tau_{\mathscr{V}}(x)(A(v_{\pi_1}), \dots, A(v_{\pi_{n-1}})), \\ &= \det A \cdot \tau_{\mathscr{V}}(x)(v_{\pi_1}, \dots, v_{\pi_{n-1}}), \\ &= \det A \cdot \tau_o(x)(v_1, \dots, v_{n-1}), \qquad \det A > 0. \end{aligned} \tag{8.48}$$

If $\det A < 0$, then A reverses orientations. Thus, if $v_{\pi_1}, \dots, v_{\pi_{n-1}}$ are positively oriented, the sequence $A(v_{\pi_1}), \dots, A(v_{\pi_{n-1}})$ is negatively oriented. Since the sequence $A(v)_{\mu_1}, \dots, A(v)_{\mu_{n-1}}$ has to be positively oriented,

$$\begin{aligned} \tau_{\mathscr{V}}(x)(A(v)_{\mu_1} \dots, A(v)_{\mu_{n-1}}) &= -\tau_{\mathscr{V}}(x)(A(v_{\pi_1}), \dots, A(v_{\pi_{n-1}})), \\ &= -\det A \cdot \tau_{\mathscr{V}}(x)(v_{\pi_1}, \dots, v_{\pi_{n-1}}), \\ &= -\det A \cdot \tau_o(v_1, \dots, v_{n-1}), \qquad \det A < 0. \end{aligned} \tag{8.49}$$

It is concluded that for both cases,

$$\tau_o(x)(A(v_1), \dots, A(v_{n-1})) = |\det A|\, \tau_o(v_1, \dots, v_{n-1}). \tag{8.50}$$

7. This last property implies that τ_o is an odd form (as in [dR84a, p. 19]), a density, or a twisted $(n-1)$-form on $\mathscr{V}$ (as in [AMR88, p. 407 and p. 486] or [Lee02, pp. 375–380]). We conclude that the differential form J induces on an orientable submanifold $\mathscr{V}$ two opposite twisted $(n-1)$-forms, one for each of its orientations. This construction is analogous to the classical definition of internal forces in mechanics where a single entity, the stress field or the flux vector field, induces on a surface two opposite densities corresponding to the two orientations of the surface. The orientation of surfaces is traditionally indicated by an "outwards"-pointing unit normal. Finally, we note that integration of densities on manifolds is well-defined (loc. cit.).

8.2.2 *Kinetic Fluxes and Kinematic Fluxes*

We describe here some of the properties of flux $(n-1)$-forms that generalize familiar properties of classical flux vector fields. First, it is shown how a volume element, a weaker geometric structure than a Riemannian metric, enables one to represent the flux form by a vector field. To that end, we use notions considered in Sects. 5.8 and 6.2.3.

Let θ be a volume element on $\mathcal{S}$. Locally, for a positively oriented chart in $\mathcal{S}$, θ is of the form

$$\theta_{1\dots n}(x^i)\mathrm{d}x^1 \wedge \cdots \wedge \mathrm{d}x^n, \tag{8.51}$$

where $\theta_{1\dots n}(x^i) > 0$. Let $v : \mathcal{S} \to T\mathcal{S}$ be a vector field in $\mathcal{S}$, represented locally by $\sum_i v^i(x)\partial_i$; then, using Example 5.4, the $(n-1)$-form $v \lrcorner \theta$, $(v \lrcorner \theta)(x) = v(x) \lrcorner \theta(x)$ is represented locally by

$$\sum_k (-1)^{k-1}\theta_{1\dots n} v^k \mathrm{d}x^1 \wedge \cdots \wedge \widehat{\mathrm{d}x^k} \wedge \dots \mathrm{d}x^n. \tag{8.52}$$

Let J be an $(n-1)$-flux form. The condition that $J = v \lrcorner \theta$ may be written for the local representation $\sum_k J_{1\dots\widehat{k}\dots n}\mathrm{d}x^1 \wedge \cdots \wedge \widehat{\mathrm{d}x^k} \wedge \dots \mathrm{d}x^n$ as

$$J_{1\dots\widehat{k}\dots n} = (-1)^{k-1}\theta_{1\dots n} v^k. \tag{8.53}$$

Since $\theta_{1\dots,n} > 0$, given J, there is a unique vector field v, given locally by

$$v^k = (-1)^{k-1}\frac{J_{1\dots\widehat{k}\dots n}}{\theta_{1\dots n}}, \tag{8.54}$$

such that $J = v \lrcorner \theta$. In other words, the volume element θ induces a vector bundle isomorphism

$$\mathcal{I}_\theta : \bigwedge^{n-1}(T^*\mathcal{S}) \longrightarrow T\mathcal{S}, \tag{8.55}$$

whose inverse is given by $\mathcal{I}_\theta^{-1}(v) = v \lrcorner \theta$.

If a positive extensive property is under consideration, such as mass or entropy, the density ρ of the property, a positive n-form, may be used as a volume element. It follows that in such a case the vector field $w = \mathcal{I}_\rho(J)$ is physically meaningful. For example, for the case of the mass property, J indicates the flux of mass, and $w = \mathcal{I}_\rho(J)$, obtained by dividing the components of the mass flux (with the appropriate sign) by the mass density, models the velocity of the mass particles. Therefore, for a case where the volume element used is the positive density of the property, the flux of which is under consideration, it is natural to refer to $w = \mathcal{I}_\rho(J)$ as the *kinematic flux*, distinguishing it from the *kinetic flux* J.

For a given volume element θ, let $w = \mathcal{I}_\theta(J)$ so that $J = w \lrcorner \theta$. Let $v_1, \dots, v_{n-1}$ be linearly independent. Since $\theta(u_1, \dots, u_n)$ does not vanish if and only if $u_1, \dots, u_n$ are linearly independent,

$$J(v_1, \dots, v_{n-1}) = \theta(w, v_1, \dots, v_{n-1}) = 0 \tag{8.56}$$

if and only if w may be represented as a linear combination of $v_1, \ldots, v_{n-1}$. We conclude that the flux through the simplex $[v_1, \ldots, v_{n-1}]$ vanishes if and only if the subspace that the vectors span contains the kinematic flux vector $w = \mathcal{I}_\theta(J)$.

8.2.3 The Flux Bundle

We observe that the dependence of $v = \mathcal{I}_\theta(J)$ on the choice of volume element θ as exhibited in Eq. (8.54) is via multiplication by a positive constant. It follows that the flux form J induces a 1-dimensional subbundle of the tangent bundle F_J such that

$$(F_J)_x = \left\{\mathcal{I}_{\theta(x)}(J(x)) \mid \theta(x) \in \bigwedge^n(T_y\mathcal{S})\right\}. \tag{8.57}$$

Using (8.54), we may write locally

$$(F_J)_x = \left\{a\sum_k(-1)^{k-1}J_{1\ldots\widehat{k}\ldots n}\partial_k \mid a \in \mathbb{R}\right\}. \tag{8.58}$$

The orientation of $\mathcal{S}$ as reflected in the sign of $\theta_{1\ldots n}$ for a positively oriented chart induces an orientation in F_J. We will refer to F_J as the *flux bundle* associated with J. It is emphasized that the flux bundle is independent of the volume element.

Since for $a \in \mathbb{R}$, $a \neq 0$, the vector aw belongs to $\operatorname{span}\{v_1, \ldots, v_{n-1}\}$ if and only if $w \in \operatorname{span}\{v_1, \ldots, v_r\}$, it follows that the flux through a simplex $[v_1, \ldots, v_{n-1}]$ vanishes if and only if $(F_J)_x$ is a subspace of $\operatorname{span}\{v_1, \ldots, v_{n-1}\}$. Furthermore, if $\mathcal{R}$ is a region, a collection of linearly independent vectors $v_1, \ldots, v_{n-1}$ span $T_x\partial\mathcal{R}$. It follows from the generalized Cauchy formula, $\tau_{\mathcal{R}} = (T\mathcal{I}_{\partial\mathcal{R}})^*(J)$, that $\tau_{\mathcal{R}}(x)$ vanishes if and only if the flux space $(F_J)_x$ is a subspace of $T_x\partial\mathcal{R}$. This is analogous to the situation in the context of the classical formulations of continuum mechanics where the flux density through a surface element vanishes if and only if the nonvanishing flux vector field is tangent to the surface element.

8.2.4 Flow Potentials and Stream Functions

For the case where $\beta = \varsigma = 0$, the differential balance law (8.39) assumes the form $\mathrm{d}J = 0$. Thus, the balance law implies that the flux form is closed. It follows from the identity $\mathrm{d}^2\psi = 0$ (6.138) that for any $(n-2)$-form ψ, $J = \mathrm{d}\psi$ satisfies the differential balance equation. In such a case, we will refer to ψ as the *potential* or *stream function* for the flux field J. This way, the Maxwell 2-form of electrodynamics induces the flux of the electric charge in the spacetime formulation (see Chap. 11).

By the Poincaré lemma, locally, every closed form is exact. Thus, locally, a flow potential always exists. It is recalled that the problem of existence of a stream function for J defined on all of $\mathcal{S}$ is related to the global topological structure of the manifold. If the manifold is simply connected, every closed form is exact.

Clearly, the potential form α is not unique. Using $\mathrm{d}^2\lambda = 0$, if an r-form ω is exact, i.e., $\omega = \mathrm{d}\alpha$ for an $(r-1)$-potential form α, then, for any $(r-2)$-form λ, $\alpha' = \alpha + \mathrm{d}\lambda$ is also a potential.

Chapter 9
Frames, Body Points, and Spacetime Structure

In the analysis of the previous chapters, we interpreted the manifold $\mathcal{S}$ as a fixed n-dimensional space manifold. Thus, assuming that time is an element of a one-dimensional manifold, $\mathcal{T}$, the event space, or equivalently, spacetime, $\mathcal{E}$, was assumed to have a Cartesian product structure, $\mathcal{E} = \mathcal{T} \times \mathcal{S}$. This is a very restrictive assumption that contradicts even the Galilean point of view of physics. In a more general situation, one may assume for the $(n+1)$-dimensional spacetime $\mathcal{E}$, the existence of local frames. A *local frame* Φ is an embedding

$$\Phi : U \subset \mathcal{E} \longrightarrow \mathcal{T} \times \mathcal{S}, \tag{9.1}$$

where U is an open subset of $\mathcal{E}$. A frame Φ will be *global* in case $U = \mathcal{E}$. The existence of a global frame clearly depends on the topological structure of $\mathcal{E}$. We will denote by $\Phi_{\mathcal{T}}$ and $\Phi_{\mathcal{S}}$ the two components of the frame and will write (t, x) for $\Phi(e)$, $e \in \mathcal{E}$. The domains of the local frames are assumed to cover $\mathcal{E}$. It follows that the foregoing analysis applies in the image of a local frame. We will assume henceforth that $\mathcal{T} = \mathbb{R}$.

9.1 Frames, Balance, and Fluxes in Spacetime

In general, a local chart (ϕ, U) on the spacetime manifold $\mathcal{E}$ provides local *world coordinates* $(z^1, \dots, z^{n+1})$. For the case where a local frame Φ is given, the local representative $\tilde{\Phi}$ of Φ is of the form $(z^1, \dots, z^{n+1}) \mapsto (t, x^1, \dots, x^n)$, where $(x^1, \dots, x^n)$ are local coordinates in $\mathcal{S}$. In the sequel, given a frame Φ, we will refer to $(z^1, \dots, z^{n+1})$ as the *frame coordinates* when $z^1 = t$ is the time coordinate, and for $i > 1$, $z^i = x^{i-1}$ are local coordinates in some chart in $\mathcal{S}$.

The tangent mapping $T_e\Phi : T_e\mathcal{E} \to T_{\Phi_{\mathcal{T}}(e)}\mathcal{T} \times T_{\Phi_{\mathcal{S}}(e)}\mathcal{S}$ has two components $T_e\Phi_{\mathcal{T}}$ and $T_e\Phi_{\mathcal{S}}$. For $u \in T_e\mathcal{E}$, we will write $u_{\mathcal{T}} = T\Phi_{\mathcal{T}}(u)$, $u_{\mathcal{S}} = T\Phi_{\mathcal{S}}(u)$.

R. Segev, *Foundations of Geometric Continuum Mechanics*, Advances in Continuum Mechanics 49, https://doi.org/10.1007/978-3-031-35655-1_9

As $T_t\mathcal{T} = \mathbb{R}$ and $T_t^*\mathcal{T} \cong \mathbb{R}$, the natural base vector of $T_t^*\mathcal{T}$, $t \in \mathcal{T}$, is the number 1, and $T_e^*\Phi_{\mathcal{T}}(1) \in T_e^*\mathcal{E}$ will be denoted by $\mathrm{d}t$. Writing $\partial_t := \partial/\partial t = (T_e\Phi)^{-1}(1,0)$, it is noted that $\mathrm{d}t(\partial_t) = 1$, $(T_e\Phi_{\mathcal{T}}(\partial_t)) = 1$, $\mathrm{d}t(T_e^{-1}\Phi(0,v)) = 0$, $(T_e\Phi_{\mathcal{T}}(T_e^{-1}\Phi(0,v))) = 0$, etc. For an alternating tensor $\omega \in \bigwedge^r T_x\mathcal{S}$, we set $\underset{\sim}{\omega} = T_e^*\Phi_{\mathcal{S}}(\omega) \in \bigwedge^r T_x\mathcal{E}$. Note that

$$
\begin{aligned}
(\partial_t \lrcorner \underset{\sim}{\omega})(v_1, \dots, v_{r-1}) &= \underset{\sim}{\omega}(\partial_t, v_1, \dots, v_{r-1}), \\
&= \omega(T_e\Phi_{\mathcal{S}}((T_e\Phi)^{-1}(1,0)), T_e\Phi_{\mathcal{S}}(v_1), \dots, T_e\Phi_{\mathcal{S}}(v_{r-1})), \\
&= 0.
\end{aligned}
\tag{9.2}
$$

Thus, given a global frame, Φ, a time-dependent r-form ω, $\omega(t) \in \Omega^r(\mathcal{S})$, induces a form $\mathfrak{f} = \mathrm{d}t \wedge \underset{\sim}{\omega} \in \Omega^{r+1}(\mathcal{E})$. Using Eq. (5.107) as well as the previous equation,

$$
\partial_t \lrcorner \mathfrak{f} = \partial_t \lrcorner (\mathrm{d}t \wedge \underset{\sim}{\omega}) = \underset{\sim}{\omega}. \tag{9.3}
$$

Let $\underset{\sim}{\rho}$, $\underset{\sim}{\beta}$, $\underset{\sim}{\varsigma}$, and $\underset{\sim}{J}$ be time-dependent differential forms on $\mathcal{E}$ corresponding, respectively, to the density, ρ, its rate of change, β, the source, ς, and the flux, J, of some extensive property P (as in Chap. 8). One may now define on $\mathcal{E}$ the forms

$$
\mathfrak{b} = \mathrm{d}t \wedge \underset{\sim}{\beta}, \qquad \mathfrak{s} = \mathrm{d}t \wedge \underset{\sim}{\varsigma}, \qquad \mathfrak{J} = -\mathrm{d}t \wedge \underset{\sim}{J} + \underset{\sim}{\rho}. \tag{9.4}
$$

Let J and ρ be represented locally as

$$
\sum_k J_{1\dots\hat{k}\dots n}\mathrm{d}x^1 \wedge \dots \wedge \widehat{\mathrm{d}x^k} \wedge \dots \wedge \mathrm{d}x^n, \qquad \text{and} \qquad \rho_{1\dots n}\mathrm{d}x^1 \wedge \dots \wedge \mathrm{d}x^n. \tag{9.5}
$$

Then, by Eq. (5.131), $\underset{\sim}{J}$ and $\underset{\sim}{\rho}$ are represented locally as

$$
\sum_k J_{1\dots\hat{k}\dots n}\underset{\sim}{\mathrm{d}x^1} \wedge \dots \wedge \widehat{\underset{\sim}{\mathrm{d}x^k}} \wedge \dots \wedge \underset{\sim}{\mathrm{d}x^n}, \qquad \text{and} \qquad \rho_{1\dots n}\underset{\sim}{\mathrm{d}x^1} \wedge \dots \wedge \underset{\sim}{\mathrm{d}x^n}, \tag{9.6}
$$

respectively. It follows that the local representation of $\mathfrak{J}$ is of the form

$$
-\sum_k J_{1\dots\hat{k}\dots n}\mathrm{d}t \wedge \underset{\sim}{\mathrm{d}x^1} \wedge \dots \wedge \widehat{\underset{\sim}{\mathrm{d}x^k}} \wedge \dots \wedge \underset{\sim}{\mathrm{d}x^n} + \rho_{1\dots n}\underset{\sim}{\mathrm{d}x^1} \wedge \dots \wedge \underset{\sim}{\mathrm{d}x^n}. \tag{9.7}
$$

In terms of frame coordinates, the local expression (9.7) for $\mathfrak{J}$ is

$$
-\sum_{k=2}^{n+1} J_{1\dots\widehat{k-1}\dots n}\mathrm{d}z^1 \wedge \mathrm{d}z^2 \wedge \dots \wedge \widehat{\mathrm{d}z^k} \wedge \dots \wedge \mathrm{d}z^{n+1} + \rho_{1\dots n}\mathrm{d}z^2 \wedge \dots \wedge \mathrm{d}z^{n+1}, \tag{9.8}
$$

so that

$$\mathfrak{J}_{2\ldots n+1} = \rho_{1\ldots n}, \quad \text{and} \quad \mathfrak{J}_{1\ldots\widehat{k}\ldots n+1} = -J_{1\ldots\widehat{k-1}\ldots n}, \quad \text{for } k > 1. \tag{9.9}$$

To compare the role played by $\mathfrak{J}$ in comparison with that of J, consider an $(n+1)$-dimensional region $\mathfrak{R} \subset \mathscr{E}$ with boundary $\partial\mathfrak{R}$. Consider $e \in \partial\mathfrak{R}$, and assume that $T\Phi\{T_e\partial\mathfrak{R}\} = \{0\} \times T_x\mathcal{S}$, i.e., the tangent space to the boundary is "spacelike." Then, for $v_1, \ldots, v_n \in T_e\partial\mathfrak{R}$,

$$\begin{aligned}\mathfrak{J}(v_1, \ldots, v_n) &= \underset{\sim}{\rho}(v_1, \ldots, v_n) - (\mathrm{d}t \wedge \underset{\sim}{J})(v_1, \ldots, v_n),\\ &= \underset{\sim}{\rho}(v_1, \ldots, v_n),\\ &= \rho(T\Phi_{\mathcal{S}}(v_1), \ldots, T\Phi_{\mathcal{S}}(v_n)),\end{aligned} \tag{9.10}$$

where the second term in the first line vanishes because $\mathrm{d}t(v_i) = 0$, for all $i = 1, \ldots, n$. On the other hand, if $T_e\partial\mathfrak{R}$ contains ∂_t, $\{T\Phi_{\mathcal{S}}(v_1), \ldots, T\Phi_{\mathcal{S}}(v_n)\}$ cannot contain n-linearly independent vectors. It follows that $\underset{\sim}{\rho}(v_1, \ldots, v_n) = 0$ and

$$\mathfrak{J}(v_1, \ldots, v_n) = -(\mathrm{d}t \wedge \underset{\sim}{J})(v_1, \ldots, v_n). \tag{9.11}$$

In case $v_i = \partial_t$,

$$\begin{aligned}\mathfrak{J}(v_1, \ldots, v_n) &= -(-1)^{i-1}(\mathrm{d}t \wedge \underset{\sim}{J})(v_i, v_1, \ldots, \widehat{v_i}, \ldots, v_n),\\ &= (-1)^i \partial_t \lrcorner (\mathrm{d}t \wedge \underset{\sim}{J})(v_1, \ldots, \widehat{v_i}, \ldots, v_n),\\ &= (-1)^i \underset{\sim}{J}(v_1, \ldots, \widehat{v_i}, \ldots, v_n),\\ &= (-1)^i J(T\Phi_{\mathcal{S}}(v_1), \ldots, \widehat{T\Phi_{\mathcal{S}}(v_i)}, \ldots, T\Phi_{\mathcal{S}}(v_n)),\end{aligned} \tag{9.12}$$

where the identity (5.107) was used to arrive at the third line. Thus, in this situation, the spacetime flux $\mathfrak{J}$ provides the flux density in accordance with the flux field J.

Next, we present the way the balance equations are written in terms of the spacetime forms. First, it is noted that using Eqs. (6.149) and (6.138),

$$\begin{aligned}\mathrm{d}\mathfrak{J} &= \mathrm{d}(-\mathrm{d}t \wedge \underset{\sim}{J}) + \mathrm{d}\underset{\sim}{\rho},\\ &= \mathrm{d}t \wedge \mathrm{d}\underset{\sim}{J} + \mathrm{d}\underset{\sim}{\rho}.\end{aligned} \tag{9.13}$$

Since $\underset{\sim}{\rho}$ is represented locally in the form $\rho_{1\ldots n}\mathrm{d}\underset{\sim}{x}^1 \wedge \cdots \wedge \mathrm{d}\underset{\sim}{x}^n$, the exterior derivative $\mathrm{d}\underset{\sim}{\rho}$ is represented by

$$\frac{\partial \rho_{1\ldots n}}{\partial t}\mathrm{d}t \wedge \mathrm{d}\underset{\sim}{x}^1 \wedge \cdots \wedge \mathrm{d}\underset{\sim}{x}^n, \tag{9.14}$$

and by the definition of β as the time derivative of ρ, we obtain

$$\mathrm{d}\,\underset{\sim}{\rho} = \mathrm{d}t \wedge \underset{\sim}{\beta} = \mathfrak{b}. \tag{9.15}$$

Finally, the differential balance equation, (8.39), yields the spacetime version

$$\mathrm{d}\mathfrak{J} = \mathfrak{s}. \tag{9.16}$$

Integrating the balance equation in spacetime over a region $\mathfrak{R} \subset \mathscr{E}$ and using Stokes' theorem, the integral version of the balance in spacetime is

$$\int_{\partial\mathfrak{R}} (\mathscr{I}_{\partial\mathfrak{R}})^*(\mathfrak{J}) = \int_{\mathfrak{R}} \mathfrak{s}. \tag{9.17}$$

It is natural therefore to set

$$\mathfrak{t}_{\mathfrak{R}} = (\mathscr{I}_{\partial\mathfrak{R}})^*(\mathfrak{J}) \tag{9.18}$$

and rewrite the integral balance equation in spacetime in the form

$$\int_{\partial\mathfrak{R}} \mathfrak{t}_{\mathfrak{R}} = \int_{\mathfrak{R}} \mathfrak{s}. \tag{9.19}$$

We conclude that for any given frame the equation of balance in spacetime has the same form as the equation of balance in space, and in fact, it is simpler in the sense that it does not contain the term β. The relations between the various spacetime forms on $\mathscr{E}$ and the corresponding forms on $\mathcal{S}$ make no sense if we do not have a specific frame. However, if one postulates the balance equation (9.19) in spacetime, the procedure of Sect. 8.1.4 may be repeated yielding the existence of the n-form $\mathfrak{J}$ satisfying Equations (9.18) and (9.16), independently of any frame. It is also noted that the various fields on $\mathscr{E}$ are naturally time-dependent.

9.2 Worldlines

Let θ be a volume element in spacetime and let $\mathfrak{J}$ be a flux field on $\mathscr{E}$. Then, the isomorphism $\mathscr{I}_\theta$ of Eq. (8.55) induces a vector field w on $\mathscr{E}$ such that $w \lrcorner\, \theta = \mathfrak{J}$. We will refer to an integral line of w as a *worldline* corresponding to the property P for which $\mathfrak{J}$ is the flux. Clearly, each worldline is a one-dimensional submanifold of $\mathscr{E}$. We want to show that as a submanifold, a worldline is independent of the choice of volume element. Let θ' be another positive volume element. Then, there is a unique positive function $a\colon \mathscr{E} \to \mathbb{R}^+$ such that $\theta'(e) = a(e)\theta(e)$ for all events $e \in \mathscr{E}$. It follows from Eq. (8.54) that the vector field $w' = \mathscr{I}_{\theta'}(\mathfrak{J})$ is related to

$w = \mathcal{J}_\theta(\mathfrak{J})$ by $a(e)w'(e) = w(e)$. The differential equations for the integral lines c and c' corresponding to the two vector fields are

$$\frac{\mathrm{d}c}{\mathrm{d}s}(s) = w(c(s)) \quad \text{and} \quad \frac{\mathrm{d}c'}{\mathrm{d}s}(s) = w'(c'(s)) = \frac{1}{a(c'(s))}w(c'(s)), \tag{9.20}$$

where we use $\mathrm{d}c/\mathrm{d}s$ (instead of a superimposed dot) to indicate the lift of the curve c parametrized by a parameter s. For a solution $c'(s)$ of the equation on the right of (9.20), consider the reparametrization $s' = s'(s)$, satisfying the condition

$$\frac{\mathrm{d}s'}{\mathrm{d}s} = \frac{1}{a(c'(s))} \tag{9.21}$$

so that the reparametrization is invertible and

$$\frac{\mathrm{d}s}{\mathrm{d}s'}(s') = a(c'(s(s'))).$$

Let $\widehat{c}$ be the reparametrized curve defined by $\hat{c}(s') = c'(s(s'))$. Using the chain rule,

$$\begin{aligned} \frac{\mathrm{d}\hat{c}}{\mathrm{d}s'}(s') &= \frac{\mathrm{d}c'}{\mathrm{d}s}(s(s'))\frac{\mathrm{d}s}{\mathrm{d}s'}(s'), \\ &= \frac{1}{a(c'(s(s')))}w(c'(s(s')))a(c'(s(s'))), \\ &= w(\hat{c}(s')). \end{aligned} \tag{9.22}$$

Thus, the reparametrization $\hat{c}$ of the integral line c' coincides with the integral line c.

We conclude that worldlines are independent of the choice of a volume element and make up invariant one-dimensional submanifolds. The tangent bundle of the worldline submanifold is clearly the flux bundle of Sect. 8.2.3. In terms of the theory of distributions—subbundles of the tangent bundle—the one-dimensional flux bundle is integrable, i.e., it is the tangent bundle of a one-dimensional manifold.

Example 9.1 Consider the case where a frame is given on $\mathscr{E}$ and assume that $J = 0$. It follows from (9.7) that $\mathfrak{J}$ is represented locally by

$$\sum_{k=1}^{n+1} \mathfrak{J}_{1\dots\widehat{k}\dots n+1}(z^i)\mathrm{d}z^1 \wedge \cdots \wedge \widehat{\mathrm{d}z^k} \wedge \cdots \wedge \mathrm{d}z^{n+1} = \rho_{1\dots n}(z^i)\mathrm{d}z^2 \wedge \cdots \wedge \mathrm{d}z^{n+1}. \tag{9.23}$$

Thus,

$$\mathfrak{J}_{1\dots\widehat{k}\dots n+1} = \begin{cases} \rho_{1\dots n}, & k = 1, \\ 0, & k > 1. \end{cases} \tag{9.24}$$

Let θ be a volume element on $\mathscr{E}$. Using Eq. (8.54), it follows that independently of the choice of a volume element, the last n local differential equations for the integral lines are

$$\frac{\mathrm{d}c^k}{\mathrm{d}s} = 0, \quad k > 1, \tag{9.25}$$

which imply that $c^k(s) = z_0^k$, $k > 1$, where z_0 is the initial condition or the event where the integral line originates.

9.3 Material Points, the Material Universe, and Material Frames

The notion of a material point is one of the basic elements of continuum mechanics. The existence of material points has been motivated by the conservation of mass. Their basic property is postulated in the principle of material impenetrability, which in its elementary version implies that two distinct material points do not occupy the same event in spacetime. (See for example [Tru77].)

The foregoing discussion enables one to generalize the notion of a material point. For a given extensive property whose balance in spacetime is reflected by the flux n-form $\mathfrak{J}$ in $\mathscr{E}$, we will identify a material point with an integral worldline of the flux bundle induced by $\mathfrak{J}$. It is implied that material points may be defined for extensive properties other than mass and that they may be considered even in cases where the source term in the balance equation does not vanish. Thus, one could define material points for the electric charge, for a volumetrically growing biological body, and for the heat flux field where the term "material point" is used only figuratively.

The collection of material points is traditionally referred to as the *material universe*. It would be desirable if the material universe had a structure of a differentiable manifold. Let θ be a volume element in $\mathscr{E}$. We first note that it follows from the result quoted in the last paragraph of Sect. 6.3.1 that for any point $e_0 \in \mathscr{E}$, such that $\mathfrak{J}(e_0) \neq 0$, there is a chart (U_0, ϕ) in a neighborhood of e_0 such that the vector field $\mathscr{J}_\theta(\mathfrak{J})$ is represented locally by $\partial/\partial x^1$. Thus, the worldlines in U_0 may be represented locally in the form $z^1(s) = z_0^1 + s$, $z^i(s) = z_0^i$, for $i > 1$, where z_0^i are the coordinates of e_0. We can assume without loss of generality that $\phi\{U_0\}$ is convex. Consider the collection F_0 of initial conditions—points in U_0 such that each $e \in F_0$ is represented locally in the form $(z_0^1, z^2, \dots, z^{n+1})$. It follows that two distinct points on a single worldline both cannot belong to F_0 and that $z^2, \dots, z^{n+1}$ parametrize the various worldlines. Thus, regarding worldlines as material points, F_0 is a local n-dimensional submanifold of $\mathscr{E}$ that contains the material points in the neighborhood. Each event in U_0 may be represented by (s, X), $s \in (a, b)$, $X \in F_0$. We may refer to the mapping $\Phi : U_0 \subset \mathscr{E} \to (a, b) \times F_0$ as a local *material frame*.

The set F_0 may be thought of as a material body, and the mapping $(a, b)\times F_0 \longrightarrow \mathscr{E}$, represented by

$$(z_0^1, z^2, \ldots, z^{n+1}) \longmapsto (s, z^2, \ldots, z^{n+1}), \tag{9.26}$$

may be thought of as a motion of that body. For each value of s, viewed as the time variable, we have a configuration, an embedding $F_0 \longrightarrow \mathscr{E}$ represented by

$$(z^2, \ldots, z^{n+1}) \longmapsto (s, z^2, \ldots, z^{n+1}). \tag{9.27}$$

Clearly, the choice of volume element does not change the material frame in any significant way. It may only change the values determining the interval (a, b).

Considering the material universe from the global point of view, we define the equivalence relation $e_1 \sim e_2$ if e_1 and e_2 are on the same worldline so that an equivalence class is a material point. Thus, the material universe may be defined as the quotient set $\mathscr{E}/\sim$. The situation discussed above need not be extendable to a product of the material universe and the real axis. In other words, global material frames do not exist necessarily. For example, consider a vector field on the two-dimensional torus that winds around the torus an irrational number of times. In such a case, the flow “mixes” the material points in such a way that a manifold structure cannot be assigned to the collection of all integral lines.

Chapter 10
Stresses

This chapter presents the generalization of the stress object of continuum mechanics to the settings of general manifolds. Only smooth distributions of stresses are considered here. The existence of the stress object as a tool for the determination of surface forces on the boundary of subbodies—the traction stress field—is proved in analogy with the proof of the existence of flux fields in Chap. 8. In distinction to the classical theory, the traction stress object does not determine the power expended. Another object—the variational stress field—is needed for that. The variational stress determines the traction stress but not vice-versa.

Non-smooth stress theory, based on a completely different approach, is presented in Part III.

10.1 Force Fields on Manifolds

In the traditional formulation of continuum mechanics, set in a Euclidean physical space, one integrates the body force field and the surface force field to obtain their resultants. This is possible because, in a Euclidean space, one can transport vectors from one point to another and then add them up in a natural way. As mentioned in Sect. 6.1.2, for the general geometry of differential manifolds, there is no natural way to compare tangent vectors belonging to two distinct tangent spaces. It follows that the basic definitions of force fields should be revised. The point of view adopted here is that the value of a force field at a point in space acts on the value of a generalized velocity field at that point to produce the power density. In accordance with classical mechanics of mass particles, the power depends linearly on the velocity. We will show how these general ideas lead to natural definitions of force fields for general manifolds.

The power density should be integrated over the appropriate manifold in order to yield the total power, a real number. The theory of integration of forms over

R. Segev, *Foundations of Geometric Continuum Mechanics*, Advances in Continuum Mechanics 49, https://doi.org/10.1007/978-3-031-35655-1_10

manifolds implies that the power density over a manifold of dimension n should be represented by an n-form over that manifold. Thus, for a region $\mathcal{R}$ in $\mathcal{S}$, a power density on $\mathcal{R}$ should be an n-form defined on $\mathcal{R}$, and a power flux density on $\partial\mathcal{R}$ should be modeled by an $(n-1)$-form on $\partial\mathcal{R}$.

Let $\mathcal{M}$ be an n-dimensional orientable manifold representing, as examples, a material body manifold $\mathcal{B}$, the physical space $\mathcal{S}$ (assuming a particular frame is given), the event space (spacetime) $\mathcal{E}$, or a submanifold thereof. As discussed in Example 6.5, we regard *generalized velocity fields* over $\mathcal{M}$ as sections of some given vector bundle $\pi : W \to \mathcal{M}$, the fibers of which are m-dimensional. In the case of standard continuum mechanics, for a given configuration $\kappa : \mathcal{B} \to \mathcal{S}$ of the body manifold in space, velocity fields are sections of

$$\pi = \kappa^* \tau_{\mathcal{S}} : W = \kappa^* T\mathcal{S} \longrightarrow \mathcal{B}. \tag{10.1}$$

For the case where $\mathcal{M}$ is either the physical space or spacetime, consider an extensive property for which the generalized Cauchy's postulates (Sect. 8.1.4) hold. Then, given a volume element, the kinematic flux, a vector field over the corresponding manifold, may serve as an example of a generalized velocity—the generalized velocity of the material points induced by the flux form as in Sect. 9.3. Furthermore, in order to abandon the dependence on the choice of a volume element, the kinetic flux form itself may be taken as an example of a section representing a generalized velocity. In this case, a generalized velocity w is a section of $W = \bigwedge^{n-1}(T^*\mathcal{M})$.

Let $\mathcal{R}$ be a compact n-dimensional admissible region (a smooth orientable submanifold or a chain) in $\mathcal{M}$ with boundary $\partial\mathcal{R}$. In particular, $\mathcal{R}$ may be the image of a configuration of a material body $\mathcal{X}$ into the physical space $\mathcal{S}$. As in Sect. 6.2, $L\big(W, \bigwedge^n T^*\mathcal{R}\big)$ denotes the vector bundle whose fiber at x is the space of linear mappings $W_x \to \bigwedge^n T_x^*\mathcal{R}$. Using the terminology of Sect. 6.2.5, $L\big(W, \bigwedge^n T^*\mathcal{R}\big)$ is the vector bundle of density-dual tensors, W'. In analogy with the case of scalar-valued fluxes discussed in Sect. 8.1.1, a *body force* field on $\mathcal{R}$ is a section $\mathbf{b}_{\mathcal{R}} : \mathcal{R} \to L(W, \bigwedge^n(T^*\mathcal{R})) = W'$, a density-covector field, so that the total power expended by the body force for a velocity field $w : \mathcal{R} \to W$ is

$$P = \int_{\mathcal{R}} \mathbf{b}_{\mathcal{R}} \cdot w. \tag{10.2}$$

Here $\mathbf{b}_{\mathcal{R}} \cdot w$ is the n-form given by $(\mathbf{b}_{\mathcal{R}} \cdot w)(x) = \mathbf{b}_{\mathcal{R}}(x)(w(x))$. A *boundary force* field, or a *surface force* field, on $\partial\mathcal{R}$ is a section

$$\mathbf{t}_{\mathcal{R}} : \partial\mathcal{R} \to L\big(W, \textstyle\bigwedge^{n-1} T^*\partial\mathcal{R}\big), \tag{10.3}$$

where we wrote W instead of $W|_{\partial\mathcal{R}}$ for the sake of simplicity. Thus, for any oriented $(n-1)$-dimensional submanifold, D, of $\partial\mathcal{R}$, the power expended by the surface force for a velocity field u defined on $\partial\mathcal{R}$ is given by

$$P = \int_D \mathbf{t}_{\mathcal{R}} \cdot u. \tag{10.4}$$

The total power expended by both the body force and surface force over the region $\mathcal{R}$ and its boundary for the virtual velocity field, w, is, therefore,

$$P = F_{\mathcal{R}}(w) = \int_{\mathcal{R}} \mathbf{b}_{\mathcal{R}} \cdot w + \int_{\partial\mathcal{R}} \mathbf{t}_{\mathcal{R}} \cdot w. \tag{10.5}$$

The total power is viewed as the result of the action of the force linear functional $F_{\mathcal{R}}$, corresponding to $\mathcal{R}$, on the virtual displacement w. The assignment $\{F_{\mathcal{R}}\}$ of a force to each subbody $\mathcal{R} \subset \mathcal{X}$ is referred to as a *force system* over $\mathcal{X}$.

Using the scheme of notation of Sect. 6.1.4, an element w of W is represented under a vector bundle chart in the forms $(x^1, \dots, x^n, w^1, \dots, w^m)$ and $\sum_\alpha w^\alpha g_\alpha$, where g_α are the local base vectors in the fibers of W induced by the vector bundle chart. A section of W is represented therefore in the forms

$$(x^1, \dots, x^n, w^1(x^i), \dots, w^m(x^i))$$

and $\sum_\alpha w^\alpha(x^i) g_\alpha$. An element $\mathbf{b}$ of $L(W, \bigwedge^n(T^*\mathcal{R}))$ is represented by $(x^i, \mathbf{b}_{\alpha 1 \dots n})$ or

$$\sum_\alpha \mathbf{b}_{\alpha 1 \dots n} g^\alpha \otimes \mathrm{d}x^1 \wedge \dots \wedge \mathrm{d}x^n,$$

so that the action, $\mathbf{b} \cdot w$, is represented by $(x^i, \sum_\alpha \mathbf{b}_{\alpha 1 \dots n} w^\alpha)$ or equivalently by

$$\sum_\alpha \mathbf{b}_{\alpha 1 \dots n} w^\alpha \mathrm{d}x^1 \wedge \dots \wedge \mathrm{d}x^n.$$

Similarly, for local coordinates y^l on $\partial\mathcal{R}$, a surface force $\mathbf{t}$ is represented in the form $(y^l, \mathbf{t}_{\alpha 1 \dots (n-1)}(y^q))$ or

$$\sum_\alpha \mathbf{t}_{\alpha 1 \dots (n-1)} g^\alpha \otimes \mathrm{d}y^1 \wedge \dots \wedge \mathrm{d}y^{n-1}.$$

The action, $\mathbf{t} \cdot u$, is represented by $(y^l, \sum_\alpha t_{\alpha 1 \dots (n-1)} u^\alpha)$ or

$$\sum_\alpha \mathbf{t}_{\alpha 1 \dots (n-1)} u^\alpha \mathrm{d}y^1 \wedge \dots \wedge \mathrm{d}y^{n-1}.$$

The definitions of Sect. 6.2.4, regarding vector-valued forms, may be applied to the body force field $\mathbf{b}_{\mathcal{R}}$, a section of $L(W, \bigwedge^n(T^*\mathcal{R}))$, so that $\mathbf{b}_{\mathcal{R}}^{\mathsf{T}}$ is a covector-valued form—a section of $\bigwedge^n(T^*\mathcal{R}, W^*)$. While, $\mathbf{b}_{\mathcal{R}}(x)(w(x)) \in \bigwedge^n T_x\mathcal{R}$ is interpreted as the power density at x, $\mathbf{b}_{\mathcal{R}}^{\mathsf{T}}(x)(v_1, \dots, v_n) \in W_x^*$ may be interpreted as the resultant of the body force acting on the infinitesimal simplex defined by the vectors $v_1, \ldots, v_n$. The analogous observations hold for the surface force $\mathbf{t}_{\mathcal{R}}$; $\mathbf{t}_{\mathcal{R}}^{\mathsf{T}}$ is a section of $\bigwedge^{n-1}(T^*\partial\mathcal{R}, W^*)$ and $\mathbf{t}_{\mathcal{R}}^{\mathsf{T}}(y)(v_1, \ldots, v_{n-1}) \in W_y^*$ is interpreted as the resultant of the surface force acting on the infinitesimal simplex on the boundary at y defined by the vectors $v_1, \ldots, v_{n-1}$.

It is noted that in general, there is no definition for the exterior derivative of differential forms valued in a vector bundle.

10.2 Traction Stresses and Cauchy's Formula on Manifolds

This section introduces the analog of the Cauchy fluxes of Sect. 8.1.2 to the case of force fields. The analog of a flux form associated with a given extensive property is a *traction stress* field. A traction stress is a section s of $L\big(W, \bigwedge^{n-1}T^*\mathcal{M}\big)$. Thus, for a section w of W, the $(n-1)$-form $s \cdot w$, defined by $(s \cdot w)(x) = s(x)(w(x))$, is a flux field representing a flux of power. Given an oriented $(n-1)$-submanifold $\mathcal{V} \subset \mathcal{M}$, the $(n-1)$-form $s \cdot w$ may be restricted to vectors tangent to $\mathcal{V}$ using the inclusion $\mathcal{I}_{\mathcal{V}} : \mathcal{V} \to \mathcal{M}$ so that

$$\rho_{\mathcal{V}} = (T\mathcal{I}_{\mathcal{V}})^* : \Omega^{n-1}(\mathcal{M}) \longrightarrow \Omega^{n-1}(\mathcal{V}), \tag{10.6}$$

with

$$(\rho_{\mathcal{V}}(s \cdot u))(y)(v_1, \ldots, v_{n-1}) = (s \cdot u)(y)(v_1, \ldots, v_{n-1}), \quad y \in \mathcal{V}. \tag{10.7}$$

The total power expended over $\mathcal{V}$ for the vector field $u : \mathcal{V} \to W|_{\mathcal{V}}$ is given by

$$P = \int_{\mathcal{V}} \rho_{\mathcal{V}}(s \cdot u). \tag{10.8}$$

The transposed s^{T}, as in (6.111), is a W^*-valued $(n-1)$-form satisfying

$$(s^{\mathsf{T}}(y)(v_1, \ldots, v_{n-1}))(w) = (s(y)(w))(v_1, \ldots, v_{n-1}), \tag{10.9}$$

where $y \in \mathcal{V}$ and $v_1, \ldots, v_{n-1} \in T_y\mathcal{V}$. It is natural to extend the definition of the restriction, $\rho_{\mathcal{V}}$, of $(n-1)$-forms to $(n-1)$-covector-valued forms such that the restriction, $\rho_{\mathcal{V}} s^{\mathsf{T}}$, of s^{T} is the covector valued form in $\Omega^{n-1}(T^*\mathcal{V}, W^*)$ given by

$$\rho_{\mathcal{V}} s^{\mathsf{T}}(y)(v_1, \ldots, v_{n-1}) = s^{\mathsf{T}}(y)(v_1, \ldots, v_{n-1}) \in W_y^*. \tag{10.10}$$

In particular, for a region $\mathcal{R}$ in $\mathcal{M}$, $\rho_{\partial\mathcal{R}} s^{\mathsf{T}}$ is a W^*-valued $(n-1)$-form on $\partial\mathcal{R}$.

We conclude that a traction stress s induces on $\partial\mathcal{R}$ a surface force field, $\mathbf{t}_{\mathcal{R}}$, by

$$\mathbf{t}_{\mathcal{R}}^{\mathsf{T}} = \rho_{\partial\mathcal{R}} s^{\mathsf{T}} \tag{10.11}$$

—the Cauchy formula for traction stresses. We note that Cauchy's formula (10.11) is equivalent to the statement that for each vector field u on $\partial\mathcal{R}$,

$$\mathbf{t}_{\mathcal{R}} \cdot u = \rho_{\partial\mathcal{R}}(s \cdot u). \tag{10.12}$$

In analogy with the situation for scalar-valued fluxes in Sect. 8.2.1, we observe that the section $\mathbf{t} : \mathcal{V} \to L\big(W, \bigwedge^{n-1} T^*\mathcal{V}\big)$ given by $\mathbf{t}^{\mathsf{T}} = \rho_{\mathcal{V}} s^{\mathsf{T}}$ is well-defined for every $(n-1)$-dimensional submanifold $\mathcal{V}$. However, the meaning of the integral of $\mathbf{t} \cdot u$, for a velocity field u over $\mathcal{V}$, depends on the orientation chosen for $\mathcal{V}$. The orientation of $\partial\mathcal{R}$ is automatically determined by the requirement that $\mathcal{R}$ inherits its orientation from $\mathcal{M}$ and by the definition of the orientation of the boundary of a region. Thus, we may extend the procedure proposed in Sect. 8.2.1 to vector-valued forms. Let o be a particular orientation of $\mathcal{V}$ and define the tensor field $\mathbf{t}_o$ on $\mathcal{V}$, a section of $W^* \otimes (T^*\mathcal{V})^{n-1}$, by

$$\mathbf{t}_o^{\mathsf{T}}(y)(v_1, \ldots, v_{n-1}) = \mathbf{t}_{\mathcal{V}}^{\mathsf{T}}(y)(v_{\pi_1}, \ldots, v_{\pi_{n-1}}) = s^{\mathsf{T}}(y)(v_{\pi_1}, \ldots, v_{\pi_{n-1}}) \tag{10.13}$$

where $v_1, \ldots, v_{n-1}$ is any collection of vectors (not necessarily oriented) and π is any permutation that rearranges the vectors so that $v_{\pi_1}, \ldots, v_{\pi_{n-1}}$ are positively oriented. It immediately follows that if o' is the opposite orientation, then, $\mathbf{t}_{o'} = -\mathbf{t}_o$. Since s^{T} is a section of $\bigwedge^{n-1} T^*\mathcal{M} \otimes W^*$, the arguments of Sect. 8.2.1 still hold, and one concludes that for every linear isomorphism $A : T_y\mathcal{M} \to T_y\mathcal{M}$,

$$\mathbf{t}_o^{\mathsf{T}}(x)(A(v_1), \ldots, A(v_{n-1})) = |\det A|\, \mathbf{t}_o^{\mathsf{T}}(v_1, \ldots, v_{n-1}). \tag{10.14}$$

Thus, $\mathbf{t}_o^{\mathsf{T}}$ and $\mathbf{t}_{o'}^{\mathsf{T}}$ are vector-valued odd forms, twisted forms, or densities. This construction is an abstraction of the long-standing tradition in solid mechanics where a surface is assigned two surface force fields depending on the orientation.

As a section of $L\big(W, \bigwedge^{n-1} T^*\mathcal{M}\big)$, a traction stress s is represented locally in the form $(x^i, s_{\alpha 1 \ldots \widehat{k} \ldots n}(x^j))$ or by

$$\sum_{k,\alpha} s_{\alpha 1 \ldots \widehat{k} \ldots n} g^{\alpha} \otimes (\mathrm{d}x^1 \wedge \cdots \wedge \widehat{\mathrm{d}x^k} \wedge \cdots \wedge \mathrm{d}x^n). \tag{10.15}$$

The transposed, s^{T}, is represented by

$$\sum_{k,\alpha} s_{\alpha 1 \ldots \widehat{k} \ldots n} (\mathrm{d}x^1 \wedge \cdots \wedge \widehat{\mathrm{d}x^k} \wedge \cdots \wedge \mathrm{d}x^n) \otimes g^{\alpha}, \tag{10.16}$$

and $s \cdot u$ is represented locally by

$$\sum_{k,\alpha} s_{\alpha 1 \ldots \widehat{k} \ldots n} u^{\alpha} \mathrm{d}x^1 \wedge \cdots \wedge \widehat{\mathrm{d}x^k} \wedge \cdots \wedge \mathrm{d}x^n. \tag{10.17}$$

10.3 The Power in Terms of the Traction Stresses

For a traction stress field s, the power of the induced surface force $\mathbf{t}_{\mathcal{R}}$ corresponding to a generalized velocity field u on $\partial\mathcal{R}$ may be written now as

$$\begin{aligned}\int_{\partial\mathcal{R}} \mathbf{t}_{\mathcal{R}} \cdot u &= \int_{\partial\mathcal{R}} \rho_{\partial\mathcal{R}}(s \cdot u), \\ &= \int_{\mathcal{R}} \mathrm{d}(s \cdot u),\end{aligned} \tag{10.18}$$

so that

$$F_{\mathcal{R}}(w) = \int_{\mathcal{R}} \mathbf{b}_{\mathcal{R}} \cdot w + \mathrm{d}(s \cdot w). \tag{10.19}$$

Using the local representation of $s \cdot w$ in (10.17), the local expression for $\mathrm{d}(s \cdot w)$ is

$$\begin{aligned}&\sum_{k,\alpha}(-1)^{k-1}\left(s_{\alpha 1\ldots\widehat{k}\ldots n} w^{\alpha}\right)_{,k} \mathrm{d}x^1 \wedge \cdots \wedge \mathrm{d}x^n \\ &\quad = \sum_{k,\alpha}(-1)^{k-1}\left(s_{\alpha 1\ldots\widehat{k}\ldots n,k} w^{\alpha} + s_{\alpha 1\ldots\widehat{k}\ldots n} w^{\alpha}_{,k}\right) \mathrm{d}x^1 \wedge \cdots \wedge \mathrm{d}x^n.\end{aligned} \tag{10.20}$$

It is observed therefore that in the expression (10.19) for the total power, the integrand depends linearly on both components of w and their derivatives. In other words, the integrand is linear in the jet, $j^1 w$, of the velocity field (see Sect. 6.1.8). In traditional continuum mechanics, one obtains that the power of the external forces is equal to the power that the stresses perform on the derivatives of the velocity fields. In the general geometry of manifolds, one cannot disassociate the values of the derivatives of the vector field from the values of the vector field itself.

10.4 Forces and Stresses for Kinematic Fluxes

Consider the case where a generalized velocity field represents the velocity field of material points in the physical space $\mathcal{S}$. For a configuration $\kappa : \mathcal{B} \to \mathcal{S}$, the field w is a section of the pullback of the tangent bundle $\kappa^*(T\mathcal{S})$ as in Example 6.5, and so $W = \kappa^*(T\mathcal{S})$. Hence, a traction stress is a section s of $L\left(\kappa^*(T\mathcal{S}), \bigwedge^{n-1} T^*\mathcal{B}\right)$ and it is represented locally in the form

$$\sum_{k,\alpha} s_{\alpha 1\ldots\widehat{k}\ldots n} \mathrm{d}y^{\alpha} \otimes (\mathrm{d}x^1 \wedge \cdots \wedge \widehat{\mathrm{d}x^k} \wedge \cdots \wedge \mathrm{d}x^n). \tag{10.21}$$

Here, $\{\mathrm{d}y^\alpha\}$ are the dual base vectors in $(\kappa^*(T\mathcal{S}))^*_x \cong T^*_{\kappa(x)}\mathcal{S}$ corresponding to the basis $\{\partial/\partial y^r\}$ of $(\kappa^*(T\mathcal{S}))_x \cong T_{\kappa(x)}\mathcal{S}$. Given a volume element θ on $\mathcal{B}$ and the isomorphism $\mathcal{J}_\theta : \bigwedge^{n-1}(T^*\mathcal{B}) \longrightarrow T\mathcal{B}$ as in (8.55), we may represent the stress s by $\mathcal{J}_\theta \circ s$, a section of $L\big(\kappa^*(T\mathcal{S}), T\mathcal{B}\big)$. Using (8.54), $\mathcal{J}_\theta \circ s$ is represented locally by

$$\sum_{k,\alpha}(-1)^{k-1}\frac{s_{\alpha 1\dots\widehat{k}\dots n}}{\theta_{1\dots n}}\mathrm{d}y^\alpha \otimes \frac{\partial}{\partial x^k}, \tag{10.22}$$

where $\theta_{1\dots n}\mathrm{d}x^1 \wedge \cdots \wedge \mathrm{d}x^n$ is the local representation of θ.

As observed by Noll [Nol59], for a Euclidean physical space, a reference configuration is a chart for the body manifold $\mathcal{B}$. If the stress were a tensor, a section of $L\big(\kappa^*(T\mathcal{S}), T\mathcal{B}\big)$, this would be reflected in its transformation rule under a change of chart. It is well known, however, that under a change of reference configuration (e.g., from the current configuration to some other reference configuration), the stress transforms using the Piola transform, which involves the Jacobian determinant. This supports the observation that the stress is not a tensor in the classical sense. A volume element θ (e.g., the one induced by a reference configuration, or one induced by a mass density distribution) enables the representation by a tensor. The last equation is analogous to the definition of the Piola-Kirchhoff stress, the counterpart of our s, on the basis of the Cauchy stress tensor, corresponding to $\mathcal{J}_\theta \circ s$.

10.5 Force Fields and Traction Stresses for Kinetic Fluxes

This section is concerned with the situation where a generalized velocity field is a kinetic flux J on $\mathcal{M}$ represented locally, as in Eq. (8.11), in the form

$$\sum_k J_{1\dots\widehat{k}\dots n}\mathrm{d}x^1 \wedge \cdots \wedge \widehat{\mathrm{d}x^k} \wedge \cdots \wedge \mathrm{d}x^n. \tag{10.23}$$

This situation is relevant if one does not assume the existence of material points a priori, but obtains the kinetic flux from the balance of some extensive property as in Chap. 9. In addition, given a volume element θ, a vector field v induces a kinetic flux $v \lrcorner \theta$. For example, in the case of classical continuum mechanics, one may use mass density as a volume element.

In light of Sect. 6.2.5, we have an isomorphism

$$\wedge_\lrcorner : \bigwedge{}^1 T^*\mathcal{M} \longrightarrow (\bigwedge{}^{n-1}T^*\mathcal{M})' := L(\bigwedge{}^{n-1}T^*\mathcal{M}, \bigwedge{}^n T^*\mathcal{M}) \tag{10.24}$$

whereby

$$\wedge_{\lrcorner}(\phi)(\omega) = \phi \wedge \omega. \tag{10.25}$$

For the case of generalized velocities modeled by kinetic fluxes, the action $\mathbf{b} \cdot J$ of a body force density $\mathbf{b}$ on a flux field is represented in the form $\mathbf{b} \cdot J = \alpha \wedge J$ for a unique 1-form α.

A traction stress will be a section of

$$L\big(\textstyle\bigwedge^{n-1} T^*\mathcal{M}, \bigwedge^{n-1} T^*\mathcal{M}\big) = \big(\textstyle\bigwedge^{n-1} T^*\mathcal{M}\big)^* \otimes \textstyle\bigwedge^{n-1} T^*\mathcal{M}. \tag{10.26}$$

Thus, the traction stress maps the flux of the extensive property, P, to the corresponding flux of power. From the discussion on multivectors, in particular Eq. (5.54), a basis of $\big(\bigwedge^{n-1} T^*\mathcal{M}\big)^*$ which is dual to the basis $\{\mathrm{d}x^1 \wedge \cdots \wedge \widehat{\mathrm{d}x^r} \wedge \cdots \wedge \mathrm{d}x^n\}_{r=1}^n$ is

$$\left\{\partial_1 \wedge \cdots \wedge \widehat{\partial_r} \wedge \cdots \wedge \partial_n\right\}_{r=1}^n. \tag{10.27}$$

Hence, s is represented locally in the form

$$\sum_{k,r} s_{1\ldots\widehat{k}\ldots n}^{1\ldots\widehat{r}\ldots,n}(\partial_1 \wedge \cdots \wedge \widehat{\partial_r} \wedge \cdots \wedge \partial_n) \otimes (\mathrm{d}x^1 \wedge \cdots \wedge \widehat{\mathrm{d}x^k} \wedge \cdots \wedge \mathrm{d}x^n), \tag{10.28}$$

and $s \cdot J$ is represented locally as

$$\sum_{k,r} s_{1\ldots\widehat{k}\ldots n}^{1\ldots\widehat{r}\ldots,n} J_{1\ldots\widehat{r}\ldots n} \mathrm{d}x^1 \wedge \cdots \wedge \widehat{\mathrm{d}x^k} \wedge \cdots \wedge \mathrm{d}x^n. \tag{10.29}$$

Let θ be a given volume element on $\mathcal{M}$. Then, θ induces an isomorphism

$$L\big(\textstyle\bigwedge^{n-1} T^*\mathcal{M}, \bigwedge^{n-1} T^*\mathcal{M}\big) \longrightarrow L\big(T\mathcal{M}, T\mathcal{M}\big) \tag{10.30}$$

given by

$$s \longmapsto \mathcal{J}_\theta \circ s \circ \mathcal{J}_\theta^{-1}, \tag{10.31}$$

where $\mathcal{J}_\theta : \bigwedge^{n-1} T^*\mathcal{M} \longrightarrow T\mathcal{M}$ is the isomorphism defined in Example 5.4. Thus, a volume element enables the representation of a traction stress that is dual to kinetic flux fields by a section of the bundle of linear mappings $T\mathcal{M} \to T\mathcal{M}$. Let $\tilde{s} = \mathcal{J}_\theta \circ s \circ \mathcal{J}_\theta^{-1}$ be expressed locally as

$$\sum_{k,r} \tilde{s}^k_{\cdot r} \mathrm{d}x^r \otimes \partial_k, \tag{10.32}$$

so that $\tilde{s}^k_{\cdot r} = \mathrm{d}x^k \cdot (\tilde{s} \cdot \partial_k)$. Using (8.54), one has

$$
\begin{aligned}
\tilde{s}^k_{\cdot r} &= \mathrm{d}x^k \cdot ((\mathcal{J}_\theta \circ s \circ \mathcal{J}_\theta^{-1}) \cdot \partial_k), \\
&= (-1)^{r-1}\theta_{1\dots n}\mathrm{d}x^k \left(\mathcal{J}_\theta \circ s(\mathrm{d}x^1 \wedge \cdots \wedge \widehat{\mathrm{d}x^r} \wedge \cdots \wedge \mathrm{d}x^n)\right), \\
&= (-1)^{r-1}\theta_{1\dots n}\mathrm{d}x^k \mathcal{J}_\theta \left(\sum_l s^{1\dots\widehat{r}\dots,n}_{1\dots\widehat{l}\dots n}\mathrm{d}x^1 \wedge \cdots \wedge \widehat{\mathrm{d}x^l} \wedge \cdots \wedge \mathrm{d}x^n\right), \\
&= (-1)^{r-1}\theta_{1\dots n}\sum_l \mathrm{d}x^k \left(\frac{(-1)^{l-1}}{\theta_{1\dots n}} s^{1\dots\widehat{r}\dots,n}_{1\dots\widehat{l}\dots n}\partial_l\right), \\
&= (-1)^{r+k} s^{1\dots\widehat{r}\dots,n}_{1\dots\widehat{k}\dots n}.
\end{aligned}
\tag{10.33}
$$

It follows that the isomorphism of Eq. (10.30) is actually independent of the choice of volume element. In other words, the equation above represents a natural isomorphism between $L\big(\bigwedge^{n-1}T^*\mathcal{M}, \bigwedge^{n-1}T^*\mathcal{M}\big)$ and $L\big(T\mathcal{M}, T\mathcal{M}\big) = T^*\mathcal{M} \otimes T\mathcal{M}$.

An isomorphism

$$
\mathcal{J} : T^*\mathcal{M} \otimes T\mathcal{M} \longrightarrow L\big(\textstyle\bigwedge^{n-1}T^*\mathcal{M}, \bigwedge^{n-1}T^*\mathcal{M}\big) \tag{10.34}
$$

can be constructed, without using provisionally a volume element, as follows: Consider $\tilde{s} \in T^*\mathcal{M} \otimes T\mathcal{M}$ represented by $\sum_{k,r} \tilde{s}^k_{\cdot r}\mathrm{d}x^r \otimes \partial_k$. For any flux form J, represented as in (10.23), set locally

$$
\mathcal{J}(\tilde{s}) \cdot J = \sum_{k,r} \tilde{s}^k_{\cdot r}\partial_k \lrcorner (\mathrm{d}x^r \wedge J). \tag{10.35}
$$

Since the exterior product is bilinear and the contraction is linear, $\mathcal{J}(\tilde{s})$ is linear in J. To show that Eq. (10.33) is indeed a representation of $\mathcal{J}$ as defined by (10.35), we use the representation (10.23) and obtain

$$
\begin{aligned}
\mathcal{J}(\tilde{s})(J) &= \sum_{k,r} \tilde{s}^k_{\cdot r}\partial_k \lrcorner \left(\mathrm{d}x^r \wedge \left(\sum_p J_{1\dots\widehat{p}\dots n}\mathrm{d}x^1 \wedge \cdots \wedge \widehat{\mathrm{d}x^p} \wedge \cdots \wedge \mathrm{d}x^n\right)\right), \\
&= \sum_{k,r} \tilde{s}^k_{\cdot r}\partial_k \lrcorner \left((-1)^{r-1}J_{1\dots\widehat{r}\dots n}\mathrm{d}x^1 \wedge \cdots \wedge \mathrm{d}x^n\right), \\
&= \sum_{k,r} \tilde{s}^k_{\cdot r}(-1)^{r-1}(-1)^{k-1}J_{1\dots\widehat{r}\dots n}\mathrm{d}x^1 \wedge \cdots \wedge \widehat{\mathrm{d}x^k} \wedge \cdots \wedge \mathrm{d}x^n, \\
&= \sum_{k,r} (-1)^{r+k}\tilde{s}^k_{\cdot r}J_{1\dots\widehat{r}\dots n}\mathrm{d}x^1 \wedge \cdots \wedge \widehat{\mathrm{d}x^k} \wedge \cdots \wedge \mathrm{d}x^n.
\end{aligned}
\tag{10.36}
$$

Comparing the last equation with (10.29), the result of Eq. (10.33) follows.

Thus, in spite of the various generalizations pertaining to the other variables, for the case where $(n-1)$-flux forms represent generalized velocities, the traction stress is represented as a tensor in the traditional sense. For example, if we replace the generalized velocity vector field of continuum mechanics with a flux form representing linear momentum, the traction stress will be represented by a tensor field transforming the flux of momentum to the flux of power.

10.6 Cauchy's Theorem for Traction Stresses

In analogy with Sect. 8.1.4, we lay down the assumptions that will be used in the proof of Cauchy's theorem for traction stresses on manifolds. Then, we state the theorem and prove it.

We consider the collection $\{(\mathbf{b}_{\mathcal{R}}, \mathbf{t}_{\mathcal{R}})\}$ for the various admissible regions $\mathcal{R} \subset \mathcal{M}$ so that each $\mathbf{b}_{\mathcal{R}}$ is a smooth section of $L(W, \bigwedge^n T^*\mathcal{R})$ and each $\mathbf{t}_{\mathcal{R}}$ is a smooth section of $L(W|_{\partial\mathcal{R}}, \bigwedge^{n-1} T^*\partial\mathcal{R})$.

Assumption 1: Boundedness
There is a section ξ of $(J^1W)' := L(J^1W, \bigwedge^n T^*\mathcal{M})$ such that for each $\mathcal{R}$,

$$\left| \int_{\partial\mathcal{R}} \mathbf{t}_{\mathcal{R}} \cdot w|_{\partial\mathcal{R}} \right| \leqslant \int_{\mathcal{R}} \left| \xi \cdot j^1 w \right|, \tag{10.37}$$

for every smooth section of W.

In the sequel, we will use the notation $\mathbf{t}_{\mathcal{R}} \cdot w$ freely and will omit the indication of the restriction to the boundary. Note that for the case of (scalar-valued) fluxes, the vector bundle is trivial, and for all $x \in \mathcal{M}$, $W_x = \mathbb{R}$. Thus, one may choose the function $w(x) = 1$ for all $x \in \mathcal{M}$, the derivative of which vanishes, so that its jet may be represented by the number 1. Therefore, the boundedness assumption for fluxes (8.6) is a special case of (10.37).

Remark 10.1 In view of Sect. 6.1.8, a section ξ of $L(J^1W, \bigwedge^n T^*\mathcal{M})$ is represented locally in the form

$$(x^i) \longmapsto (x^i, \xi_{\alpha 1\dots n}(x^i), \Xi^j_{\alpha 1\dots n}(x^i)) \tag{10.38}$$

so that the n-form $\xi \cdot j^1 w$ is represented locally as

$$\sum_{\alpha, j} (\xi_{\alpha 1\dots n} w^\alpha + \Xi^j_{\alpha 1\dots n} w^\alpha_{,j}) \mathrm{d}x^1 \wedge \cdots \wedge \mathrm{d}x^n. \tag{10.39}$$

Cauchy's postulate of locality for stresses is a straightforward analog of the one corresponding to fluxes.

Assumption 2: Cauchy's Postulate of Locality
(*i*) Let $x \in \mathcal{M}$, and let $v_1, \dots, v_{n-1} \in T_x\mathcal{M}$ be a collection of $n-1$ vectors. Let $\mathcal{R}$ be any region such that $x \in \partial\mathcal{R}$, $v_1, \dots, v_{n-1} \in T_x\partial\mathcal{R}$, and the

collection of vectors is positively oriented relative to the orientation of $\partial\mathcal{R}$. Then, $\mathbf{t}^{\mathsf{T}}_{\mathcal{R}}(x)(v_1, \ldots, v_{n-1})$ is independent of $\mathcal{R}$.

(*ii*) The second locality postulate pertains to the body force. It is assumed that $\mathbf{b}_{\mathcal{R}}(x)$, $x \in \mathcal{M}$ is independent of $\mathcal{R}$.

Cauchy's postulate implies that for the given point $x \in \mathcal{M}$, $\mathbf{t}_{\mathcal{R}}(x)$ depends on $\mathcal{R}$ only through the tangent space $T_x\partial\mathcal{R}$ and the orientation induced on it.

This assumption makes it possible to define a mapping

$$\Sigma_x : (T_x\mathcal{M})^{n-1} \longrightarrow W^*_x, \tag{10.40}$$

such that for any collection $v_1, \ldots, v_{n-1} \in T_x\mathcal{M}$,

$$\Sigma_x(v_1, \ldots, v_{n-1}) = \mathbf{t}^{\mathsf{T}}_{\mathcal{R}}(x)(v_1, \ldots, v_{n-1}) \tag{10.41}$$

for any region $\mathcal{R}$ satisfying the condition that the collection $(v_1, \ldots, v_{n-1}) \in (T_x\partial\mathcal{R})^{n-1}$ is positively oriented. Thus, the order in which the collection is written specifies the orientation for which the action of Σ_x applies. At this point, for an odd permutation π, there is no relation between $\Sigma_x(v_1, \ldots, v_{n-1})$ and $\Sigma_x(v_{\pi_1}, \ldots, v_{\pi_{n-1}})$. It is noted that $\Sigma_x(v_1, \ldots, v_{n-1})$ depends only on the multivector $\mathfrak{v} = v_1 \wedge \cdots \wedge v_{n-1}$. Thus, in analogy with (8.19) for fluxes, we have a fiber preserving mapping between vector bundles,

$$\Sigma : T^{n-1}\mathcal{M} \longrightarrow W^*, \qquad \text{such that} \quad \Sigma|_{(T_x\mathcal{M})^{n-1}} = \Sigma_x. \tag{10.42}$$

In view of the locality assumption (*ii*), we will simply use $\mathbf{b}$ for the body force in what follows. This postulate may be somewhat weakened by assuming that the dependence on $\mathcal{R}$ is bounded, but we will not explore this generalization any further.

Assumption 3: Regularity

The mapping Σ is smooth.

We refer to the collection $\{\mathbf{t}_{\mathcal{R}}\}$ for the various regions $\mathcal{R}$ satisfying the foregoing assumptions as a *system of Cauchy surface forces*. These assumptions make it possible to prove Cauchy's theorem for traction stresses on manifolds, i.e., to prove that there exists a unique traction stress s such that (10.11) and the equivalent (10.12) hold.

Consider a generic point $x_0 \in \mathcal{M}$ and a vector bundle chart (U, φ, Φ) with coordinates (x^i, u^α), $i = 1, \ldots, n$, $\alpha = 1, \ldots, m$, in an open neighborhood U of x_0. The local base vectors in $\pi^{-1}\{U\}$ will be denoted as $\{g_\alpha\}$ so that the restriction of a section w of W to U may be written in the form $\sum_\alpha w^\alpha(x^i)g_\alpha$.

We want to prove that Σ_{x_0} is multilinear and alternating, i.e., that $\mathbf{t}^{\mathsf{T}}_{\mathcal{R}}(x_0)$ is the restriction to $(T_{x_0}\partial\mathcal{R})^{n-1}$ of a multilinear alternating mapping on $(T_{x_0}\mathcal{M})^{n-1}$. With this objective in mind, we construct for each $\alpha = 1, \ldots, m$ a Cauchy flux system $\{\tau_{\alpha\mathcal{R}}\}$ for which Cauchy's theorem for fluxes of Sect. 8.1.4 may be applied.

Let $\psi : U \to [0, 1]$ be a smooth function having the following properties: (a) ψ is compactly supported in U. (b) There is an open neighborhood V of x_0 such that

$V \subset U$, and for each $x \in V$, $\psi(x) = 1$. Thus, the cutoff function ψ decreases from 1 to 0 in $U \setminus V$. For each $\alpha = 1, \ldots, m$, we define the section $w_{(\alpha)}$ of W by

$$w_{(\alpha)}(x) = \begin{cases} \psi(x) g_\alpha, & \text{for } x \in U, \\ 0, & \text{for } x \notin U. \end{cases} \tag{10.43}$$

By the definition of the cutoff function ψ, $w_{(\alpha)}$ is evidently a smooth field. Letting $\tilde{\psi}$ be the local representative of ψ in U, then, locally,

$$w^\beta_{(\alpha)}(x^i) = \tilde{\psi}(x^i)\delta^\beta_\alpha, \qquad w^\beta_{(\alpha),j} = \tilde{\psi}_{,j}\delta^\beta_\alpha. \tag{10.44}$$

For each region $\mathcal{R}$ and $\alpha = 1, \ldots, m$, consider the flux distribution $\tau_{\alpha\mathcal{R}}$ defined by

$$\tau_{\alpha\mathcal{R}} = \mathbf{t}_\mathcal{R} \cdot w_{(\alpha)}, \tag{10.45}$$

so that for a positively oriented collection $v_1, \ldots, v_{n-1}$ of vectors, tangent to $\partial\mathcal{R}$, one has

$$\begin{aligned} \tau_{\alpha\mathcal{R}}(x)(v_1, \ldots, v_{n-1}) &= \mathbf{t}_\mathcal{R}(x)(w_{(\alpha)}(x))(v_1, \ldots, v_{n-1}), \\ &= \Sigma_x(v_1, \ldots, v_{n-1})(w_{(\alpha)}(x)). \end{aligned} \tag{10.46}$$

Clearly, $\tau_{\alpha\mathcal{R}}(x)$ vanishes outside U.

To show that for each $\alpha = 1, \ldots, m$, $\{\tau_{\alpha\mathcal{R}}\}$ satisfies the boundedness condition of Sect. 8.1.4, we observe that the boundedness assumption for forces, (10.37), implies

$$\begin{aligned} \left|\int_{\partial\mathcal{R}} \tau_{\alpha\mathcal{R}}\right| &= \left|\int_{\partial\mathcal{R}} \mathbf{t}_\mathcal{R} \cdot w_{(\alpha)}\right|, \\ &\leqslant \int_\mathcal{R} \left|\xi \cdot j^1 w_{(\alpha)}\right|, \\ &= \int_{\mathcal{R}\cap U} \left|\xi \cdot j^1 w_{(\alpha)}\right|, \\ &= \int_{\varphi(\mathcal{R}\cap U)} \left|\xi_{\alpha 1\ldots n}\tilde{\psi} + \textstyle\sum_j \Xi^j_{\alpha 1\ldots n}\tilde{\psi}_{,j}\right| \mathrm{d}x^1 \cdots \mathrm{d}x^n, \\ &\leqslant \int_{\varphi(\mathcal{R}\cap U)} C\Big(|\xi_{\alpha 1\ldots n}| + \textstyle\sum_j \left|\Xi^j_{\alpha 1\ldots n}\right|\Big) \mathrm{d}x^1 \cdots \mathrm{d}x^n, \\ &\leqslant \int_{\varphi(\mathcal{R}\cap U)} \varsigma_{(\alpha)1\ldots n}\mathrm{d}x^1 \cdots \mathrm{d}x^n, \\ &= \int_\mathcal{R} \varsigma_{(\alpha)}, \end{aligned} \tag{10.47}$$

where in the fifth line, the constant C exists because $\tilde{\psi}_{,i}$ is different from zero in a compact set and $\max_x \psi(x) = 1$. In the sixth line,

$$\varsigma_{(\alpha)1\dots n}(x^i) := C\Big(\Big|\xi_{\alpha 1\dots n}(x^i)\Big| + \sum_j \big|\Xi^j_{\alpha 1\dots n}(x^i)\big|\Big), \qquad x \in U, \tag{10.48}$$

and $\varsigma_{(\alpha)}$ is the n-form on $\mathscr{M}$ such that $\varsigma(x)$ is represented by

$$\begin{cases} \varsigma_{(\alpha)1\dots n}(x^i)\mathrm{d}x^1 \wedge \cdots \wedge \mathrm{d}x^n, & x \in U, \\ \varsigma_{(\alpha)1\dots n}(x^i) = 0, & x \notin U. \end{cases} \tag{10.49}$$

The flux system $\{\tau_{\alpha\mathscr{R}}\}$ satisfies the locality assumption because

$$\begin{aligned} \tau_{\alpha\mathscr{R}}(x)(v_1, \dots, v_{n-1}) &= (\mathbf{t}_{\mathscr{R}}(x)(w_{(\alpha)}))(v_1, \dots, v_{n-1}), \\ &= (\mathbf{t}^{\mathsf{T}}_{\mathscr{R}}(x)(v_1, \dots, v_{n-1}))(w_{(\alpha)}), \end{aligned} \tag{10.50}$$

which, by the locality assumption for surface forces, is independent of $\mathscr{R}$ as long as $(v_1, \dots, v_{n-1})$ is a positively oriented collection of vectors in $T_x\partial\mathscr{R}$.

Let $\mathrm{t}_\alpha : T^{n-1}\mathscr{M} \longrightarrow \mathbb{R}$ be the Cauchy-Whitney mapping corresponding to the flux system $\{\tau_{\alpha\mathscr{R}}\}$ as in Eq. (8.19). We have to show that t_α is smooth. Independently of the region $\mathscr{R}$, as long as the collection of tangent vectors $(v_1, \dots, v_{n-1}) \in (T_x\mathscr{M})^{n-1}$ is positively oriented in $\partial\mathscr{R}$, one has

$$\begin{aligned} \mathrm{t}_\alpha(v_1, \dots, v_{n-1}) &= \tau_{\alpha\mathscr{R}}(x)(v_1, \dots, v_{n-1}), \\ &= \mathbf{t}^{\mathsf{T}}_{\mathscr{R}}(x)(v_1, \dots, v_{n-1})(w_{(\alpha)}(x)), \\ &= \Sigma(v_1, \dots, v_{n-1})(w_{(\alpha)}(x)). \end{aligned} \tag{10.51}$$

Since Σ is assumed to be smooth and $w_{(\alpha)}$ is smooth by its construction, so is t_α. (This may be easily verified using local expressions.)

Cauchy's theorem for fluxes implies now that for each α there is a flux form $s_\alpha \in \Omega^{n-1}(\mathscr{M})$ such that $\tau_{\alpha\mathscr{R}}(x_0) = \rho_{\partial\mathscr{R}}(s_\alpha(x_0))$. Using the definition (10.45),

$$\begin{aligned} \mathbf{t}_{\mathscr{R}}(x_0)(g_\alpha) &= \mathbf{t}_{\mathscr{R}}(x_0)(w_{(\alpha)}(x_0)), \\ &= \rho_{\partial\mathscr{R}}(s_\alpha(x_0)), \end{aligned} \tag{10.52}$$

and it follows that for $w_0 = w(x_0)$, represented by $\sum_\alpha w_0^\alpha g_\alpha$,

$$\begin{aligned}\mathbf{t}_{\mathcal{R}}(x_0)(w_0) &= \mathbf{t}_{\mathcal{R}}(x_0)\big(\textstyle\sum_\alpha w_0^\alpha g_\alpha\big),\\ &= \sum_\alpha w_0^\alpha \mathbf{t}_{\mathcal{R}}(x_0)(g_\alpha),\\ &= \sum_\alpha w_0^\alpha \rho_{\partial\mathcal{R}}(s_\alpha(x_0)).\end{aligned} \tag{10.53}$$

Define the section s of $L\big(W, \bigwedge^{n-1}T^*\mathcal{M}\big)$ by setting, for $x_0 \in U$,

$$(s \cdot w)(x_0) := s(x_0)(w(x_0)) := \sum_\alpha w^\alpha(x_0)s_\alpha(x_0), \tag{10.54}$$

which is a valid expression because s_α is supported in U. Hence, we obtain for any $w_0 \in W_{x_0}$,

$$\begin{aligned}\rho_{\partial\mathcal{R}}(s^{\mathsf{T}}(x_0))(v_1,\dots,v_{n-1})(w_0) &= s(x_0)(w_0)(T\mathcal{I}_{\partial\mathcal{R}}(v_1),\dots,T\mathcal{I}_{\partial\mathcal{R}}(v_{n-1})),\\ &= \sum_\alpha w_0^\alpha \rho_{\partial\mathcal{R}}(s_\alpha(x_0))(v_1,\dots,v_{n-1}),\\ &= \mathbf{t}^{\mathsf{T}}_{\mathcal{R}}(x_0)(v_1,\dots,v_{n-1})(w_0),\end{aligned} \tag{10.55}$$

which is the required Cauchy formula (10.11) and is independent of the chart. It is noted that although the foregoing construction depends on x_0 and the chart, the locality assumption makes it valid. Finally, the uniqueness of flux forms (cf. the beginning of Sect. 8.1.4) implies that the traction stress induced by a system of Cauchy surface forces is unique.

10.7 Variational Stresses

Continuing the observations made in Sect. 10.3, we conclude that the total power of the body force and surface force for a region $\mathcal{R}$ is given by

$$F_{\mathcal{R}}(w) = \int_{\mathcal{R}} \mathbf{b} \cdot w + \mathrm{d}(s \cdot w). \tag{10.56}$$

The local expression for the integrand is

$$\sum_{k,\alpha}\big[\big(\mathbf{b}_{\alpha 1\dots n} + (-1)^{k-1}s_{\alpha 1\dots\widehat{k}\dots n,k}\big)w^\alpha + (-1)^{k-1}s_{\alpha 1\dots\widehat{k}\dots n}w^\alpha_{,k}\big]\mathrm{d}x^1 \wedge\dots\wedge \mathrm{d}x^n. \tag{10.57}$$

It is noted that the values of the n-form in the expression above are linear in the local representatives $(w^k, w^\alpha_{,i})$ (cf. Sect. 6.1.8) of j^1w. Thus, recalling the definition of density-dual spaces in Sect. 6.2.5, there is a section S of $(J^1W)' := L(J^1W, \bigwedge^n T^*\mathcal{M})$ such that

$$\mathbf{b} \cdot w + \mathrm{d}(s \cdot w) = S \cdot j^1 w. \tag{10.58}$$

We will refer to S as the *variational stress*, or, more accurately, as the *variational stress density field*. This may be summarized by

$$F_{\mathcal{R}}(w) = \int_{\mathcal{R}} \mathbf{b} \cdot w + \mathrm{d}(s \cdot w) = \int_{\mathcal{R}} \mathbf{b} \cdot w + \int_{\partial\mathcal{R}} \mathbf{t}_{\mathcal{R}} \cdot w = \int_{\mathcal{R}} S \cdot j^1 w, \tag{10.59}$$

which is the metric independent version of the *principle of virtual power* (or *virtual work*). (Compare the principle of virtual power with the boundedness assumption in Eq. (10.37).) In contrast with the classical version of the principle of virtual work, for the metric independent analysis, one has to make a distinction between the traction stress s that induces the surface traction using the generalized Cauchy formula and the variational stress which determines the density of the power via the principle of virtual power.

The last equation exhibits the major role played by the stress fields. A variational stress field is a means for restricting a force from a body to its subbodies. In other words, a variational stress field induces a force system over the body.

In view of Remark 10.1, S is represented locally in the form $(x^i, R_{\alpha 1 \dots n}, S^k_{\alpha 1 \dots n})$, or equivalently

$$\left(\sum_{\alpha} R_{\alpha 1 \dots n} g^{\alpha} + \sum_{k,\alpha} S^k_{\alpha 1 \dots n} \partial_k \otimes g^{\alpha}\right) \otimes (\mathrm{d}x^1 \wedge \cdots \wedge \mathrm{d}x^n), \tag{10.60}$$

so that $S \cdot j^1 w$ is represented locally by

$$\left(\sum_{\alpha} R_{\alpha 1 \dots n} w^{\alpha} + \sum_{k,\alpha} S^k_{\alpha 1 \dots n} w^{\alpha}_{,k}\right) \mathrm{d}x^1 \wedge \cdots \wedge \mathrm{d}x^n. \tag{10.61}$$

Since the principle of virtual power holds for any region $\mathcal{R}$ and virtual velocity field w, it follows from the smoothness of the various fields and the local representation of (10.57) and (10.61) that

$$\begin{aligned} R_{\alpha 1 \dots n} &= \mathbf{b}_{\alpha 1 \dots n} + \sum_{k} (-1)^{k-1} s_{\alpha 1 \dots \widehat{k} \dots n, k}, \\ S^k_{\alpha 1 \dots n} &= (-1)^{k-1} s_{\alpha 1 \dots \widehat{k} \dots n}. \end{aligned} \tag{10.62}$$

Using the second equation above, the first equation may be written in the form

$$\mathbf{b}_{\alpha 1 \dots n} = R_{\alpha 1 \dots n} - \sum_{k} S^k_{\alpha 1 \dots n, k}. \tag{10.63}$$

As a section of $(J^1W)'$, a variational stress S induces the vector-valued form $S^{\mathsf{T}} \in \Omega^n(T^*\mathcal{M}, (J^1W)^*)$, which assumes values in the dual of the jet bundle, and is defined by

$$(S^{\mathsf{T}} \cdot (v_1, \ldots, v_n)) \cdot \chi = (S \cdot \chi) \cdot (v_1, \ldots, v_n). \tag{10.64}$$

It is noted that for a velocity field w, the components $w^\alpha_{,k}$ of the jet j^1w do not represent an invariant geometric object unless they are in conjunction with the components w^i. On the other hand, the second of Eqs. (10.62) and the fact that the traction stress is an invariant object imply that the dual components $S^k_{\alpha 1\ldots n}$ do represent an invariant object which we describe below. In the terminology of [Pal68], the components $S^k_{\alpha 1,\ldots,n}$ represent the symbol of S, where S is viewed as a linear differential operator.

To present the relations without recourse to local representation, we construct a surjective vector bundle morphism

$$p_s : (J^1W)' \longrightarrow L\Big(W, \textstyle\bigwedge^{n-1}T^*\mathcal{M}\Big), \tag{10.65}$$

which associates with every variational stress S a traction stress $s = p_s \circ S$ that we also write as $p_s(S)$.

Using the jet projection mapping $\pi^1_0 : J^1W \to W$, consider the *vertical subbundle*, $V^1_0J^1W = \operatorname{Kernel}\pi^1_0$. Since π^1_0 is represented locally by $(x^i, u^\alpha, A^\beta_k) \mapsto (x^i, u^\alpha)$, an element of $V^1_0J^1W$ is represented locally in the form $(x^i, 0, A^\beta_k)$. In other words, the fiber $(V^1_0J^*W)_{x_0}$ contains jets of sections of W that vanish at x_0. It is noted that there is a natural vector bundle isomorphism

$$V^1_0J^1W \cong L\big(T\mathcal{M}, W\big) \cong T^*\mathcal{M} \otimes W, \tag{10.66}$$

by which an element of $V^1_0J^1W$ is represented in the form (x^i, A^β_k). We use $\mathcal{I}_{V^1_0} : V^1_0J^1W \hookrightarrow J^1W$ to denote the inclusion of the vertical subbundle—a vector bundle morphism over $\mathcal{M}$ represented by $(x^i, A^\beta_k) \mapsto (x^i, 0, A^\beta_k)$. Then, the dual vector bundle morphism

$$\mathcal{I}^*_{V^1_0} : (J^1W)^* \longrightarrow (V^1_0J^1W)^* \cong T\mathcal{M} \otimes W, \tag{10.67}$$

is a projection represented locally in the form $(x^i, \xi_\alpha, \Xi^i_\beta) \mapsto (x^i, \Xi^i_\beta)$—the restriction of $\xi \in (J^1W)^*$ to vertical elements of the jet bundle. Thus, $\mathcal{I}^*_{V^1_0}(\xi)(A)$, $A \in V^1_0J^1W$ is represented by $\sum_{i,\beta} \Xi^i_\beta A^\beta_i$. Similarly, recalling the definition of a density-dual mapping in (6.122),

$$\mathcal{I}'_{V^1_0} : (J^1W)' \longrightarrow (V^1_0J^1W)', \tag{10.68}$$

and for a section S of $(J^1W)'$, $\mathcal{I}'_{V_0^1}(S)$, a section of $(V_0^1J^1W)'$, is given by $\mathcal{I}'_{V_0^1}(S(x))(A) = S(x)(\mathcal{I}_{V_0^1}(A)) \in \bigwedge^n T_x^*\mathcal{M}$. The evaluation $\mathcal{I}'_{V_0^1}(S)(x)(A)$ is represented by $\sum_{k,\alpha} S^k_{\alpha 1\ldots n}(x)A^\alpha_k \mathrm{d}x^1 \wedge \cdots \wedge \mathrm{d}x^n$ and so $\mathcal{I}'_{V_0^1}(S)$ is represented in the form

$$\sum_{k,\alpha} S^k_{\alpha 1\ldots n}\partial_k \otimes g^\alpha \otimes (\mathrm{d}x^1 \wedge \cdots \wedge \mathrm{d}x^n). \tag{10.69}$$

For a given variational stress distribution S, we will refer to $S^+ := \mathcal{I}'_{V_0^1} \circ S$, represented locally as in the previous equation, as the *principal component* of S. As mentioned above, the principal component $S^+ := \mathcal{I}'_{V_0^1}(S)$ is the symbol of the linear differential operator S as defined in [Pal68].

Using the isomorphism $V_0^1J^1W \cong L(T\mathcal{M}, W)$, we regard $S^+ = \mathcal{I}'_{V_0^1}(S)$ as a section of

$$\begin{aligned} L(T\mathcal{M}, W)' &\cong L(T\mathcal{M}, W)^* \otimes \textstyle\bigwedge^n T^*\mathcal{M}, \\ &\cong (T^*M \otimes W)^* \otimes \textstyle\bigwedge^n T^*\mathcal{M}, \\ &\cong T\mathcal{M} \otimes W^* \otimes \textstyle\bigwedge^n T^*\mathcal{M}. \end{aligned} \tag{10.70}$$

It follows that a section of $L(T\mathcal{M}, W)'$ may be represented locally in the form $\sum_a v_a \otimes \phi^a \otimes \theta$ for an n-form θ and pairs v_a, ϕ^a of sections of $T\mathcal{M}$ and W^*, respectively. We can use the contraction of the first and third factors in the product to obtain $\sum_a \phi^a \otimes (v_a \lrcorner \theta)$. Thus, we have a natural mapping

$$\begin{aligned} \mathsf{C} : L(T\mathcal{M}, W)' &\longrightarrow L(W, \textstyle\bigwedge^{n-1} T^*\mathcal{M}) \\ &\cong W^* \otimes \textstyle\bigwedge^{n-1}(T^*\mathcal{M}). \end{aligned} \tag{10.71}$$

It follows that

$$p_s := \mathsf{C} \circ \mathcal{I}'_{V_0^1} : (J^1W)' \longrightarrow L(W, \textstyle\bigwedge^{n-1} T^*\mathcal{M}) \tag{10.72}$$

associates a traction stress $s = p_s \circ S$ with any variational stress S. Using (10.69) and Sect. 5.8, C is represented locally by

$$\begin{aligned} \sum_{k,\alpha} S^k_{\alpha 1\ldots n}\partial_k \otimes g^\alpha \otimes (\mathrm{d}x^1 \wedge \cdots \wedge \mathrm{d}x^n) &\longmapsto \sum_{k,\alpha} S^k_{\alpha 1\ldots n} g^\alpha \otimes (\partial_k \lrcorner (\mathrm{d}x^1 \wedge \cdots \wedge \mathrm{d}x^n)), \\ &= \sum_{k,\alpha} (-1)^{k-1} S^k_{\alpha 1\ldots n} g^\alpha \otimes (\mathrm{d}x^1 \wedge \cdots \wedge \widehat{\mathrm{d}x^k} \wedge \cdots \wedge \mathrm{d}x^n). \end{aligned} \tag{10.73}$$

Note that the local expression implies immediately that C is an isomorphism of vector bundles over $\mathscr{M}$.

We conclude that

$$s = p_s(S) = \mathsf{C}(S^+), \tag{10.74}$$

is represented as in (10.62), as expected.

10.8 The Divergence of a Variational Stress Field

We have seen in the previous section that a given variational stress S induces a unique traction stress field $s = p_s(S)$ which, in turn, induces a surface force field on the boundary of each region using the Cauchy formula (10.11). In this section, we define a generalization of the divergence operator of classical vector analysis that may be applied to a variational stress field to determine a body force field. As indicated in the first of (10.62) and again below, the traction stress field does not contain, in the general geometric setting considered here, all the information needed in order to determine the body force.

The *divergence*, $\operatorname{div} S$, of the variational stress field S is a section of $W' := L\big(W, \bigwedge^n T^*\mathscr{M}\big)$ which is defined invariantly by

$$\operatorname{div} S \cdot w = \mathrm{d}(p_s(S) \cdot w) - S \cdot j^1 w, \tag{10.75}$$

for every differentiable vector field w.

We now study the divergence in further detail. Define the mapping

$$\mathfrak{d} : C^\infty(L(W, \textstyle\bigwedge^{n-1} T^*\mathscr{M})) \longrightarrow C^\infty((J^1 W)'), \tag{10.76}$$

between the spaces of smooth sections of the corresponding bundles, by

$$(\mathfrak{d}(s))(j^1 w) := \mathrm{d}(s \cdot w). \tag{10.77}$$

The mapping $\mathfrak{d}$ acts on traction stress fields to produce variational stress fields. Then, we may rewrite the definition of the divergence as

$$\operatorname{div} S = \mathfrak{d}(p_s(S)) - S \circ j^1. \tag{10.78}$$

To show that the mapping $\mathfrak{d}$ and the divergence are well-defined, we first note that $\mathrm{d}(s \cdot w)$ is represented locally by

$$\sum_{k,\alpha} \mathrm{d}(s_{\alpha 1 \dots \widehat{k} \dots n} w^\alpha) \wedge \mathrm{d}x^1 \wedge \dots \wedge \widehat{\mathrm{d}x^k} \wedge \dots \wedge \mathrm{d}x^n$$

$$
\begin{aligned}
&= \sum_{k,\alpha} (s_{\alpha 1 \dots \widehat{k} \dots n} w^\alpha)_{,k} \mathrm{d}x^k \wedge \mathrm{d}x^1 \wedge \cdots \wedge \widehat{\mathrm{d}x^k} \wedge \cdots \wedge \mathrm{d}x^n, \\
&= \sum_{k,\alpha} (s_{\alpha 1 \dots \widehat{k} \dots n} w^\alpha)_{,k} (-1)^{k-1} \mathrm{d}x^1 \wedge \cdots \wedge \mathrm{d}x^n, \\
&= \sum_{k,\alpha} (s_{\alpha 1 \dots \widehat{k} \dots n,k} w^\alpha + s_{\alpha 1 \dots \widehat{k} \dots n} w^\alpha_{,k}) (-1)^{k-1} \mathrm{d}x.
\end{aligned}
\tag{10.79}
$$

Hence, $\mathfrak{d}(s)$ is indeed a section of $(J^1 W)'$, represented by $(s_{\alpha 1 \dots \widehat{k} \dots n,k}, s_{\alpha 1 \dots \widehat{k} \dots n})$. In addition, in view of the local expression, the action of $\mathfrak{d}s$ may be factored as

$$
w \xmapsto{j^1} j^1 w \xmapsto{\widetilde{\mathfrak{d}s}} \mathrm{d}(s \cdot w), \tag{10.80}
$$

where $\widetilde{\mathfrak{d}s}$ is a vector bundle morphism $J^1 W \to \bigwedge^n T^* \mathcal{M}$, over $\mathcal{M}$. Moreover, if $s = p_s(S)$, then, by Eq. (10.62), locally,

$$
\mathrm{d}(s \cdot w) = \sum_{k,\alpha} (S^k_{\alpha 1 \dots n} w^\alpha)_{,k} \mathrm{d}x \tag{10.81}
$$

Using Eq. (10.61), the local expression for div $S \cdot w$ is, therefore,

$$
\begin{aligned}
&\left[\sum_{k,\alpha} (S^k_{\alpha 1 \dots n} w^\alpha)_{,k} - \left(\sum_{\alpha} R_{\alpha 1 \dots n} w^\alpha + \sum_{k,\alpha} S^k_{\alpha 1 \dots n} w^\alpha_{,k} \right) \right] \mathrm{d}x^1 \wedge \cdots \wedge \mathrm{d}x^n \\
&\qquad = \sum_{k,\alpha} (S^k_{\alpha 1 \dots n,k} - R_{\alpha 1 \dots n}) w^\alpha \mathrm{d}x^1 \wedge \cdots \wedge \mathrm{d}x^n.
\end{aligned}
\tag{10.82}
$$

We conclude that div S is indeed a section of W' that is represented locally by

$$
\sum_{k,\alpha} (S^k_{\alpha 1 \dots n,k} - R_{\alpha 1 \dots n}) g^\alpha \otimes \mathrm{d}x. \tag{10.83}
$$

It is noted that in the case where $R_{\alpha 1 \dots n} = 0$ locally, the expression for the divergence reduces to the traditional expression for the divergence of a tensor field in a Euclidean space. However, such a condition is not invariant.

Given a variational stress S, Eq. (10.63) implies that

$$
\mathbf{b} = -\operatorname{div} S \tag{10.84}
$$

is the body force that satisfies the principle of virtual power (10.59)—an obvious generalization of the differential equation for the stress tensor in continuum mechanics.

Summarizing, the conditions of compatibility of the stress field with the external loading for any region, $\mathscr{R}$, are

$$\operatorname{div} S + \mathbf{b} = 0, \qquad \text{in } \mathscr{R}, \tag{10.85}$$

$$\rho_{\partial \mathscr{R}}(p_s S)^{\mathsf{T}} = \mathbf{t}_{\mathscr{R}}^{\mathsf{T}}, \qquad \text{on } \partial \mathscr{R}.$$

These conditions exhibit explicitly the way the stress field induces a force system in the body. Conversely, given a force system satisfying Cauchy's postulate, the variational stress representing it is unique. As mentioned in the last paragraph of Sect. 10.6, the force system corresponds to a unique traction stress field. The traction stress field and the body force determine a unique variational stress field given locally by (10.62). Alternatively, to determine $R_{\alpha 1 \ldots n}$ uniquely, one can apply the representation of the action of forces on subbodies by variational stress (as in (10.59)) to subbodies contained in charts and consider vector fields for which the local representatives are "uniform."

10.9 The Case Where a Connection Is Given

In this section, we study the consequences of introducing a linear connection of the vector bundle W. First, we review the geometric structure involved with a linear connection, and then we consider the implications for stress theory. In short, a connection enables one to decompose the jet of a section w of W into the pair $(w, \nabla w)$, where ∇w is the covariant derivative of w, a section of $T^*\mathscr{M} \otimes W$. This decomposition induces a dual decomposition of the variational stress into two components, and the one dual to ∇w is the analog of the classical stress tensor, a section of $T\mathscr{M} \otimes W^*$.

10.9.1 The Tangent Bundle of a Vector Bundle

As a special case of Sect. 6.1.6 concerning the tangent bundle to a fiber bundle, we consider first the various structures on the tangent bundle $\tau_W : TW \to W$ of the vector bundle $\pi : W \to \mathscr{M}$. Elements of TW are represented in the form $(x^i, w^\alpha, \dot{x}^j, \dot{w}^\beta)$ or $\sum_j \dot{x}^j \partial_j + \sum_\beta \dot{w}^\beta g_\beta$ for a local basis $\{g_b\}$ of the fiber bundle. The tangent mapping, $T\pi : TW \to T\mathscr{M}$, is given locally by $(x^i, w^\alpha, \dot{x}^j, \dot{w}^\beta) \mapsto (x^i, \dot{x}^j)$. The vertical subbundle $VW := \operatorname{Kernel} T\pi \subset TW$ contains elements of the form $(x^i, w^\alpha, 0, \dot{w}^\beta)$. Consider the pullback structure

$$\begin{array}{ccc} \pi^*W & \xrightarrow{\pi'^*\pi} & W \\ {\scriptstyle \pi^*\pi'}\downarrow & & \downarrow{\scriptstyle \pi'} \\ W & \xrightarrow{\pi} & \mathcal{M} \end{array} \tag{10.86}$$

where the fiber $(\pi^*W)_w$ of π^*W at $w \in W$ is identified with $W_{\pi(w)}$. Note that we make an artificial distinction between the projection π which pulls back the fiber bundle and the projection π' of the fiber bundle which is pulled back. Elements of π^*W are of the form (x^i, w^α, v^β). Locally, $\pi^*\pi'$ is represented by $(x^i, w^\alpha, v^\beta) \mapsto (x^i, w^\alpha)$ and $\pi'^*\pi$ is represented by $(x^i, w^\alpha, v^\beta) \mapsto (x^i, v^\beta)$. There is a natural isomorphism

$$\pi^*W \cong VW, \qquad \text{given locally by} \qquad (x^i, w^\alpha, \dot{w}^\beta) \longmapsto (x^i, w^\alpha, 0, \dot{w}^\beta). \tag{10.87}$$

The inclusion of $\pi^*W \to TW$ may also be written without using components. For each $x \in \mathcal{M}$, let $\mathcal{I}_x : W_x \to W$ be the inclusion of the fiber at x in the vector bundle. Then, the tangent mapping $T(\mathcal{I}_x) : T(W_x) \cong W_x \times W_x \to TW$ is the inclusion of vectors tangent to the fiber at x into TW. (Note that the tangent space at each point in a vector space is naturally isomorphic to the vector space itself, W_x in our case.) Letting x vary, we have an inclusion of vector bundles $\tilde{\mathcal{I}} : \pi^*W \to TW$, the image of which is VW.

Consider the isomorphism,

$$(VW)_{w_0} = T_{w_0}(W_{\pi(w_0)}) \cong W_{\pi(w_0)}, \qquad w_0 \in W. \tag{10.88}$$

It implies that for a section $w : \mathcal{M} \to W$, and the pullback bundle

$$w^*\tau_W : w^*VW \longrightarrow \mathcal{M}, \tag{10.89}$$

one has

$$(w^*VW)_x = (VW)_{w(x)} \cong W_x. \tag{10.90}$$

We conclude that for a vector bundle as above and a section w,

$$w^*VW \cong W. \tag{10.91}$$

An additional pullback structure is

$$\begin{array}{ccc} \pi^*T\mathcal{M} \cong \tau^*W & \xrightarrow{\tau^*\pi} & T\mathcal{M} \\ {\scriptstyle \pi^*\tau}\downarrow & & \downarrow{\scriptstyle \tau} \\ W & \xrightarrow{\pi} & \mathcal{M} \end{array} \tag{10.92}$$

where the fiber $(\pi^* T\mathcal{M})_w$, $w \in \pi^{-1}\{x\}$, is identified with $T_x\mathcal{M}$. An element of $\pi^* T\mathcal{M}$ is represented locally by $(x^i, w^\alpha, \dot{x}^j)$. As indicated in the diagram, there is a fiber bundle isomorphism $\pi^* T\mathcal{M} \cong \tau^* W$ over $\mathcal{M}$, where the fiber $(\tau^* W)_v$, $v \in T\mathcal{M}$, is identified with $W_{\tau(v)}$. The isomorphism is given locally by $(x^i, w^\alpha, \dot{x}^j) \mapsto (x^i, \dot{x}^j, w^\alpha)$.

Consider the following diagram:

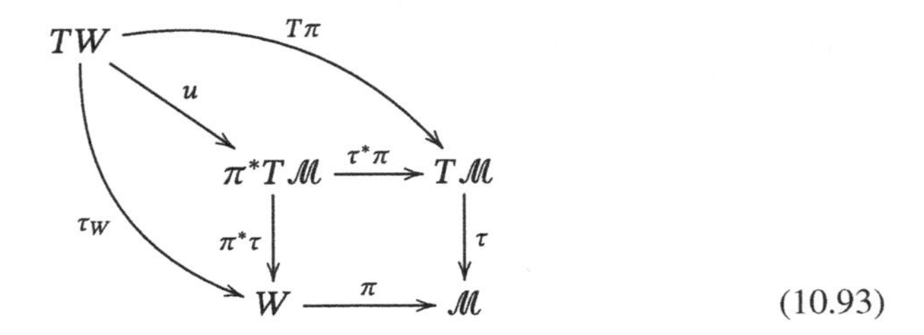

(10.93)

The universal property of the pullback structure implies that there is a unique vector bundle morphism

$$u : TW \longrightarrow \pi^* T\mathcal{M}, \tag{10.94}$$

over W, such that

$$\tau^*\pi \circ u = T\pi, \quad \text{and} \quad \pi^*\tau \circ u = \tau_W. \tag{10.95}$$

Locally, the mapping u is given by

$$(x^i, w^\alpha, \dot{x}^j, \dot{w}^\beta) \longmapsto (x^i, w^\alpha, \dot{x}^j). \tag{10.96}$$

In particular, it follows that $VW = \text{Kernel}\, u$. In other words, one has an exact sequence of vector bundle morphisms

$$\begin{array}{ccccccccc} 0 & \longrightarrow & \pi^* W & \xrightarrow{\tilde{\mathcal{J}}} & TW & \xrightarrow{u} & \pi^* T\mathcal{M} & \longrightarrow & 0 \\ & & & \searrow^{\pi^*\pi'} & \downarrow{\scriptstyle \tau_W} & \swarrow_{\pi^*\tau} & & & \\ & & & & W & & & & \end{array} \tag{10.97}$$

over W.

10.9.2 Linear Connections on a Vector Bundle and Covariant Derivatives

Although VW is a subbundle of TW, there is no natural complementary subbundle HW such that $TW = VW \oplus HW$. Such a splitting of TW is given by the additional structure of a connection.

A *linear connection* is specified by an inverse

$$\Gamma : \pi^* T\mathcal{M} \cong \tau^* W \longrightarrow TW \tag{10.98}$$

of the mapping $u : TW \to \pi^* T\mathcal{M}$, which is a vector bundle morphism relative to both vector bundle structures of $\pi^* T\mathcal{M} \cong \tau^* W$. Locally, Γ is represented using an array field (not a tensor field) $\Gamma^{\alpha}_{\beta i}(x^j)$, $i = 1, \ldots, n$, $\alpha, \beta = 1, \ldots, m$, in the form

$$(x^i, w^\gamma, \dot{x}^j) \longmapsto (x^i, w^\gamma, \dot{x}^j, -\textstyle\sum_{i,\beta} \Gamma^{\alpha}_{\beta j}(x^i) w^\beta \dot{x}^j). \tag{10.99}$$

Evidently $u \circ \Gamma = \mathrm{Id}_{\pi^* T\mathcal{M}}$ and $\Gamma(v)$ is not vertical as long as $\tau^*\pi(v) \in T\mathcal{M}$ does not vanish. Thus, Image$(\Gamma \circ u)$, the *horizontal distribution*, is a subbundle that is complementary to VW. For an element $\dot{w} \in TW$, $\Gamma \circ u(\dot{w})$ is the *horizontal component* of $\dot{w}$. The *vertical component* of $\dot{w}$ is the complement of the horizontal component. It is given by $C(\dot{w}) := \dot{w} - \Gamma \circ u(\dot{w})$ so that C is represented locally by

$$(x^i, w^\gamma, \dot{x}^j, \dot{w}^\alpha) \longmapsto (x^i, w^\gamma, 0, \dot{w}^\alpha + \textstyle\sum_{i,\beta} \Gamma^{\alpha}_{\beta j}(x^i) w^\beta \dot{x}^j). \tag{10.100}$$

Thus, C is a vector bundle morphism $TW \to VW \cong \pi^* W$—the vertical projection induced by the connection.

Finally, using $\pi'^*\pi : \pi^* W \to W$, the vertical component of $\dot{w} \in TW$ may be represented as an element of W. In other words, we have a mapping

$$K := \pi'^*\pi \circ C : TW \to W, \tag{10.101}$$

whereby, locally,

$$(x^i, w^\gamma, \dot{x}^j, \dot{w}^\alpha) \longmapsto (x^i, \dot{w}^\alpha + \textstyle\sum_{i,\beta} \Gamma^{\alpha}_{\beta j}(x^i) w^\beta \dot{x}^j). \tag{10.102}$$

The various mappings are illustrated in the following diagram:

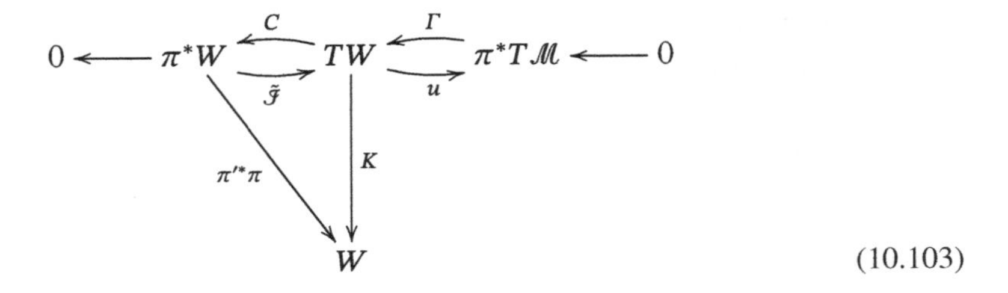

(10.103)

where $(u, C) : TW \to \pi^*T\mathcal{M} \oplus \pi^*W$ is the required decomposition.

Let $w : \mathcal{M} \to W$ be a section of π. The tangent to the field is

$$Tw : T\mathcal{M} \longrightarrow TW, \tag{10.104}$$

given locally by

$$(x^i, \dot{x}^j) \longmapsto (x^i, w^\alpha(x^k), \dot{x}^j, \textstyle\sum_l w^\beta_{,l}(x^k)\dot{x}^l). \tag{10.105}$$

It is noted that $Tw|_{T_x\mathcal{M}} : T_x\mathcal{M} \to TW$ contains the same data as $j^1_x w$.

The *covariant derivative of* w is the mapping induced by the connection and the tangent mapping Tw by

$$\nabla w := K \circ Tw : T\mathcal{M} \longrightarrow W. \tag{10.106}$$

Using the Christoffel symbol, Γ, the action of the covariant derivative is given by

$$(x^i, \dot{x}^j) \longmapsto \big(x^i, \textstyle\sum_l \big(w^\beta_{,l}(x^k) + \sum_\alpha \Gamma^\beta_{\alpha l}(x^k) w^\alpha(x^k)\big)\dot{x}^l\big). \tag{10.107}$$

Alternatively, since ∇w is a vector bundle morphism, it may be viewed as a section of $L(T\mathcal{M}, W) \cong T^*\mathcal{M} \otimes W$. As such, one may write its local representative as

$$\sum_{\beta,l} (\nabla w)^\beta_l \mathrm{d}x^l \otimes g_\beta = \sum_{\alpha,\beta,l} (w^\beta_{,l} + \Gamma^\beta_{\alpha l} w^\alpha) \mathrm{d}x^l \otimes g_\beta. \tag{10.108}$$

Thus, the connection enables one to store the information concerning Tw and $j^1 w$ by the pair $(w, \nabla w)$.

10.9.3 The Covariant Divergence of the Stress

In this section, it is assumed that a connection is given on the vector bundle W, and we want to study the new covariant objects that result from the additional structure. To that end, we consider the integrand $S \cdot j^1 w$, the density of power, as follows from the expression

$$F_{\mathcal{R}}(w) = \int_{\mathcal{R}} S \cdot j^1 w. \tag{10.109}$$

Locally, $S \cdot j^1 w$ is represented by $\left(\sum_\alpha R_{\alpha 1,\dots,n} w^\alpha + \sum_{\alpha,i} S^i_{\alpha 1,\dots,n} w^\alpha_{,i}\right)\mathrm{d}x$. Since by Eq. (10.108), $w^\alpha_{,l} = (\nabla w)^\alpha_l - \Gamma^\alpha_{\beta i} w^\beta$, the local expression $S \cdot j^1 w$ may be written in terms of the components of the covariant derivative as

$$\sum_{\alpha,\beta,i} \left((R_{\alpha 1,\dots,n} - S^i_{\beta 1,\dots,n}\Gamma^\beta_{\alpha i})w^\alpha + S^i_{\alpha 1,\dots,n}(\nabla w)^\alpha_i\right)\mathrm{d}x. \tag{10.110}$$

It is observed that the term $S^i_{\alpha 1,\dots,n}(\nabla w)^\alpha_i \mathrm{d}x$ above is the covariant field $S^+ \cdot \nabla w$, where $S^+ := \mathcal{J}'_{V^1_0}(S)$ is the principal component of the stress. Since the equation above represents the invariant object $S \cdot j^1 w$, we conclude that

$$\tilde{R}_{\alpha 1,\dots,n} := R_{\alpha 1,\dots,n} - S^i_{\beta 1,\dots,n}\Gamma^\beta_{\alpha i} \tag{10.111}$$

represent a covariant section $\tilde{R}$ of W' to which we will refer as the *self-force*. These observations give rise to a covariant decomposition

$$S \cdot j^1 w = \tilde{R} \cdot w + S^+ \cdot \nabla w. \tag{10.112}$$

For example, one may postulate that the self-force, $\tilde{R}$, vanishes identically. We conclude that

$$F_{\mathcal{R}}(w) = \int_{\mathcal{R}} \tilde{R} \cdot w + \int_{\mathcal{R}} S^+ \cdot \nabla w. \tag{10.113}$$

Locally, one has

$$\begin{aligned}
S^+ \cdot \nabla w &= \sum_{\alpha,\beta,i} S^i_{\alpha 1,\dots,n}(w^\alpha_{,i} + \Gamma^\alpha_{\beta i} w^\beta)\mathrm{d}x, \\
&= \sum_{\alpha,i} (S^i_{\alpha 1,\dots,n} w^\alpha)_{,i}\mathrm{d}x - \sum_{\alpha,i} S^i_{\alpha 1,\dots,n,i} w^\alpha \mathrm{d}x + \sum_{\alpha,\beta,i} S^i_{\alpha 1,\dots,n}\Gamma^\alpha_{\beta i} w^\beta \mathrm{d}x, \\
&= \mathrm{d}(p_s S \cdot w) - \sum_{\alpha,\beta,i} (S^i_{\beta 1,\dots,n,i} - S^i_{\alpha 1,\dots,n}\Gamma^\alpha_{\beta i})w^\beta \mathrm{d}x,
\end{aligned} \tag{10.114}$$

where we used (10.81) for the first term in the last line. As both $S^+ \cdot \nabla w$ and $\mathrm{d}(p_s S \cdot w)$ are covariant terms, the same holds for the second term on the right. Thus, there is a section $\nabla \cdot S^+$ of W', given locally by

$$\sum_{\alpha,\beta,i} (S^i_{\beta 1,\dots,n,i} - S^i_{\alpha 1,\dots,n}\Gamma^\alpha_{\beta i})g^\beta \otimes \mathrm{d}x, \tag{10.115}$$

to which we will refer as the *covariant divergence* of the stress. It follows that

$$S^+ \cdot \nabla w = \mathrm{d}(p_s S \cdot w) - (\nabla \cdot S^+) \cdot w, \tag{10.116}$$

and

$$\begin{aligned} F_{\mathcal{R}}(w) &= \int_{\mathcal{R}} \tilde{R} \cdot w + \int_{\mathcal{R}} \mathrm{d}(p_s S \cdot w) - \int_{\mathcal{R}} (\nabla \cdot S^+) \cdot w, \\ &= \int_{\mathcal{R}} \tilde{R} \cdot w + \int_{\partial\mathcal{R}} p_s S \cdot w - \int_{\mathcal{R}} (\nabla \cdot S^+) \cdot w. \end{aligned} \tag{10.117}$$

Comparing the last expression to Eq. (10.59) for the virtual power, it follows immediately that

$$\mathbf{b} = \tilde{R} - \nabla \cdot S^+. \tag{10.118}$$

In particular, if one assumes that the self-force $\tilde{R}$ vanishes identically, we obtain the analog of the classical equilibrium conditions

$$\begin{aligned} \nabla \cdot S^+ + \mathbf{b} &= 0, && \text{in } \mathcal{R}, \\ \rho_{\partial\mathcal{R}}(\mathsf{C}(S^+))^{\mathsf{T}} &= \mathbf{t}_{\mathcal{R}}^{\mathsf{T}}, && \text{on } \partial\mathcal{R}. \end{aligned} \tag{10.119}$$

Chapter 11
Smooth Electromagnetism on Manifolds

The foregoing theory of forces and stresses is meant, naturally, to generalize classical mechanics of solid bodies. However, the generality of the proposed setting allows one to apply it to other fields. In this chapter, we formulate notions from the classical theory of electromagnetism as examples of the constructions developed above. In the first section, we consider the metric-independent (pre-metric) Maxwell equations in a Lorentz frame in four-dimensional spacetime. In addition, we consider the corresponding Lorentz force and energy-momentum tensor (cf. [HIO06, HO03]). Then, in the second section, we show how the Maxwell equations and their generalizations to p-form electrodynamics (as in [HT86, HT88]) in an n-dimensional manifold may be presented as a special case of geometric stress theory.

11.1 Electromagnetism in a Lorentz Frame

We first review the basics of electromagnetism in a Lorentz frame in spacetime. This will illustrate the relation between the Maxwell equations, as written traditionally in terms of vector fields, and the metric-independent formulation using differential forms.

Metric-independent, or pre-metric, aspects of electrodynamics have been studied since the beginning of the twentieth century. Whittaker [Whi53, pp. 192–196] attributes the first work in this direction to Kottler [Kot22], while Truesdell and Toupin [TT60, Section F] attribute the main contribution to van Dantzig [vD34]. In recent decades, renewed interest in the subject led to further work in which notions of modern differential geometry have been utilized (see, e.g., [HO, Kai04, HIO06]).

R. Segev, *Foundations of Geometric Continuum Mechanics*, Advances in Continuum Mechanics 49, https://doi.org/10.1007/978-3-031-35655-1_11

11.1.1 The Metric-Independent Maxwell Equations in Four-Dimensional Spacetime

The first fundamental quantity of electromagnetism that we consider is the 3-form $\mathfrak{J}$ in spacetime, $\mathscr{E}$, modeling the 4-current density. The representation of $\mathfrak{J}$ as a 3-form may be motivated by the theory of fluxes outlined in Sect. 8.1.4. In frame coordinates we may use Eq. (9.8) and write locally

$$\begin{aligned}\mathfrak{J} &= \sum_k \mathfrak{J}_{1\ldots\widehat{k}\ldots 4}\mathrm{d}z^1 \wedge \cdots \wedge \widehat{\mathrm{d}z^k} \wedge \cdots \wedge \mathrm{d}z^4, \\ &= \rho_{123}\mathrm{d}z^2 \wedge \mathrm{d}z^3 \wedge \mathrm{d}z^4 - J_{23}\mathrm{d}z^1 \wedge \mathrm{d}z^3 \wedge \mathrm{d}z^4 \\ &\quad - J_{13}\mathrm{d}z^1 \wedge \mathrm{d}z^2 \wedge \mathrm{d}z^4 - J_{12}\mathrm{d}z^1 \wedge \mathrm{d}z^2 \wedge \mathrm{d}z^3,\end{aligned} \tag{11.1}$$

where $\rho = \rho_{123}\mathrm{d}x^1 \wedge \mathrm{d}x^2 \wedge \mathrm{d}x^3$ is the space charge density and

$$J = \sum_i J_{1\ldots\widehat{i}\ldots 3}\mathrm{d}x^1 \wedge \cdots \wedge \widehat{\mathrm{d}x^i} \wedge \cdots \wedge \mathrm{d}x^3$$

is the 2-form in space describing the current density.

In case a volume element $\theta = \vartheta \mathrm{d}z^1 \wedge \cdots \wedge \mathrm{d}z^4$ is given on spacetime and a volume element $\theta_0 = \vartheta_0 \mathrm{d}x^1 \wedge \mathrm{d}x^2 \wedge \mathrm{d}x^3$ is given in $\mathcal{S}$, one may use the isomorphism $\mathcal{J}_\theta$ defined in Sect. 8.2.2 to represent the flux forms by tangent vectors $\mathbf{j} = \sum_{i=1}^{3} \mathbf{j}^i \partial/\partial x^i$ and $w = \sum_{k=1}^{4} w^k \partial/\partial z^k$, such that $\mathfrak{J} = w \lrcorner \theta$ and $J = \mathbf{j} \lrcorner \theta_0$. It follows from Eq. (8.54) that

$$\mathbf{j}^i = (-1)^{i-1}\frac{J_{1\ldots\widehat{i}\ldots 3}}{\vartheta_0}, \qquad w^k = (-1)^{k-1}\frac{\mathfrak{J}_{1\ldots\widehat{k}\ldots 4}}{\vartheta}. \tag{11.2}$$

Using Eq. (9.9), for $k > 1$,

$$\mathfrak{J}_{1\ldots\widehat{k}\ldots 4} = (-1)^{k-1}w^k\vartheta = -J_{1\ldots\widehat{k-1}\ldots 3} = -(-1)^{k-2}\mathbf{j}^{k-1}\vartheta_0, \tag{11.3}$$

and so,

$$w^1 = \frac{\rho_{123}}{\vartheta}, \quad w^2 = \frac{\vartheta_0}{\vartheta}\mathbf{j}^1, \quad w^3 = \frac{\vartheta_0}{\vartheta}\mathbf{j}^2, \quad w^4 = \frac{\vartheta_0}{\vartheta}\mathbf{j}^3 \tag{11.4}$$

(*cf.* [MTW73, p. 81 and p. 113], [LL95, pp. 60–71]).

A basic assumption of electromagnetism is conservation of charge, i.e., for any four-dimensional region, $\mathfrak{R}$, of spacetime

$$\int_{\partial\mathfrak{R}} \mathfrak{J} = 0. \tag{11.5}$$

Thus, it follows from Stokes's theorem that $\mathrm{d}\mathfrak{J} = 0$.

The Maxwell 2-form, $\mathfrak{g}$, is a flow potential for the flux, so that $\mathfrak{J} = \mathrm{d}\mathfrak{g}$. The local representation of the Maxwell form in Lorentzian frame coordinates is given by

$$\begin{aligned}\mathfrak{g} &= \sum_{k<l} \mathfrak{g}_{kl} \mathrm{d}z^k \wedge \mathrm{d}z^l, \\ &= H_1 \mathrm{d}z^1 \wedge \mathrm{d}z^2 + H_2 \mathrm{d}z^1 \wedge \mathrm{d}z^3 + H_3 \mathrm{d}z^1 \wedge \mathrm{d}z^4 \\ &\quad + D_3 \mathrm{d}z^2 \wedge \mathrm{d}z^3 - D_2 \mathrm{d}z^2 \wedge \mathrm{d}z^4 + D_1 \mathrm{d}z^3 \wedge \mathrm{d}z^4\end{aligned} \tag{11.6}$$

where D_i are the component of the electric displacement and H_i are the components of the magnetic field intensity. The matrix representing $\mathfrak{g}$ is therefore

$$\begin{pmatrix} 0 & H_1 & H_2 & H_3 \\ -H_1 & 0 & D_3 & -D_2 \\ -H_2 & -D_3 & 0 & D_1 \\ -H_3 & D_2 & -D_1 & 0 \end{pmatrix}. \tag{11.7}$$

The components of the equation $\mathfrak{J} = \mathrm{d}\mathfrak{g}$ are

$$\begin{aligned}-\mathbf{j}^3\vartheta_0 &= -J_{12} = \mathfrak{J}_{123} = \mathfrak{g}_{12,3} - \mathfrak{g}_{13,2} + \mathfrak{g}_{23,1} = \frac{\partial H_1}{\partial x^2} - \frac{\partial H_2}{\partial x^1} + \frac{\partial D_3}{\partial t}, \\ -\mathbf{j}^1\vartheta_0 &= -J_{23} = \mathfrak{J}_{134} = \mathfrak{g}_{13,4} - \mathfrak{g}_{14,3} + \mathfrak{g}_{34,1} = \frac{\partial H_2}{\partial x^3} - \frac{\partial H_3}{\partial x^2} + \frac{\partial D_1}{\partial t}, \\ \mathbf{j}^2\vartheta_0 &= -J_{13} = \mathfrak{J}_{124} = \mathfrak{g}_{12,4} - \mathfrak{g}_{14,2} + \mathfrak{g}_{24,1} = \frac{\partial H_1}{\partial x^3} - \frac{\partial H_3}{\partial x^1} - \frac{\partial D_2}{\partial t}, \\ \rho_{1\ldots4} &= \mathfrak{J}_{234} = \mathfrak{g}_{23,4} - \mathfrak{g}_{24,3} + \mathfrak{g}_{34,2} = \frac{\partial D_3}{\partial x^3} + \frac{\partial D_2}{\partial x^2} + \frac{\partial D_1}{\partial x^1}\end{aligned} \tag{11.8}$$

—the first part of Maxwell's equations.

The next fundamental object of electromagnetism is the Faraday 2-form $\mathfrak{f}$. Using frame coordinates, $\mathfrak{f}$ is represented locally by

$$\begin{aligned}\mathfrak{f} &= \sum_{k<l} \mathfrak{f}_{kl} \mathrm{d}z^k \wedge \mathrm{d}z^l, \\ &= -E_1 \mathrm{d}z^1 \wedge \mathrm{d}z^2 - E_2 \mathrm{d}z^1 \wedge \mathrm{d}z^3 - E_3 \mathrm{d}z^1 \wedge \mathrm{d}z^4 \\ &\quad + B_3 \mathrm{d}z^2 \wedge \mathrm{d}z^3 - B_2 \mathrm{d}z^2 \wedge \mathrm{d}z^4 + B_1 \mathrm{d}z^3 \wedge \mathrm{d}z^4\end{aligned} \tag{11.9}$$

where E_i are the components of the electric field in space and B_i are the components of the magnetic flux density. Thus, the matrix representation of $\mathfrak{f}$ is

$$\begin{pmatrix} 0 & -E_1 & -E_2 & -E_3 \\ E_1 & 0 & B_3 & -B_2 \\ E_2 & -B_3 & 0 & B_1 \\ E_3 & B_2 & -B_1 & 0 \end{pmatrix}. \tag{11.10}$$

The other set of Maxwell's equations states that the Faraday 2-form is closed, i.e., $\mathrm{d}\mathfrak{f} = 0$. The local expression in frame components is

$$\begin{aligned} 0 = &(-E_{1,3} + E_{2,2} + B_{3,1})\mathrm{d}z^1 \wedge \mathrm{d}z^2 \wedge \mathrm{d}z^3 \\ &+ (-E_{1,4} + E_{3,2} - B_{2,1})\mathrm{d}z^1 \wedge \mathrm{d}z^2 \wedge \mathrm{d}z^4 \\ &+ (-E_{2,4} + E_{3,3} - B_{1,1})\mathrm{d}z^1 \wedge \mathrm{d}z^3 \wedge \mathrm{d}z^4 \\ &+ (-B_{3,4} + B_{2,3} + B_{1,2})\mathrm{d}z^2 \wedge \mathrm{d}z^3 \wedge \mathrm{d}z^4, \end{aligned} \tag{11.11}$$

where the indices for the partial derivatives indicate differentiation with respect to the world coordinates z^k (as opposed to x^i). Thus, in terms of time-space coordinates (t, x^i), the four components of this equation are

$$\begin{aligned} &2\frac{\partial E_2}{\partial x^1} - \frac{\partial E_1}{\partial x^2} + \frac{\partial B_3}{\partial t} = 0, \qquad -\frac{\partial E_1}{\partial x^3} + \frac{\partial E_3}{\partial x^1} - \frac{\partial B_2}{\partial t} = 0, \\ &-\frac{\partial E_2}{\partial x^3} + \frac{\partial E_3}{\partial x^2} + \frac{\partial B_1}{\partial t} = 0, \qquad \frac{\partial B_1}{\partial x^1} + \frac{\partial B_2}{\partial x^2} + \frac{\partial B_3}{\partial x^3} = 0. \end{aligned} \tag{11.12}$$

Depending on the topology of spacetime, one may replace the second set of the Maxwell equations with a possibly stronger assumption that $\mathfrak{f}$ is an exact form. That is, we assume that there is a covector potential 1-form $\mathfrak{a}$ over spacetime such that

$$\mathfrak{f} = \mathrm{d}\mathfrak{a}. \tag{11.13}$$

11.1.2 The Lorentz Force

Let a Lorentzian frame be given in spacetime. For a particle of charge Q traveling in space with velocity $v = \sum_i v^i \partial/\partial x^i$, we consider the 4-velocity $\mathfrak{v}$ given locally by $\mathfrak{v} = \partial/\partial z^1 + v^1\partial/\partial z^2 + v^2\partial/\partial z^3 + v^3\partial/\partial z^4$. The 4-Lorentz force acting on the particle is given by $\mathfrak{F} = -Q\,\mathfrak{v} \lrcorner \mathfrak{f}$. Using Example 5.6, one has

$$\begin{aligned} \frac{\mathfrak{F}}{Q} = &-(E_1v^1 + E_2v^2 + E_3v^3)\mathrm{d}z^1 + (E_1 + v^2B_3 - v^3B_2)\mathrm{d}z^2 \\ &+ (E_2 - v^1B_3 + v^3B_1)\mathrm{d}z^3 + (E_3 + v^1B_2 - v^2B_1)\mathrm{d}z^4, \end{aligned} \tag{11.14}$$

so that the time-like component is minus the power expanded by the electromagnetic field and the three space-like components make up the traditional expression for the Lorentz force, as expected. The virtual power performed by the 4-Lorentz force for a virtual 4-velocity u is

$$\mathfrak{F}(u) = -(Q\mathfrak{v} \lrcorner \mathfrak{f})(u) = -Q\,\mathfrak{f}(\mathfrak{v}, u) \tag{11.15}$$

and so the skew symmetry of the Faraday 2-form implies that

$$\mathfrak{F}(u) = \mathfrak{f}(u, Q\mathfrak{v}) = (u \lrcorner \mathfrak{f})(Q\mathfrak{v}). \tag{11.16}$$

For the case of a continuum, we want to write an analogous expression for the Lorentz force density, a body force, acting on the charge, the motion of which is specified by a charge flux $\mathfrak{J}$. Given a volume element θ, one may replace the 4-velocity above by the vector $\mathcal{I}_\theta(\mathfrak{J})$ and consider the 1-form $-\mathcal{I}_\theta(\mathfrak{J}) \lrcorner \mathfrak{f}$. However, as a body force, the Lorentz force density should be a section of $L\big(T\mathcal{E}, \bigwedge^4 T^*\mathcal{E}\big)$. We can use the inverse $\mathcal{I}_\theta^{-1} : T\mathcal{E} \to \bigwedge^3 T^*\mathcal{E}$ and the isomorphism $\mathcal{I}_\wedge : \bigwedge^1 T^*\mathcal{E} \to L\big(\bigwedge^3 T^*\mathcal{E}, \bigwedge^4 T^*\mathcal{E}\big)$ to define the Lorentz force density $\mathfrak{b}$ by $\mathfrak{b}(w) = -(\mathcal{I}_\theta(\mathfrak{J}) \lrcorner \mathfrak{f}) \wedge (\mathcal{I}_\theta^{-1}(w))$. Using Eq. (5.127), we obtain the simpler expression

$$\mathfrak{b} \cdot w = (w \lrcorner \mathfrak{f}) \wedge \mathfrak{J} \tag{11.17}$$

for any virtual 4-velocity field w. In addition, by Eq. (5.118) in Example 5.9, the last equation may also be written as

$$\mathfrak{b} \cdot w = -\mathfrak{f} \wedge (w \lrcorner \mathfrak{J}). \tag{11.18}$$

It is noted that the Lorentz force density does not depend on the choice of a volume element.

The local representative of $\mathfrak{b}$ is of the form $\sum_p \mathfrak{b}_{1\dots 4p} \mathrm{d}z^p \otimes (\mathrm{d}z^1 \wedge \cdots \wedge \mathrm{d}z^4)$, with $\mathfrak{b}(w)$ represented by $\sum_p \mathfrak{b}_{1\dots 4p} w^p \mathrm{d}z^1 \wedge \cdots \wedge \mathrm{d}z^4$. It follows from the expression for the power density for the Lorentz force, and from Eq. (5.125), that

$$\begin{aligned}
\mathfrak{b}_{1\dots 41} &= E_1\mathfrak{J}_{134} - E_2\mathfrak{J}_{124} + E_3\mathfrak{J}_{123},\\
\mathfrak{b}_{1\dots 42} &= E_1\mathfrak{J}_{234} + B_3\mathfrak{J}_{124} + B_2\mathfrak{J}_{123},\\
\mathfrak{b}_{1\dots 43} &= E_2\mathfrak{J}_{234} + B_3\mathfrak{J}_{134} - B_1\mathfrak{J}_{123},\\
\mathfrak{b}_{1\dots 44} &= E_3\mathfrak{J}_{234} - B_2\mathfrak{J}_{134} - B_1\mathfrak{J}_{124}.
\end{aligned} \tag{11.19}$$

In terms of the three-dimensional flux vector field $\mathbf{j}$, these may be written as

$$\mathfrak{b}_{1\dots 41} = -E \cdot \mathbf{j}\vartheta_0, \qquad \mathfrak{b}_{1\dots 4i} = \vartheta_0[\rho_{123}E + (\mathbf{j} \times B)]_{i-1}, \qquad \text{for } i > 1, \tag{11.20}$$

where $E \cdot \mathbf{j}$ denotes the inner product of vectors in $\mathbb{R}^3$, and $\mathbf{j} \times B$ denotes the vector product, as expected.

11.1.3 Metric-Invariant Maxwell Stress Tensor

We are now in a position to introduce a metric-invariant version of the Maxwell stress-energy tensor. The Maxwell stress tensor is a traction stress field, i.e., a section s_M of $L(T\mathscr{E}, \bigwedge^3 T^*\mathscr{E})$ defined by

$$s_M \cdot w = (w \lrcorner \mathfrak{f}) \wedge \mathfrak{g} - (w \lrcorner \mathfrak{g}) \wedge \mathfrak{f} \tag{11.21}$$

Note that we use here the opposite sign in comparison with the standard practice (cf. [Seg02, HIO06, HO03]).

Using Eqs. (5.107) and (5.89), we may also write

$$\begin{aligned} s_M \cdot w &= \mathfrak{g} \wedge (w \lrcorner \mathfrak{f}) - (w \lrcorner \mathfrak{g}) \wedge \mathfrak{f}, \\ &= 2(w \lrcorner \mathfrak{f}) \wedge \mathfrak{g} - w \lrcorner (\mathfrak{g} \wedge \mathfrak{f}), \end{aligned} \tag{11.22}$$

etc.

By the definition (11.21) of the Maxwell stress, the matrix of components $[s_{M1\ldots\widehat{k}\ldots4r}]$ for the local representation

$$\sum_{k,r} s_{M1\ldots\widehat{k}\ldots4r} \mathrm{d}z^r \otimes (\mathrm{d}z^1 \wedge \cdots \wedge \widehat{\mathrm{d}z^k} \wedge \cdots \wedge \mathrm{d}z^4), \tag{11.23}$$

cf. (10.15), is given by

$$-\begin{bmatrix} B \cdot H + D \cdot E & 2(B \times D)_1 & 2(B \times D)_2 & 2(B \times D)_3 \\ 2(H \times E)_1 & P_{11} & P_{12} & P_{13} \\ -2(H \times E)_2 & P_{21} & P_{22} & P_{23} \\ 2(H \times E)_3 & P_{31} & P_{32} & P_{33} \end{bmatrix}, \tag{11.24}$$

where $P_{ij} = (-1)^{i-1}[(B \cdot H + D \cdot E)\delta_{ij} - 2(B \otimes H + D \otimes E)_{ij}]$.

Having an expression for a traction stress and the body force, we can compute the corresponding variational stress using $\mathfrak{b} \cdot w + \mathrm{d}(s_M \cdot w) = S_M \cdot j^1 w$, as in (10.58). First,

$$\mathrm{d}(s_M \cdot w) = \mathrm{d}(w \lrcorner \mathfrak{f}) \wedge \mathfrak{g} - (w \lrcorner \mathfrak{f}) \wedge \mathrm{d}\mathfrak{g} - [\mathrm{d}(w \lrcorner \mathfrak{g}) \wedge \mathfrak{f} - (w \lrcorner \mathfrak{g}) \wedge \mathrm{d}\mathfrak{f}]\,. \tag{11.25}$$

Next, by the Maxwell equations and the expression (11.17) for the Lorentz force, we arrive at

$$\mathrm{d}(s_M \cdot w) = \mathrm{d}(w \lrcorner \mathfrak{f}) \wedge \mathfrak{g} - \mathrm{d}(w \lrcorner \mathfrak{g}) \wedge \mathfrak{f} - \mathfrak{b} \cdot w. \tag{11.26}$$

We may now use the definition of the variational stress, as in (10.58), to obtain the power density as

$$S_M \cdot j^1 w = \mathrm{d}(w \lrcorner \mathfrak{f}) \wedge \mathfrak{g} - \mathrm{d}(w \lrcorner \mathfrak{g}) \wedge \mathfrak{f}. \tag{11.27}$$

It may be shown that for the case where the aether relations are used in a Lorentzian spacetime, the two terms on the right-hand side cancel, so that $S_M \cdot j^1 w = 0$ and $\mathfrak{b} \cdot w = -\mathrm{d}(s_M \cdot w)$.

11.2 Metric-Independent p-Form Electrodynamics

In this section, we show how the fundamental mathematical objects of electromagnetism, considered in the preceding section, may be exhibited as a particular case of stress theory for geometric continuum mechanics as developed in Chap. 10. Moreover, the general setting we have used makes it possible to extend the four-dimensional nature of electromagnetism to a spacetime of an arbitrary dimension n. In fact, for $n > 4$, there may be several generalized electromagnetic theories, depending basically on the degree, p, of the potential form. Each such theory is therefore termed p-form electrodynamics (see [HT86, HT88, NS12]). In addition to the extension of the Maxwell equations, we will consider the corresponding generalization of the Lorentz force and Maxwell stress.

11.2.1 The Maxwell Equations for p-Form Electrodynamics

We consider a generalization of the Maxwell equations

$$\mathfrak{f} = \mathrm{d}\mathfrak{a}, \qquad \mathrm{d}\mathfrak{g} = \mathfrak{J}, \tag{11.28}$$

where the Maxwell and Faraday forms are related by constitutive relations. The extended theory will be presented as an example of stress theory. The basic variable is the potential form, or a variation thereof, which is represented as a p-form in an n-dimensional spacetime. The stress for the case of generalized electrodynamics is assumed to be represented by a differential $(n-p-1)$-form, a generalization of the Maxwell 2-form. This setting will be generalized in Part III, to the case of non-smooth fields, making use of the notion of de Rham currents.

In accordance with the notation scheme of the Chap. 10, we make the identification

$$\mathcal{M} = \mathcal{E}, \qquad W = \bigwedge^p T^*\mathcal{E}, \qquad 0 \leqslant p < n. \tag{11.29}$$

A section of W, a p-form, $\mathfrak{a}$, over spacetime, is thus interpreted as a potential field. A variational stress field is therefore a smooth section

$$S : \mathcal{E} \longrightarrow L(J^1(\bigwedge^p T^*\mathcal{E}), \bigwedge^n T^*\mathcal{E}). \tag{11.30}$$

The total power of the variational stress is

$$F(\mathfrak{a}) = \int_{\mathcal{E}} S \cdot j^1\mathfrak{a}, \tag{11.31}$$

and for the integral to make sense, we assume that at least one of the fields, S or $\mathfrak{a}$, has a compact support.

A traction stress is a section

$$s : \mathscr{E} \longrightarrow L(\bigwedge^p T^*\mathscr{E}, \bigwedge^{n-1} T^*\mathscr{E}). \tag{11.32}$$

For an n-dimensional region $\mathscr{R} \subset \mathscr{E}$, the total flux of power through $\partial\mathscr{R}$ is

$$\Phi = \int_{\partial\mathscr{R}} s \cdot \mathfrak{a}.$$

The first assumption that we make is that

$$\mathbf{b} = -\operatorname{div} S = 0. \tag{11.33}$$

Let $s = p_s S$. It follows immediately from Eq. (10.75) that

$$S \cdot j^1\mathfrak{a} = \mathrm{d}(s \cdot \mathfrak{a}). \tag{11.34}$$

The second basic assumption is concerned with a particular form of the traction stress. It is assumed that there is an $(n - p - 1)$-form,

$$\mathfrak{g} : \mathscr{E} \longrightarrow \bigwedge^{n-p-1} T^*\mathscr{E}, \tag{11.35}$$

such that

$$s \cdot \mathfrak{a} = \mathfrak{g} \wedge \mathfrak{a}. \tag{11.36}$$

It follows immediately that for any n-dimensional region $\mathscr{R} \subset \mathscr{E}$, the induced force $F_{\mathscr{R}}$ satisfies

$$\begin{aligned} F_{\mathscr{R}}(\mathfrak{a}) &= \int_{\mathscr{R}} S \cdot j^1\mathfrak{a}, \\ &= \int_{\mathscr{R}} \mathrm{d}(s \cdot \mathfrak{a}), \\ &= \int_{\mathscr{R}} \mathrm{d}(\mathfrak{g} \wedge \mathfrak{a}), \\ &= \int_{\partial\mathscr{R}} \mathfrak{g} \wedge \mathfrak{a}, \\ &= \int_{\mathscr{R}} \mathrm{d}\mathfrak{g} \wedge \mathfrak{a} + (-1)^{n-p-1} \int_{\mathscr{R}} \mathfrak{g} \wedge \mathrm{d}\mathfrak{a}. \end{aligned} \tag{11.37}$$

In particular, if $\mathfrak{g}$ or $\mathfrak{a}$ is compactly supported in $\mathscr{R}$, then $F_{\mathscr{R}}(\mathfrak{a}) = 0$. The simple derivation above implies that the action of the variational stress, S, is given by

$$S \cdot j^1\mathfrak{a} = \mathrm{d}\mathfrak{g} \wedge \mathfrak{a} + (-1)^{n-p-1}\mathfrak{g} \wedge \mathrm{d}\mathfrak{a}. \tag{11.38}$$

We now set

$$\mathfrak{f} := \mathrm{d}\mathfrak{a}, \qquad \mathfrak{J} := \mathrm{d}\mathfrak{g}, \tag{11.39}$$

so that $\mathfrak{f}$ is a $(p+1)$-form and $\mathfrak{J}$ is an $(n-p)$-form. Evidently,

$$\mathrm{d}\mathfrak{f} = 0, \qquad \mathrm{d}\mathfrak{J} = 0. \tag{11.40}$$

Thus, we naturally view $\mathfrak{f}$ as a generalization of the Faraday form and view $\mathfrak{J}$ as a generalization of the 4-charge-current density of electrodynamics. Consequently, the equations above generalize the Maxwell equations. The power may be rewritten now as

$$F_{\mathscr{R}}(\mathfrak{a}) = \int_{\mathscr{R}} \mathfrak{J} \wedge \mathfrak{a} + (-1)^{n-p-1} \int_{\mathscr{R}} \mathfrak{g} \wedge \mathfrak{f}. \tag{11.41}$$

Remark 11.1 One may consider another example of forces that are dual to differential p-forms, maybe even more natural than p-form electrodynamics. Let

$$\mathfrak{A}_J : J^1\left(\bigwedge^p T^*\mathscr{E}\right) \longrightarrow \bigwedge^{p+1} T^*\mathscr{E} \tag{11.42}$$

be the natural mapping given by

$$\mathfrak{A}_J(j^1_x\mathfrak{a}) := \mathrm{d}\mathfrak{a}(x), \tag{11.43}$$

or $A \mapsto \mathrm{d}\mathfrak{a}(x)$, where $\mathfrak{a}$ is any p-form representing $A \in (J^1(\bigwedge^p T^*\mathscr{E}))_x$. Thus, the density dual,

$$\mathfrak{A}'_J : (\bigwedge^{p+1} T^*\mathscr{E})' \longrightarrow (J^1(\bigwedge^p T^*\mathscr{E}))', \tag{11.44}$$

may be used to consider particular variational stress densities represented in the form

$$S = \mathfrak{A}'_J(S'), \qquad S' \in (\bigwedge^{p+1} T^*\mathscr{E})'. \tag{11.45}$$

Using the notation presented in Sect. 6.2.3, we recall that

$$\wedge_\lrcorner : \bigwedge^{n-p} T^*\mathscr{X} \longrightarrow L(\bigwedge^p T^*\mathscr{X}, \bigwedge^n T^*\mathscr{X}), \qquad \wedge_\lrcorner(\mathfrak{g})(\mathfrak{a}) = \mathfrak{g} \wedge \mathfrak{a}, \tag{11.46}$$

is an isomorphism. Thus, for each section S' of $(\bigwedge^{p+1} T^*\mathscr{E})'$, there is an $(n-p-1)$-form $\mathfrak{g}$ such that

$$S' \cdot \beta = \mathfrak{g} \wedge \beta \tag{11.47}$$

for every $(n-p-1)$-form β. Hence, for a variational stress density field S,

$$S \cdot j^1\mathfrak{a} = S' \cdot \mathrm{d}\mathfrak{a} = \mathfrak{g} \wedge \mathrm{d}\mathfrak{a}. \tag{11.48}$$

It follows that

$$\begin{aligned} F_{\mathcal{R}}(\mathfrak{a}) &= \int_{\mathcal{R}} S \cdot j^1\mathfrak{a}, \\ &= \int_{\mathcal{R}} \mathfrak{g} \wedge \mathrm{d}\mathfrak{a}, \\ &= (-1)^{n-p-1} \int_{\mathcal{R}} \mathrm{d}(\mathfrak{g} \wedge \mathfrak{a}) + (-1)^{n-p} \int_{\mathcal{R}} \mathrm{d}\mathfrak{g} \wedge \mathfrak{a}, \\ &= (-1)^{n-p-1} \int_{\partial\mathcal{R}} \mathfrak{g} \wedge \mathfrak{a} + (-1)^{n-p} \int_{\mathcal{R}} \mathrm{d}\mathfrak{g} \wedge \mathfrak{a}. \end{aligned} \tag{11.49}$$

In particular, if $\mathfrak{g}$ or $\mathfrak{a}$ is compactly supported in $\mathcal{R}$, then,

$$F_{\mathcal{R}}(\mathfrak{a}) = (-1)^{n-p} \int_{\mathcal{R}} \mathrm{d}\mathfrak{g} \wedge \mathfrak{a}. \tag{11.50}$$

11.2.2 The Lorentz Force and Maxwell Stress for p-Form Electrodynamics

The Lorentz force density, $\mathfrak{b}$, (not to be confused with $\mathbf{b} = 0$, as above) for standard electromagnetism, $p = 1$, $n = 4$, is given by Eq. (11.17). For $p > 1$, we slightly generalize the expression and set

$$\mathfrak{b} \cdot w = (-1)^{p-1} (w \lrcorner \mathfrak{f}) \wedge \mathfrak{J}, \tag{11.51}$$

which evidently includes the classical case.

Leaving the expression for the Maxwell traction stress unchanged from (11.21), we note that $w \lrcorner \mathfrak{f}$ is a p-form and $w \lrcorner \mathfrak{g}$ is an $(n-p-2)$-form. Thus,

$$
\begin{aligned}
\mathrm{d}(s_M \cdot w) &= \mathrm{d}(w \lrcorner \mathfrak{f}) \wedge \mathfrak{g} + (-1)^p (w \lrcorner \mathfrak{f}) \wedge \mathrm{d}\mathfrak{g} \\
&\quad - \left[\mathrm{d}(w \lrcorner \mathfrak{g}) \wedge \mathfrak{f} + (-1)^{n-p-2} (w \lrcorner \mathfrak{g}) \wedge \mathrm{d}\mathfrak{f}\right], \\
&= \mathrm{d}(w \lrcorner \mathfrak{f}) \wedge \mathfrak{g} - \mathrm{d}(w \lrcorner \mathfrak{g}) \wedge \mathfrak{f} - \mathfrak{b} \cdot w,
\end{aligned}
\tag{11.52}
$$

where the Maxwell equations and the expression for the Lorentz force were used in arriving at the second line. Hence, the expression for $\mathrm{d}(s_M \cdot w)$ remains unchanged.

Finally, using $S_M \cdot j^1 w = \mathrm{d}(s_M \cdot w) + \mathfrak{b} \cdot w$, we immediately recover

$$
S_M \cdot j^1 w = \mathrm{d}(w \lrcorner \mathfrak{f}) \wedge \mathfrak{g} - \mathrm{d}(w \lrcorner \mathfrak{g}) \wedge \mathfrak{f} \tag{11.53}
$$

for the representation of the Maxwell variational stress.

Chapter 12
The Elasticity Problem

The equilibrium problem of continuum mechanics is concerned with the stress field distribution on a given body, $\mathcal{X}$, resulting from a given external loading. It is well known that continuum mechanics is statically indeterminate. That is, there is no unique solution to the system of equations (10.85).

This inherent nonuniqueness is resolved by the introduction of constitutive relations that couple the statics with the kinematics of the body $\mathcal{X}$ in the physical space $\mathcal{S}$. The constitutive relation for an elastic body provides the stress distribution in the body as a function of its configuration in space. As a result, the configuration of the body, $\kappa : \mathcal{X} \to \mathcal{S}$ as in Example 6.1, becomes an additional unknown of the problem.

In the previous chapter, the vector bundle W, used in the formulation of force and stress theory, was fixed. In the context of the elasticity problem, $W := \kappa^* T\mathcal{S}$, in accordance with Example 6.5 and Eq. (10.1). Now that the configuration varies, the vector bundle W is no longer fixed. Thus, the introduction of constitutive relations and the formulation of the elasticity problem require additional study of kinematics and the corresponding modifications to the formulation of force and stress theory. It turns out that the constructions involved in the formulation apply naturally in the more general situation where a configuration is viewed as a section $\kappa : \mathcal{X} \to \mathcal{Y}$ of a general fiber bundle $\pi : \mathcal{Y} \to \mathcal{X}$ (see [KOS17a], on which much of this chapter is based).

12.1 Kinematics

Configurations As mentioned in Example 6.1, up to this point, a configuration is assumed to be an embedding

$$\kappa : \mathcal{X} \longrightarrow \mathcal{S}. \tag{12.1}$$

R. Segev, *Foundations of Geometric Continuum Mechanics*, Advances in Continuum Mechanics 49, https://doi.org/10.1007/978-3-031-35655-1_12

The significant consequences of the assumption that κ is an embedding will be studied in the third part of the book. For now, the relevant property is that κ is assumed to be a smooth mapping.

This structure may be generalized somewhat as follows: Let

$$\mathrm{pr}_1 : \mathcal{Y} := \mathcal{X} \times \mathcal{S} \longrightarrow \mathcal{X} \tag{12.2}$$

be the trivial, product bundle. Then, any smooth mapping $\kappa : \mathcal{X} \to \mathcal{S}$ induces a unique section $\tilde{\kappa} : \mathcal{X} \to \mathcal{Y}$ of the fiber bundle $\mathcal{Y}$ given by $x \mapsto (x, \kappa(x))$. Moreover, one may consider a general fiber bundle

$$\pi : \mathcal{Y} \longrightarrow \mathcal{X}, \tag{12.3}$$

and define a configuration to be a section,

$$\kappa : \mathcal{X} \longrightarrow \mathcal{Y}, \tag{12.4}$$

of this fiber bundle. A configuration will be represented locally by

$$(x^i) \longmapsto (x^i, y^\alpha = \kappa^\alpha(x^j)), \qquad i = 1, \dots, n,\ \alpha = 1, \dots, m, \tag{12.5}$$

or by

$$\boldsymbol{x} \longmapsto (\boldsymbol{x}, \boldsymbol{y} = \boldsymbol{\kappa}(\boldsymbol{x})), \qquad \boldsymbol{x} \in U \subset \mathbb{R}^n,\ \boldsymbol{y} \in \mathbb{R}^m, \tag{12.6}$$

where $\boldsymbol{\kappa}$ denotes the local representative of the section.

Generalized (Virtual) Velocity Fields For the general setting where a configuration is a section κ of a fiber bundle $\pi : \mathcal{Y} \to \mathcal{X}$, we introduce below the notion of a general velocity field as a section $\mathcal{X} \to V\mathcal{Y}$, where $V\mathcal{Y}$ is the vertical subbundle of $T\mathcal{Y}$ defined in Sect. 6.1.6.

We start with the case where a configuration is viewed as an embedding $\mathcal{X} \to \mathcal{S}$ and assume that spacetime has the structure of a product $\mathbb{R} \times \mathcal{S}$. A motion is viewed as a smooth mapping

$$M : \mathbb{R} \times \mathcal{X} \longrightarrow \mathcal{S} \tag{12.7}$$

such that for each time $t \in \mathbb{R}$,

$$\kappa_t := M|_{\{t\}\times\mathcal{X}} : \mathcal{X} \longrightarrow \mathcal{S} \tag{12.8}$$

is a configuration. Noting that for each $x \in \mathcal{X}$, $M_x := M|_{\mathbb{R}\times\{x\}} : \mathbb{R} \to \mathcal{S}$ is a curve in space, its tangent, the value of its lift, $\dot{M}_x$, at time t, is interpreted as the velocity of the material point x at t. Letting x vary, one obtains, at each time t, a mapping

$$w_t := D_t M(t,\cdot) := \frac{\partial M}{\partial t}(t,\cdot) : \mathcal{X} \longrightarrow T\mathcal{S}, \qquad x \longmapsto \dot{M}_x(t). \tag{12.9}$$

Let M be a motion of a body and consider $\kappa := \kappa_0$ and $w := w_0$. As in Example 6.5, the generalized velocity, w, is a smooth vector field along κ. That is, it is a mapping $w : \mathcal{X} \to T\mathcal{S}$, such that $\tau_{\mathcal{S}} \circ w = \kappa$, where $\tau_{\mathcal{S}} : T\mathcal{S} \to \mathcal{S}$ is the tangent bundle projection for $\mathcal{S}$. In such a case, we say that w is a velocity field at the configuration κ. Note that unlike the traditional formulation in a Euclidean physical space, for a general space manifold, the value, $w(x)$, of a velocity of a body point has no meaning independently of the point $\kappa(x)$ in space where that body point is located.

For a given configuration κ, any velocity field, w, at κ, can be identified with a section $\tilde{w}$ of $\kappa^*\tau_{\mathcal{S}} : \kappa^*T\mathcal{S} \to \mathcal{X}$ because of the universal property of the pullback bundle as in the following diagram (where Id denotes the identity mapping).

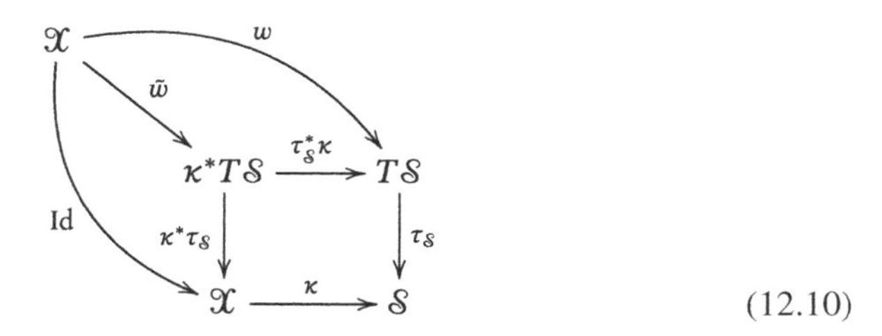

(12.10)

Indeed, we will use the same notation, w, for both.

Returning to the case where κ is no longer fixed, a velocity field is modeled by a mapping $w : \mathcal{X} \to T\mathcal{S}$ such that $\kappa := \tau_{\mathcal{S}} \circ w$ is a configuration.

Locally, a motion is represented in the form

$$(t, \boldsymbol{x}) \longmapsto \boldsymbol{y} = \boldsymbol{M}(t, \boldsymbol{x}) \in \mathbb{R}^m, \tag{12.11}$$

for a function $\boldsymbol{M} : \mathbb{R} \times U \subset \mathbb{R} \times \mathbb{R}^n \to \mathbb{R}^m$, where U is the image of a chart in $\mathcal{X}$. It follows from (12.9) that a velocity field w_t, induced by the motion, is represented locally by

$$\boldsymbol{x} \longmapsto (\boldsymbol{M}(t, \boldsymbol{x}), D_t \boldsymbol{M}(t, \boldsymbol{x})). \tag{12.12}$$

In particular, $w := w_0$ is represented in the form

$$\boldsymbol{x} \longmapsto (\boldsymbol{\kappa}(\boldsymbol{x}), \dot{\boldsymbol{\kappa}}(\boldsymbol{x})), \quad \text{where} \quad \boldsymbol{\kappa} := \boldsymbol{M}(0,\cdot),\ \dot{\boldsymbol{\kappa}} := D_t \boldsymbol{M}(0,\cdot) : U \subset \mathbb{R}^n \longrightarrow \mathbb{R}^m. \tag{12.13}$$

Consider the tangent bundle

$$T\mathcal{Y} := T(\mathcal{X} \times \mathcal{S}) \cong T\mathcal{X} \times T\mathcal{S} \xrightarrow{\tau_{\mathcal{Y}}} \mathcal{Y} := \mathcal{X} \times \mathcal{S} \xrightarrow{\mathrm{pr}_1} \mathcal{X}, \tag{12.14}$$

where the isomorphism $T(\mathcal{X} \times \mathcal{S}) \longrightarrow T\mathcal{X} \times T\mathcal{S}$ is given locally by $(\boldsymbol{x}, \boldsymbol{y}, \dot{\boldsymbol{x}}, \dot{\boldsymbol{y}}) \mapsto (\boldsymbol{x}, \dot{\boldsymbol{x}}, \boldsymbol{y}, \dot{\boldsymbol{y}})$. A mapping $w : \mathcal{X} \to T\mathcal{S}$ induces uniquely a mapping $\tilde{w} : \mathcal{X} \to T\mathcal{Y}$ represented locally by

$$\boldsymbol{x} \longmapsto (\boldsymbol{x}, \boldsymbol{\kappa}(\boldsymbol{x}), \boldsymbol{0}, \dot{\boldsymbol{\kappa}}(\boldsymbol{x})). \tag{12.15}$$

The natural choice $\dot{\boldsymbol{x}} = \boldsymbol{0} \in \mathbb{R}^n$ may be written invariantly as follows: Let $T\mathrm{pr}_1 : T\mathcal{Y} = T(\mathcal{X} \times \mathcal{S}) \to T\mathcal{X}$ be the tangent of the product bundle projection, which is represented locally by $(\boldsymbol{x}, \boldsymbol{y}, \dot{\boldsymbol{x}}, \dot{\boldsymbol{y}}) \mapsto (\boldsymbol{x}, \dot{\boldsymbol{x}})$. Then, the condition $\dot{\boldsymbol{x}} = \boldsymbol{0}$ may be written as

$$T\mathrm{pr}_1 \circ \tilde{w} = 0, \tag{12.16}$$

where, here, 0 is the zero section in $T\mathcal{X}$. Letting $V\mathcal{Y} := \mathrm{Kernel} T\mathrm{pr}_1$, we may identify a velocity field with a section $\mathcal{X} \to V\mathcal{Y}$, and we represent it locally in the form

$$\boldsymbol{x} \longmapsto (\boldsymbol{x}, \boldsymbol{\kappa}(\boldsymbol{x}), \dot{\boldsymbol{\kappa}}(\boldsymbol{x})). \tag{12.17}$$

As for the case of configurations above, the last observation may be generalized naturally to the case where $\pi : \mathcal{Y} \to \mathcal{X}$ is a general fiber bundle. For this general case, a motion is a mapping

$$M : \mathbb{R} \times \mathcal{X} \longrightarrow \mathcal{Y}, \tag{12.18}$$

such that for every $t \in \mathbb{R}$, $\kappa_t := M|_{\{t\}\times\mathcal{X}} : \mathcal{X} \to \mathcal{Y}$ is a section of π. For each $x \in \mathcal{X}$, $M_x := M|_{\mathbb{R}\times\{x\}} : \mathbb{R} \to \mathcal{Y}_x$ is a curve in the fiber over x due to the requirement that

$$\kappa_t := M|_{\{t\}\times\mathcal{X}} : \mathcal{X} \longrightarrow \mathcal{Y} \tag{12.19}$$

is a section for all t. Thus, the lift $\dot{M}_x : \mathbb{R} \to T(\mathcal{Y}_x)$ contains vectors that are tangent to the fibers. Letting x vary, and recalling the definition of the vertical subbundle $V\mathcal{Y} \subset T\mathcal{Y}$ in Sect. 6.1.6, one obtains the mapping

$$\dot{M} := D_t M : \mathbb{R} \times \mathcal{X} \longrightarrow V\mathcal{Y}, \qquad \dot{M}(t, x) := \dot{M}_x(t). \tag{12.20}$$

In analogy with (12.9), we set

$$w_t := \frac{\mathrm{d}\kappa_t}{\mathrm{d}t} := D_t M(t, \cdot) := \frac{\partial M}{\partial t}(t, \cdot) : \mathcal{X} \longrightarrow V\mathcal{Y}, \qquad x \longmapsto \dot{M}_x(t). \tag{12.21}$$

Particularly, the section

$$w = \dot{M}(0, \cdot) : \mathcal{X} \longrightarrow V\mathcal{Y} \tag{12.22}$$

is the natural velocity field induced by the motion. Therefore, we extend the definition of a *generalized velocity field* to be a section $w : \mathcal{X} \to V\mathcal{Y}$ of $\pi \circ \tau_{\mathcal{Y}} : V\mathcal{Y} \to \mathcal{X}$. The resulting structure is illustrated in the following diagram, where $\mathcal{J}_V$ denotes the natural inclusion of the vertical bundle.

$$V\mathcal{Y} \xrightarrow{\mathcal{J}_V} T\mathcal{Y} \xrightarrow{\tau_{\mathcal{Y}}} \mathcal{Y} \xrightarrow{\pi} \mathcal{X}, \qquad w : \mathcal{X} \to V\mathcal{Y}. \tag{12.23}$$

Let $\kappa_0 : \mathcal{X} \to \mathcal{Y}$ be a given configuration. Then, a *generalized velocity field at the configuration* κ_0 is a generalized velocity $w : \mathcal{X} \to V\mathcal{Y}$ such that $\tau_{\mathcal{Y}} \circ w = \kappa_0$. We note that a generalized velocity field at the configuration κ_0 may be identified with a section of the pullback bundle $\kappa_0^* VY \to \mathcal{X}$ as in the following diagram:

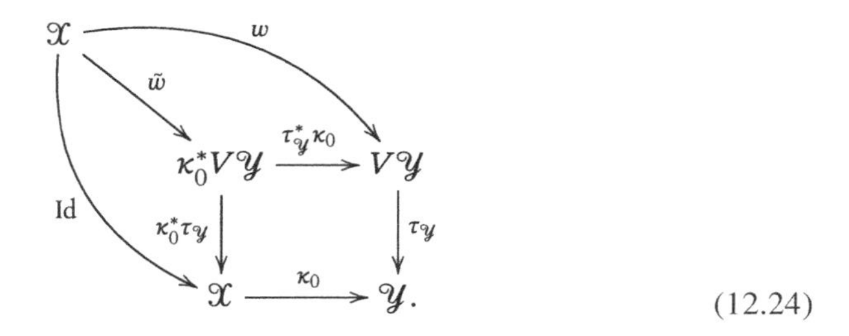

(12.24)

Here and in what follows, we simplify the notation and use $\tau_{\mathcal{Y}}$ to denote its restriction, $\tau_{\mathcal{Y}} \circ \mathcal{J}_V$, to the vertical subbundle.

Locally, a motion is represented in the form

$$(t, \boldsymbol{x}) \longmapsto (\boldsymbol{x}, \boldsymbol{y} = \boldsymbol{M}(t, \boldsymbol{x})), \tag{12.25}$$

for a function $\boldsymbol{M} : \mathbb{R} \times U \subset \mathbb{R} \times \mathbb{R}^n \to \mathbb{R}^m$, where U is the image of a chart in $\mathcal{X}$. A velocity field, w_t, induced by the motion, is represented locally by

$$\boldsymbol{x} \longmapsto (\boldsymbol{x}, \boldsymbol{M}(t, \boldsymbol{x}), D_t \boldsymbol{M}(t, \boldsymbol{x})). \tag{12.26}$$

In particular, $w := w_0$ is represented in the form

$$\boldsymbol{x} \longmapsto (\boldsymbol{x}, \boldsymbol{\kappa}(\boldsymbol{x}), \dot{\boldsymbol{\kappa}}(\boldsymbol{x})), \tag{12.27}$$

where $\boldsymbol{\kappa} := \boldsymbol{M}(0, \cdot)$ and $\dot{\boldsymbol{\kappa}} := D_t \boldsymbol{M}(0, \cdot) : U \subset \mathbb{R}^n \longrightarrow \mathbb{R}^m$.

Locally, a velocity field at the configuration κ_0 is represented in the form

$$\boldsymbol{x} \longmapsto (\boldsymbol{x}, \boldsymbol{\kappa}_0(\boldsymbol{x}), \dot{\boldsymbol{y}}(\boldsymbol{x})), \qquad \text{or} \qquad \textstyle\sum_\alpha \dot{y}^a(x^j)\partial_\alpha, \tag{12.28}$$

where for short we write ∂_α for what is actually the pullback $\kappa_{0\tau_{\mathcal{Y}}}^* \partial_\alpha$.

Tangent Mappings and Deformation Jets As in Example 6.4, the deformation gradient of classical continuum mechanics is generalized by defining it to be the tangent mapping to the configuration mapping as illustrated by the diagram

$$\begin{array}{ccc} T\mathcal{X} & \xrightarrow{T\kappa} & T\mathcal{S} \\ {\scriptstyle \tau_{\mathcal{X}}}\downarrow & & \downarrow{\scriptstyle \tau_{\mathcal{S}}} \\ \mathcal{X} & \xrightarrow{\kappa} & \mathcal{S}. \end{array} \tag{12.29}$$

Locally the tangent mapping is represented by

$$(\boldsymbol{x}, \boldsymbol{v}) \longmapsto (\boldsymbol{\kappa}(\boldsymbol{x}), D\boldsymbol{\kappa}(\boldsymbol{x})(\boldsymbol{v})), \tag{12.30}$$

where $D\boldsymbol{\kappa}$ is the derivative mapping, so that $D\boldsymbol{\kappa}(\boldsymbol{x})$ is a matrix of dimensions $m \times n$. The values of the deformation gradient at a point x are represented by $\boldsymbol{\kappa}(\boldsymbol{x})$ and $D\boldsymbol{\kappa}(\boldsymbol{x})$. It is emphasized that unlike the case of Euclidean geometry, for the general setting considered here, one cannot disassociate invariantly the value of the derivative $D\boldsymbol{\kappa}(\boldsymbol{x})$ from the value of the configuration $\boldsymbol{\kappa}(\boldsymbol{x})$. Thus, the values of the deformation gradient at $x \in \mathcal{X}$ belong to $J^1(\mathcal{X}, \mathcal{S})_x$. As mentioned in Sect. 6.1.7, $J^1(\mathcal{X}, \mathcal{S})$ is identified with

$$J^1\pi := J^1(\mathrm{pr}_1) : J^1\mathcal{Y} := J^1(\mathcal{X} \times \mathcal{S}) \longrightarrow \mathcal{X}. \tag{12.31}$$

This may be immediately generalized to the case where $\pi : \mathcal{Y} \to \mathcal{X}$ is a general fiber bundle. For a general fiber bundle, we interpret elements of $J^1\mathcal{Y}$ as possible values of the deformation gradient of a configuration. For a configuration κ, a section of π, the *deformation gradient corresponding to* κ is the jet extension,

$$j^1\kappa : \mathcal{X} \longrightarrow J^1\mathcal{Y}, \tag{12.32}$$

which is represented locally by

$$\boldsymbol{x} \longmapsto (\boldsymbol{x}, \boldsymbol{\kappa}(\boldsymbol{x}), D\boldsymbol{\kappa}(\boldsymbol{x})). \tag{12.33}$$

It is noted that a general section, $\chi : \mathcal{X} \to J^1\mathcal{Y}$, is represented locally by $\boldsymbol{x} \longmapsto (\boldsymbol{x}, \boldsymbol{\chi}_0(\boldsymbol{x}), \boldsymbol{\chi}_1(\boldsymbol{x}))$ for smooth mappings $\boldsymbol{\chi}_0 : U \subset \mathbb{R}^n \to \mathbb{R}^m$ and $\boldsymbol{\chi}_1 : U \subset \mathbb{R}^n \to L(\mathbb{R}^n, \mathbb{R}^m)$. Such a section, a *jet field*, is not necessarily integrable—compatible, holonomic—as χ need not be the jet extension, $j^1\kappa$, of a section $\kappa : \mathcal{X} \to \mathcal{Y}$. For example, the local representatives, $\boldsymbol{\chi}_1$, need not be the derivative of some $\boldsymbol{\chi}_0$. This is analogous to the situation in classical continuum mechanics.

Velocity Jets The principle of virtual power, as presented in Eq. (10.59), demonstrates the major role played by jets of velocity fields. Since we now view a velocity field as a section $w : \mathcal{X} \to V\mathcal{Y}$ of the vertical bundle as in (12.23), jets of velocity fields are valued in the jet bundle, $J^1V\mathcal{Y}$. Thus, for a velocity field w

at the configuration κ, we have

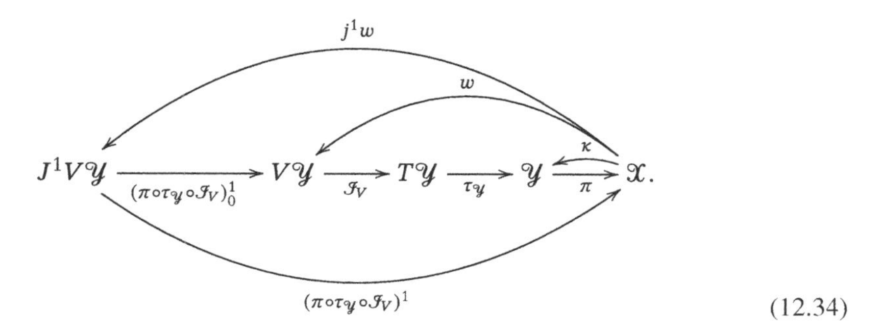

(12.34)

Remark 12.1 Notation: It is noted that $V\mathcal{Y}$ is also a fiber (vector) bundle over $\mathcal{Y}$, and the notation $J^1 V\mathcal{Y}$ is ambiguous, as it could also mean jets of sections $\mathcal{Y} \to V\mathcal{Y}$. This ambiguity may be removed by writing $J^1(\mathcal{X}, V\mathcal{Y})$, in distinction with $J^1(\mathcal{Y}, V\mathcal{Y})$. As a convention, for the sake of simplicity, when we have a sequence of fiber bundles $\mathcal{Y} \to \mathcal{Z} \to \cdots \to \mathcal{X}$, $J^1\mathcal{Y} := J^1(\mathcal{X}, \mathcal{Y})$. With this convention, the notation in the diagram above is well-defined.

In view of our notation conventions, the foregoing diagram can be represented in short by

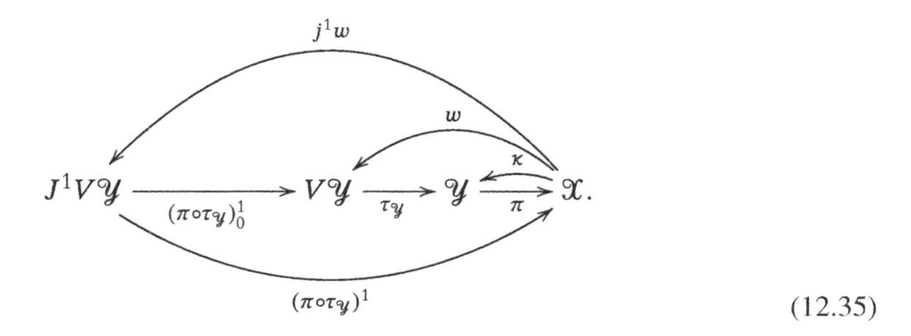

(12.35)

An element of $J^1 V\mathcal{Y}$ is represented locally in the form $(\boldsymbol{x}, \boldsymbol{y}, \dot{\boldsymbol{y}}, \boldsymbol{y}', \dot{\boldsymbol{y}}')$, where $\boldsymbol{y}', \dot{\boldsymbol{y}}' \in L(\mathbb{R}^n, \mathbb{R}^m)$. If a velocity field w is represented locally as in Eq. (12.15), then $j^1 w$ is represented locally by

$$\boldsymbol{x} \longmapsto (\boldsymbol{x}, \boldsymbol{\kappa}(\boldsymbol{x}), \dot{\boldsymbol{\kappa}}(\boldsymbol{x}), D\boldsymbol{\kappa}(\boldsymbol{x}), D\dot{\boldsymbol{\kappa}}(\boldsymbol{x})). \tag{12.36}$$

Similarly to the case of deformation jets, it is noted that, in general, a section $\mathcal{X} \to J^1 V\mathcal{Y}$ need not be holonomic. That is, if the section is represented by

$$\boldsymbol{x} \longmapsto (\boldsymbol{x}, \boldsymbol{\kappa}(\boldsymbol{x}), \dot{\boldsymbol{\kappa}}(\boldsymbol{x}), \boldsymbol{\chi}(\boldsymbol{x}), \dot{\boldsymbol{\chi}}(\boldsymbol{x})), \tag{12.37}$$

then, $\boldsymbol{\chi}$ and $\dot{\boldsymbol{\chi}}$ need not be the derivatives of $\boldsymbol{\kappa}$ and $\dot{\boldsymbol{\kappa}}$, respectively.

To study the properties of jets of velocity field further, let $\varepsilon \in J^1VY$ and let $(\pi \circ \tau_{\mathcal{Y}})^1(\varepsilon) = x_0$. Then, there is a velocity field $w : \mathcal{X} \to V\mathcal{Y}$ such that $\varepsilon = j^1_{x_0} w$. Consequently, there is a motion $M : \mathbb{R} \times \mathcal{X} \to \mathcal{Y}$ such that for every $t \in \mathbb{R}$, $\kappa_t := M|_{\{t\}\times\mathcal{X}} : \mathcal{X} \to \mathcal{Y}$ is a section of π. Moreover, for

$$\dot{M} := D_t M : \mathbb{R} \times \mathcal{X} \longrightarrow V\mathcal{Y}, \tag{12.38}$$

we have $w = \dot{M}(0, \cdot)$, and $j^1 w = j^1(\dot{M}(0, \cdot))$.

Let M be represented locally in the form

$$(t, \boldsymbol{x}) \longmapsto (\boldsymbol{x}, \boldsymbol{M}(t, \boldsymbol{x})), \qquad \boldsymbol{M}(t, \boldsymbol{x}) \in \mathbb{R}^m. \tag{12.39}$$

Then, using $D_t \boldsymbol{M}$ and $D_{\boldsymbol{x}} \boldsymbol{M}$ to denote the respective partial derivatives,

$$\begin{aligned}
\dot{M} &\text{ is represented by } (t, \boldsymbol{x}) \longmapsto (\boldsymbol{x}, \boldsymbol{M}(t, \boldsymbol{x}), D_t \boldsymbol{M}(t, \boldsymbol{x})),\\
w &\text{ is represented by } \quad \boldsymbol{x} \longmapsto (\boldsymbol{x}, \boldsymbol{M}(0, \boldsymbol{x}), D_t \boldsymbol{M}(0, \boldsymbol{x})),\\
j^1 w &\text{ is represented by } \quad \boldsymbol{x} \longmapsto (\boldsymbol{x}, \boldsymbol{M}(0, \boldsymbol{x}), D_t \boldsymbol{M}(0, \boldsymbol{x}),\\
&\qquad\qquad D_{\boldsymbol{x}} \boldsymbol{M}(0, \boldsymbol{x}), D_{\boldsymbol{x}}(D_t \boldsymbol{M}(0, \boldsymbol{x}))).
\end{aligned} \tag{12.40}$$

Jet Velocities A time-dependent jet field is a smooth mapping

$$M^J : \mathbb{R} \times \mathcal{X} \longrightarrow J^1\mathcal{Y} \tag{12.41}$$

such that for every $t \in \mathbb{R}$, $\chi_t := M^J|_{\{t\}\times\mathcal{X}} : \mathcal{X} \to J^1\mathcal{Y}$ is a jet field—a section of π^1. In analogy with (12.20), one can define

$$\dot{M}^J := D_t M^J : \mathbb{R} \times \mathcal{X} \longrightarrow VJ^1\mathcal{Y}, \qquad \dot{M}^J(t, x) := \dot{M}^J_x(t), \tag{12.42}$$

where $M^J_x := M^J|_{\mathbb{R}\times\{x\}} : \mathbb{R} \to (J^1\mathcal{Y})_x$. A *rate of change of a jet field*, or, for short, a *jet velocity field*, is a section $\eta : \mathcal{X} \longrightarrow VJ^1\mathcal{Y}$. Thus, $\eta = \dot{M}^J(0, \cdot)$, for some motion M^J, is a jet velocity field.

Locally,

$$\begin{aligned}
M^J &\text{ is represented by } (t, \boldsymbol{x}) \longmapsto (\boldsymbol{x}, \boldsymbol{M}^J_0(t, \boldsymbol{x}), \boldsymbol{M}^J_1(t, \boldsymbol{x})),\\
\dot{M}^J &\text{ is represented by } (t, \boldsymbol{x}) \longmapsto (\boldsymbol{x}, \boldsymbol{M}^J_0(t, \boldsymbol{x}), \boldsymbol{M}^J_1(t, \boldsymbol{x}),\\
&\qquad\qquad D_t \boldsymbol{M}^J_0(t, \boldsymbol{x}), D_t \boldsymbol{M}^J_1(t, \boldsymbol{x})),\\
\eta &\text{ is represented by } \quad \boldsymbol{x} \longmapsto (\boldsymbol{x}, \boldsymbol{M}^J_0(0, \boldsymbol{x}), \boldsymbol{M}^J_1(0, \boldsymbol{x}),\\
&\qquad\qquad D_t \boldsymbol{M}^J_0(0, \boldsymbol{x}), D_t \boldsymbol{M}^J_1(0, \boldsymbol{x})).
\end{aligned} \tag{12.43}$$

Let $\chi_0 : \mathcal{X} \to J^1\mathcal{Y}$ be a jet field. We will say that a jet velocity field $\eta : \mathcal{X} \to VJ^1\mathcal{Y}$ is a *jet velocity field at* χ_0 if $\pi_0^1 \circ \eta = \chi_0$. A jet velocity field at χ_0 is represented in the form

$$\boldsymbol{x} \longmapsto (\boldsymbol{x}, (\boldsymbol{\chi}_0)_0(\boldsymbol{x}), (\boldsymbol{\chi}_0)_1(\boldsymbol{x}), \dot{\boldsymbol{y}}(\boldsymbol{x}), \dot{\boldsymbol{y}}'(\boldsymbol{x})), \tag{12.44}$$

or

$$(x^j) \longmapsto \textstyle\sum_\alpha \dot{y}^\alpha(x^j)\partial_\alpha + \sum_{\alpha,i}(\dot{y}')_i^\alpha(x^j)\mathrm{d}x^i \otimes \partial_\alpha. \tag{12.45}$$

For the case where the motion $M^J(t,\cdot)$ is compatible for each $t \in \mathbb{R}$, $\boldsymbol{M}_1^J = D_{\boldsymbol{x}}\boldsymbol{M}_0^J$, and η is represented by

$$\boldsymbol{x} \longmapsto (\boldsymbol{x}, \boldsymbol{M}_0^J(0,\boldsymbol{x}), D_{\boldsymbol{x}}\boldsymbol{M}_0^J(0,\boldsymbol{x}), D_t\boldsymbol{M}_0^J(0,\boldsymbol{x}), D_t(D_{\boldsymbol{x}}\boldsymbol{M}_0^J)(0,\boldsymbol{x})). \tag{12.46}$$

Note that for every $x \in \mathcal{X}$, one can construct a compatible motion as above. Hence comparing the last equation the last line of (12.40) and observing that $D_t \circ D_{\boldsymbol{x}} = D_{\boldsymbol{x}} \circ D_t$, we have constructed a fiber bundle diffeomorphism (see [Pal68, pp. 82–83], [Sau89, p.125]),

$$\imath : J^1V\mathcal{Y} \longrightarrow VJ^1\mathcal{Y}, \tag{12.47}$$

whereby, locally,

$$(\boldsymbol{x}, \boldsymbol{y}, \dot{\boldsymbol{y}}, \boldsymbol{y}', \dot{\boldsymbol{y}}') \longmapsto (\boldsymbol{x}, \boldsymbol{y}, \boldsymbol{y}', \dot{\boldsymbol{y}}, \dot{\boldsymbol{y}}'). \tag{12.48}$$

Roughly speaking, the value of a velocity jet at $x \in \mathcal{X}$ may be identified with a value of a jet velocity at that point. In case $\varepsilon \in J^1V\mathcal{Y}$ is represented by a motion M as $\varepsilon = j_x^1(\dot{M}(0,\cdot))$, then,

$$\imath(\varepsilon) = (j^1(M(t,\cdot)))^{\cdot}(0,x). \tag{12.49}$$

The various mappings associated with $J^1V\mathcal{Y}$ and VJ^1Y are illustrated in the following diagram: Note that in the diagram we have made use of the fact that tangent mappings to fiber bundle morphisms map vertical vectors to vertical vectors. We also use the definition of a jet of a fiber bundle morphism, (6.48), to obtain $J^1\tau_{\mathcal{Y}}$ from the fiber bundle morphism $\tau_{\mathcal{Y}}$ over the identity $\mathcal{X} \to \mathcal{X}$.

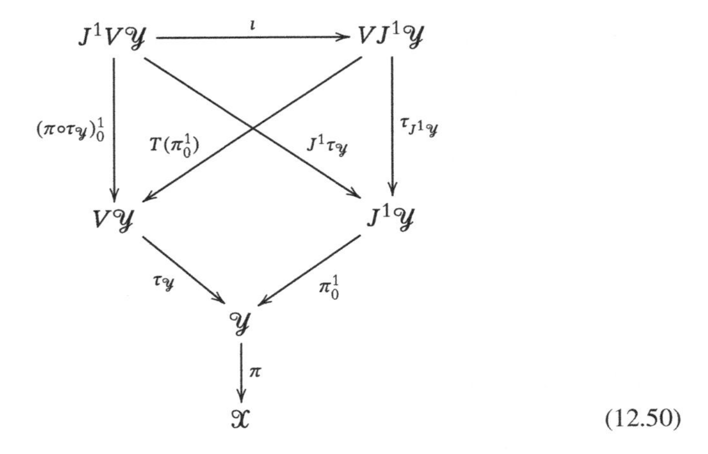

(12.50)

Let $\kappa : \mathcal{X} \to \mathcal{Y}$ be a configuration and let $\chi = j^1\kappa$. We want to show that as fiber bundles over $\mathcal{X}$,

$$J^1(\kappa^* V\mathcal{Y}) \cong \chi^*(VJ^1Y) \tag{12.51}$$

(see [Pal68, p. 82]). Consider the following commutative diagrams:

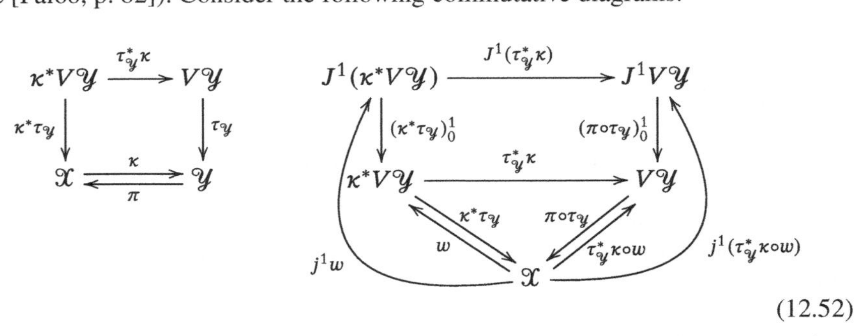

(12.52)

The next commutative diagram illustrates the actions of the local representatives of the various maps. (For simplicity, the maps are indicated on the arrows, but the actions are those of the corresponding local representatives.)

$$\begin{array}{ccc}
(\boldsymbol{x}, \dot{\boldsymbol{\kappa}}(\boldsymbol{x}), D\dot{\boldsymbol{\kappa}}(\boldsymbol{x})) & \xmapsto{J^1(\tau^*_{\mathcal{Y}}\kappa)} & (\boldsymbol{x}, \boldsymbol{\kappa}(\boldsymbol{x}), \dot{\boldsymbol{\kappa}}(\boldsymbol{x}), D\boldsymbol{\kappa}(\boldsymbol{x}), D\dot{\boldsymbol{\kappa}}(\boldsymbol{x})) \\
\downarrow (\kappa^*\tau_{\mathcal{Y}})^1_0 & & \downarrow (\pi\circ\tau_{\mathcal{Y}})^1_0 \\
(\boldsymbol{x}, \dot{\boldsymbol{\kappa}}(\boldsymbol{x})) & \xmapsto{\tau^*_{\mathcal{Y}}\kappa} & (\boldsymbol{x}, \boldsymbol{\kappa}(\boldsymbol{x}), \dot{\boldsymbol{\kappa}}(\boldsymbol{x})) \\
\kappa^*\tau_{\mathcal{Y}} \searrow \nwarrow w & & \pi\circ\tau_{\mathcal{Y}} \swarrow \nearrow \tau^*_{\mathcal{Y}}\kappa\circ w \\
& \boldsymbol{x} &
\end{array}$$

with $j^1 w : \boldsymbol{x} \mapsto (\boldsymbol{x}, \dot{\boldsymbol{\kappa}}(\boldsymbol{x}), D\dot{\boldsymbol{\kappa}}(\boldsymbol{x}))$ and $j^1(\tau^*_{\mathcal{Y}}\kappa\circ w) : \boldsymbol{x} \mapsto (\boldsymbol{x}, \boldsymbol{\kappa}(\boldsymbol{x}), \dot{\boldsymbol{\kappa}}(\boldsymbol{x}), D\boldsymbol{\kappa}(\boldsymbol{x}), D\dot{\boldsymbol{\kappa}}(\boldsymbol{x}))$

(12.53)

Composing $J^1(\tau^*_{\mathcal{Y}}\kappa)$ with the involution ι above, we have

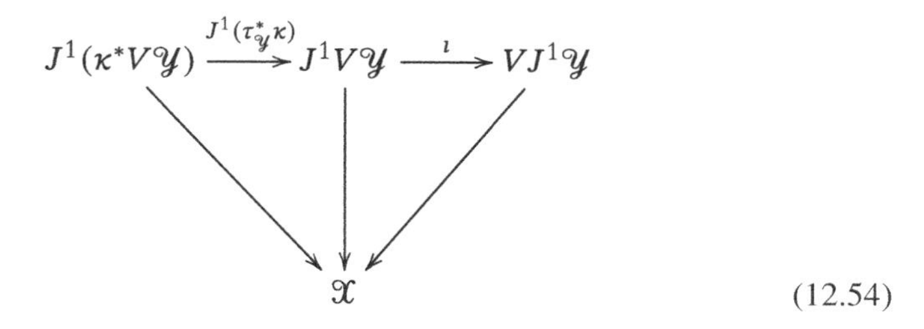

$$\begin{array}{ccccc} J^1(\kappa^* V\mathcal{Y}) & \xrightarrow{J^1(\tau^*_{\mathcal{Y}}\kappa)} & J^1 V\mathcal{Y} & \xrightarrow{\iota} & V J^1\mathcal{Y} \\ & \searrow & \downarrow & \swarrow & \\ & & \mathcal{X} & & \end{array} \tag{12.54}$$

where, in view of (12.48), $\iota \circ j^1(\tau^*_{\mathcal{Y}}\kappa)$ is represented locally by

$$(\boldsymbol{x}, \dot{\boldsymbol{\kappa}}(\boldsymbol{x}), D\dot{\boldsymbol{\kappa}}(\boldsymbol{x})) \longmapsto (\boldsymbol{x}, \boldsymbol{\kappa}(\boldsymbol{x}), D\boldsymbol{\kappa}(\boldsymbol{x}), \dot{\boldsymbol{\kappa}}(\boldsymbol{x}), D\dot{\boldsymbol{\kappa}}(\boldsymbol{x})). \tag{12.55}$$

Since $(\boldsymbol{x}, \boldsymbol{\kappa}(\boldsymbol{x}), D\boldsymbol{\kappa}(\boldsymbol{x}))$ represents the fixed $\chi(x) := j^1\kappa(x)$, we conclude that image $\iota \circ J^1(\tau^*_{\mathcal{Y}}\kappa)$ is $(j^1\kappa)^* V J^1\mathcal{Y}$.

In particular, we note that for a motion $M : \mathbb{R} \times \mathcal{X} \to \mathcal{Y}$, and the induced time-dependent jet field

$$M^J : \mathbb{R} \times \mathcal{X} \longrightarrow J^1\mathcal{Y}, \qquad (t, x) \longmapsto j^1\kappa_t(x), \tag{12.56}$$

one may make the identification,

$$\dot{M}^J(t, x) := D_t M^J(t, x) := D_t(j^1\kappa_t)(t, x) = j^1(D_t\kappa_t)(x) =: (j^1\dot{M})(x, t). \tag{12.57}$$

12.2 Statics

In this section, using the terminology introduced above, we adapt the notions of force fields and stresses to the case where the configuration of the body in space is variable.

12.2.1 Preliminaries on Vector Bundle Morphisms and Pullbacks

We consider here various structures involving vector bundle morphisms and pullbacks of bundles that will be used below.

It is first noted that a vector bundle morphism over $\mathcal{X}$,

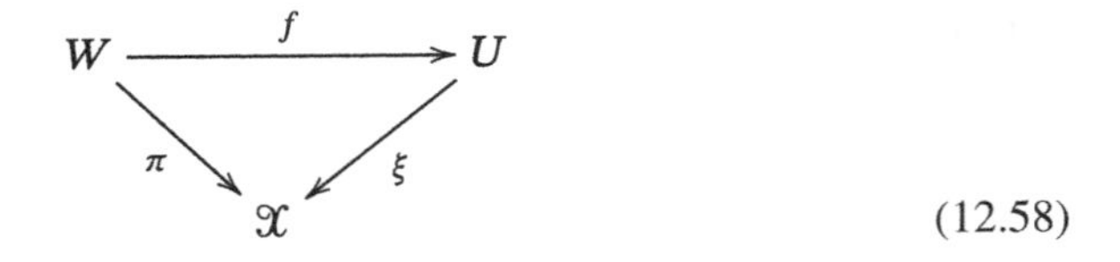

(12.58)

may be identified with a smooth section $\tilde{f}$ of $L(W, U) = W^* \otimes U$. Given f, one sets $\tilde{f}(x) := f|_{W_x}$. Conversely, given $\tilde{f}$ one sets $f(w) := \tilde{f}(\pi(w))(w)$. Thus, in the sequel, we will identify $\tilde{f}$ and f.

A slightly more general situation will be encountered below. There are two base manifolds, $\mathfrak{X}, \mathcal{X}$, and there is a mapping $\mu : \mathfrak{X} \to \mathcal{X}$, such that the following diagram is commutative.

$$
\begin{array}{ccc}
W & \xrightarrow{f} & U \\
{\scriptstyle \xi}\downarrow & & \downarrow{\scriptstyle \pi} \\
\mathfrak{X} & \xrightarrow{\mu} & \mathcal{X}
\end{array}
\tag{12.59}
$$

In such a case, one may use the universality of the pullback bundle $\mu^*\pi$ to construct the mapping f_U that makes the following diagram commutative.

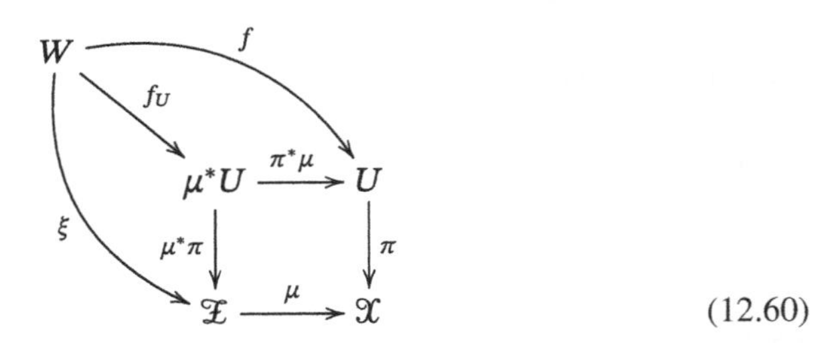

(12.60)

As a vector bundle morphism over $\mathfrak{X}$, for a fixed base mapping μ, f_U may be identified with a section of $L(W, \mu^*U)$. Thus, by universality, the vector bundle morphism f may also be viewed as a section of $L(W, \mu^*U)$. For example, given $\mu : \mathfrak{X} \to \mathcal{X}$, then, the tangent mapping $T\mu : T\mathfrak{X} \to T\mathcal{X}$ induces a unique section

$$
\mathrm{d}\mu : \mathfrak{X} \longrightarrow L(T\mathfrak{X}, \mu^*T\mathcal{X}), \tag{12.61}
$$

as in

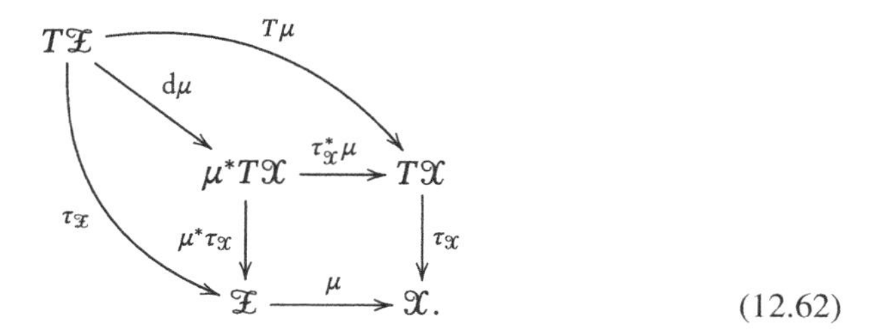

(12.62)

Next, we recall the definition of a pullback of a bundle morphism. Let $\pi_{\mathcal{Y}} : \mathcal{Y} \to \mathcal{X}$ and $\pi_{\mathcal{Z}} : \mathcal{Z} \to \mathcal{X}$ be two fiber bundles. Let $\phi : \mathcal{Y} \to \mathcal{Z}$ be a bundle morphism and let $f : \mathcal{M} \to \mathcal{X}$ be a smooth mapping defined on the manifold $\mathcal{M}$. Then, the pullback of the bundle morphism ϕ by the mapping f is the bundle morphism, $f^*\phi : f^*\mathcal{Y} \to f^*\mathcal{Z}$, defined so as to make the following diagram commutative.

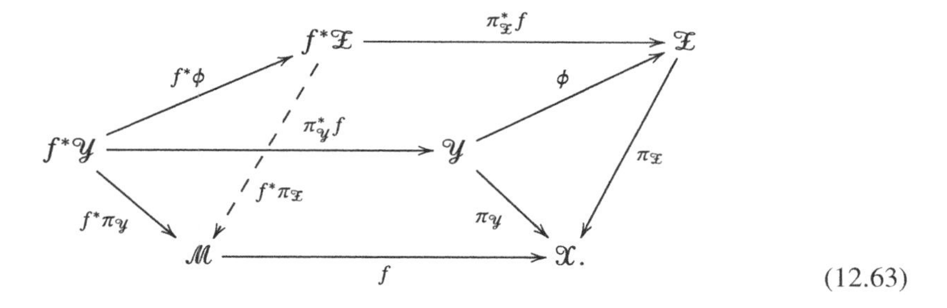

(12.63)

Consider the special case where W and U are vector bundles over $\mathcal{X}$ and ϕ is a vector bundle morphism, as in the following diagram:

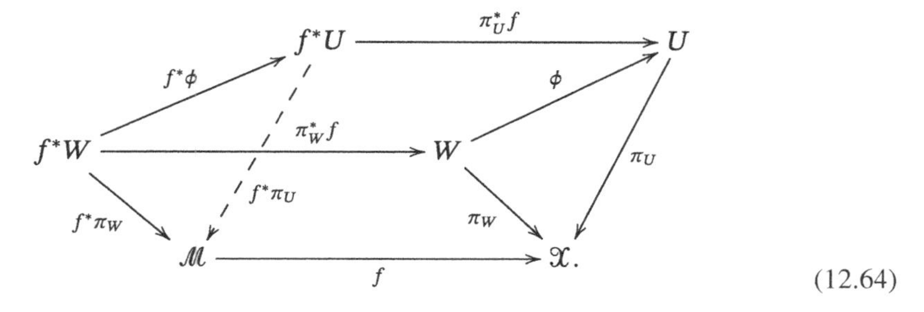

(12.64)

Then, we may identify ϕ with a section of $\pi_L : L(W, U) \to \mathcal{X}$ and identify $f^*\phi$ with a section of

$$L(f^*W, f^*U). \tag{12.65}$$

Since, for any $m \in \mathcal{M}$,

$$
\begin{aligned}
L(f^*W, f^*U)_m &= L((f^*W)_m, (f^*U)_m), \\
&= L(W_{f(m)}, U_{f(m)}), \\
&= L(W, U)_{f(m)}, \\
&= (f^*L(W, U))_m,
\end{aligned}
\tag{12.66}
$$

we conclude that

$$
L(f^*W, f^*U) = f^*L(W, U), \qquad f^*\pi_L : f^*L(W, U) \longrightarrow \mathcal{M}. \tag{12.67}
$$

It follows that $f^*\phi$ may be identified with a section $\widetilde{f^*\phi}$ of $f^*L(W, U)$, which is naturally the pullback of the section $\tilde{\phi}$ representing ϕ, as in the following diagram:

$$
\begin{array}{ccc}
f^*L(W,U) & \xrightarrow{\pi_L^* f} & L(W,U) \\
{\scriptstyle f^*_{\pi_L}\tilde{\phi}=\widetilde{f^*\phi}}\uparrow\downarrow{\scriptstyle f^*\pi_L} & & {\scriptstyle \pi_L}\downarrow\uparrow{\scriptstyle \tilde{\phi}} \\
\mathcal{M} & \xrightarrow[f]{} & \mathcal{X}.
\end{array}
\tag{12.68}
$$

12.2.2 *Force Fields*

Body Forces Identifying the vector bundle W of Chap. 10 with the pullback $\kappa^*V\mathcal{Y}$ for some configuration $\kappa : \mathcal{X} \to \mathcal{Y}$, as we did above, a body force field $\mathbf{b}$ at the configuration κ is a section of $L(\kappa^*V\mathcal{Y}, \bigwedge^n T^*\mathcal{X})$, which one may identify with a vector bundle morphism $\kappa^*V\mathcal{Y} \to \bigwedge^n T^*\mathcal{X}$ over $\mathcal{X}$.

When one lets κ vary, a *body force field* $\mathbf{b}$ is modeled as a fiber bundle morphism

$$
\begin{array}{ccc}
V\mathcal{Y} & \xrightarrow{\mathbf{b}} & \bigwedge^n T^*\mathcal{X} \\
{\scriptstyle \tau_{\mathcal{Y}}}\downarrow & & \downarrow{\scriptstyle \tau^{*n}_{\mathcal{X}}} \\
\mathcal{Y} & \xrightarrow{\pi} & \mathcal{X}.
\end{array}
\tag{12.69}
$$

Then, as noted above, and in view of the diagram

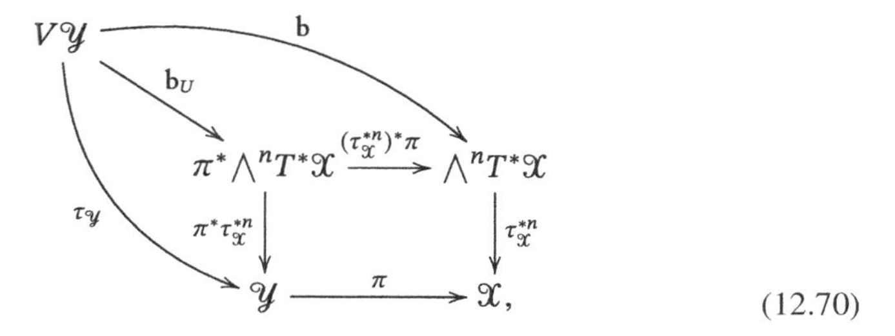

(12.70)

$\mathbf{b}$ may be represented as a section of $L(V\mathcal{Y}, \pi^*\bigwedge^n T^*\mathcal{X})$. Accordingly, we will refer to elements of $L(V\mathcal{Y}, \pi^*\bigwedge^n T^*\mathcal{X})$ as body forces.

We want to show how the definition of $\mathbf{b}$, as a section of

$$\xi_{\mathbf{b}} : L(V\mathcal{Y}, \pi^*\textstyle\bigwedge^n T^*\mathcal{X}) \to \mathcal{Y}, \tag{12.71}$$

agrees with the definition of a body force at a particular configuration κ. Consider the following commutative diagram for a body force field $\mathbf{b}$ and a configuration κ.

$$\begin{array}{ccc} \kappa^* L(V\mathcal{Y}, \pi^*\bigwedge^n T^*\mathcal{X}) & \xrightarrow{\xi_{\mathbf{b}}^*\kappa} & L(V\mathcal{Y}, \pi^*\bigwedge^n T^*\mathcal{X}) \\ \kappa_{\xi_{\mathbf{b}}}^*\mathbf{b} \uparrow\downarrow \kappa^*\xi_{\mathbf{b}} & & \xi_{\mathbf{b}} \downarrow\uparrow \mathbf{b} \\ \mathcal{X} & \underset{\kappa}{\overset{\pi}{\rightleftarrows}} & \mathcal{Y}, \end{array} \tag{12.72}$$

where $\kappa_{\xi_{\mathbf{b}}}^*\mathbf{b}$ is the pullback of the section $\mathbf{b}$ by κ as in Sect. 6.1.9. Next, it is observed that

$$\begin{aligned} (\kappa^* L(V\mathcal{Y}, \pi^*\textstyle\bigwedge^n T^*\mathcal{X}))_x &= L(V\mathcal{Y}, \pi^*\textstyle\bigwedge^n T^*\mathcal{X})_{\kappa(x)}, \\ &= L((V\mathcal{Y})_{\kappa(x)}, (\pi^*\textstyle\bigwedge^n T^*\mathcal{X})_{\kappa(x)}), \\ &= L((\kappa^* V\mathcal{Y})_x, (\textstyle\bigwedge^n T^*\mathcal{X})_{\pi(\kappa(x))}), \\ &= L(\kappa^* V\mathcal{Y}, \textstyle\bigwedge^n T^*\mathcal{X})_x, \end{aligned} \tag{12.73}$$

so that $\kappa^* L(V\mathcal{Y}, \pi^*\bigwedge^n T^*\mathcal{X}) = L(\kappa^* V\mathcal{Y}, \bigwedge^n T^*\mathcal{X})$. Moreover, by the definition of the pullback of a section in Sect. 6.1.9,

$$\mathbf{b}_\kappa := \kappa_{\xi_{\mathbf{b}}}^*\mathbf{b} = \mathbf{b} \circ \kappa : \mathcal{X} \longrightarrow L(\kappa^* V\mathcal{Y}, \textstyle\bigwedge^n T^*\mathcal{X}), \tag{12.74}$$

which is the induced *body force at the configuration* κ.

In order to represent the local form of a body force, we recall (see (12.28)) that a vector u in $V\mathcal{Y}$ may be represented in the form $u = \sum_\alpha u^\alpha \partial_\alpha$, where $\partial_\alpha := \partial/\partial y^\alpha$. Accordingly, an element ϕ of $(V\mathcal{Y})^*$ may be represented in the form $\phi = \sum_\alpha \phi_\alpha \mathrm{d}y^\alpha$. Next, we consider $\pi^*\bigwedge^n T^*\mathcal{X}$ as represented in the following diagram:

$$\begin{array}{ccc} \pi^*\bigwedge^n T^*\mathcal{X} & \xrightarrow{(\tau^*)^*\pi} & \bigwedge^n T^*\mathcal{X} \\ \pi^*_{\tau^*}\mathrm{d}x \uparrow\downarrow \pi^*\tau^* & & \tau^* \downarrow\uparrow \mathrm{d}x \\ \mathcal{Y} & \xrightarrow[\pi]{} & \mathcal{X}, \end{array} \tag{12.75}$$

where we wrote τ^* for $\tau^{*n}_{\mathcal{X}}$ and ignored the fact that $\mathrm{d}x$ is defined only locally. In what follows, we will simply write $\mathrm{d}x$ for the pullback form $\pi^*_{\tau^{*n}_{\mathcal{X}}}\mathrm{d}x$. Thus, a body force field may be represented locally in the form

$$\sum_\alpha \mathbf{b}_{\alpha 1\ldots n}(x^j, y^\beta)\mathrm{d}y^\alpha \otimes \mathrm{d}x. \tag{12.76}$$

Note that the function $\mathbf{b}_{\alpha 1\ldots n}(x^j, y^\beta)$ contains constitutive information as to the dependence of the body force on the values of a configuration. A body force field at the configuration κ, the pullback section $\kappa^*_{\xi_\mathbf{b}}\mathbf{b}$, is represented locally by

$$\sum_\alpha \mathbf{b}_{\alpha 1\ldots n}(x^j, \kappa^\beta(x^i))g^\alpha \otimes \mathrm{d}x, \tag{12.77}$$

where $g^\alpha = \kappa^*_{\xi_\mathbf{b}}\mathrm{d}y^\alpha$.

For a given velocity field, $w : \mathcal{X} \to V\mathcal{Y}$, we will use $\mathbf{b} \cdot w$ to denote the n-form $(\kappa^*_{\xi_\mathbf{b}}\mathbf{b}) \cdot w$ as an extension of the notation used in Sect. 10.1.

Surface Forces We consider surface forces acting on the boundary $\partial\mathcal{X}$ of the body $\mathcal{X}$. As the body is fixed, the indication of the region, as in Eq. (10.3), will be omitted and we will simply use $\mathbf{t}$ for $\mathbf{t}_\mathcal{X}$. In addition, to simplify the notation, we will use $\pi : W \to \partial\mathcal{X}$ and $\kappa : \partial\mathcal{X} \to \mathcal{Y}$, rather than the more accurate $\pi|_{\pi^{-1}(\partial\mathcal{X})} : W|_{\partial\mathcal{X}} \to \partial\mathcal{X}$ and $\kappa|_{\partial\mathcal{X}} : \partial\mathcal{X} \to \mathcal{Y}$, respectively. Thus, a surface force on $\partial\mathcal{X}$ at the configuration κ is a section $\partial\mathcal{X} \longrightarrow L(\kappa^* V\mathcal{Y}, \bigwedge^{n-1}T^*\partial\mathcal{X})$.

The modifications needed in order to allow variable configurations are analogous to those made for body forces. Thus, a surface force field is modeled by a vector bundle morphism

$$\begin{array}{ccc} V\mathcal{Y} & \xrightarrow{\mathbf{t}} & \bigwedge^{n-1}T^*\partial\mathcal{X} \\ \tau_\mathcal{Y} \downarrow & & \downarrow \tau^{*n-1}_{\partial\mathcal{X}} \\ \mathcal{Y} & \xrightarrow{\pi} & \partial\mathcal{X}. \end{array} \tag{12.78}$$

Alternatively and similarly to the situation for body forces, using the universal property, a *surface force field* may be defined as a section

$$\mathbf{t} : \mathcal{Y} \longrightarrow L(V\mathcal{Y}, \pi^*\textstyle\bigwedge^{n-1}T^*\partial\mathcal{X}), \tag{12.79}$$

of the fiber bundle $\xi_{\mathbf{t}} : L(V\mathcal{Y}, \pi^* \bigwedge^{n-1} T^* \partial \mathcal{X}) \longrightarrow \mathcal{Y}$. For a configuration, κ, the induced surface force—the *surface force at the configuration* κ—is the pullback section

$$\mathbf{t}_\kappa = \kappa^*_{\xi_{\mathbf{t}}} \mathbf{t} = \mathbf{t} \circ \kappa : \partial \mathcal{X} \longrightarrow L(\kappa^* V\mathcal{Y}, \bigwedge^{n-1} T^* \partial \mathcal{X}). \tag{12.80}$$

Locally, $\mathbf{t}$ is represented in the form

$$\sum_\alpha \mathbf{t}_{\alpha 1 \ldots (n-1)}(z^b, y^\beta) \mathrm{d}y^\alpha \otimes \mathrm{d}z, \tag{12.81}$$

where z^b, $b = 1, \ldots, n-1$, are local coordinates in $\partial \mathcal{X}$, $\mathrm{d}z$ is the induced $(n-1)$-form on $\partial \mathcal{X}$, and as above, we use $\mathrm{d}z$, for short, to denote what is actually $\pi^*_{\tau^{*n-1}_{\mathcal{X}}} \mathrm{d}z$. The induced field, $\kappa^*_{\xi_{\mathbf{t}}} \mathbf{t}$, at the configuration κ is represented by

$$\sum_\alpha \mathbf{t}_{\alpha 1 \ldots (n-1)}(z^b, \boldsymbol{\kappa}^\beta (x^i(z^a))) g^\alpha \otimes \mathrm{d}z, \tag{12.82}$$

where $x^i(z^a)$ denote the representative of the inclusion $\partial \mathcal{X} \to \mathcal{X}$. It is noted that here also, the field contains constitutive information on the dependence of the surface force on the configuration.

Loadings For a given body $\mathcal{X}$, a pair $(\mathbf{b}, \mathbf{t})$ containing a body force and a surface force is a *loading*. Given a loading $(\mathbf{b}, \mathbf{t})$, and a configuration κ, the loading corresponding to a configuration κ is the pair

$$(\mathbf{b}_\kappa, \mathbf{t}_\kappa) := (\kappa^*_{\xi_{\mathbf{b}}} \mathbf{b}, \kappa^*_{\xi_{\mathbf{t}}} \mathbf{t}) = (\mathbf{b} \circ \kappa, \mathbf{t} \circ \kappa). \tag{12.83}$$

Thus, given a loading as above and a configuration κ, a force, F_κ, is induced so that

$$F_\kappa(w) = \int_{\mathcal{X}} \mathbf{b}_\kappa \cdot w + \int_{\partial \mathcal{X}} \mathbf{t}_\kappa \cdot w \tag{12.84}$$

for every generalized velocity field $w : \mathcal{X} \to \kappa^* V\mathcal{Y}$.

12.2.3 Traction Stresses

Traction Stress A *traction stress field* is defined as a vector bundle morphism

$$\begin{array}{ccc} V\mathcal{Y} & \xrightarrow{s} & \bigwedge^{n-1}T^*\mathcal{X} \\ \tau_{\mathcal{Y}}\downarrow & & \downarrow \tau_{\mathcal{X}}^{*n-1} \\ \mathcal{Y} & \xrightarrow{\pi} & \mathcal{X}. \end{array} \tag{12.85}$$

Thus, a traction stress field may be represented as a section of the vector bundle,

$$\xi_s : L(V\mathcal{Y}, \pi^*\bigwedge\nolimits^{n-1}T^*\mathcal{X}) \longrightarrow \mathcal{Y}, \tag{12.86}$$

the elements of which represent possible values of traction stresses. A *traction stress at a configuration* $\kappa : \mathcal{X} \to \mathcal{Y}$ is an element of $\kappa^*L(V\mathcal{Y}, \pi^*\bigwedge^{n-1}T^*\mathcal{X}) = L(\kappa^*V\mathcal{Y}, \bigwedge^{n-1}T^*\mathcal{X})$. A traction stress field $s : \mathcal{Y} \longrightarrow L(V\mathcal{Y}, \pi^*\bigwedge^{n-1}T^*\mathcal{X})$ induces a traction stress field at the configuration κ by

$$s_\kappa := \kappa^*_{\xi_s} s = s \circ \kappa : \mathcal{X} \longrightarrow L(\kappa^*V\mathcal{Y}, \bigwedge\nolimits^{n-1}T^*\mathcal{X}). \tag{12.87}$$

The local representation of a traction stress field is in the form

$$\sum_{k,\alpha} s_{\alpha 1\dots\widehat{k}\dots n}(x^j, y^\beta)\mathrm{d}y^\alpha \otimes (\mathrm{d}x^1 \wedge \cdots \wedge \widehat{\mathrm{d}x^k} \wedge \cdots \wedge \mathrm{d}x^n), \tag{12.88}$$

and for a given configuration, κ, the induced field, $s_\kappa = \kappa^*_{\xi_s} s$, is represented by

$$\sum_{k,\alpha} s_{\alpha 1\dots\widehat{k}\dots n}(x^j, \kappa^\beta(x^i))g^\alpha \otimes (\mathrm{d}x^1 \wedge \cdots \wedge \widehat{\mathrm{d}x^k} \wedge \cdots \wedge \mathrm{d}x^n). \tag{12.89}$$

In analogy with the notation convention described above, in the last two equations, we have written $\mathrm{d}x^1 \wedge \cdots \wedge \widehat{\mathrm{d}x^k} \wedge \cdots \wedge \mathrm{d}x^n$ for what is actually the local section $\pi^*_{\tau_{\mathcal{X}}^{*n-1}}(\mathrm{d}x^1 \wedge \cdots \wedge \widehat{\mathrm{d}x^k} \wedge \cdots \wedge \mathrm{d}x^n)$ of $\pi^*\bigwedge^{n-1}T^*\mathcal{X} \to \mathcal{Y}$.

The Cauchy Formula, Revisited We want to show now that the relation between the traction stress and the surface force on an oriented $(n-1)$-dimensional submanifold $\mathcal{V} \subset \mathcal{X}$, using restriction of forms as in Sect. 10.2, still applies in the current setting. We first note that the fiber bundle $\mathcal{Y}$ and any vector bundle $\xi : W \to \mathcal{Y}$ may be restricted to $\mathcal{V}$ using the pullback relative to the inclusion $\mathcal{I}_{\mathcal{V}} : \mathcal{V} \to \mathcal{X}$. Thus, using the notation

$$\pi|_{\mathcal{V}} := \mathcal{I}^*_{\mathcal{V}}\pi, \quad \mathcal{Y}|_{\mathcal{V}} := \mathcal{I}^*_{\mathcal{V}}\mathcal{Y}, \quad \xi|_{\mathcal{V}} := (\pi^*\mathcal{I}_{\mathcal{V}})^*\xi, \quad W|_{\mathcal{V}} := (\pi^*\mathcal{I}_{\mathcal{V}})^*W, \tag{12.90}$$

one has the following diagram:

$$\begin{array}{ccc} W|_{\mathscr{V}} & \xrightarrow{\xi^*(\pi^*\mathcal{I}_{\mathscr{V}})} & W \\ \Big\downarrow{\scriptstyle \xi|_{\mathscr{V}}} & & \Big\downarrow{\scriptstyle \xi} \\ \mathscr{Y}|_{\mathscr{V}} & \xrightarrow{\pi^*\mathcal{I}_{\mathscr{V}}} & \mathscr{Y} \\ \Big\downarrow{\scriptstyle \pi|_{\mathscr{V}}} & & \Big\downarrow{\scriptstyle \pi} \\ \mathscr{V} & \xrightarrow{\mathcal{I}_{\mathscr{V}}} & \mathscr{X}. \end{array} \tag{12.91}$$

Consider

$$\tau^*|_{\mathscr{V}} : (\textstyle\bigwedge^{n-1} T^*\mathscr{X})|_{\mathscr{V}} \longrightarrow \mathscr{V}, \tag{12.92}$$

where we write τ^* for short instead or $\tau_{\mathscr{X}}^{*n-1}$ and define the restriction,

$$\rho_{\mathscr{V}} : (\textstyle\bigwedge^{n-1} T^*\mathscr{X})|_{\mathscr{V}} \longrightarrow \textstyle\bigwedge^{n-1} T^*\mathscr{V}, \tag{12.93}$$

by

$$\rho_{\mathscr{V}}(\omega)(v_1, \ldots, v_{n-1}) := \omega(v_1, \ldots, v_{n-1}). \tag{12.94}$$

Evidently, $\rho_{\mathscr{V}}$ is a vector bundle morphism over $\mathscr{V}$. Thus, considering the diagram

$$\begin{array}{ccccc} V\mathscr{Y}|_{\mathscr{V}} & \xrightarrow{s|_{\mathscr{V}}} & (\bigwedge^{n-1} T^*\mathscr{X})|_{\mathscr{V}} & \xrightarrow{\rho_{\mathscr{V}}} & \bigwedge^{n-1} T^*\mathscr{V} \\ & {\scriptstyle \pi|_{\mathscr{V}} \circ \tau_{\mathscr{Y}}|_{\mathscr{V}}} \searrow & \Big\downarrow{\scriptstyle \tau^*_{\mathscr{X}}|_{\mathscr{V}}} & \swarrow {\scriptstyle \tau^*_{\mathscr{V}}} & \\ & & \mathscr{V}, & & \end{array} \tag{12.95}$$

we have a vector bundle morphism

$$\mathbf{t}_{\mathscr{V}} := \rho_{\mathscr{V}} \circ s|_{\mathscr{V}}, \tag{12.96}$$

which is the required form of the Cauchy formula.

In (10.6) we defined $\rho_{\mathscr{V}}$ as a mapping between sections over $\mathscr{V}$ and in (12.93) as a vector bundle morphism over $\mathscr{V}$. Next, we describe the action it induces on elements of the vector bundle

$$L(V\mathscr{Y}, \pi^*\textstyle\bigwedge^{n-1} T^*\mathscr{X})|_{\mathscr{V}} \longrightarrow \mathscr{Y}|_{\mathscr{V}}, \quad \text{or} \quad (\pi^*\mathcal{I}_{\mathscr{V}})^* L(V\mathscr{Y}, \pi^*\textstyle\bigwedge^{n-1} T^*\mathscr{X}) \longrightarrow \mathcal{I}^*_{\mathscr{V}}\mathscr{Y}, \tag{12.97}$$

where traction stresses over $\mathscr{V}$ assume their values.

We first consider the following diagram that describes the various morphisms associated with the restriction of $\pi^* \bigwedge^{n-1} T^*\mathcal{X}$ to $\mathcal{Y}|_{\mathcal{V}}$.

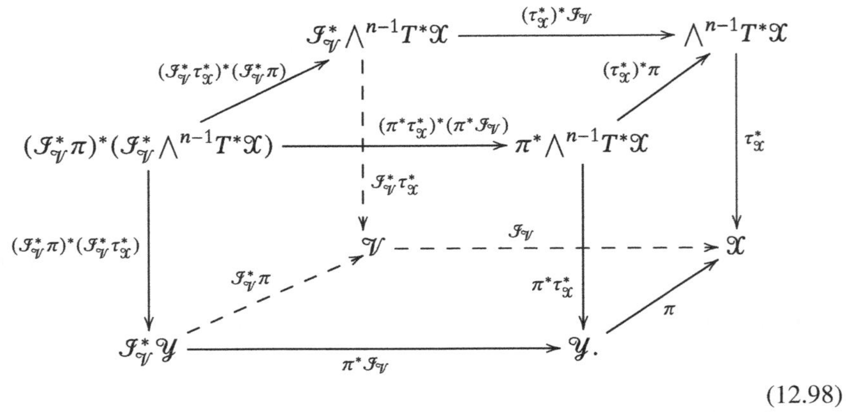

(12.98)

Note that the definition of pullbacks of bundles implies that the foregoing diagram is equivalent to

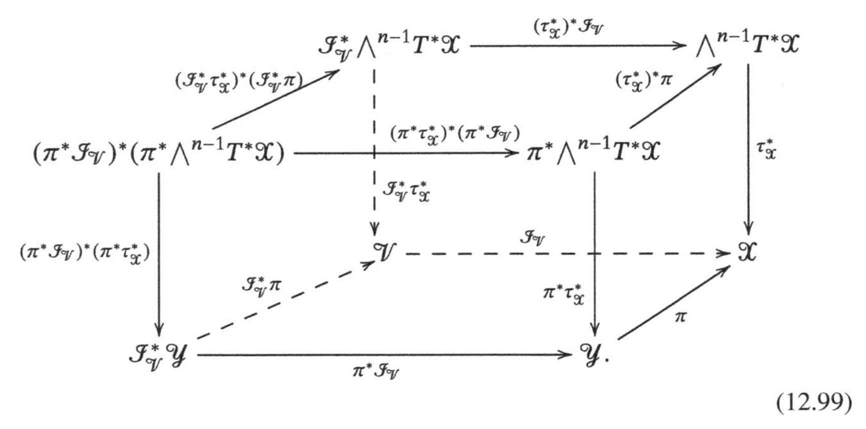

(12.99)

The pullback of a vector bundle morphism as in (12.64) may be applied to the vector bundle morphism $\rho_{\mathcal{V}}$ and its pullback by the mapping $\pi|_{\mathcal{V}}$ as in the following diagram:

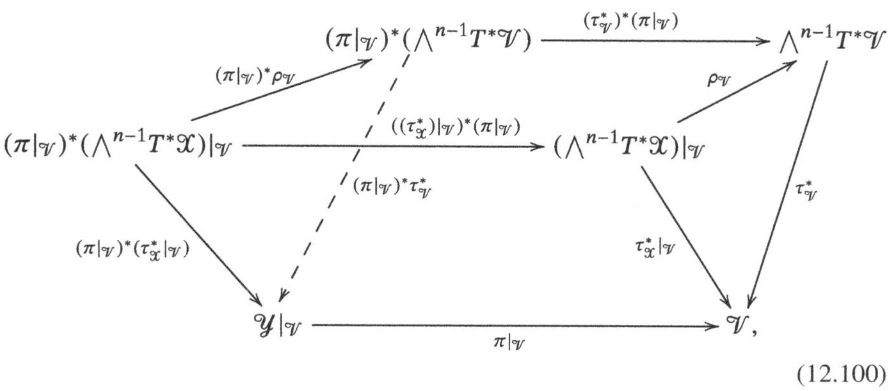

(12.100)

or, equivalently,

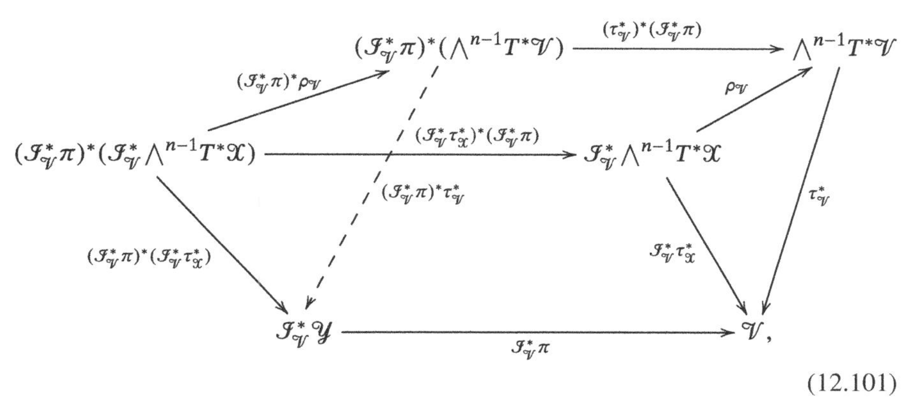

(12.101)

where the mapping $(\mathcal{J}^*_{\mathcal{V}}\pi)^*\rho_{\mathcal{V}} = (\pi|_{\mathcal{V}})^*\rho_{\mathcal{V}}$ is defined so as to make the diagrams commutative.

Considering the restriction of the bundle of traction stresses, it follows from (12.67) that

$$\begin{aligned}(\pi^*\mathcal{J}_{\mathcal{V}})^*L(V\mathcal{Y},\pi^*\textstyle\bigwedge^{n-1}T^*\mathcal{X}) &\cong L((\pi^*\mathcal{J}_{\mathcal{V}})^*V\mathcal{Y},(\pi^*\mathcal{J}_{\mathcal{V}})^*(\pi^*\textstyle\bigwedge^{n-1}T^*\mathcal{X})),\\ &\cong L((\pi^*\mathcal{J}_{\mathcal{V}})^*V\mathcal{Y},(\mathcal{J}^*_{\mathcal{V}}\pi)^*(\mathcal{J}^*_{\mathcal{V}}\textstyle\bigwedge^{n-1}T^*\mathcal{X})),\end{aligned} \tag{12.102}$$

as vector bundles over $\mathcal{J}^*_{\mathcal{V}}\mathcal{Y}$, where in the second line we used the equivalence of the pullbacks in diagrams (12.98,12.99). Thus, we may identify naturally any $s_0 \in (\pi^*\mathcal{J}_{\mathcal{V}})^*L(V\mathcal{Y},\pi^*\bigwedge^{n-1}T^*\mathcal{X})$ with an element of $L((\pi^*\mathcal{J}_{\mathcal{V}})^*V\mathcal{Y},(\mathcal{J}^*_{\mathcal{V}}\pi)^*(\mathcal{J}^*_{\mathcal{V}}\bigwedge^{n-1}T^*\mathcal{X}))$. In view of (12.101), we may compose s_0 with $(\mathcal{J}^*_{\mathcal{V}}\pi)^*\rho_{\mathcal{V}}$, and we obtain an element

$$((\mathcal{J}^*_{\mathcal{V}}\pi)^*\rho_{\mathcal{V}})\circ s_0 \in L((\pi^*\mathcal{J}_{\mathcal{V}})^*V\mathcal{Y},(\mathcal{J}^*_{\mathcal{V}}\pi)^*\textstyle\bigwedge^{n-1}T^*\mathcal{V}). \tag{12.103}$$

Here, $(\pi^*\mathcal{J}_{\mathcal{V}})^*V\mathcal{Y}$ is the vector bundle over $\mathcal{Y}|_{\mathcal{V}}$ induced by $V\mathcal{Y}$ as in diagram (12.91), and $(\mathcal{J}^*_{\mathcal{V}}\pi)^*\bigwedge^{n-1}T^*\mathcal{V}$ is the pullback of $\bigwedge^{n-1}T^*\mathcal{V}$ onto $\mathcal{Y}|_{\mathcal{V}}$ as illustrated in the following diagram:

$$
\begin{array}{ccc}
(\mathcal{J}^*\pi)^*\bigwedge^{n-1}T^*\mathcal{V} & \xrightarrow{(\tau^*_{\mathcal{V}})^*(\mathcal{J}^*\pi)} & \bigwedge^{n-1}T^*\mathcal{V} \\
\Big\downarrow {\scriptstyle (\mathcal{J}^*_{\mathcal{V}}\pi)^*\tau^*_{\mathcal{V}}} & & \Big\downarrow {\scriptstyle \tau^*_{\mathcal{V}}} \\
\mathcal{J}^*_{\mathcal{V}}\mathcal{Y} =: \mathcal{Y}|_{\mathcal{V}} & \xrightarrow{\mathcal{J}^*_{\mathcal{V}}\pi} & \mathcal{V} \\
\Big\downarrow {\scriptstyle \pi^*\mathcal{J}_{\mathcal{V}}} & & \Big\downarrow {\scriptstyle \mathcal{J}_{\mathcal{V}}} \\
\mathcal{Y} & \xrightarrow{\pi} & \mathcal{X}.
\end{array} \tag{12.104}
$$

We conclude that the traction $\mathbf{t}_0 \in L((\pi^*\mathcal{J}_{\mathcal{V}})^*V\mathcal{Y}, (\mathcal{J}^*_{\mathcal{V}}\pi)^*\bigwedge^{n-1}T^*\mathcal{V})$ induced by the element s_0, as above, is given by

$$
\mathbf{t}_0 = ((\mathcal{J}^*_{\mathcal{V}}\pi)^*\rho_{\mathcal{V}}) \circ s_0, \tag{12.105}
$$

giving us another version of the Cauchy formula.

12.2.4 Variational Stresses and Constitutive Relations

Variational Stresses A *variational stress field, Φ*, is defined to be a vector bundle morphism

$$
\begin{array}{ccc}
VJ^1\mathcal{Y} \cong J^1V\mathcal{Y} & \xrightarrow{\Phi} & \bigwedge^n T^*\mathcal{X} \\
\Big\downarrow {\scriptstyle \tau_{J^1\mathcal{Y}}} & & \Big\downarrow {\scriptstyle \tau^{*n}_{\mathcal{X}}} \\
J^1\mathcal{Y} & \xrightarrow{\pi^1} & \mathcal{X}.
\end{array} \tag{12.106}
$$

In analogy with the foregoing objects, Φ may be viewed also as a section of the vector bundle

$$
\xi_S : L(VJ^1\mathcal{Y}, \pi^{1*}\textstyle\bigwedge^n T^*\mathcal{X}) \longrightarrow J^1\mathcal{Y}. \tag{12.107}
$$

Thus, a variational stress S_0 is an element of $L(VJ^1\mathcal{Y}, \pi^{1*}\bigwedge^n T^*\mathcal{X})$.

Locally, a variational stress field is represented in the form

$$
\Big(\sum_{\alpha} R_{\alpha 1\dots n}(x^i, y^\beta, y'^\gamma_j)\mathrm{d}y^\alpha + \sum_{k,\alpha} S^k_{\alpha 1\dots n}(x^i, y^\beta, y'^\gamma_j)\partial_k \otimes \mathrm{d}y^\alpha\Big) \otimes \mathrm{d}x. \tag{12.108}
$$

If $S_0 \in L(VJ^1\mathcal{Y}, \pi^{1*}\bigwedge^n T^*\mathcal{X})_{\chi_0}$, $\chi_0 \in J^1\mathcal{Y}$, is represented by

$$\left(\sum_{\alpha} R_{\alpha 1\dots n}\mathrm{d}y^{\alpha} + \sum_{k,\alpha} S^{k}_{\alpha 1\dots n}\partial_k \otimes \mathrm{d}y^{\alpha}\right) \otimes \mathrm{d}x, \tag{12.109}$$

and $\varepsilon_0 \in (VJ^1\mathcal{Y})_{\chi_0}$ is represented by

$$\textstyle\sum_{\alpha} \dot{y}^{\alpha}\partial_{\alpha} + \sum_{\alpha,i}(\dot{y}')^{\alpha}_{i}\mathrm{d}x^{i} \otimes \partial_{\alpha}, \tag{12.110}$$

then, $S_0(\varepsilon_0)$ is represented by

$$\left(\sum_{\alpha} R_{\alpha 1\dots n}\dot{y}^{\alpha} + \sum_{k,\alpha} S^{k}_{\alpha 1\dots n}(\dot{y}')^{\alpha}_{k}\right)\mathrm{d}x. \tag{12.111}$$

For a variational stress field, Φ, and a jet velocity field. ε, we set $\Phi \cdot \varepsilon$ to be the section of $\pi^{1*}\tau^{*}_{\mathcal{X}} : \pi^{1*}\bigwedge^{n} T^{*}\mathcal{X} \to J^1\mathcal{Y}$ given by

$$(\Phi \cdot \varepsilon)(\chi_0) := \Phi(\chi_0)(\varepsilon(\chi_0)), \qquad \chi_0 \in J^1\mathcal{Y}.$$

It is emphasized that a variational stress field, defined on $J^1\mathcal{Y}$ as above, generalizes the notion of an elastic constitutive relation to the current geometric setting. The stress at a point is determined by the jet—the value and the derivative of the configuration at that point. Thus, we will also refer to a variational stress field as a *constitutive relation.* Clearly, at this level of generality, the dependence on the value of the configuration cannot be omitted on a basis of an analog of the principle of material frame indifference, as this principle cannot be formulated on general manifolds.

A *variational stress field at the deformation jet field* $\chi : \mathcal{X} \to J^1\mathcal{Y}$ is a section of

$$\chi^{*}\xi_S : \chi^{*}L(VJ^1\mathcal{Y}, \pi^{1*}\textstyle\bigwedge^{n} T^{*}\mathcal{X}) \longrightarrow \mathcal{X}. \tag{12.112}$$

For a point $x \in \mathcal{X}$,

$$\begin{aligned}(\chi^{*}L(VJ^1\mathcal{Y}, \pi^{1*}\textstyle\bigwedge^{n} T^{*}\mathcal{X}))_x &= L((\chi^{*}VJ^1\mathcal{Y})_x, \chi^{*}(\pi^{1*}\textstyle\bigwedge^{n} T^{*}\mathcal{X})_x)\\ &= L((VJ^1\mathcal{Y})_{\chi(x)}, (\textstyle\bigwedge^{n} T^{*}\mathcal{X})_{\pi^1(\chi(x))}),\\ &= L((VJ^1\mathcal{Y})_{\chi(x)}, (\textstyle\bigwedge^{n} T^{*}\mathcal{X})_x).\end{aligned} \tag{12.113}$$

Particularly, in case $\chi = j^1\kappa$, for a section $\kappa : \mathcal{X} \to \mathcal{Y}$, the isomorphism $J^1(\kappa^{*}V\mathcal{Y}) \cong \chi^{*}(VJ^1Y)$, (12.51), gives

$$(\chi^{*}L(VJ^1\mathcal{Y}, \pi^{1*}\textstyle\bigwedge^{n} T^{*}\mathcal{X}))_x = L((J^1(\kappa^{*}V\mathcal{Y}))_x, (\textstyle\bigwedge^{n} T^{*}\mathcal{X})_x), \tag{12.114}$$

so that

$$\chi^* L(VJ^1\mathcal{Y}, \pi^{1*}\bigwedge^n T^*\mathcal{X}) = L(J^1(\kappa^* V\mathcal{Y}), \bigwedge^n T^*\mathcal{X}), \tag{12.115}$$

in agreement with the definition in Chap. 10 for the case $W := \kappa^* V\mathcal{Y}$.

For a given configuration κ, a *variational stress field at the configuration* κ is a section

$$S : \mathcal{X} \longrightarrow (j^1\kappa)^* L(VJ^1\mathcal{Y}, \pi^{1*}\bigwedge^n T^*\mathcal{X}) = L(J^1(\kappa^* V\mathcal{Y}), \bigwedge^n T^*\mathcal{X}). \tag{12.116}$$

A section

$$\Phi : J^1\mathcal{Y} \longrightarrow L(VJ^1\mathcal{Y}, \pi^{1*}\bigwedge^n T^*\mathcal{X}), \tag{12.117}$$

a constitutive relation, induces for any configuration κ the variational stress field S on $\mathcal{X}$ that corresponds to κ by

$$S = \Phi_\kappa := (j^1\kappa)^*_{\xi_S}\Phi : \mathcal{X} \longrightarrow L(J^1(\kappa^* V\mathcal{Y}), \bigwedge^n T^*\mathcal{X}), \tag{12.118}$$

as in the following diagram:

$$\begin{array}{ccc} (j^1\kappa)^* L(VJ^1\mathcal{Y}, \pi^{1*}\bigwedge^n T^*\mathcal{X}) & \xrightarrow{\xi_S^* j^1\kappa} & L(VJ^1\mathcal{Y}, \pi^{1*}\bigwedge^n T^*\mathcal{X}) \\ {\scriptstyle S=(j^1\kappa)^*_{\xi_S}\Phi}\uparrow\ \downarrow{\scriptstyle (j^1\kappa)^*\xi_S} & & {\scriptstyle \xi_S}\downarrow\ \uparrow{\scriptstyle \Phi} \\ \mathcal{X} & \xrightarrow[j^1\kappa]{} & J^1\mathcal{Y} \end{array} \tag{12.119}$$

The stress field, S, corresponding to κ is given locally by

$$\left(\sum_\alpha R_{\alpha 1\ldots n}(\boldsymbol{x}, \boldsymbol{\kappa}(\boldsymbol{x}), D\boldsymbol{\kappa}(\boldsymbol{x}))g^\alpha + \sum_{k,\alpha} S^k_{\alpha 1\ldots n}(\boldsymbol{x}, \boldsymbol{\kappa}(\boldsymbol{x}), D\boldsymbol{\kappa}(\boldsymbol{x}))\partial_k \otimes g^\alpha\right) \otimes \mathrm{d}x. \tag{12.120}$$

The Traction Stress Induced by a Variational Stress In analogy with Sect. 10.7, we want to extract from a given variational stress,

$$S_0 \in L(VJ^1\mathcal{Y}, \pi^{1*}\bigwedge^n T^*\mathcal{X}),$$

a traction stress, $s_0 = p_s(S_0) \in L(V\mathcal{Y}, \pi^*\bigwedge^{n-1} T^*\mathcal{X})$, so that we have a vector bundle morphism

$$\begin{array}{ccc} L(VJ^1\mathcal{Y}, \pi^{1*}\bigwedge^n T^*\mathcal{X}) & \xrightarrow{p_s} & L(V\mathcal{Y}, \pi^*\bigwedge^{n-1} T^*\mathcal{X}) \\ & {\scriptstyle \pi^1_0\circ\xi_S}\searrow \quad \swarrow{\scriptstyle \xi_s} & \\ & \mathcal{Y}. & \end{array} \tag{12.121}$$

We first consider the vertical subbundle,

$$V_0^1 J^1 \mathcal{Y} := \operatorname{Kernel} T\pi_0^1 \subset VJ^1\mathcal{Y} \subset TJ^1\mathcal{Y}. \tag{12.122}$$

If $(\boldsymbol{x}, \boldsymbol{y}, \boldsymbol{y}', \dot{\boldsymbol{x}}, \dot{\boldsymbol{y}}, \dot{\boldsymbol{y}}')$ represents a generic element of $TJ^1\mathcal{Y}$, the action of $T\pi_0^1$ is represented by

$$(\boldsymbol{x}, \boldsymbol{y}, \boldsymbol{y}', \dot{\boldsymbol{x}}, \dot{\boldsymbol{y}}, \dot{\boldsymbol{y}}') \longmapsto (\boldsymbol{x}, \boldsymbol{y}, \dot{\boldsymbol{x}}, \dot{\boldsymbol{y}}), \tag{12.123}$$

so that an element of $V_0^1 J^1\mathcal{Y}$ is represented in the form $(\boldsymbol{x}, \boldsymbol{y}, \boldsymbol{y}', \boldsymbol{0}, \boldsymbol{0}, \dot{\boldsymbol{y}}')$, whereas an element of $VJ^1\mathcal{Y}$ is of the form $(\boldsymbol{x}, \boldsymbol{y}, \boldsymbol{y}', \boldsymbol{0}, \dot{\boldsymbol{y}}, \dot{\boldsymbol{y}}')$. We will use $\mathcal{I}_{V_0^1}$ to denote the natural inclusion as illustrated below.

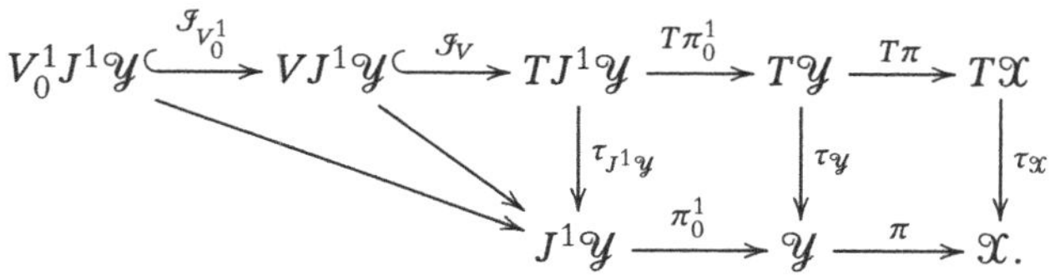

$$\begin{array}{ccccccccc} V_0^1 J^1\mathcal{Y} & \overset{\mathcal{I}_{V_0^1}}{\hookrightarrow} & VJ^1\mathcal{Y} & \overset{\mathcal{I}_V}{\hookrightarrow} & TJ^1\mathcal{Y} & \xrightarrow{T\pi_0^1} & T\mathcal{Y} & \xrightarrow{T\pi} & T\mathcal{X} \\ & \searrow & & \searrow & \downarrow \tau_{J^1\mathcal{Y}} & & \downarrow \tau_{\mathcal{Y}} & & \downarrow \tau_{\mathcal{X}} \\ & & & & J^1\mathcal{Y} & \xrightarrow{\pi_0^1} & \mathcal{Y} & \xrightarrow{\pi} & \mathcal{X}. \end{array} \tag{12.124}$$

In what follows, to simplify the notation, we will use $\tau_{J^1\mathcal{Y}}$ to denote the restriction of the tangent bundle projection to the vertical subbundles.

Let $y_0 \in \mathcal{Y}$, $x_0 = \pi(y_0)$, and consider the inclusion,

$$\mathcal{I}_{y_0} : (J^1\mathcal{Y})_{y_0} = (\pi_0^1)^{-1}\{y_0\} \longrightarrow J^1\mathcal{Y}, \tag{12.125}$$

of the fiber of the jet bundle at y_0 into the jet bundle. It is noted that an element $\chi_0 \in (J^1\mathcal{Y})_{y_0}$, with $x_0 = \pi(y_0)$, is represented locally as $(\boldsymbol{x}_0, \boldsymbol{y}_0, \boldsymbol{y}')$ for fixed values of $\boldsymbol{x}_0, \boldsymbol{y}_0$. In addition, since y_0 is given, χ_0 is identified uniquely with an element of

$$L(T_{x_0}\mathcal{X}, T_{y_0}(\mathcal{Y}_{x_0})) = L(T_{x_0}\mathcal{X}, (V\mathcal{Y})_{y_0}), \tag{12.126}$$

so that we have the natural isomorphism

$$(J^1\mathcal{Y})_{y_0} \cong T^*_{\pi(y_0)}\mathcal{X} \otimes (V\mathcal{Y})_{y_0} \tag{12.127}$$

(see [Sau89, p. 97]).

The tangent mapping at a point $\chi_0 \in (J^1\mathcal{Y})_{y_0}$,

$$T_{\chi_0}\mathcal{I}_{y_0} : T_{\chi_0}((J^1\mathcal{Y})_{y_0}) \longrightarrow T_{\chi_0}(J^1\mathcal{Y}), \tag{12.128}$$

is represented locally in the form

$$(\boldsymbol{x}_0, \boldsymbol{y}_0, \boldsymbol{\chi}_0, \dot{\boldsymbol{\chi}}) \longmapsto (\boldsymbol{x}_0, \boldsymbol{y}_0, \boldsymbol{\chi}_0, \boldsymbol{0}, \boldsymbol{0}, \dot{\boldsymbol{\chi}}). \tag{12.129}$$

Thus,

$$T_{\chi_0}\mathcal{J}_{y_0} : T_{\chi_0}((J^1\mathcal{Y})_{y_0}) \longrightarrow (V_0^1 J^1\mathcal{Y})_{\chi_0} \tag{12.130}$$

is an isomorphism, and $V_0^1 J^1\mathcal{Y}$ may be viewed as the subbundle containing vectors that are tangent to the fibers of π_0^1. In addition, by (12.127), $(J^1\mathcal{Y})_{y_0}$ is identified with a vector space. Hence, its tangent space at each $\chi_0 \in (J^1\mathcal{Y})_{y_0}$ is identified with the same vector space, and we have the isomorphisms

$$(V_0^1 J^1\mathcal{Y})_{\chi_0} \cong T_{\chi_0}((J^1\mathcal{Y})_{y_0}) \cong T^*_{\pi(y_0)}\mathcal{X} \otimes (V\mathcal{Y})_{y_0} = (\pi^*T^*\mathcal{X} \otimes V\mathcal{Y})_{y_0}, \tag{12.131}$$

independently of $\chi_0 \in (J^1\mathcal{Y})_{y_0}$. We may view the vertical subbundle as the pullback of a vector bundle over $\mathcal{Y}$, as illustrated in the diagram

$$\begin{array}{ccc} V_0^1 J^1\mathcal{Y} = \pi_0^{1*}(\pi^*T^*\mathcal{X} \otimes V\mathcal{Y}) & \longrightarrow & \pi^*T^*\mathcal{X} \otimes V\mathcal{Y} \\ \tau_{J^1\mathcal{Y}} \downarrow & & \downarrow \\ J^1\mathcal{Y} & \xrightarrow{\quad \pi_0^1 \quad} & \mathcal{Y}. \end{array} \tag{12.132}$$

In Sect. 10.7 we defined the vertical jet bundle $V_0^1 J^1 W := \text{Kernel}\pi_0^1$ for a vector bundle $\pi : W \to \mathcal{X}$. We verify next that this is a special case of the foregoing definition, $V_0^1 J^1\mathcal{Y} = \text{Kernel}T\pi_0^1$, when $W = \kappa^* V\mathcal{Y}$, for a section $\kappa : \mathcal{X} \to \mathcal{Y}$, as is the situation in the current context. It is noted that the meaning of the prefix V_0^1 depends on following jet bundle, either the jet of a fixed vector bundle, W, or a fiber bundle, $\mathcal{Y}$. Given a section κ of $\mathcal{Y}$,

$$(j^1\kappa)^* V_0^1 J^1\mathcal{Y} \cong V_0^1 J^1(\kappa^* V\mathcal{Y}). \tag{12.133}$$

Indeed, (12.51) states that $(j^1\kappa)^* V J^1\mathcal{Y} \to J^1(\kappa^* V\mathcal{Y})$ is a natural isomorphism represented locally by

$$(\boldsymbol{x}, \kappa(\boldsymbol{x}), D\kappa(\boldsymbol{x}), \boldsymbol{0}, \dot{\boldsymbol{y}}, \dot{\boldsymbol{y}}') \longmapsto (\boldsymbol{x}, \kappa(\boldsymbol{x}), \boldsymbol{0}, \dot{\boldsymbol{y}}, D\kappa(\boldsymbol{x}), \dot{\boldsymbol{y}}'). \tag{12.134}$$

For an element of the subbundle $(j^1\kappa)^* V_0^1 J^1\mathcal{Y}$, $\dot{\boldsymbol{y}} = \boldsymbol{0}$, hence,

$$(\boldsymbol{x}, \kappa(\boldsymbol{x}), D\kappa(\boldsymbol{x}), \boldsymbol{0}, \boldsymbol{0}, \dot{\boldsymbol{y}}') \longmapsto (\boldsymbol{x}, \kappa(\boldsymbol{x}), \boldsymbol{0}, \boldsymbol{0}, D\kappa(\boldsymbol{x}), \dot{\boldsymbol{y}}'), \tag{12.135}$$

which represents an element of $V_0^1 J^1(\kappa^* V\mathcal{Y})$.

The inclusion $\mathcal{J}_{V_0^1} : V_0^1 J^1\mathcal{Y} \to V J^1\mathcal{Y}$ induces a vertical projection on the space of variational stresses. Specifically,

$$\mathcal{J}'_{V_0^1} : L(V J^1\mathcal{Y}, \pi^{1*} \textstyle\bigwedge^n T^*\mathcal{X}) \longrightarrow L(V_0^1 J^1\mathcal{Y}, \pi^{1*} \textstyle\bigwedge^n T^*\mathcal{X}), \tag{12.136}$$

is given by

$$S_0 \longmapsto S_0 \circ \mathcal{I}_{V_0^1}. \tag{12.137}$$

We also observe that for any vector bundle $U \to \mathcal{X}$,

$$\begin{aligned}(\pi^{1*}U)_{\chi_0} &= U_{\pi(\pi_0^1(\chi_0))}, \\ &= (\pi^* U)_{\pi_0^1(\chi_0)}, \\ &= (\pi_0^{1*}(\pi^* U))_{\chi_0}.\end{aligned} \tag{12.138}$$

Hence,

$$\pi^{1*}U = (\pi \circ \pi_0^1)^* U = \pi_0^{1*}\pi^* U, \tag{12.139}$$

and in particular,

$$\pi^{1*}\textstyle\bigwedge^n T^*\mathcal{X} = \pi_0^{1*}\pi^*\bigwedge^n T^*\mathcal{X}, \tag{12.140}$$

as illustrated in the diagram

$$\begin{array}{ccccc}\pi^{1*}\bigwedge^n T^*\mathcal{X} = \pi_0^{1*}\pi^*\bigwedge^n T^*\mathcal{X} & \xrightarrow{(\pi^*\tau_{\mathcal{X}}^*)^*\pi_0^1} & \pi^*\bigwedge^d T^*\mathcal{X} & \xrightarrow{\tau_{\mathcal{X}}^*\pi^*} & \bigwedge^d T^*\mathcal{X} \\ {\scriptstyle \pi_0^{1*}\pi^*\tau_{\mathcal{X}}^*}\Big\downarrow{\scriptstyle \pi^{1*}\tau_{\mathcal{X}}^*} & & \Big\downarrow{\scriptstyle \pi^*\tau_{\mathcal{X}}^*} & & \Big\downarrow{\scriptstyle \tau_{\mathcal{X}}^*} \\ J^1\mathcal{Y} & \xrightarrow{\pi_0^1} & \mathcal{Y} & \xrightarrow{\pi} & \mathcal{X}.\end{array} \tag{12.141}$$

In view of (12.111), for a variational stress S_0 represented as in (12.109), the action of $\mathcal{I}'_{V_0^1}$ is represented locally by

$$\Big(\sum_\alpha R_{\alpha 1\dots n}\mathrm{d}y^\alpha + \sum_{k,\alpha} S^k_{\alpha 1\dots n}\partial_k \otimes \mathrm{d}y^\alpha\Big) \otimes \mathrm{d}x \longmapsto \sum_{k,\alpha} S^k_{\alpha 1\dots n}\partial_k \otimes \mathrm{d}y^\alpha \otimes \mathrm{d}x, \tag{12.142}$$

where again, we make no distinction in the notation between the pullbacks of the base vectors and the base vectors themselves.

Now, using the identification (12.132), we may rewrite (12.136) as

$$\mathcal{I}'_{V_0^1} : L(VJ^1\mathcal{Y}, \pi^{1*}\textstyle\bigwedge^n T^*\mathcal{X}) \longrightarrow \pi_0^{1*}(\pi^*T\mathcal{X} \otimes (V\mathcal{Y})^*) \otimes \pi_0^{1*}\pi^*\bigwedge^n T^*\mathcal{X}. \tag{12.143}$$

Setting

$$\xi_{\mathcal{P}} : \mathcal{P} := \pi^* T\mathcal{X} \otimes (V\mathcal{Y})^* \otimes \pi^* \textstyle\bigwedge^n T^*\mathcal{X} \longrightarrow \mathcal{Y},$$

one has (*cf.* (10.70))

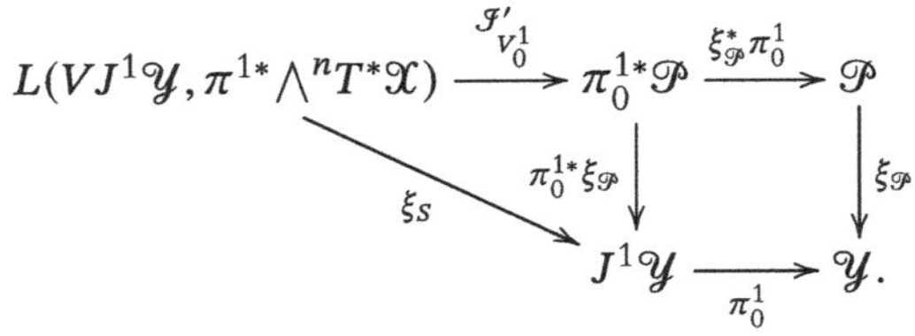

$$\begin{array}{ccccc} L(VJ^1\mathcal{Y}, \pi^{1*}\bigwedge^n T^*\mathcal{X}) & \xrightarrow{\mathcal{I}'_{V_0^1}} & \pi_0^{1*}\mathcal{P} & \xrightarrow{\xi^*_{\mathcal{P}}\pi_0^1} & \mathcal{P} \\ & \searrow^{\xi_S} & \Big\downarrow{\scriptstyle \pi_0^{1*}\xi_{\mathcal{P}}} & & \Big\downarrow{\scriptstyle \xi_{\mathcal{P}}} \\ & & J^1\mathcal{Y} & \xrightarrow[\pi_0^1]{} & \mathcal{Y}. \end{array} \tag{12.144}$$

For an element $S_0 \in L(VJ^1\mathcal{Y}, \pi^{1*}\bigwedge^n T^*\mathcal{X})$, we will refer to $\mathcal{I}'_{V_0^1}(S_0)$ as the *vertical component* of the stress.

Once we have defined the vertical component of a variational stress, we may consider *horizontal variational stresses*. Thus, we say that a variational stress S_0 is horizontal if $S_0 \in \text{Kernel}\,\mathcal{I}'_{V_0^1}$. It follows immediately from (12.142) that S is horizontal if for its local representation, $S^k_{\alpha 1 \dots n} = 0$ for all $\alpha = 1, \dots, m$, and $k = 1, \dots, n$, and S is represented by

$$\left(\sum_\alpha R_{\alpha 1 \dots n} \mathrm{d}y^\alpha \right) \otimes \mathrm{d}x. \tag{12.145}$$

It follows that a horizontal variational stress may be viewed as a body force in $L(V\mathcal{Y}, \pi^* \bigwedge^n T^*\mathcal{X})$.

In analogy with (10.71), one can consider the contraction operation

$$\begin{aligned} \mathsf{C} : \mathcal{P} &= \pi^* T\mathcal{X} \otimes (V\mathcal{Y})^* \otimes \pi^* \textstyle\bigwedge^n T^*\mathcal{X} \\ &\longrightarrow L(V\mathcal{Y}, \pi^* \textstyle\bigwedge^{n-1} T^*\mathcal{X}) \cong (V\mathcal{Y})^* \otimes \pi^* \textstyle\bigwedge^{n-1} T^*\mathcal{X}, \end{aligned} \tag{12.146}$$

a vector bundle morphism over $\mathcal{Y}$, whereby

$$v \otimes \phi \otimes \theta \longmapsto \phi \otimes (v \lrcorner \theta), \tag{12.147}$$

$v \in (\pi^* T\mathcal{X})_y = (T\mathcal{X})_{\pi(y)}$, $\phi \in (V\mathcal{Y})_y$, $\theta \in (\pi^* \bigwedge^n T^*\mathcal{X})_y = (\bigwedge T^*\mathcal{X})_{\pi(y)}$, $y \in \mathcal{Y}$. Thus, the foregoing diagram may be extended, giving

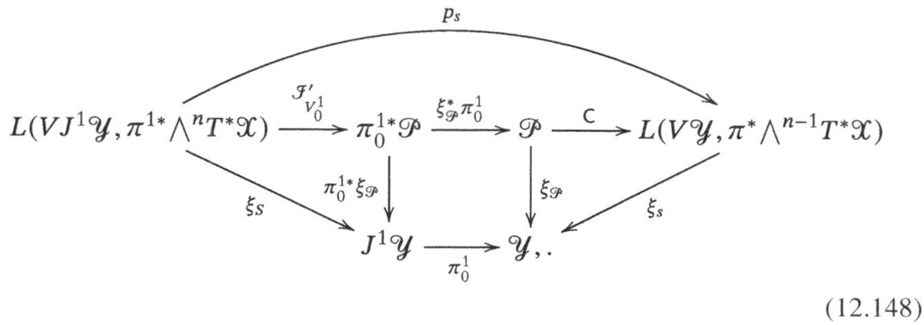

(12.148)

and we set

$$p_s := \mathsf{C} \circ \xi^*_{\mathcal{P}}\pi^1_0 \circ \mathcal{F}'_{V_0^1}, \tag{12.149}$$

determining the traction stress associated with a variational stress.

Applying contraction to the right-hand side of (12.142), p_s is represented locally by

$$\begin{aligned}\Big(\sum_{\alpha} R_{\alpha 1\dots n}\mathrm{d}y^\alpha + \sum_{k,\alpha} S^k_{\alpha 1\dots n}\partial_k \otimes \mathrm{d}y^\alpha\Big) \otimes \mathrm{d}x \longmapsto \sum_{k,\alpha} S^k_{\alpha 1\dots n}\mathrm{d}y^\alpha \otimes (\partial_k \lrcorner \mathrm{d}x)\\ = \sum_{k,\alpha}(-1)^{k-1} S^k_{\alpha 1\dots n}\mathrm{d}y^\alpha \otimes (\mathrm{d}x^1 \wedge \dots \wedge \widehat{\mathrm{d}x^k} \wedge \dots \wedge \mathrm{d}x^n),\end{aligned} \tag{12.150}$$

in analogy with (10.73).

The Divergence of a Variational Stress Field over $\mathcal{X}$ We proceed in analogy with Sect. 10.8. Let $\kappa : \mathcal{X} \to \mathcal{Y}$ be a configuration and let

$$s_\kappa : \mathcal{X} \to \kappa^* L(V\mathcal{Y}, \pi^*\bigwedge^{n-1} T^*\mathcal{X}) = L(\kappa^* V\mathcal{Y}, \bigwedge^{n-1} T^*\mathcal{X}) \tag{12.151}$$

be a traction stress field at κ. Thus, for a section w of $\kappa^* V\mathcal{Y}$, $s_\kappa \cdot w$ is an $(n-1)$-form, and its exterior derivative $\mathrm{d}(s_\kappa \cdot w)$ is an n-form over $\mathcal{X}$. If s_κ is represented as in (12.89), then, $s_\kappa \cdot w$ is represented by

$$\sum_{k,\alpha} s_{\alpha 1\dots\widehat{k}\dots n} w^\alpha (\mathrm{d}x^1 \wedge \dots \wedge \widehat{\mathrm{d}x^k} \wedge \dots \wedge \mathrm{d}x^n),$$

and $\mathrm{d}(s_\kappa \cdot w)$ is represented by

$$\sum_{k,\alpha}(-1)^{k-1}(s_{\alpha 1\dots\widehat{k}\dots n} w^\alpha)_{,k}\mathrm{d}x = \sum_{k,\alpha}(-1)^{k-1}(s_{\alpha 1\dots\widehat{k}\dots n,k} w^\alpha + s_{\alpha 1\dots\widehat{k}\dots n} w^\alpha_{,k})\mathrm{d}x. \tag{12.152}$$

It is noted that the value of the n-form $\mathrm{d}(s_\kappa \cdot w)$, at each point $x \in \mathcal{X}$, depends linearly on $j^1 w$. In other words, the action $w \mapsto \mathrm{d}(s_\kappa \cdot w)$ may be factored as follows:

$$w \overset{j^1}{\longmapsto} j^1 w \overset{\widetilde{\mathfrak{d}s}_\kappa}{\longmapsto} \mathrm{d}(s_\kappa \cdot w), \tag{12.153}$$

where $\widetilde{\mathfrak{d}s}_\kappa$ is a vector bundle morphism over $\mathcal{X}$ determined by the first jet of s_κ, represented by $(s_{\alpha 1\ldots\widehat{k}\ldots n}, s_{\alpha 1\ldots\widehat{k}\ldots n,k})$,

$$\begin{array}{ccc} J^1(\kappa^* V\mathcal{Y}) & \xrightarrow{\widetilde{\mathfrak{d}s}_\kappa} & \bigwedge^n T^*\mathcal{X} \\ & {\scriptstyle (\kappa^*\tau_\mathcal{Y})^1} \searrow \quad \swarrow {\scriptstyle \tau^*_\mathcal{X}} & \\ & \mathcal{X} & \end{array} \tag{12.154}$$

That is,

$$\widetilde{\mathfrak{d}s}_\kappa \cdot j^1 w := \mathrm{d}(s_\kappa \cdot w) \tag{12.155}$$

in accordance with Sect. 10.8. The factorization makes the action $w \mapsto \mathrm{d}(s_\kappa \cdot w)$ a linear differential operator of order one (see [Pal68]). As a vector bundle morphism, $\widetilde{\mathfrak{d}s}_\kappa$ may be identified with a section of

$$L(J^1(\kappa^* V\mathcal{Y}), \textstyle\bigwedge^n T^*\mathcal{X}) = L((j^1\kappa)^* V J^1\mathcal{Y}, \textstyle\bigwedge^n T^*\mathcal{X}),$$

identifying it with a variational stress field at the configuration κ.

The mapping

$$\widetilde{\mathfrak{d}} : s_\kappa \longmapsto \widetilde{\mathfrak{d}s}_\kappa, \tag{12.156}$$

is also a differential operator, as it may be factored into a vector bundle morphism composed with a jet

$$s_\kappa \overset{j^1}{\longmapsto} j^1 s_\kappa \overset{\widetilde{\mathfrak{d}}}{\longmapsto} \widetilde{\mathfrak{d}s}_\kappa. \tag{12.157}$$

Here,

$$\begin{aligned} \widetilde{\mathfrak{d}} : J^1(L(\kappa^* V\mathcal{Y}, \textstyle\bigwedge^{n-1} T^*\mathcal{X})) \longrightarrow\ & L(J^1(\kappa^* V\mathcal{Y}), \textstyle\bigwedge^n T^*\mathcal{X}) \\ &= L(j^1\kappa)^* V J^1\mathcal{Y}, \textstyle\bigwedge^n T^*\mathcal{X}), \end{aligned} \tag{12.158}$$

is represented locally by

$$\sum_{k,\alpha,i}\left(s_{\alpha 1\ldots\widehat{k}\ldots n}g^{\alpha}+s_{\alpha 1\ldots\widehat{k}\ldots n,i}\mathrm{d}x^{i}\otimes g^{\alpha}\right)\otimes(\mathrm{d}x^{1}\wedge\cdots\wedge\widehat{\mathrm{d}x^{k}}\wedge\cdots\wedge\mathrm{d}x^{n})$$

$$\longmapsto\sum_{k,\alpha}\left((-1)^{k-1}s_{\alpha 1\ldots\widehat{k}\ldots n}\partial_{k}\otimes g^{\alpha}+(-1)^{k-1}s_{\alpha 1\ldots\widehat{k}\ldots n,k}g^{\alpha}\right)\otimes\mathrm{d}x. \tag{12.159}$$

Let $S:\mathcal{X}\longrightarrow(j^{1}\kappa)^{*}L(VJ^{1}\mathcal{Y},\pi^{1*}\bigwedge^{n}T^{*}\mathcal{X})$ be a variational stress field at the configuration κ, represented as in (12.120), and let $s_{\kappa}=p_{s}\circ S$. By the local representation of p_{s} in (12.150), $\widetilde{\mathfrak{d}s_{\kappa}}=\mathfrak{d}\widetilde{(p_{s}\circ S)}$ is represented by

$$\left(\sum_{k,\alpha}S^{k}_{\alpha 1\ldots n}\partial_{k}\otimes g^{\alpha}+\sum_{k,\alpha}S^{k}_{\alpha 1\ldots n,k}g^{\alpha}\right)\otimes\mathrm{d}x. \tag{12.160}$$

For a variational stress field, S, represented by (12.120), consider the *generalized divergence* defined as

$$\mathrm{div}S:=\mathfrak{d}\widetilde{(p_{s}\circ S)}-S\qquad\text{or}\qquad\mathrm{div}S(w)=\mathrm{d}(p_{s}S\cdot w)-S\cdot j^{1}w. \tag{12.161}$$

We have seen that $\mathfrak{d}\widetilde{(p_{s}\circ S)}$ is a variational stress field so that the definition is valid. Locally, $\mathrm{div}S$ is represented in the form

$$\left(\sum_{k,\alpha}S^{k}_{\alpha 1\ldots n}\partial_{k}\otimes g^{\alpha}+\sum_{k,\alpha}S^{k}_{\alpha 1\ldots n,k}g^{\alpha}-\sum_{\alpha}R_{\alpha 1\ldots n}g^{\alpha}-\sum_{k,\alpha}S^{k}_{\alpha 1\ldots n}\partial_{k}\otimes g^{\alpha}\right)\otimes\mathrm{d}x.$$

$$=\sum_{k,\alpha}\left(S^{k}_{\alpha 1\ldots n,k}-R_{\alpha 1\ldots n}\right)g^{\alpha}\otimes\mathrm{d}x. \tag{12.162}$$

Comparing the last expression with (12.145), it follows immediately that $\mathrm{div}S$ is horizontal and may be identified with a section of $L(\kappa^{*}V\mathcal{Y},\bigwedge^{n}T^{*}\mathcal{X})$. Thus, we may write

$$\mathrm{div}S\cdot w=\mathrm{d}(p_{s}S\cdot w)-S\cdot j^{1}w. \tag{12.163}$$

12.2.5 The Problem of Elasticity

In this section, we use the structure and terminology introduced above in order to present the balance equation for continuum mechanics in terms of constitutive relations. Then, we consider the case where the external forces and stresses are given in terms of potential functions.

The Balance Equation and the Elasticity Problem Using the definition of the divergence, we can write, for a variational stress field S at the configuration κ and every generalized velocity field w,

$$\begin{aligned} F_\kappa(w) &= \int_{\mathcal{X}} S \cdot j^1 w = \int_{\mathcal{X}} \mathrm{d}(p_s S \cdot w) - \int_{\mathcal{X}} \operatorname{div} S \cdot w, \\ &= \int_{\partial\mathcal{X}} p_s S \cdot w - \int_{\mathcal{X}} \operatorname{div} S \cdot w, \end{aligned} \tag{12.164}$$

where Stokes's theorem was used to obtain the second line. Given a loading, as in Sect. 12.2.2 and Eq. (12.84), one has, for every generalized velocity w,

$$\int_{\mathcal{X}} \mathbf{b}_\kappa \cdot w + \int_{\partial\mathcal{X}} \mathbf{t}_\kappa \cdot w = \int_{\mathcal{X}} S \cdot j^1 w = \int_{\partial\mathcal{X}} p_s S \cdot w - \int_{\mathcal{X}} \operatorname{div} S \cdot w. \tag{12.165}$$

As the equality above holds for every field w, and using (12.96), one obtains the differential balance equation and boundary conditions in the form

$$\operatorname{div} S + \mathbf{b}_\kappa = 0, \qquad \rho_{\partial\mathcal{X}} \circ p_s S|_{\partial\mathcal{X}} = \mathbf{t}_\kappa. \tag{12.166}$$

It is now recalled that a variational stress field

$$\Phi : J^1\mathcal{Y} \longrightarrow L(VJ^1\mathcal{Y}, \pi^{1*}\textstyle\bigwedge^n T^*\mathcal{X}) \tag{12.167}$$

contains an elastic constitutive relation between the jet of a configuration and the resulting stress. Specifically, for any given configuration κ, one has the stress field S_κ over $\mathcal{X}$, given by

$$S_\kappa = \Phi \circ j^1\kappa = j^1\kappa^*_{\xi S}\Phi, \tag{12.168}$$

as in (12.118). Thus, given a loading, $(\mathbf{b}, \mathbf{t})$,

$$\mathbf{b} : \mathcal{Y} \longrightarrow L(V\mathcal{Y}, \pi^*\textstyle\bigwedge^n T^*\mathcal{X}), \qquad \mathbf{t} : \mathcal{Y} \longrightarrow L(V\mathcal{Y}, \pi^*\textstyle\bigwedge^{n-1} T^*\partial\mathcal{X}), \tag{12.169}$$

for any configuration κ, a corresponding pair $(\mathbf{b}_\kappa, \mathbf{t}_\kappa) = (\mathbf{b} \circ \kappa, \mathbf{t} \circ \kappa)$ is induced. The problem of elasticity is formulated as follows: Find a configuration κ such that the fields $\mathbf{b}_\kappa$, $\mathbf{t}_\kappa$, and S_κ, induced by the loading and constitutive relations, satisfy the differential equation and boundary condition (12.166), that is,

$$\operatorname{div}(\Phi \circ j^1\kappa) + \mathbf{b} \circ \kappa = 0, \qquad \rho_{\partial\mathcal{X}} \circ p_s(\Phi \circ j^1\kappa)|_{\partial\mathcal{X}} = \mathbf{t} \circ \kappa. \tag{12.170}$$

Potential Functions and Hyperelasticity The derivations of the various static fields from their corresponding potential functions follow the same pattern. Let $\pi : \mathfrak{Z} \to \mathfrak{X}$ be a fiber bundle, and let $\xi : U \to \mathfrak{X}$ be a vector bundle. Then, one may consider the structure described in the following commutative diagram:

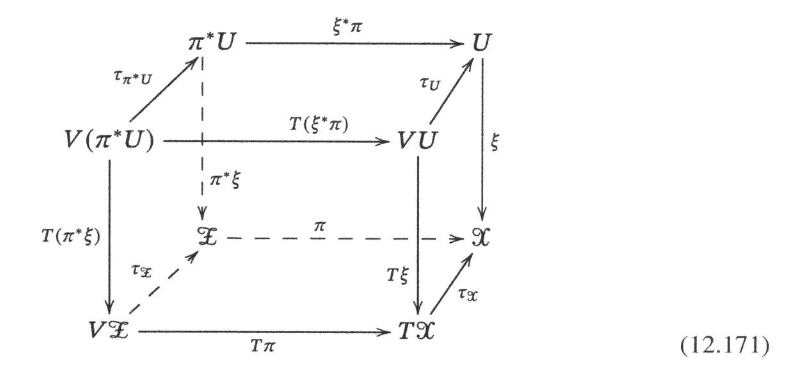

(12.171)

The local representatives of elements in the various objects are mapped as follows:

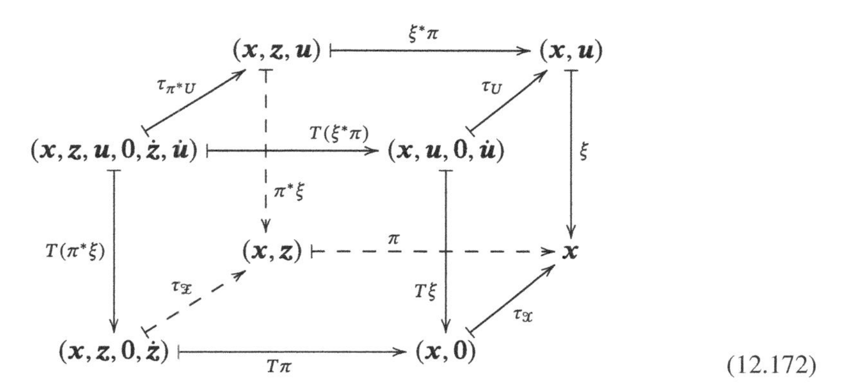

(12.172)

We will be concerned with a section

$$\psi : \mathfrak{Z} \longrightarrow \pi^*U \tag{12.173}$$

and its tangent mapping as illustrated below, where it is recalled that tangent mappings preserve vertical subbundles.

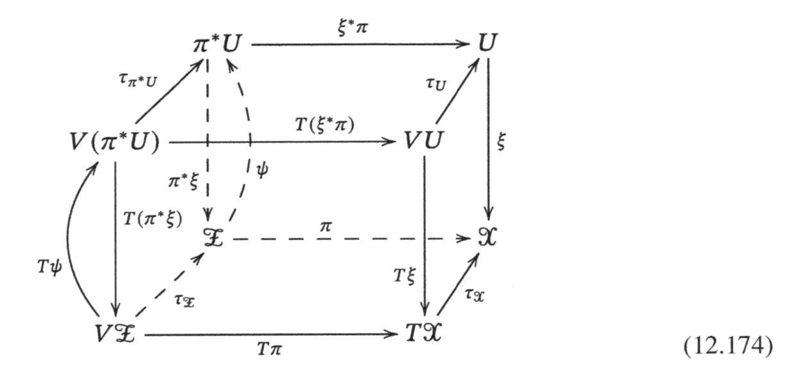

(12.174)

Note that one has the following isomorphism

$$\mathcal{J} : V(\pi^*U) \longrightarrow (T\pi)^*VU, \tag{12.175}$$

whereby, locally,

$$(\boldsymbol{x}, z, \boldsymbol{u}, \boldsymbol{0}, \dot{z}, \dot{\boldsymbol{u}}) \longmapsto (\boldsymbol{x}, z, \boldsymbol{0}, \dot{z}, \boldsymbol{u}, \dot{\boldsymbol{u}}). \tag{12.176}$$

Thus, we may consider the diagram

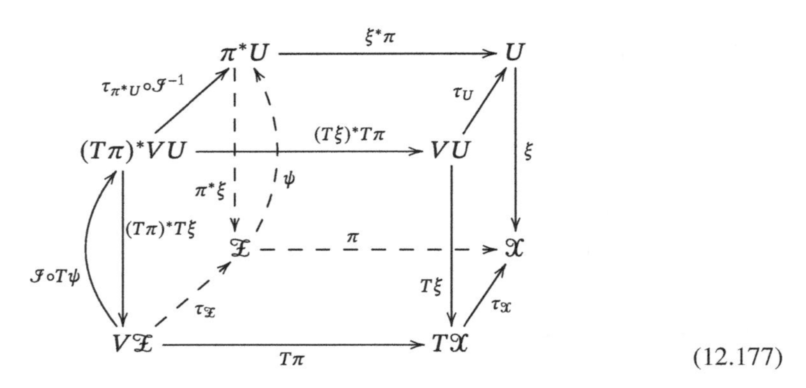

(12.177)

Now, in view of the definition of the section $\mathrm{d}\psi$ in terms of the tangent mapping as in (12.62), and the isomorphism $\psi^*(V(\pi^*U)) \cong \pi^*U$ as in (10.91), we have the commutative diagram

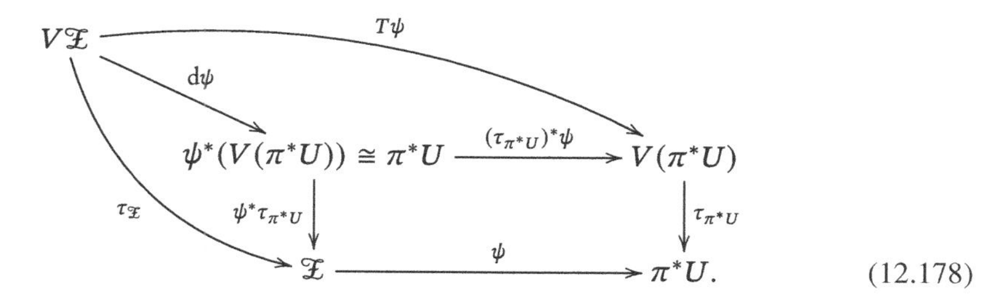

(12.178)

It is concluded that the section $\psi : \mathfrak{X} \to \pi^*U$ induces a section

$$\mathrm{d}\psi : \mathfrak{X} \longrightarrow L(V\mathfrak{X}, \pi^*U). \tag{12.179}$$

As a first example of the foregoing structure, we consider a potential function for the body force. Thus, for a case where the fiber bundle $\mathfrak{X}$ is replaced by $\pi : \mathcal{Y} \to \mathcal{X}$ and the vector bundle U is taken as $\tau^*_{\mathcal{X}} : \bigwedge^n T^*\mathcal{X} \to \mathcal{X}$, a *potential function for the body force* **b** is a section

$$\psi_{\mathbf{b}} : \mathcal{Y} \longrightarrow \pi^* \bigwedge^n T^*\mathcal{X}. \tag{12.180}$$

For a given potential function $\psi_{\mathbf{b}}$, a configuration $\kappa : \mathcal{X} \to \mathcal{Y}$ induces an n-form $\psi_{\mathbf{b}\kappa} := \kappa^*_{\pi^*\tau^*_{\mathcal{X}}} \psi_{\mathbf{b}} := \psi_{\mathbf{b}} \circ \kappa$ on $\mathcal{X}$ by pulling back the potential

$$\begin{array}{ccc}
\pi^* \bigwedge^n T^*\mathcal{X} & \xrightarrow{(\tau^*_{\mathcal{X}})^*\pi} & \bigwedge^n T^*\mathcal{X} \\
\psi_{\mathbf{b}} \uparrow \downarrow \pi^*\tau^*_{\mathcal{X}} & & \tau^*_{\mathcal{X}} \downarrow \uparrow \psi_{\mathbf{b}\kappa} = \kappa^*_{\pi^*\tau^*_{\mathcal{X}}} \psi_{\mathbf{b}} \\
\mathcal{Y} & \underset{\kappa}{\overset{\pi}{\rightleftarrows}} & \mathcal{X},
\end{array} \tag{12.181}$$

where, evidently, $\kappa^*(\pi^* \bigwedge^n T^*\mathcal{X}) = \bigwedge^n T^*\mathcal{X}$. The *total potential energy of the body force at the configuration* κ can be defined now naturally as

$$\Psi_{\mathbf{b}\kappa} := \int_{\mathcal{X}} \psi_{\mathbf{b}\kappa}. \tag{12.182}$$

A potential function induces a body force distribution, a section,

$$\mathbf{b} = -\mathrm{d}\psi_{\mathbf{b}} : \mathcal{Y} \longrightarrow L(V\mathcal{Y}, \pi^* \bigwedge^n T^*\mathcal{X}), \tag{12.183}$$

where the minus sign has been introduced due to the traditional convention in mechanics.

Similarly, we may substitute, for $\pi : \mathfrak{X} \to \mathcal{X}$, the fiber bundle $\pi|_{\partial\mathcal{X}} : \mathcal{Y}|_{\partial\mathcal{X}} \to \partial\mathcal{X}$, and for the vector bundle $\xi : U \to \mathcal{X}$, we substitute $\tau^*_{\partial\mathcal{X}} : \bigwedge^{n-1} T^*\partial\mathcal{X} \to \partial\mathcal{X}$. Thus, a *potential function for the surface force* **t** is a section

$$\psi_{\mathbf{t}} : \mathcal{Y}|_{\partial\mathcal{X}} \longrightarrow (\pi|_{\partial\mathcal{X}})^* \bigwedge^{n-1} T^*\partial\mathcal{X}. \tag{12.184}$$

Given a configuration, κ, by pulling back the potential function of the surface force using $\kappa|_{\partial\mathcal{X}}$, one obtains an $(n-1)$-form $\psi_{\mathbf{t}\kappa}$ in analogy with the case of a body force. Thus, the *total potential energy of the surface force at* κ is

$$\Psi_{\mathbf{t}\kappa} := \int_{\partial\mathcal{X}} \psi_{\mathbf{t}\kappa}. \tag{12.185}$$

A potential function $\psi_{\mathbf{t}}$ induces a surface force distribution, a section

$$\mathbf{t} = -\mathrm{d}\psi_{\mathbf{t}} : \mathcal{Y}|_{\partial\mathcal{X}} \longrightarrow L(V(\mathcal{Y}|_{\partial\mathcal{X}}), (\pi|_{\partial\mathcal{X}})^* \bigwedge^{n-1} T^*\partial\mathcal{X}). \tag{12.186}$$

Finally, for a *hyperelastic material* ,there is a potential function for the stress. That is, for the fiber bundle $\mathfrak{X}$, we substitute the jet bundle $\pi^1 : J^1\mathcal{Y} \to \mathcal{X}$, and for the vector bundle U, we substitute $\bigwedge^n T^*\mathcal{X}$. An *elastic energy function* is a section

$$\psi_S : J^1\mathcal{Y} \longrightarrow \pi^{1*} \bigwedge^n T^*\mathcal{X}. \tag{12.187}$$

For a given configuration, κ, one may pull back the elastic energy function by $j^1\kappa$ to obtain an n-form $\psi_{S\kappa} = \psi_S \circ j^1\kappa$ over $\mathcal{X}$ in analogy with the above. The *total elastic energy at the configuration* κ is

$$\Psi_{S\kappa} := \int_{\mathcal{X}} \psi_{S\kappa}.$$

As expected, an elastic energy function ψ_S induces a constitutive equation

$$\Phi = \mathrm{d}\psi_S : J^1\mathcal{Y} \longrightarrow L(VJ^1\mathcal{Y}, \pi^{1*} \bigwedge^n T^*\mathcal{X}). \tag{12.188}$$

Let $M : \mathbb{R} \times \mathcal{X} \to \mathcal{Y}$ be a motion of a body. Using the notation of Sect. 12.1, we consider the rates of change of the total potential energies. For the body forces, using the chain rule,

$$\begin{aligned}
\frac{\mathrm{d}\Psi_{\mathbf{b}\kappa_t}}{\mathrm{d}t}(t) &:= \frac{\mathrm{d}}{\mathrm{d}t}\int_{\mathcal{X}} \psi_{\mathbf{b}\kappa_t}(t), \\
&= \int_{\mathcal{X}} \frac{\partial}{\partial t}((\kappa_t)^*_{\pi^*\tau^*_{\mathcal{X}}}\psi_{\mathbf{b}})(t), \\
&= \int_{\mathcal{X}} \frac{\partial}{\partial t}(\psi_{\mathbf{b}}\circ\kappa_t)(t), \\
&= \int_{\mathcal{X}} \mathrm{d}\psi_{\mathbf{b}}\cdot\frac{\mathrm{d}\kappa_t}{\mathrm{d}t}(t), \\
&= -\int_{\mathcal{X}} \mathbf{b}\cdot w_t.
\end{aligned} \tag{12.189}$$

In analogy, for the total potential energy of the surface forces,

$$\frac{\mathrm{d}\Psi_{\mathbf{t}\kappa_t}}{\mathrm{d}t}(t) = -\int_{\mathcal{X}} \mathbf{t}\cdot w_t. \tag{12.190}$$

For the total elastic energy, one has

$$\begin{aligned}
\frac{\mathrm{d}\Psi_{S\kappa_t}}{\mathrm{d}t}(t) &:= \frac{\mathrm{d}}{\mathrm{d}t}\int_{\mathcal{X}} \psi_{S\kappa_t}(t), \\
&= \int_{\mathcal{X}} \frac{\partial}{\partial t}(\psi_S\circ j^1\kappa_t)(t), \\
&= \int_{\mathcal{X}} \mathrm{d}\psi_S\cdot\frac{\mathrm{d}(j^1\kappa_t)}{\mathrm{d}t}(t), \\
&= \int_{\mathcal{X}} \Phi\cdot j^1 w_t,
\end{aligned} \tag{12.191}$$

where (12.57) was used in the last line above.

Thus, using the balance (12.165), we conclude that the total energy is conserved, that is,

$$\frac{\mathrm{d}}{\mathrm{d}t}(\Psi_{S\kappa_t} + \Psi_{\mathbf{b}\kappa_t} + \Psi_{\mathbf{t}\kappa_t}) = 0. \tag{12.192}$$

12.3 Eulerian Fields

So far, most relevant fields were defined on the body/base manifold $\mathcal{X}$. As such, they are traditionally referred to as Lagrangian. Given a configuration $\kappa : \mathcal{X} \to \mathcal{Y}$, the Eulerian counterparts of these fields may be defined as follows:

Consider $\kappa(\mathcal{X}) := \text{Image}\kappa \subset \mathcal{Y}$, the inclusion $\mathcal{J}_\kappa : \kappa(\mathcal{X}) \longrightarrow \mathcal{Y}$, and the restricted mapping

$$\hat{\kappa} : \mathcal{X} \longrightarrow \kappa(\mathcal{X}), \qquad \hat{\kappa}(x) := \kappa(x), \qquad x \in \mathcal{X}. \tag{12.193}$$

Evidently, $\hat{\kappa}$ is a diffeomorphism, the inverse of which is the restriction of the fiber bundle projection

$$\pi|_{\kappa(\mathcal{X})} : \kappa(\mathcal{X}) \to \mathcal{X}. \tag{12.194}$$

Consider the following diagram:

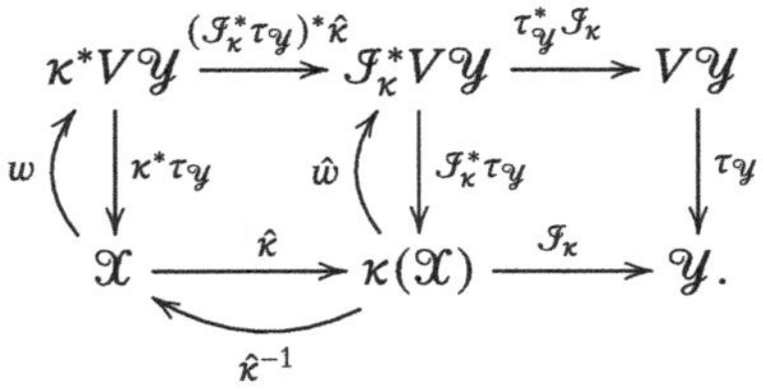

$$\begin{array}{ccccc}
\kappa^* V\mathcal{Y} & \xrightarrow{(\mathcal{J}_\kappa^* \tau_{\mathcal{Y}})^* \hat{\kappa}} & \mathcal{J}_\kappa^* V\mathcal{Y} & \xrightarrow{\tau_{\mathcal{Y}}^* \mathcal{J}_\kappa} & V\mathcal{Y} \\
w \uparrow \downarrow \kappa^* \tau_{\mathcal{Y}} & & \hat{w} \uparrow \downarrow \mathcal{J}_\kappa^* \tau_{\mathcal{Y}} & & \downarrow \tau_{\mathcal{Y}} \\
\mathcal{X} & \underset{\hat{\kappa}^{-1}}{\overset{\hat{\kappa}}{\rightleftarrows}} & \kappa(\mathcal{X}) & \xrightarrow{\mathcal{J}_\kappa} & \mathcal{Y}.
\end{array} \tag{12.195}$$

It is recalled that $(\kappa^* V\mathcal{Y})_x = (V\mathcal{Y})_{\kappa(x)} = (\mathcal{J}_\kappa^* V\mathcal{Y})_{\kappa(x)}$, so that $[(\mathcal{J}_\kappa^* \tau_{\mathcal{Y}})^* \hat{\kappa}]|_x$ is the identity on each fiber. Given a generalized velocity field $w : \mathcal{X} \to \kappa^* V\mathcal{Y}$, one can define its Eulerian counterpart

$$\hat{w} := (\hat{\kappa}^{-1}_{\kappa^* \tau_{\mathcal{Y}}})^* w : \kappa(\mathcal{X}) \longrightarrow \hat{\kappa}^{-1*}(\kappa^* V\mathcal{Y}) = \mathcal{J}_\kappa^* V\mathcal{Y}, \tag{12.196}$$

by

$$\hat{w} := (\mathcal{J}_\kappa^* \tau_{\mathcal{Y}})^* \hat{\kappa} \circ w \circ \hat{\kappa}^{-1}, \qquad \text{so that} \qquad \hat{w}(\kappa(x)) := w(x). \tag{12.197}$$

Evidently, given a Eulerian vector field $\hat{w} : \kappa(\mathcal{X}) \longrightarrow \mathcal{J}^* V\mathcal{Y}$, the corresponding Lagrangian counterpart is obtained by switching the roles of κ and $\hat{\kappa}^{-1}$ in the preceding relation.

For the jets of the velocity fields, one has

$$j^1 \hat{w} = j^1((\mathcal{J}_\kappa^* \tau_{\mathcal{Y}})^* \hat{\kappa} \circ w \circ \hat{\kappa}^{-1}) = J^1((\mathcal{J}_\kappa^* \tau_{\mathcal{Y}})^* \hat{\kappa}) \circ j^1 w \circ \hat{\kappa}^{-1}, \tag{12.198}$$

where we applied the definition of a jet of a fiber bundle morphism in (6.48) to the current settings as in the following diagram.

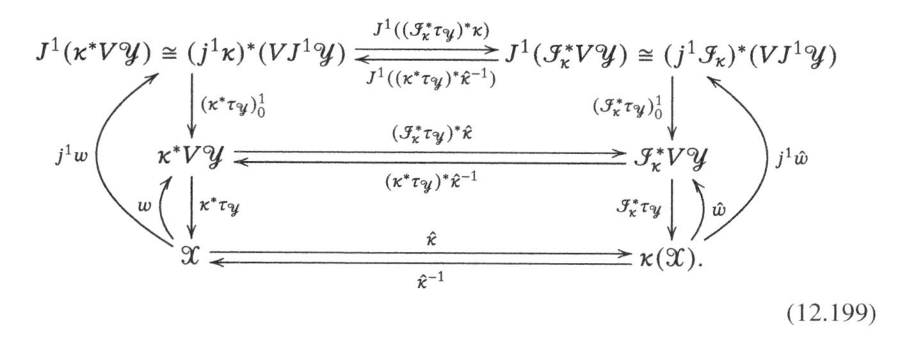

(12.199)

Let $\hat{\kappa}$ be represented locally by $\boldsymbol{x} \mapsto \boldsymbol{y} = \hat{\boldsymbol{\kappa}}(\boldsymbol{x})$, and let $w_0 \in \kappa^* V\mathcal{Y}$ be represented by $(\boldsymbol{x}, \boldsymbol{w}_0)$. Then, the natural mapping $(\mathcal{J}_\kappa^*\tau_\mathcal{Y})^*\hat{\kappa}$ is represented locally by $(\boldsymbol{x}, \boldsymbol{w}_0) \mapsto (\hat{\boldsymbol{\kappa}}(\boldsymbol{x}), \boldsymbol{w}_0)$. If $w : \mathcal{X} \to \kappa^* V\mathcal{Y}$ is a vector field, represented locally by $\boldsymbol{x} \mapsto (\boldsymbol{x}, \boldsymbol{w}(\boldsymbol{x}))$, the corresponding Eulerian vector field, $\hat{w} : \kappa(\mathcal{X}) \to \mathcal{J}_\kappa^* V\mathcal{Y}$, is represented locally by $\boldsymbol{y} \mapsto (\boldsymbol{y}, \boldsymbol{w}(\hat{\boldsymbol{\kappa}}^{-1}(\boldsymbol{y})))$. It follows that

$$j^1\hat{w} = j^1((\mathcal{J}_\kappa^*\tau_\mathcal{Y})^*\hat{\kappa} \circ w \circ \hat{\kappa}^{-1}) = J^1((\mathcal{J}_\kappa^*\tau_\mathcal{Y})^*\hat{\kappa}) \circ j^1 w \circ \hat{\kappa}^{-1} \tag{12.200}$$

is represented by

$$\begin{aligned} \boldsymbol{y} \longmapsto (\boldsymbol{y}, \boldsymbol{w}(\hat{\boldsymbol{\kappa}}^{-1}(\boldsymbol{y})), D_{\boldsymbol{y}}\boldsymbol{w}(\hat{\boldsymbol{\kappa}}^{-1}(\boldsymbol{y}))) \\ = (\boldsymbol{y}, \boldsymbol{w}(\hat{\boldsymbol{\kappa}}^{-1}(\boldsymbol{y})), D_{\boldsymbol{x}}\boldsymbol{w}(\hat{\boldsymbol{\kappa}}^{-1}(\boldsymbol{y})) \circ D_{\boldsymbol{y}}\hat{\boldsymbol{\kappa}}^{-1}(\boldsymbol{y})). \end{aligned} \tag{12.201}$$

To represent $J^1((\mathcal{J}_\kappa^*\tau_\mathcal{Y})^*\hat{\kappa})$ locally, consider $\eta \in J^1(\kappa^* V\mathcal{Y})$, given by $\eta = j_x^1 w$, $w : \mathcal{X} \to \kappa^* V\mathcal{Y}$, and represented by

$$\boldsymbol{\eta} = (\boldsymbol{x}, \boldsymbol{w}_0, \boldsymbol{w}_0') = (\boldsymbol{x}, \boldsymbol{w}(\boldsymbol{x}), D_{\boldsymbol{x}}\boldsymbol{w}(\boldsymbol{x})). \tag{12.202}$$

Then, using $J^1((\mathcal{J}_\kappa^*\tau_\mathcal{Y})^*\hat{\kappa})(\eta) := j^1((\mathcal{J}_\kappa^*\tau_\mathcal{Y})^*\hat{\kappa} \circ w \circ \hat{\kappa}^{-1})$, $J^1((\mathcal{J}_\kappa^*\tau_\mathcal{Y})^*\hat{\kappa})$ is represented by

$$(\boldsymbol{x}, \boldsymbol{w}_0, \boldsymbol{w}_0') \longmapsto (\boldsymbol{y} = \hat{\boldsymbol{\kappa}}(\boldsymbol{x}), \boldsymbol{w}_0, \boldsymbol{w}_0' \circ D_{\boldsymbol{y}}\hat{\boldsymbol{\kappa}}^{-1}(\boldsymbol{y})). \tag{12.203}$$

Interchanging the roles of $\hat{\kappa}$ and $\hat{\kappa}^{-1}$, as illustrated in the diagram above, we can consider, for a given Eulerian field $\hat{w} : \kappa(\mathcal{X}) \to \mathcal{J}_\kappa^* V\mathcal{Y}$, the Lagrangian counterpart

$$w = (\hat{\kappa}_{\mathcal{J}_\kappa^*\tau_\mathcal{Y}})^*\hat{w} : \mathcal{X} \longrightarrow \kappa^* V\mathcal{Y}. \tag{12.204}$$

Representing $\hat{w}$ in the form $\boldsymbol{y} \mapsto (\boldsymbol{y}, \hat{\boldsymbol{w}}(\boldsymbol{y}))$, w is represented in the form $\boldsymbol{x} \mapsto (\boldsymbol{x}, \hat{\boldsymbol{w}}(\hat{\boldsymbol{\kappa}}(\boldsymbol{x}))$, and $j^1 w$ is represented in the form

$$\boldsymbol{x} \longmapsto (\boldsymbol{x}, \hat{\boldsymbol{w}}(\hat{\boldsymbol{\kappa}}(\boldsymbol{x}), D_{\boldsymbol{y}}\hat{\boldsymbol{w}}(\hat{\boldsymbol{\kappa}}(\boldsymbol{x})) \circ D_{\boldsymbol{x}}\hat{\boldsymbol{\kappa}}(\boldsymbol{x})). \tag{12.205}$$

The jet morphism $J^1((\kappa^* \tau_{\mathcal{Y}})^* \hat{\kappa}^{-1})$ is represented by

Cauchy Variational Stresses The variational stress we have been considering so far is defined on the body/base manifold, and it expends power when acting on the rates of change of deformation jets, or, equivalently, on jets of Lagrangian velocity fields. As such, it is analogous to the first Piola-Kirchhoff stress of standard continuum mechanics. Having defined Eulerian velocity fields and their corresponding jets, we may now define the analog of the Cauchy stress, that is, a field defined on the image, $\kappa(\mathcal{X})$, of a configuration and acting on jets of Eulerian velocity fields to produce power densities.

We define, therefore, in analogy with a variational stress field, a Cauchy variational stress field[1], $\hat{S}$, to be a section

$$\hat{S} : \kappa(\mathcal{X}) \longrightarrow L(J^1(\mathcal{J}_\kappa^* V\mathcal{Y}, \bigwedge^n T^*(\kappa(\mathcal{X}))). \tag{12.206}$$

A variational stress field, S, at a configuration κ induces a Cauchy variational stress, $\hat{S}$, and vice-versa. The natural condition we impose is

$$(T\hat{\kappa})^*(x)[\hat{S}(\kappa(x))(j^1\hat{w}(\kappa(x)))] = S(x)(j^1 w(x)), \tag{12.207}$$

for every $x \in \mathcal{X}$ and every pair of fields w, $\hat{w}$ with $\hat{w} = (\hat{\kappa}^{-1}_{\kappa^* \tau_{\mathcal{Y}}})^* w$. Here, $(T\hat{\kappa})^*$ is the pullback of forms as in (6.78). Using (12.200), the condition may be represented by the following commutative diagram:

$$\begin{array}{ccc} J^1(\kappa^* V\mathcal{Y})_x & \xrightarrow{J^1((\mathcal{J}_\kappa^* \tau_{\mathcal{Y}})^* \hat{\kappa})} & J^1(\mathcal{J}_\kappa^* V\mathcal{Y}) \\ {\scriptstyle S(x)}\downarrow & & \downarrow{\scriptstyle \hat{S}(\kappa(x))} \\ \bigwedge^n T_x^* \mathcal{X} & \xleftarrow{(T_x \kappa)^*} & \bigwedge^n T^*_{\kappa(x)} \kappa(\mathcal{X}), \end{array} \tag{12.208}$$

or

$$S(x) = (T_x \kappa)^* \circ \hat{S}(\kappa(x)) \circ J^1((\mathcal{J}_\kappa^* \tau_{\mathcal{Y}})^* \hat{\kappa})|_x. \tag{12.209}$$

It follows from (6.155) that

$$\int_{\mathcal{X}} S \cdot j^1 w = \int_{\mathcal{X}} (T\hat{\kappa})^* (\hat{S} \cdot j^1 \hat{w}) = \int_{\kappa(\mathcal{X})} \hat{S} \cdot j^1 \hat{w}. \tag{12.210}$$

Let S be represented locally by

[1] Note that in [Seg02] we used the term "Cauchy stress" to what is referred to here as the "traction stress."

$$(x^i) \longmapsto (x^i, R_{\alpha 1 \dots n}(x^i), S^k_{\alpha 1 \dots n}(x^i)) \tag{12.211}$$

as in (10.61), and let $\hat{S}$ be represented locally by

$$(y^{i'}) \longmapsto (y^{i'}, \hat{R}_{\alpha 1 \dots n}(y^{i'}), \hat{S}^{k'}_{\alpha 1 \dots n}(y^{i'})), \tag{12.212}$$

where $y^{i'}$ are local coordinates in $\kappa(\mathcal{X})$. Then, condition (12.207) is represented locally by

$$\begin{aligned}\sum_{\alpha,k'} \left| D_x \hat{\kappa}(x) \right| \left[\hat{R}_{\alpha 1 \dots n}(\hat{\kappa}(x)) \hat{w}^\alpha(\hat{\kappa}(x)) + \hat{S}^{k'}_{\alpha 1 \dots n}(\hat{\kappa}(x)) \hat{w}^\alpha_{,k'}(\hat{\kappa}(x)) \right] \mathrm{d}y \\ = \sum_{\alpha,k} \left[R_{\alpha 1 \dots n}(x) w^\alpha(x) + S^k_{\alpha 1 \dots n}(x) w^\alpha_{,k}(x) \right] \mathrm{d}x.\end{aligned} \tag{12.213}$$

Using the local relations (12.197, 12.201) between the Eulerian field and their jets to the corresponding Lagrangian fields, the last condition is rewritten as

$$\begin{aligned}\sum_{\alpha,k,k'} \left| D_x \hat{\kappa}(x) \right| \left[\hat{R}_{\alpha 1 \dots n}(\hat{\kappa}(x)) w^\alpha(x) + \hat{S}^{k'}_{\alpha 1 \dots n}(\hat{\kappa}(x)) w^\alpha_{,k}(x) (\hat{\kappa}^{-1})^k_{,k'}(\hat{\kappa}(x)) \right] \mathrm{d}y \\ = \sum_{\alpha,k} \left[R_{\alpha 1 \dots n}(x) w^\alpha(x) + S^k_{\alpha 1 \dots n}(x) w^\alpha_{,k}(x) \right] \mathrm{d}x.\end{aligned} \tag{12.214}$$

Since this condition holds for arbitrary values of $w^\alpha(x)$ and $w^\alpha_{,k}(x)$,

$$\begin{aligned} R_{\alpha 1 \dots n}(x) \mathrm{d}x &= \left| D_x \hat{\kappa}(x) \right| \hat{R}_{\alpha 1 \dots n}(\hat{\kappa}(x)) \mathrm{d}y. \\ S^k_{\alpha 1 \dots n}(x) \mathrm{d}x &= \left| D_x \hat{\kappa}(x) \right| \hat{S}^{k'}_{\alpha 1 \dots n}(\hat{\kappa}(x)) (\kappa^{-1})^k_{,k'}(\hat{\kappa}(x)) \mathrm{d}y. \end{aligned} \tag{12.215}$$

These relations are analogous to the relation between the first Piola-Kirchhoff stress and the Cauchy stress in standard continuum mechanics.

Appendix: Notation Used in This Chapter

In Table 12.1 below, we summarize the terminology and notation used in this chapter.

Table 12.1 Terminology and notation

Object	Description	Local representation
Point in the domain/body (base manifold)	$x \in \mathcal{X}$	$(x^i), i = 1, \dots, n, \boldsymbol{x} \in \mathbb{R}^n$
Value of a field/configuration	$y \in \mathcal{Y}, \pi : \mathcal{Y} \to \mathcal{X}$	$(y^\alpha), \alpha = 1, \dots, m, \boldsymbol{y} \in \mathbb{R}^m$
Configuration	$\kappa : \mathcal{X} \to \mathcal{Y}$	$\kappa^\alpha(x^i), \boldsymbol{\kappa}(\boldsymbol{x})$
Motion	$M : \mathbb{R} \times \mathcal{X} \to \mathcal{Y}$	$(\boldsymbol{x}, \boldsymbol{y} = \boldsymbol{M}(t, \boldsymbol{x}))$
Vertical subbundle	$V\mathcal{Y} := \text{Kernel}T\pi$	$(\boldsymbol{x}, \boldsymbol{y}, \boldsymbol{0}, \dot{\boldsymbol{y}})$
Generalized velocity	$w : \mathcal{X} \to V\mathcal{Y}$	$\boldsymbol{x} \mapsto (\boldsymbol{x}, \boldsymbol{\kappa}(\boldsymbol{x}), \dot{\boldsymbol{\kappa}}(\boldsymbol{x}))$
Velocities at κ	$w : \mathcal{X} \to \kappa^* V\mathcal{Y}$	$\boldsymbol{x} \mapsto (\boldsymbol{x}, \dot{\boldsymbol{\kappa}}(\boldsymbol{x}))$
Inclusion of $\kappa(\mathcal{X}) := \text{Image}\kappa$	$\mathcal{J}_\kappa : \kappa(\mathcal{X}) \to \mathcal{Y}$	
Restriction of κ to its image	$\hat{\kappa} : \mathcal{X} \to \kappa(\mathcal{X})$	
Eulerian field corresponding to w	$\hat{w} : w \circ \hat{\kappa}^{-1} : \kappa(\mathcal{X}) \to \mathcal{J}_\kappa^* V\mathcal{Y}$	$\boldsymbol{y} = \boldsymbol{\kappa}(\boldsymbol{x}) \mapsto (\boldsymbol{y}, \dot{\boldsymbol{\kappa}}(\boldsymbol{x}))$
Jet field	$\chi : \mathcal{X} \to J^! \mathcal{Y}$	$\boldsymbol{x} \longmapsto (\boldsymbol{x}, \boldsymbol{\kappa}(\boldsymbol{x}), \boldsymbol{\chi}(\boldsymbol{x}))$
Deformation jet/gradient	$\chi = j^1\kappa : \mathcal{X} \to J^! \mathcal{Y}$	$\boldsymbol{x} \longmapsto (\boldsymbol{x}, \boldsymbol{\kappa}(\boldsymbol{x}), \dot{\boldsymbol{\kappa}}(\boldsymbol{x}),$ $D\boldsymbol{\kappa}(\boldsymbol{x}), D\dot{\boldsymbol{\kappa}}(\boldsymbol{x}))$
Velocity jet	$j^1 w : \mathcal{X} \to J^! V\mathcal{Y}$	$\boldsymbol{x} \longmapsto (\boldsymbol{x}, \boldsymbol{\kappa}(\boldsymbol{x}), D\boldsymbol{\kappa}(\boldsymbol{x}))$
Time-dependent jet field	$M^J : \mathbb{R} \times \mathcal{X} \longrightarrow J^1\mathcal{Y}$	$(t, \boldsymbol{x}) \mapsto (\boldsymbol{x}, \boldsymbol{M}_0^J(t, \boldsymbol{x}),$ $\boldsymbol{M}_1^J(t, \boldsymbol{x}))$
Jet velocity field	$\eta : \mathcal{X} \longrightarrow VJ^1\mathcal{Y}$	$(\boldsymbol{x}, \boldsymbol{M}_0^J(0, \boldsymbol{x}), \boldsymbol{M}_1^J(0, \boldsymbol{x}),$ $D_t\boldsymbol{M}_0^J(0, \boldsymbol{x}), D_t\boldsymbol{M}_1^J(0, \boldsymbol{x}))$
Natural isomorphism	$\iota : J^1 V\mathcal{Y} \longrightarrow VJ^1\mathcal{Y},$	$(\boldsymbol{x}, \boldsymbol{y}, \dot{\boldsymbol{y}}, \boldsymbol{y}', \dot{\boldsymbol{y}}')$ $\longmapsto (\boldsymbol{x}, \boldsymbol{y}, \boldsymbol{y}', \dot{\boldsymbol{y}}, \dot{\boldsymbol{y}}')$
Body forces	$\xi_{\mathbf{b}} : L(V\mathcal{Y}, \pi^* \bigwedge^n T^*\mathcal{X}) \to \mathcal{Y}$	$\sum_\alpha \mathbf{b}_{\alpha 1 \dots n} \mathrm{d}y^\alpha \otimes \mathrm{d}x$
Body force field	$\mathbf{b} : \mathcal{Y} \to L(V\mathcal{Y}, \pi^* \bigwedge^n T^*\mathcal{X})$	$\sum_\alpha \mathbf{b}_{\alpha 1 \dots n}(x^j, y^\beta) \mathrm{d}y^\alpha \otimes \mathrm{d}x$
Body force field at configuration κ	$\mathbf{b}_\kappa : \mathcal{X} \to L(\kappa^* V\mathcal{Y}, \bigwedge^n T^*\mathcal{X})$	
Surface forces	$\xi_{\mathbf{t}} : L(V\mathcal{Y}, \pi^* \bigwedge^{n-1} T^*\partial\mathcal{X}) \to \mathcal{Y}$	$\sum_\alpha \mathbf{t}_{\alpha 1 \dots n-1} \mathrm{d}y^\alpha \otimes \mathrm{d}z$
Surface force field	$\mathbf{t} : \mathcal{Y} \to L(V\mathcal{Y}, \pi^* \bigwedge^{n-1} T^*\partial\mathcal{X})$	$\sum_\alpha \mathbf{t}_{\alpha 1 \dots n-1}(z^b, y^\beta) \mathrm{d}y^\alpha \otimes \mathrm{d}z$
Surface force field at configuration κ	$\mathbf{t}_\kappa : \partial\mathcal{X} \to L(\kappa^* V\mathcal{Y}, \bigwedge^{n-1} T^*\partial\mathcal{X})$	
Loading	$(\mathbf{b}, \mathbf{t})$	
Traction stresses	$\xi_s : L(V\mathcal{Y}, \pi^* \bigwedge^{n-1} T^*\mathcal{X}) \longrightarrow \mathcal{Y}$	$\sum_{\alpha,k} (-1)^{k-1} s_{\alpha 1 \dots \widehat{k} \dots n}$ $\mathrm{d}y^\alpha \otimes (\partial_k \lrcorner \mathrm{d}x)$
Traction stress field	$s : \mathcal{Y} \longrightarrow L(V\mathcal{Y}, \pi^* \bigwedge^{n-1} T^*\mathcal{X})$	$(\boldsymbol{x}, \boldsymbol{y}) \longmapsto s_{\alpha 1 \dots \widehat{k} \dots n}(\boldsymbol{x}, \boldsymbol{y})$
Traction stress field at κ	$s_\kappa : \mathcal{X} \longrightarrow L(\kappa^* V\mathcal{Y}, \bigwedge^{n-1} T^*\mathcal{X})$	$\boldsymbol{x} \longmapsto s_{\alpha 1 \dots \widehat{k} \dots n}(x^j, \kappa^\beta(x^i))$
$(n-1)$-dimensional submanifold	$\mathcal{J}_\mathcal{V} : \mathcal{V} \to \mathcal{X}$	
Restriction of $(n-1)$-forms	$\rho_\mathcal{V} : (\bigwedge^{n-1} T^*\mathcal{X})\vert_\mathcal{V} \longrightarrow \bigwedge^{n-1} T^*\mathcal{V}$	
Variational stresses	$\xi_S : L(VJ^1\mathcal{Y}, \pi^{1*} \bigwedge^n T^*\mathcal{X}) \to J^1\mathcal{Y}$	$(R_{\alpha 1 \dots n}, S^k_{\alpha 1 \dots n})$

(continued)

Table 12.1 (continued)

Object	Description	Local representation
Variational stress field	$\Phi : J^1\mathcal{Y} \to L(VJ^1\mathcal{Y}, \pi^{1*}\bigwedge^n T^*\mathcal{X})$	$(\boldsymbol{x}, \boldsymbol{y}) \longmapsto (R_{\alpha 1\ldots n}, S^k_{\alpha 1\ldots n})(\boldsymbol{x}, \boldsymbol{y}, \boldsymbol{y}')$
Variational stress field at κ	$S : \mathcal{X} \longrightarrow L(J^1(\kappa^* V\mathcal{Y}), \bigwedge^n T^*\mathcal{X})$	$\boldsymbol{x} \longmapsto$ $(R_{\alpha 1\ldots n}, S^k_{\alpha 1\ldots n})(\boldsymbol{x}, \boldsymbol{\kappa}(\boldsymbol{x}), D\boldsymbol{\kappa}(\boldsymbol{x})$
Traction stress by a variational stress	$p_s : S_0 \longmapsto s_0$	$(R_{\alpha 1\ldots n}, S^k_{\alpha 1\ldots n})$ $\longmapsto ((-1)^{k-1} S^k_{\alpha 1\ldots n})$
Cauchy (Eulerian) variational stress	$\hat{S} : \kappa(\mathcal{X}) \longrightarrow$ $L(J^1(\mathcal{J}^*_\kappa V\mathcal{Y}, \bigwedge^n T^*\kappa(\mathcal{X}))$	
Differential operator	$\widetilde{\mathfrak{d}s}_\kappa \cdot j^1 w := \mathrm{d}(s_\kappa \cdot w)$	
Divergence of a variation stress field	$\mathrm{div} S \cdot w = \mathrm{d}(p_s S \cdot w) - S \cdot j^1 w$	$(R_{\alpha 1\ldots n}, S^k_{\alpha 1\ldots n})$ $\longmapsto (S^k_{\alpha 1\ldots n,k} - R_{\alpha 1\ldots n})$
Potential function for body forces	$\psi_{\mathbf{b}} : \mathcal{Y} \longrightarrow \pi^* \bigwedge^n T^*\mathcal{X}$	
Total potential energy of body forces	$\Psi_{\mathbf{b}\kappa} := \int_{\mathcal{X}} \psi_{\mathbf{b}\kappa}$	
Potential function for surface forces	$\psi_{\mathbf{t}} : \mathcal{Y}\vert_{\partial\mathcal{X}} \longrightarrow (\pi\vert_{\partial\mathcal{X}})^* \bigwedge^{n-1} T^*\partial\mathcal{X}$	
Total potential energy of surface forces	$\Psi_{\mathbf{t}\kappa} := \int_{\partial\mathcal{X}} \psi_{\mathbf{t}\kappa}$	
Elastic energy function	$\psi_S : J^1\mathcal{Y} \longrightarrow \pi^{1*} \bigwedge^n T^*\mathcal{X}$	
Total elastic energy	$\Psi_{S\kappa} := \int_{\mathcal{X}} \psi_{S\kappa}$	
Generic potential function	$\psi : \mathcal{Z} \to \pi^* U$	$\pi : \mathcal{Z} \to \mathcal{X},\ \xi : U \to \mathcal{X}$
Differential of a potential function	$\mathrm{d}\psi : \mathcal{Z} \longrightarrow L(V\mathcal{Z}, \pi^* U)$	

Chapter 13
Symmetry and Dynamics

In this chapter, introducing additional structure on the space manifold $\mathcal{S}$, we can develop the theory further. First, we consider force resultants and invariance properties of forces and stresses by assuming a group action on $\mathcal{S}$. Next, in the spirit of classical mechanics, we adopt a general model of spacetime whereby time is absolute. Using the real line as a model for the time axis, spacetime is represented as a fiber bundle

$$\vartheta_{\mathscr{E}} : \mathscr{E} \longrightarrow \mathbb{R}.$$

Spacetime with this fiber bundle structure is referred to as a proto-Galilean spacetime (see [SE80, SE22]). Consequently, a motion of a particle is a section of this fiber bundle. Using the constructions of the previous chapter, we formulate the dynamics of a particle and a continuous body. In the proposed framework, the dynamic law, the analog of Newton's second law, is naturally contained in the constitutive relation.

Remark 13.1 Notation. In what follows, unless otherwise indicated, summation is implied for repeated indices and multi-indices. Naturally, when the notation does not comply with the summation convention, e.g., when an index is repeated three times, summation is not implied (unless explicitly indicated).

13.1 Symmetry Group Action: Totals and Invariance

At the current level of generality, classical notions of continuum mechanics, such as the total force and total torque acting on a subbody, cannot be formulated. Thus, one cannot formulate conditions of equilibrium for forces and stresses. For example, a symmetry property of stresses, which follows from a balance of moments, is meaningless.

R. Segev, *Foundations of Geometric Continuum Mechanics*, Advances in Continuum Mechanics 49, https://doi.org/10.1007/978-3-031-35655-1_13

It is recalled that for standard continuum mechanics in a three-dimensional Euclidean space, equilibrium may be imposed by requiring that for any subbody, the total power will vanish for rigid vector fields—those representing translation and rigid rotations of space. Since we do not assume that the space manifold has a Euclidean structure, we introduce symmetry explicitly through a symmetry group action on the space manifold. The rigid vector fields of standard continuum mechanics will be replaced by vector fields on the space manifold induced by the group action.

For the kinematic framework, a configuration is taken as a smooth embedding $\kappa : \mathcal{X} \to \mathcal{S}$ for a fixed space manifold $\mathcal{S}$. The symmetry properties will correspond to the space manifold. Thus, we assume that there is a Lie group G that acts on $\mathcal{S}$ on the left, so that there is a smooth mapping

$$A : G \times \mathcal{S} \longrightarrow \mathcal{S}, \tag{13.1}$$

having the following properties. Let us denote the group multiplication by $(\gamma_1, \gamma_2) \mapsto \gamma_1\gamma_2$, and let e be the identity of the group. We will also use the notation

$$\gamma \cdot y := A(\gamma, y), \qquad A_\gamma := A|_{\{\gamma\}\times\mathcal{S}} : \mathcal{S} \longrightarrow \mathcal{S}. \tag{13.2}$$

The standard assumptions regarding a group action on a manifold require that

$$A_e : \mathcal{S} \longrightarrow \mathcal{S} \tag{13.3}$$

is the identity mapping $\mathrm{Id}_{\mathcal{S}} : \mathcal{S} \to \mathcal{S}$, and for every $\gamma_1, \gamma_2 \in G$, $y \in \mathcal{S}$,

$$\gamma_2 \cdot (\gamma_1 \cdot y) = (\gamma_2\gamma_1) \cdot y \qquad \text{or} \qquad A_{\gamma_2} \circ A_{\gamma_1} = A_{\gamma_2\gamma_1}. \tag{13.4}$$

These properties imply immediately that

$$A_{\gamma^{-1}} \circ A_\gamma = \mathrm{Id}_{\mathcal{S}},$$

which in turn implies that A_γ is a diffeomorphism for every $\gamma \in G$ and

$$(A_\gamma)^{-1} = A_{\gamma^{-1}}. \tag{13.5}$$

The action A induces naturally a mapping, the partial tangent mapping,

$$T_1 A : TG \times \mathcal{S} \longrightarrow T\mathcal{S}. \tag{13.6}$$

For an element $v \in T_\gamma G$, represented by a curve

$$c : \mathbb{R} \longrightarrow G, \qquad c(0) = \gamma, \tag{13.7}$$

we have

$$T_1 A(\upsilon, y) = (A_{c(t)} y)^{\cdot}(t = 0) = (c(t) \cdot y)^{\cdot}(t = 0). \tag{13.8}$$

Evidently,

$$\tau_{\mathcal{S}}(T_1 A(\upsilon, y)) = \gamma \cdot y. \tag{13.9}$$

The mapping $T_1 A$ may be restricted to the Lie algebra, $\mathfrak{g} = T_e G$, and we obtain a mapping

$$DA := T_1 A|_{T_e G \times \mathcal{S}} : \mathfrak{g} \times \mathcal{S} \longrightarrow T\mathcal{S}, \tag{13.10}$$

which is linear in $\mathfrak{g}$. Since the tangent space $T_\gamma G$, for any $\gamma \in G$, is naturally isomorphic to the Lie algebra, DA contains all the information included in $T_1 A$. Since $A_e = \mathrm{Id}_{\mathcal{S}}$, $\tau_{\mathcal{S}}(DA(\upsilon, y)) = y$ for all $\upsilon \in \mathfrak{g}$. Thus, any element $\upsilon \in \mathfrak{g}$ determines, in turn, a vector field

$$X_\upsilon := DA|_{\{\upsilon\} \times \mathcal{S}} = T_1 A|_{\{\upsilon\} \times \mathcal{S}} : \mathcal{S} \longrightarrow T\mathcal{S}. \tag{13.11}$$

The action of G on $\mathcal{S}$ induces an action of G on configurations, whereby for $\gamma \in G$,

$$\kappa \longmapsto \gamma \cdot \kappa := A_\gamma \circ \kappa. \tag{13.12}$$

This action is well-defined since A_γ is a diffeomorphism.

Given a configuration κ, a curve $c : \mathbb{R} \to G$ induces a motion of the body

$$M : \mathbb{R} \times \mathcal{X} \longrightarrow \mathcal{S}, \qquad M(t, x) = c(t) \cdot \kappa = A_{c(t)} \circ \kappa. \tag{13.13}$$

Assume that $c(0) = e$ and let $\upsilon \in \mathfrak{g}$ be the corresponding tangent vector. Then, for the generalized velocity field corresponding to the motion, as in (12.9), we have

$$\begin{aligned} w_\upsilon(x) &= \frac{\partial M}{\partial t}(0, x), \\ &= \frac{\partial}{\partial t}(c(t) \cdot \kappa(x))|_{t=0}, \\ &= T_1 A(\upsilon, \kappa(x)), \\ &= DA(\upsilon, \kappa(x)), \\ &= X_\upsilon(\kappa(x)). \end{aligned} \tag{13.14}$$

Therefore, the induced generalized velocity field is

$$w_{\upsilon} = X_{\upsilon} \circ \kappa : \mathcal{X} \longrightarrow T\mathcal{S}, \qquad \tau_{\mathcal{S}} \circ w_{\upsilon} = \kappa. \tag{13.15}$$

In view of the definition of a pullback of a section in (6.56), and the diagram

$$\begin{array}{ccc} \kappa^* T\mathcal{S} & \xrightarrow{\tau_{\mathcal{S}}^*\kappa} & T\mathcal{S} \\ \kappa_{\tau_{\mathcal{S}}}^* X_{\upsilon} \uparrow \downarrow \kappa^*\tau_{\mathcal{S}}^* & & \tau_{\mathcal{S}} \downarrow \uparrow X_{\upsilon} \\ \mathcal{X} & \xrightarrow[\kappa]{} & \mathcal{S}, \end{array} \tag{13.16}$$

we may make the identification,

$$w_{\upsilon} = \kappa_{\tau_{\mathcal{S}}}^* X_{\upsilon}. \tag{13.17}$$

The foregoing structure may be somewhat generalized for the case where $\pi : \mathcal{Y} \to \mathcal{X}$ is a general fiber bundle, rather than a product bundle, $\mathcal{X} \times \mathcal{S}$, acted upon by a Lie group G. Thus, it is assumed that we have a smooth action

$$A : G \times \mathcal{Y} \to \mathcal{Y} \tag{13.18}$$

that satisfies the following conditions:

1. The action is assumed to preserve the fibers of $\mathcal{Y}$ so that

$$\pi(\gamma \cdot y) = \pi(y), \quad \text{for all} \quad \gamma \in G, \ y \in \mathcal{Y}. \tag{13.19}$$

2. The properties below imply that the restriction of A to a fiber is a group action on a manifold. Thus,

$$A|_{\{\gamma\}\times\mathcal{Y}_x} : \mathcal{Y}_x \longrightarrow \mathcal{Y}_x \tag{13.20}$$

is a diffeomorphism for all $\gamma \in G, \ x \in \mathcal{X}$.
3. Using the notation $A_{\gamma} := A|_{\{\gamma\}\times\mathcal{Y}} : \mathcal{Y} \longrightarrow \mathcal{Y}$, it is assumed that $A_e = \mathrm{Id}_{\mathcal{Y}}$.
4. Finally,

$$\gamma_2 \cdot (\gamma_1 \cdot y) = (\gamma_2\gamma_1) \cdot y \qquad \text{or} \qquad A_{\gamma_2} \circ A_{\gamma_1} = A_{\gamma_2\gamma_1}. \tag{13.21}$$

The Lie algebra generates vector fields by

$$DA := T_1 A|_{T_e G \times \mathcal{Y}} : \mathfrak{g} \times \mathcal{Y} \longrightarrow V\mathcal{Y}. \tag{13.22}$$

Indeed, if $c : \mathbb{R} \to G$ is a curve with $c(0) = e$, and $y \in \mathcal{Y}_x$, then, for each $t \in \mathbb{R}$, $c(t) \cdot y \in \mathcal{Y}_x$. Hence, the tangent to the curve $t \mapsto c(t) \cdot y$ at $t = 0$ belongs to $(V\mathcal{Y})_y$. Under restriction, any $\upsilon \in \mathfrak{g}$ induces, linearly, a vertical vector field

$$X_\upsilon := DA|_{\{\upsilon\}\times\mathcal{Y}} = T_1 A|_{\{\upsilon\}\times\mathcal{Y}} : \mathcal{Y} \longrightarrow V\mathcal{Y}. \tag{13.23}$$

It is noted that X_υ is defined now on $\mathcal{Y}$ and not on $\mathcal{S}$, since the fiber bundle is no longer trivial. Henceforth, we will keep this notation even for the special case of a trivial fiber bundle. When $\mathcal{Y} = \mathcal{X} \times \mathcal{S}$, X_υ is evidently independent of $x \in \mathcal{X}$.

The group action on $\mathcal{Y}$ induces an action of G on configurations whereby, for $\gamma \in G$, $\kappa \mapsto \gamma \cdot \kappa := A_\gamma \circ \kappa$. Given a configuration κ, an element $\upsilon \in \mathfrak{g}$ induces a generalized velocity w_υ at the configuration κ by

$$w_\upsilon := X_\upsilon \circ \kappa : \mathcal{X} \longrightarrow V\mathcal{Y}, \qquad \tau_{\mathcal{Y}} \circ w_\upsilon = \kappa. \tag{13.24}$$

As before, w_υ is viewed as the pullback section $\kappa^*_{\tau_{\mathcal{Y}}} X_\upsilon$ of the pullback bundle $\kappa^* V\mathcal{Y}$ as in

$$\begin{array}{ccc} \kappa^* V\mathcal{Y} & \xrightarrow{\tau^*_{\mathcal{Y}}\kappa} & V\mathcal{Y} \\ \kappa^*_{\tau_{\mathcal{Y}}} X_\upsilon \uparrow\downarrow \kappa^*\tau^*_{\mathcal{Y}} & & \tau_{\mathcal{Y}} \downarrow\uparrow X_\upsilon \\ \mathcal{X} & \xrightarrow[\kappa]{} & \mathcal{Y}. \end{array} \tag{13.25}$$

The operation $\upsilon \mapsto w_\upsilon = X_\upsilon \circ \kappa$, defined on the elements of the Lie algebra, induces a dual operation defined on force functionals. Specifically, a force F at the configuration κ determines an element $\varphi_F \in \mathfrak{g}^* = T^*_e G$ by

$$\varphi_F(\upsilon) := F(w_\upsilon) = F(X_\upsilon \circ \kappa). \tag{13.26}$$

If $F \mapsto \varphi_F \in \mathfrak{g}^*$, we will refer to φ_F as the *resultant* or *total* of the force F. For example, for a region $\mathcal{R}$, the resultants of the body force $\mathbf{b}$ and the surface force $\mathbf{t}_{\mathcal{R}}$ over $\mathcal{R}$ satisfy, respectively,

$$\varphi_{\mathbf{b}\mathcal{R}}(\upsilon) = \int_{\mathcal{R}} \mathbf{b} \cdot w_\upsilon, \quad \varphi_{\mathbf{t}\mathcal{R}}(\upsilon) = \int_{\partial\mathcal{R}} \mathbf{t} \cdot w_\upsilon, \quad \text{for all} \quad \upsilon \in \mathfrak{g}. \tag{13.27}$$

If a variational stress field S at the configuration κ induces the force $F_{\mathcal{R}}$ as in (10.59), then, the resultant of the induced force on $\mathcal{R}$ is given by

$$\varphi_{S\mathcal{R}}(\upsilon) = \int_{\mathcal{R}} S \cdot j^1(X_\upsilon \circ \kappa) = \int_{\mathcal{R}} S \cdot j^1 w_\upsilon. \tag{13.28}$$

Alternatively, the resultant force may be expressed in terms of the corresponding Cauchy variational stress. Using (12.210), one has

$$\varphi_{S\mathcal{R}}(\upsilon) = \int_{\kappa(\mathcal{R})} \hat{S} \cdot j^1(X_\upsilon \circ \kappa \circ \hat{\kappa}^{-1}) = \int_{\kappa(\mathcal{R})} \hat{S} \cdot j^1(X_\upsilon|_{\kappa(\mathcal{R})}) = \int_{\kappa(\mathcal{R})} \hat{S} \cdot j^1 \hat{w}_\upsilon. \tag{13.29}$$

We will say that a force $F_{\mathcal{R}}$ acting on a region $\mathcal{R} \subset \mathcal{X}$ at a configuration κ is *balanced* (relative to G) if its resultant $\varphi_{F_{\mathcal{R}}}$ vanishes. A system of forces $\{F_{\mathcal{R}}\}$, $\mathcal{R} \subset \mathcal{X}$ is balanced if $F_{\mathcal{R}}$ is balanced for each region $\mathcal{R}$. In particular, the force system $\{\varphi_{S\mathcal{R}}\}$ at the configuration κ induced by a variational stress field S, as in (13.28 and 13.29), is balanced if

$$\int_{\mathcal{R}} S \cdot j^1(X_\upsilon \circ \kappa) = \int_{\kappa(\mathcal{R})} \hat{S} \cdot j^1(X_\upsilon|_{\kappa(\mathcal{R})}) = 0, \quad \text{for all} \quad \upsilon \in \mathfrak{g},\ \mathcal{R} \subset \mathcal{X}. \tag{13.30}$$

The continuity of the integrand and the arbitrariness of $\mathcal{R}$ imply that for each $x \in \mathcal{X}$,

$$S(x)((j^1(X_\upsilon \circ \kappa))(x)) = \hat{S}(\kappa(x))(j^1(X_\upsilon|_{\kappa(\mathcal{R})})(\kappa(x))) = 0, \ \text{for all } \upsilon \in \mathfrak{g},\ x \in \mathcal{X}. \tag{13.31}$$

If the foregoing condition holds, we will say that S is a *symmetric variational stress* field.

To represent the condition above locally, consider a section $u : \mathcal{Y} \to V\mathcal{Y}$. The field

$$w := \kappa^*_{\tau_{\mathcal{Y}}} u = u \circ \kappa : \mathcal{X} \longrightarrow \kappa^* V\mathcal{Y}, \tag{13.32}$$

is represented by

$$\boldsymbol{x} \longmapsto (\boldsymbol{x}, \boldsymbol{u}(\boldsymbol{x}, \boldsymbol{\kappa}(\boldsymbol{x}))). \tag{13.33}$$

Here, $\boldsymbol{u}$ is a representative of u and $\boldsymbol{x} \mapsto (\boldsymbol{x}, \boldsymbol{z}) = (\boldsymbol{x}, \boldsymbol{\kappa}(\boldsymbol{x}))$ represents κ. It is noted that $\boldsymbol{y} \in \mathbb{R}^n$ is a typical collection of coordinates in $\kappa(\mathcal{X})$, while $(\boldsymbol{x}, \boldsymbol{z}) = (\boldsymbol{x}, \boldsymbol{\kappa}(\boldsymbol{x})) \in \mathbb{R}^n \times \mathbb{R}^m$ are coordinates in $\mathcal{Y}$. Thus, $j^1 w$ is represented by

$$\boldsymbol{x} \longmapsto (\boldsymbol{x}, \boldsymbol{u}(\boldsymbol{x}, \boldsymbol{\kappa}(\boldsymbol{x})), D_{\boldsymbol{x}}\boldsymbol{u}(\boldsymbol{x}, \boldsymbol{\kappa}(\boldsymbol{x})) + D_{\boldsymbol{z}}(\boldsymbol{x}, \boldsymbol{\kappa}(\boldsymbol{x})) \circ D_{\boldsymbol{x}}\boldsymbol{\kappa}(\boldsymbol{x})). \tag{13.34}$$

Note that for the special case, where $\mathcal{Y} = \mathcal{X} \times \mathcal{S}$, $u : \mathcal{X} \times \mathcal{S} \to T\mathcal{S}$ is independent of the first factor (see the remark following Eq. (13.23)). Hence, in this classical case, $j^1 w$ is represented by

$$\boldsymbol{x} \longmapsto (\boldsymbol{x}, \boldsymbol{u}(\boldsymbol{x}, \boldsymbol{\kappa}(\boldsymbol{x})), D_{\boldsymbol{z}}(\boldsymbol{x}, \boldsymbol{\kappa}(\boldsymbol{x})) \circ D_{\boldsymbol{x}}\boldsymbol{\kappa}(\boldsymbol{x})). \tag{13.35}$$

Using the relation between jets of Eulerian vector fields and the corresponding Lagrangian fields in (12.201), and setting $\boldsymbol{x} := \hat{\boldsymbol{\kappa}}^{-1}(\boldsymbol{y}) \in \mathbb{R}^n$, $\boldsymbol{z} := \boldsymbol{\kappa}(\hat{\boldsymbol{\kappa}}^{-1}(\boldsymbol{y})) \in \mathbb{R}^m$,

$$j^1 \hat{w} = j^1(u \circ \kappa \circ \hat{\kappa}^{-1}) = j^1(u|_{\kappa(\mathcal{X})}) : \kappa(\mathcal{X}) \longrightarrow J^1(\mathcal{I}_\kappa^* V\mathcal{Y}) \tag{13.36}$$

is represented by

$$
\begin{aligned}
&y = \hat{\kappa}(x) \longmapsto \\
&(y, u(x, z), D_x(u \circ \kappa)(\hat{\kappa}^{-1}(y)) \circ D_y\hat{\kappa}^{-1}(y)), \\
&= (y, u(x, z)), D_x u(x, z) \circ D_y\hat{\kappa}^{-1}(y)), \\
&= (y, u(x, z)), (D_x u(x, z) + D_z u(x, z) \circ D_x\kappa(x)) \circ D_y\hat{\kappa}^{-1}(y)), \\
&= (y, u(x, z)), D_x u(x, z) \circ D_y\hat{\kappa}^{-1}(y) + D_z u(x, z) \circ D_x\kappa(x)) \circ D_y\hat{\kappa}^{-1}(y)).
\end{aligned} \tag{13.37}
$$

It is observed that $D_x\kappa(x)) \circ D_y\hat{\kappa}^{-1}(y)$ represents the tangent mapping to the inclusion $\mathcal{J}_\kappa : \kappa(\mathcal{X}) \to \mathcal{Y}$, and it may be written using components as $z^\alpha_{,k'} = \partial z^\alpha / \partial y^{k'}$.

Using (12.212) for the Cauchy variational stress, and substituting X_υ for the field u, the condition for the symmetry of the Cauchy variational stress is expressed by

$$
\begin{aligned}
&\hat{R}_{\alpha 1 \ldots n}(y)(X_\upsilon)^\alpha(x, z) \\
&\qquad + \hat{S}^{k'}_{\alpha 1 \ldots n}(y)[(X_\upsilon)^\alpha_{,k}(x, z)(\hat{\kappa}^{-1})^k_{,k'}(y) + (X_\upsilon)^\alpha_{,\beta}(x, z) z^\beta_{,k'}(y)] \\
&\qquad\qquad = 0,
\end{aligned} \tag{13.38}
$$

for every element υ in the Lie algebra.

We note that for the case where $\mathcal{Y} = \mathcal{X} \times \mathcal{S}$, $(X_\upsilon)^\alpha_{,k} = 0$. If, in addition, $\dim \mathcal{X} = \dim \mathcal{S}$, and we use the same coordinates (z^α) on $\mathcal{S}$ and on $\kappa(\mathcal{X}) \subset \mathcal{S}$ (so that k' is identified with α), the condition for symmetry assumes the simpler form

$$
\hat{R}_{\alpha 1 \ldots n}(z)(X_\upsilon)^\alpha(z) + \hat{S}^\beta_{\alpha 1 \ldots n}(z)(X_\upsilon)^\alpha_{,\beta}(z) = 0. \tag{13.39}
$$

13.2 Dynamics

In the formulation of dynamics that we consider here, we retain the assumption of absolute time, which characterizes classical mechanics. Yet, we model spacetime as a general fiber bundle over the time axis.

By a *proto-Galilean spacetime* (see [SE80, SE22]), we mean a fiber bundle

$$
\vartheta_{\mathcal{E}} : \mathcal{E} \longrightarrow \mathbb{R} \tag{13.40}
$$

where $\mathcal{E}$ is the event space of Chap. 9, $\mathbb{R}$ represents an absolute time axis, and $\vartheta_{\mathcal{E}}$ associate an absolute time $t = \vartheta_{\mathcal{E}}(e) \in \mathbb{R}$, with any event $e \in \mathcal{E}$. Thus, $\vartheta_{\mathcal{E}}^{-1}\{t\}$ is diffeomorphic to a space manifold, $\mathcal{S}$—the typical fiber of the fiber bundle. The manifold $\vartheta_{\mathcal{E}}^{-1}\{t\}$ is interpreted as the collection of simultaneous events occurring at time t. It is observed that a proto-Galilean spacetime lacks the affine structure of a Galilean spacetime.

13.2.1 Dynamics of a Particle in a Proto-Galilean Spacetime

In the formulation of dynamics, we view a motion as a configuration of a body-time manifold in spacetime. For a single particle, a motion is simply the configuration of the time axis in spacetime. In the foregoing material, we viewed generalized velocity fields as time derivatives of motions of a body in space. This interpretation does not fit the approach we present below. Here, it will be convenient to view generalized velocity fields as virtual displacements—infinitesimal variations of placements of material points in the physical space. These are independent of the velocities of the material points in an actual motion. This distinction will be clearly apparent in the definitions below.

The dynamics of a particle in a proto-Galilean spacetime will be presented as a simple example of the elasticity problem of Chap. 12. We identify the fiber bundle, $\pi : \mathcal{Y} \to \mathcal{X}$, of Chap. 12, with $\vartheta_{\mathscr{E}} : \mathscr{E} \to \mathbb{R}$. We view the time axis, $\mathbb{R}$, as the base manifold, $\mathcal{X}$. A configuration, a section $\kappa : \mathcal{X} = \mathbb{R} \to \mathscr{E}$, is interpreted as a motion of a particle. A motion is represented locally by $t \mapsto (t, y^\alpha = \kappa^\alpha(t))$. Geometrically, as the base manifold is identified with the real numbers, a configuration may be viewed also as a curve in $\mathscr{E}$.

A section $w : \mathcal{X} = \mathbb{R} \to V\mathscr{E}$ is interpreted as a virtual displacement field of analytical mechanics. It represents an infinitesimal variation of the motion. The configuration associated with the virtual displacement w is $\kappa = \tau_{\mathscr{E}} \circ w$. In particular, a virtual displacement at the configuration κ is a section $w : \mathcal{X} = \mathbb{R} \to \kappa^* V\mathscr{E}$. In the sequel, for short, we set $W := \kappa^* V\mathscr{E}$. Locally, a virtual displacement is represented by

$$t \longmapsto (t, \kappa^\alpha(t), w^\alpha(t)). \tag{13.41}$$

Let $\kappa : \mathcal{X} \to \mathscr{E}$ be a motion. The jet

$$j^1\kappa : \mathbb{R} \longrightarrow J^1\mathscr{E}, \tag{13.42}$$

is interpreted as the velocity associated with the motion κ. Since the configuration κ may be viewed as a curve in $\mathscr{E}$, the jet, $j^1_t\kappa$ may be viewed as an element of $T\mathscr{E}$. It is emphasized that virtual displacements and velocities are separate types of objects. In particular, while both a virtual displacement and a velocity are tangent vectors, a virtual displacement is necessarily vertical, while a velocity cannot be vertical for a C^1-section κ (see Fig. 13.1). Evidently, while w may be changed independently of κ, $j^1\kappa$ is determined by κ. In view of the local representation of a motion, the local representative of a velocity is

$$t \longmapsto (t, \kappa^\alpha(t), \dot{\kappa}^\beta(t)), \tag{13.43}$$

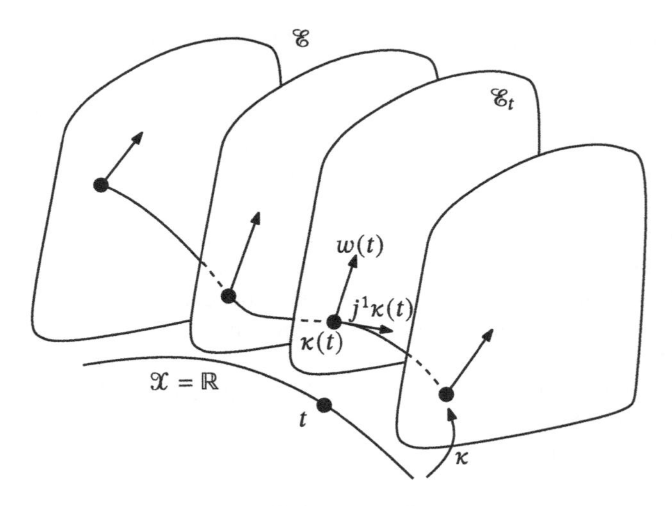

Fig. 13.1 Particle kinematics in a proto-Galilean spacetime

where a superimposed dot indicates differentiation relative to the time variable, and it replaces the differentiation relative to the variable in the base manifold $\partial/\partial x^i$. Thus, the velocity is analogous to the deformation gradient in continuum mechanics.

Let $w : \mathbb{R} = \mathscr{X} \to \kappa^* V\mathscr{C}$ be a virtual displacement at the motion κ. In analogy with the jet of a configuration, the jet of the virtual displacement

$$j^1 w : \mathbb{R} \longrightarrow J^1 W = J^1(\kappa^* V\mathscr{C}), \tag{13.44}$$

is interpreted at the rate at which the virtual displacement changes. Note that by the isomorphism $J^1(\kappa^* V\mathscr{C}) \cong (j^1\kappa)^*(VJ^1\mathscr{C})$, it may be viewed also as the variation of the velocity. Locally, the jet of a virtual displacement at the configuration κ is of the form

$$t \longmapsto (t, w^\alpha(t), \dot{w}^\beta(t)). \tag{13.45}$$

The observations made above for $j^1\kappa$ have analogs in the case of $j^1 w$. Since w may be viewed as a curve $\mathbb{R} \to W$, $j^1 w$ may be identified with the lift $\dot{w}$ of this curve, taking values in TW.

A body force field at the motion κ is a section

$$\mathbf{b} : \mathbb{R} \longrightarrow L(V\mathscr{C}, \pi^* \textstyle\bigwedge^n T^*\mathbb{R}) \cong L(V\mathscr{C}, \mathbb{R}) = (V\mathscr{C})^*. \tag{13.46}$$

Thus, a body force is represented in the form $t \mapsto (t, \mathbf{b}_\alpha)$, where it is noted that the motion κ is implicit in this representation. A body force is interpreted as a "static" force acting on the particle in the classical sense. That is, it is applied to a virtual displacement superimposed on the motion to produce virtual power.

In this setting, a variational stress field at the configuration κ is a section

$$S : \mathbb{R} \longrightarrow L(J^1W, \textstyle\bigwedge^n T^*\mathbb{R}) \cong L(J^1W, \mathbb{R}) = (J^1W)^*. \tag{13.47}$$

Thus, a variational stress field is represented locally by

$$t \longmapsto (t, R_\alpha, S_\beta), \tag{13.48}$$

so that $S \cdot j^1 w$ is represented by

$$R_\alpha w^\alpha + S_\beta \dot{w}^\beta. \tag{13.49}$$

The variational stress in this setting is interpreted as minus the generalized momentum of the particle.

It is recalled that the equation of equilibrium is

$$\operatorname{div} S + \mathbf{b} = 0, \tag{13.50}$$

where div$S$, a section of $L(V\mathcal{E}, \bigwedge^n T^*\mathcal{X}) \cong L(V\mathcal{E}, \mathbb{R}) = (V\mathcal{E})^*$, is represented locally by

$$S^k_{1\ldots n\alpha,k} - R_{1\ldots n\alpha}. \tag{13.51}$$

In the present settings, div$S$ is represented, therefore, by

$$\dot{S}_\alpha - R_\alpha, \tag{13.52}$$

and the local expression for the equation of equilibrium, or, in our interpretation, the equation of motion, is

$$\dot{S}_\alpha - R_\alpha = -\mathbf{b}_\alpha. \tag{13.53}$$

A constitutive relation is a mapping

$$m : J^1\mathcal{E} \longrightarrow L(VJ^1\mathcal{E}, \pi^{1*}\textstyle\bigwedge^n T^*\mathbb{R}) = (VJ^1\mathcal{E})^*, \tag{13.54}$$

given locally by functions $m_{0\alpha}$, and $m_{1\beta}$, $\alpha, \beta = 1, \ldots, m$, in the form

$$(t, \kappa^\alpha, \dot{\kappa}^\beta) \longmapsto (t, R_\alpha = m_{0\alpha}(t, \kappa^\gamma, \dot{\kappa}^\delta), S_\beta = m_{1\beta}(t, \kappa^\gamma, \dot{\kappa}^\delta)). \tag{13.55}$$

Consequently, given a constitutive relation, specifying the inertial properties of the particle, the equation of motion (equilibrium equation) assumes the local form

$$\frac{\mathrm{d}m_{1\alpha}(t, \kappa^\gamma(t), \dot{\kappa}^\delta(t))}{\mathrm{d}t} + m_{0\alpha}(t, \kappa^\gamma(t), \dot{\kappa}^\delta(t)) = -\mathbf{b}_\alpha(t, \kappa^\gamma(t)). \tag{13.56}$$

The second term of the left is naturally interpreted as the "centrifugal" or "Coriolis" force. These terms appear naturally in the current setting. However, it should be noted that the constitutive relations, determining the 4-momentum in terms of the velocity, should be specified. At this level of generality, they are not restricted by any symmetry or invariance conditions.

Remark 13.2 Given the equation of motion (13.56), there may be a trivialization such that the term $m_{0\alpha}(t, \kappa^\gamma(t), \dot\kappa^\delta(t))$ vanishes. Such a trivialization may be interpreted as an inertial frame. Thus, in an inertial frame, the equation of motion assumes the form

$$\frac{\mathrm{d}m_{1\alpha}(t, \kappa^\gamma(t), \dot\kappa^\delta(t))}{\mathrm{d}t} = -\mathbf{b}_\alpha(t, \kappa^\gamma(t)). \tag{13.57}$$

In an inertial frame, a free motion is a solution of the equation

$$\frac{\mathrm{d}m_{1\alpha}(t, \kappa^\gamma(t), \dot\kappa^\delta(t))}{\mathrm{d}t} = 0. \tag{13.58}$$

In the particular case where a connection is given on spacetime, one can decompose the stress object uniquely and invariantly in the form

$$S \cdot j^1 w = \boldsymbol{R} \cdot w + \boldsymbol{S} \cdot \nabla_t w, \tag{13.59}$$

where ∇_t denoted the covariant derivative and $\boldsymbol{R}$ and $\boldsymbol{S}$ are sections of $(V\mathscr{E})^*$. The corresponding equation of motion will be

$$\frac{\mathrm{d}\boldsymbol{S}}{\mathrm{d}t}(t, \dot\kappa(t)) + \boldsymbol{R}(t, \dot\kappa(t)) = -\boldsymbol{b}(t, \kappa(t)). \tag{13.60}$$

Thus, the two terms on the left of Eq. (13.56) have invariant counterparts. Given the constitutive relation, there may be some connection for which the $\boldsymbol{R}$ vanishes. In this case, the inertial frames induced by such a connection are defined globally.

13.2.2 Dynamics of a Body in a Proto-Galilean Spacetime

The kinematics of a continuous body in a proto-Galilean spacetime may be described with increasing levels of generality.

In the simplest situation, we consider the spacetime bundle

$$\vartheta_{\mathscr{E}} : \mathscr{E} \longrightarrow \mathbb{R} \tag{13.61}$$

where the projection $\vartheta_{\mathscr{E}}$ signifies the assumption that time is absolute. The manifold, $\mathscr{X}$, of Chap. 12, is interpreted as a body-time manifold. It is modeled as a Cartesian product of the time axis and a material body manifold, that is, $\mathcal{X} = \mathbb{R} \times \mathcal{B}$. We view a motion as a fiber bundle morphism over the time axis that makes the following diagram commutative.

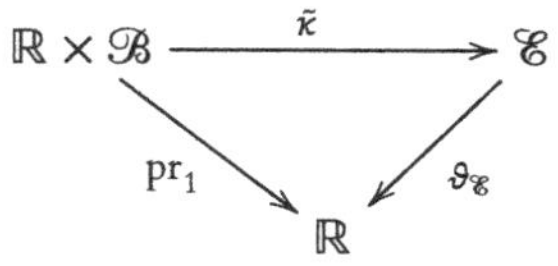

$$\begin{array}{ccc} \mathbb{R} \times \mathcal{B} & \xrightarrow{\tilde{\kappa}} & \mathcal{E} \\ & \mathrm{pr}_1 \searrow \quad \swarrow \vartheta_{\mathcal{E}} & \\ & \mathbb{R} & \end{array} \tag{13.62}$$

If an element of $\mathcal{E}$ is represented locally by (t, y^α) and an element of $\mathcal{B}$ is represented locally by (x^i), then, $\tilde{\kappa}$ is represented locally in the form

$$(z, x^i) \longmapsto (t = z, y^\alpha = \kappa^\alpha(z, x^i)). \tag{13.63}$$

In this setting, the motion can also be viewed as a section

$$\kappa : \mathcal{X} := \mathbb{R} \times \mathcal{B} \longrightarrow \mathcal{Y} := (\mathbb{R} \times \mathcal{B}) \times \mathcal{E} \tag{13.64}$$

of the fiber bundle

$$\pi : \mathcal{Y} := \mathcal{X} \times \mathcal{E} \longrightarrow \mathcal{X} := \mathbb{R} \times \mathcal{B}. \tag{13.65}$$

Observing that one has a projection

$$\vartheta_{\mathcal{Y}} := \vartheta_{\mathcal{E}} \circ \mathrm{pr}_2 : \mathcal{Y} \to \mathbb{R}, \tag{13.66}$$

the condition that $\tilde{\kappa}$ is a fiber bundle morphism over the identity, implies that the diagram

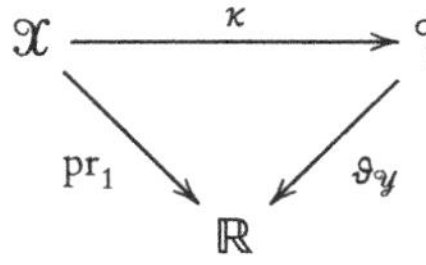

$$\begin{array}{ccc} \mathcal{X} & \xrightarrow{\kappa} & \mathcal{Y} \\ & \mathrm{pr}_1 \searrow \quad \swarrow \vartheta_{\mathcal{Y}} & \\ & \mathbb{R} & \end{array} \tag{13.67}$$

is commutative (see Fig. 13.2).

Locally, κ is of the form

$$(z, x^i) \longmapsto (z, x^i, t = z, y^\alpha = \kappa^\alpha(z, x^i)). \tag{13.68}$$

This setting may be generalized by removing the condition that $\vartheta_{\mathcal{Y}} \circ \kappa = \mathrm{pr}_1$ and modeling a motion by a fiber bundle morphism

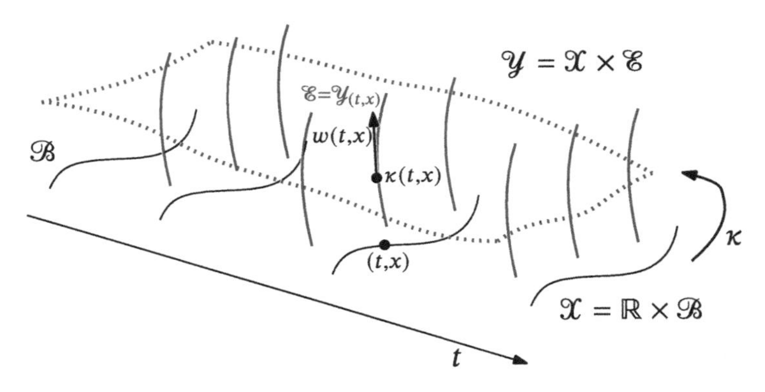

Fig. 13.2 Illustrating the kinematics of body-time $\mathcal{X} = \mathbb{R} \times \mathcal{B}$ in a proto-Galilean spacetime $\mathcal{E}$, where a motion is represented by a section of the fiber bundle $\mathcal{Y}$

$$\begin{array}{ccc} \mathbb{R} \times \mathcal{B} & \xrightarrow{\tilde{\kappa}} & \mathcal{E} \\ \downarrow \mathrm{pr}_1 & & \downarrow \vartheta_{\mathcal{E}} \\ \mathbb{R} & \xrightarrow{\hat{\kappa}_0} & \mathbb{R}. \end{array} \tag{13.69}$$

Thus, locally, $\tilde{\kappa}$ will assume the form

$$(z, x^i) \longmapsto (t = \kappa^0(z), y^\alpha = \kappa^\alpha(z, x^i)), \tag{13.70}$$

where, evidently, $\kappa^0 = \hat{\kappa}_0$ is the base mapping of the fiber bundle morphism.

Let $\mathrm{pr}_1^2 : \mathbb{R}^2 \to \mathbb{R}$ be a trivial fiber bundle structure on $\mathbb{R}^2$, whereby $(z, t) \mapsto z$. Thus, $\hat{\kappa}_0$ may be represented by a section $\kappa_0 : \mathbb{R} \to \mathbb{R}^2$ of this fiber bundle, given by $z \mapsto (z, t)$. Let

$$\vartheta : \mathcal{Y} \longrightarrow \mathbb{R}^2 \tag{13.71}$$

be the fiber bundle morphism given by

$$\vartheta(y) = (\mathrm{pr}_1(\mathrm{pr}_1(y)), \vartheta_{\mathcal{Y}}(y)). \tag{13.72}$$

Locally, ϑ is represented by

$$(z, x^i, t, y^\alpha) \longmapsto (z, t), \tag{13.73}$$

viewed as a fiber bundle morphism over the identity in $\mathbb{R}$ as in the diagram

$$\begin{array}{ccc} \mathcal{Y} & \xrightarrow{\vartheta} & \mathbb{R}^2 \\ {\scriptstyle \mathrm{pr}_1 \circ \mathrm{pr}_1}\downarrow & & \downarrow{\scriptstyle \mathrm{pr}_1^2} \\ \mathbb{R} & \xrightarrow{\mathrm{Id}} & \mathbb{R}. \end{array} \tag{13.74}$$

The foregoing definitions make it possible to represent $\hat{\kappa}$ by a section $\kappa : \mathcal{X} \to \mathcal{Y}$, as in the diagram

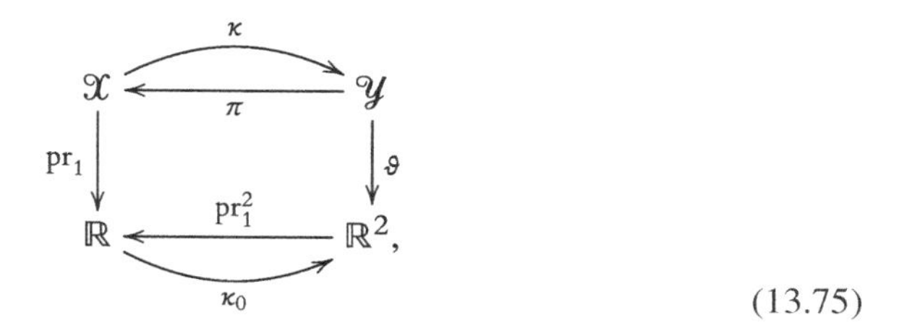

(13.75)

which has the local form

$$(z, x^i) \longmapsto (z, x^i, t = \kappa^0(z), y^\alpha = \kappa^\alpha(z, x^i)). \tag{13.76}$$

Here, κ, a section of π, is a fiber bundle morphism over κ_0, which in turn is viewed as a section of pr_1^2.

An additional generalization may be considered where the body-time manifold, $\mathcal{X}$, is a general (not necessarily a Cartesian product) $(n+1)$-dimensional fiber bundle

$$\vartheta_{\mathcal{X}} : \mathcal{X} \longrightarrow \mathbb{R}, \tag{13.77}$$

and the fiber bundle $\pi : \mathcal{Y} \to \mathcal{X}$ is not necessarily a product bundle. It is still assumed that there is a time projection

$$\vartheta_{\mathcal{Y}} : \mathcal{Y} \longrightarrow \mathbb{R} \tag{13.78}$$

and we can consider

$$\vartheta : \mathcal{Y} \longrightarrow \mathbb{R}^2, \qquad \vartheta(y) = (\vartheta_{\mathcal{X}} \circ \pi(y), \vartheta_{\mathcal{Y}}(y)). \tag{13.79}$$

The typical fiber $\pi^{-1}\{x\}$ of $\mathcal{Y}$ is diffeomorphic to $\mathcal{E}$. In other words, each element of $\mathcal{X}$ "sees" a different copy of spacetime. An element $y \in \mathcal{Y}$ is represented locally in the form (z, x^i, t, y^α) where $\pi(y)$ is represented by (z, x^i), $t = \vartheta_{\mathcal{Y}}(y)$, $z = \vartheta_{\mathcal{X}} \circ \pi(y)$.

A motion of a body-time manifold $\mathcal{X}$ is a section of π which is a fiber bundle morphism as in the following diagram:

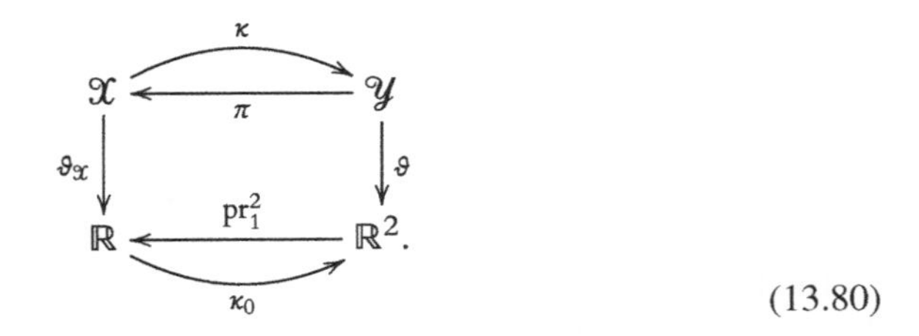

(13.80)

It follows immediately from the commutativity of the diagram that

$$\kappa\{\vartheta_{\mathcal{X}}^{-1}\{z\}\} = \vartheta^{-1}\{\kappa_0(z)\} \tag{13.81}$$

A motion is still represented locally in the form (13.76). The following kinematic and static variables will correspond to this general setting, and the reduction to product structures is straightforward.

The jet,

$$j^1\kappa : \mathcal{X} \longrightarrow J^1(\mathcal{X}, \mathcal{Y}), \tag{13.82}$$

of a motion, κ, is represented locally in the form

$$\begin{aligned}(z, x^i) \longmapsto (z, x^i, t = \kappa^0(z), y^\alpha = \kappa^\alpha(z, x^i), \kappa^0_{,0}(z), y^\beta_0 = \kappa^\beta_{,0}(z, x^i), y^\gamma_j = \kappa^\gamma_{,j}(z, x^i)),\\ = (z, x^i, t = \kappa^0(z), y^\alpha = \kappa^\alpha(z, x^i), \kappa^0_{,0}(z), y^\gamma_{\mathrm{j}} = \kappa^\gamma_{,\mathrm{j}}(z, x^i)),\end{aligned} \tag{13.83}$$

where $(\cdot)_{,0}$ indicates $\partial(\cdot)/\partial z$, $\alpha, \beta, \gamma = 1, \ldots, m$, and $\mathrm{j} = 0, 1, \ldots, m$.

In accordance with the paradigm of Sect. 12.1, a virtual displacement at the motion κ, representing a variation of the motion in the classical sense, is a section $w : \mathcal{X} \to V\mathcal{Y}$. Here, we keep the notation $V\mathcal{Y} = \mathrm{Kernel}T\xi$ in spite of the other fiber bundle structures of $\mathcal{Y}$. Locally an element of $V\mathcal{Y}$ is in the form $(z, x^i, t, y^\alpha, \delta t, \delta y^\beta)$, and w is represented locally in the form

$$(z, x^i) \longmapsto (z, x^i, t = \kappa^0(z), y^\alpha = \kappa^\alpha(z, x^i), \delta t = w^0(z), \delta y^\beta = w^\beta(z, x^i)). \tag{13.84}$$

A body force field $\mathbf{b}$ at the motion κ is a section of $L(\kappa^* V\mathcal{Y}, \bigwedge^{n+1} T^*\mathcal{X})$, which is given locally in the form

$$(z, x^i) \longmapsto (z, x^i, (\mathbf{b}_{00\ldots n}(z, x^i)\mathrm{d}t + \mathbf{b}_{\alpha 0\ldots n}(z, x^i)\mathrm{d}y^\alpha) \otimes (\mathrm{d}z \wedge \mathrm{d}x)). \tag{13.85}$$

The integrand of action $\int_{\mathcal{X}} \mathbf{b} \cdot w$, of a body force field at κ, an $(n+1)$-form on $\mathcal{X}$, is represented therefore as

$$[\mathbf{b}_{00\ldots n}(z, x^i)w^0(z) + \mathbf{b}_{\alpha 0\ldots n}(z, x^i)w^\alpha)]\mathrm{d}z \wedge \mathrm{d}x. \tag{13.86}$$

Thus, in this general setting, a body force has a component that acts on the time variation. This component may be interpreted as power supply density.

Let $w : \mathcal{X} \to V\mathcal{Y}$ be a virtual displacement field at the configuration κ. Then, $j^1 w : \mathcal{X} \to J^1(\kappa^* V\mathcal{Y})$ is represented locally in the form

$$(z, x^i) \longmapsto (z, x^i, w^0(z), w^\alpha(z, x^i), w^0_{,0}(z), w^\beta_{,0}(z, x^i), w^\gamma_{,j}(z, x^i)). \tag{13.87}$$

A variational stress field $S : \mathcal{X} \to L(J^1(\kappa^* V\mathcal{Y}), \bigwedge^{n+1} T^*\mathcal{X})$ is represented locally as

$$\begin{aligned}(z, x^i) \longmapsto (z, x^i, (&R_{00\dots n}(z, x^i)\mathrm{d}t + R_{\alpha 0\dots n}(z, x^i)\mathrm{d}y^\alpha,\\ &S^0_{00\dots n}(z, x^i)\partial_z \otimes \mathrm{d}t + S^0_{\alpha 0\dots n}(z, x^i)\partial_z \otimes \mathrm{d}y^\alpha + S^j_{\alpha 0\dots n}(z, x^i)\partial_j \otimes \mathrm{d}y^\alpha)\\ &\otimes (\mathrm{d}z \wedge \mathrm{d}x)).\end{aligned} \tag{13.88}$$

The integrand of the action $\int_{\mathcal{X}} S \cdot j^1 w$ is represented locally by

$$\begin{aligned}(&R_{00\dots n}(z, x^i)w^0(z) + R_{\alpha 0\dots n}(z, x^i)w^\alpha(z, x^i)\\ &+ S^0_{00\dots n}(z, x^i)w^0_{,0}(z) + S^0_{\alpha 0\dots n}(z, x^i)w^\alpha_{,j}(z, x^i) + S^j_{\alpha 0\dots n}(z, x^i)w^\alpha_{,j}(z, x^i))\\ &\otimes (\mathrm{d}z \wedge \mathrm{d}x)).\end{aligned} \tag{13.89}$$

Finally, the field equation $\operatorname{div} S + \mathbf{b} = 0$ is thus represented locally by

$$\begin{aligned} S^0_{00\dots n,0} - R_{00\dots n} + \mathbf{b}_{00\dots n} &= 0,\\ S^0_{\alpha 0\dots n,0} + S^j_{\alpha 0\dots n,j} - R_{\alpha 0\dots n} + \mathbf{b}_{\alpha 0\dots n} &= 0.\end{aligned} \tag{13.90}$$

The first of these equations may be viewed as a balance of power. The term $S^0_{\alpha 0\dots n,0}$ generalizes the classical time derivative of the momentum density. The equation in the second line generalizes the momentum balance of continuum mechanics, where the term $S^0_{\alpha 0\dots n,0}$ contains the time derivative of the momentum.

Part III
Non-smooth, Global Theories

In this part of the book, function space constructions, infinite-dimensional topological vector spaces, and the corresponding dual spaces are used. It is somewhat uncommon that technicalities of functional analysis are used in theoretical formulations of classical continuum mechanics. Still, except for Chap. 24, which is concerned with computational aspects of continuum mechanics, the main message of this part is that one can formulate continuum theories using the infinite-dimensional counterpart of the simplified Chap. 2. Moreover, technical considerations, related to the infinite-dimensional character of functional analysis, have far-reaching consequences in continuum mechanics.

Thus, in Chap. 15, we describe the configuration space of a body manifold in the space manifold as an infinite-dimensional Banach manifold. Following the paradigm of Chap. 2, forces are defined as elements of the cotangent bundle of the configuration space. In Chap. 21, stresses and hyper-stresses emerge as measures, valued in the dual of a jet bundle, that represent force functionals.

As described below, the basic construction may be outlined for the classical case where the body, $\mathscr{B}$, is assumed to be a three-dimensional manifold with boundary of $\mathbb{R}^3$, and the space manifold $\mathcal{S}$ is identified with $\mathbb{R}^3$.

The principle of material impenetrability implies that the configuration space is a collection of embeddings of the body into space. Assuming that configurations are sufficiently regular, the configuration space is a subset

$$\mathbb{Q} \subset C^k(\mathscr{B}, \mathbb{R}^3),$$

for some integer $k \geqslant 0$. Indeed, as shown in Chap. 15 for the general case, the configuration space of C^r-embeddings is an open subset of the space of all C^r-mappings for $r \geqslant 1$. Thus, the value $r = 1$ is the natural value to use in the formulation. For this choice, $\mathbb{Q}$ is an open subset of the Banach space $C^1(\mathscr{B}, \mathbb{R}^3)$. As such, for each $\kappa \in \mathbb{Q}$, the tangent space at a generic configuration κ is

$$T_\kappa\mathbb{Q} = C^1(\mathscr{B}, \mathbb{R}^3).$$

It is recalled that the norm on $C^1(\mathscr{B}, \mathbb{R}^3)$ is given by

$$\|w\| = \sup\{|w^i(x)|, |w^j_{,k}(x)| \mid i, j, k = 1, 2, 3, \ x \in \mathscr{B}\}.$$

The general paradigm implies that a force is an element

$$F \in T^*_\kappa \mathbb{Q} = C^1(\mathscr{B}, \mathbb{R}^3)^*,$$

that is, a linear functional

$$F : C^1(\mathscr{B}, \mathbb{R}^3) \longrightarrow \mathbb{R}, \qquad |F(w)| \leqslant C_F \|w\|,$$

for some constant C_F and every field w.

Consider the jet extension mapping

$$j^1 : C^1(\mathscr{B}, \mathbb{R}^3) \longrightarrow C^0(\mathscr{B}, \mathbb{R}^{12}), \qquad (w^i) \longmapsto (w^i, w^j_{,k}),$$

where in $C^0(\mathscr{B}, \mathbb{R}^{12})$, one uses the norm

$$\|\chi\| = \sup\{|\chi^\alpha(x)| \mid \alpha = 1, \ldots, 12\}.$$

Evidently, j^1 is a linear, injective isometry with an inverse

$$(j^1)^{-1} : \text{Image } j^1 \longrightarrow C^1(\mathscr{B}, \mathbb{R}^3),$$

which is an isometric isomorphism.

Let $F \in C^1(\mathscr{B}, \mathbb{R}^3)^*$ be a force. Then,

$$\varsigma_0 := F \circ (J^1)^{-1} : \text{Image } j^1 \longrightarrow \mathbb{R}$$

is a continuous linear mapping. By the Hahn–Banach theorem, the bounded functional ς_0 may be extended to some continuous linear functional $\varsigma \in C^0(\mathscr{B}, \mathbb{R}^{12})^*$. The resulting structure is illustrated in the following diagram, where $\mathcal{J}$ denotes the natural inclusion of the subspace Image j^1.

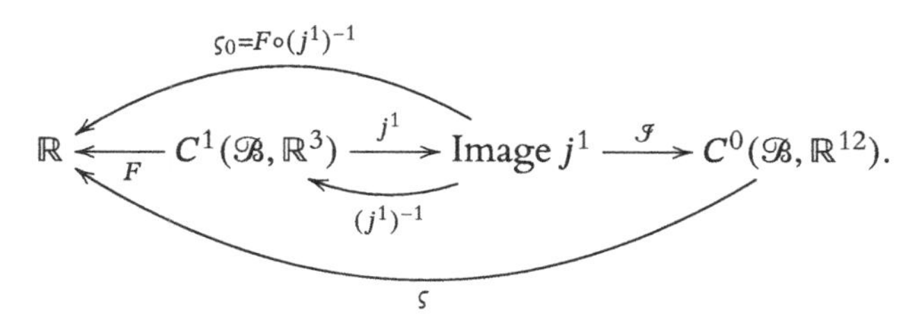

Being an extension of ς_0, the linear functional ς satisfies, for each $j^1 w \in$ Image j^1,

$$\begin{aligned} \varsigma(j^1 w) &= \varsigma_0(j^1 w), \\ &= F \circ (j^1)^{-1}(j^1 w), \\ &= F(w). \end{aligned}$$

It follows that each force $F \in C^1(\mathscr{B}, \mathbb{R}^3)^*$, is represented by some element $\varsigma \in C^0(\mathscr{B}, \mathbb{R}^{12})^*$ in the form

$$F(w) = \varsigma(j^1 w), \quad \text{for all} \quad w \in C^1(\mathscr{B}, \mathbb{R}^3).$$

An element $\varsigma \in C^0(\mathscr{B}, \mathbb{R}^{12})^*$ is a generalization of the variational stress distribution introduced in Sect. 10.7. The condition that a variational stress represents a given force is a generalization of the principle of virtual work, or virtual power, because it implies that the power expended by the force for any virtual velocity field is equal to the power of the stress for the jet of the virtual velocity, as in Eq. (10.59). Using the definition of a dual mapping, we may write this condition also as

$$F = j^{1*}(\varsigma).$$

The condition for the representation of a force by a stress, as formulated in the last two equations, applies just the same for continuum mechanics on general manifolds (see Chap. 21). This elegant and general form of the "equilibrium equation" was obtained naturally from the general paradigm and the principle of material impenetrability. The following should be noted:

1. Since $\varsigma \in C^0(\mathscr{B}, \mathbb{R}^{12})^* = (C^0(\mathscr{B}, \mathbb{R})^*)^{12}$, and since the Riesz representation theorem implies that each element of $C^0(\mathscr{B}, \mathbb{R})^*$ is represented by a Borel measure, a stress is represented by a collection of 12 measures $(\varsigma_i, \varsigma_j^k)$, $i, j, k = 1, 2, 3$. Therefore, the action of the stress may be expressed in the form

   $$\varsigma(j^1 w) = \int_{\mathscr{B}} w^i \mathrm{d}\varsigma_i + \int_{\mathscr{B}} w^j_{,k} \mathrm{d}\varsigma_j^k.$$

   For example, this non-smooth formulation applies in the case where the measures ς_j^k are Dirac measures at a point in the body, describing a stress distribution that is concentrated at a single point (sometimes referred to as a force dipole).
2. In case the stress measures are represented by smooth densities, (S_i, S_j^k), so that

   $$\mathrm{d}\varsigma_i = S_i \,\mathrm{d}V, \qquad \mathrm{d}\varsigma_j^k = S_j^k \,\mathrm{d}V,$$

   integration by parts may be carried out and one obtains a representation of the force in the form

   $$F(w) = \int_{\mathscr{B}} w^i S_i \,\mathrm{d}V + \int_{\mathscr{B}} w^j_{,k} S_j^k \,\mathrm{d}V,$$

$$= \int_{\mathscr{B}} w^i (S_i - S^k_{i,k})\, \mathrm{d}V + \int_{\partial\mathscr{B}} w^i S^k_i n_k\, \mathrm{d}A,$$

$$= \int_{\mathscr{B}} w^i \mathbf{b}_i\, \mathrm{d}V + \int_{\partial\mathscr{B}} w^i \mathbf{t}_i\, \mathrm{d}A,$$

where

$$S^k_{i,k} + \mathbf{b}_i = S_i, \qquad \mathbf{t}_i = S^k_i n_k.$$

Thus, a force is represented by a body force and a surface force in case the stress measures are smooth.

3. Since Image j^1 is a subset of $C^0(\mathscr{B}, \mathbb{R}^{12})$, which is not dense, the extension, ς, of ς_0 is not unique. This is the source of the nonuniqueness relation between stresses and forces, the inherent static indeterminacy, in continuum mechanics.
4. Since the subset of embeddings is open in $C^r(\mathscr{B}, \mathbb{R}^3)$ for integers $r \geqslant 1$, one may also consider the case $r > 1$, strictly. This will lead to r-th order continuum mechanics and hyper-stresses.
5. Let ς^{hb} be the norm-preserving extension of ς_0, as guaranteed by the Hahn–Banach theorem. It follows that

$$\begin{aligned} \|\varsigma^{\mathrm{hb}}\| &= \sup\left\{\frac{\varsigma^{\mathrm{hb}}(\chi)}{\|\chi\|} \mid \chi \in C^0(\mathscr{B}, \mathbb{R}^{12})\right\}, \\ &= \|\varsigma_0\| = \sup\left\{\frac{\varsigma_0(j^1 w)}{\|w\|} \mid w \in C^0(\mathscr{B}, \mathbb{R}^3)\right\}, \\ &= \sup\left\{\frac{F(w)}{\|w\|} \mid w \in C^0(\mathscr{B}, \mathbb{R}^3)\right\}, \\ &= \|F\|. \end{aligned}$$

Since in general, for any extension, ς, of ς_0, $\|\varsigma\| \geqslant \|\varsigma_0\|$, it is concluded that

$$\|F\| = \inf\left\{\|\varsigma\| \mid F = j^{1*}(\varsigma)\right\}.$$

Accordingly, results related to optimal stresses and load capacity of bodies, which are analogous to those considered for the finite-dimensional case in Chap. 2, apply to continuum mechanics. These and related subjects are considered in Chap. 24.

Chapters 14–22 develop this paradigm to continuum mechanics on manifolds and fiber bundles. Background material is presented concerning manifolds of sections and embeddings, generalized sections on manifolds with corners, and de Rham currents. Then, the representation of forces by hyper-stresses and stresses is considered.

Another non-smooth theory of continuum mechanics, which applies to fractal bodies and non-smooth velocity fields, is presented in Chap. 23. In accordance with the theme of this part of the book, the constructions demonstrating the relevance of Whitney's geometric integration theory (see [Whi57]) to continuum mechanics in Euclidean spaces are founded on the notion of duality.

Chapters 23 and 24 may be read independently of the preceding material.

Chapter 14
Banachable Spaces of Sections of Vector Bundles over Compact Manifolds

For two manifolds $\mathcal{X}$ and $\mathcal{S}$, $C^r(\mathcal{X}, \mathcal{S})$ will denote the collection of C^r-mappings from $\mathcal{X}$ to $\mathcal{S}$. If $\xi : \mathcal{Y} \to \mathcal{X}$ is a fiber bundle, $C^r(\xi)$ is the space of C^r-sections $\mathcal{X} \to \mathcal{Y}$. In the following few chapters, where the Lagrangian point of view of continuum mechanics is adopted, $\mathcal{X}$ is interpreted as the body manifold and $\mathcal{S}$ is interpreted as the physical space manifold. Thus, a generic point in $\mathcal{X}$ will be denoted by X (rather than x as in the preceding chapters).

If $\mathcal{X}$ is compact, $C^r(\mathcal{X}, \mathcal{S})$ and $C^r(\xi)$ may be given structures of infinite-dimensional Banach manifolds. These Banach manifolds are modeled by Banachable spaces of sections of vector bundles, as will be described in the next chapter. In this chapter, we describe the Banachable structure on spaces of differentiable sections of a vector bundle and make some related observations. Thus, we consider a vector bundle $\pi : W \to \mathcal{X}$, where $\mathcal{X}$ is a smooth compact n-dimensional manifold with corners, and the typical fiber of W is an m-dimensional vector space. The space of C^r-sections $w : \mathcal{X} \to W$, $r \geq 0$, will be denoted by $C^r(\pi)$, or by $C^r(W)$ if no ambiguity may arise. A natural real vector space structure is induced on $C^r(\pi)$ by setting $(w_1 + w_2)(X) = w_1(X) + w_2(X)$ and $(cw)(X) = cw(X)$, $c \in \mathbb{R}$.

14.1 The C^r-Topology on $C^r(\pi)$

Let K_a, $a = 1, \ldots, A$, be a finite collection of compact subsets, the interiors of which cover $\mathcal{X}$, such that for each a, K_a is a subset of a domain of a chart $\varphi_a : U_a \to \mathbb{R}^n$ on $\mathcal{X}$, and

$$(\varphi_a, \Phi_a) : \pi^{-1}(U_a) \longrightarrow \mathbb{R}^n \times \mathbb{R}^m, \quad v \longmapsto (X^i, v^\alpha), \tag{14.1}$$

is some given vector bundle chart on W. Such a covering may always be found by the compactness of $\mathcal{X}$ (using coordinate balls as, for example, in [Lee02, p. 16] or

R. Segev, *Foundations of Geometric Continuum Mechanics*, Advances in Continuum Mechanics 49, https://doi.org/10.1007/978-3-031-35655-1_14

[Pal68, p. 10]). We will refer to such a structure as a *precompact atlas*. The same terminology will apply, with the necessary adaptations, for the case of a fiber bundle over $\mathcal{X}$.

For a section w of π and each $a = 1, \dots, A$, let

$$w_a : \varphi_a(K_a) \longrightarrow \mathbb{R}^m, \tag{14.2}$$

satisfying

$$w_a(\varphi_a(X)) = \Phi_a(w(X)), \quad \text{for all} \quad X \in K_a, \tag{14.3}$$

be a local representative of w.

Such a choice of a precompact vector bundle atlas and subsets K_a makes it possible to define, for a section w,

$$\|w\|^r = \sup_{a,\alpha,|\boldsymbol{I}|\leq r} \left\{ \sup_{X\in\varphi_a(K_a)} \left\{ \left|(w_a^\alpha)_{,\boldsymbol{I}}(X)\right| \right\} \right\}. \tag{14.4}$$

It is recalled that in Sect. 6.1.7, $\boldsymbol{I}$ is defined as a non-decreasing multi-index. However, as we take the supremum, we could use all symmetric multi-indices.

Palais [Pal68, in particular, Chapter 4] shows that $\|\cdot\|^r$ is indeed a norm endowing $C^r(\pi)$ with a Banach space structure. The dependence of this norm on the particular choice of atlas and sets K_a makes the resulting topological vector space Banachable, rather than a Banach space. Other choices will correspond to different norms. However, norms induced by different choices will induce equivalent topological vector space structures on $C^r(\pi)$ [Mic20, Section 5].

Next, one observes that the foregoing may be applied, in particular, to the vector space $C^0(\pi^r) = C^0(J^r W)$ of continuous sections of the r-jet bundle $\pi^r : J^r W \to \mathcal{X}$ of π. As a continuous section B of π^r is locally of the form

$$(X^i) \longmapsto (X^i, B_{\boldsymbol{I}}^\alpha(X^i)), \qquad |\boldsymbol{I}| \leq r, \tag{14.5}$$

the analogous expression for the norm induced by a choice of a precompact vector bundle atlas is

$$\|B\|^0 = \sup_{a,\alpha,|\boldsymbol{I}|\leq r} \left\{ \sup_{X\in\varphi_a(K_a)} \left\{ \left|B_{a\boldsymbol{I}}^\alpha(X)\right| \right\} \right\}. \tag{14.6}$$

Once the topologies of $C^r(\pi)$ and $C^0(\pi^r)$ have been defined, one may consider the jet extension mapping

$$j^r : C^r(\pi) \longrightarrow C^0(\pi^r). \tag{14.7}$$

For a section $w \in C^r(\pi)$, with local representatives w_a^α, $j^r w$ induces a section $B \in C^0(\pi^k)$, the local representatives of which satisfy,

$$B_{a\boldsymbol{I}}^\alpha = w_{a,\boldsymbol{I}}^\alpha. \tag{14.8}$$

Clearly, the mapping j^r is injective and linear. Furthermore, it follows that

$$\|j^r w\|^0 = \sup_{a,\alpha,|\boldsymbol{I}|\le r} \left\{ \sup_{X\in\varphi_a(K_a)} \left\{ \left|w_{a,\boldsymbol{I}}^\alpha(X)\right| \right\} \right\}. \tag{14.9}$$

Thus, in view of Eq. (14.4),

$$\|j^r w\|^0 = \|w\|^r, \tag{14.10}$$

and we conclude that j^r is a linear isometric embedding of $C^r(\pi)$ into $C^0(\pi^r)$. Evidently, j^r is not surjective as a section A of π^k need not be compatible, i.e., it need not satisfy (14.8), for some section w of π. As a result of the above observations, j^r has a continuous right inverse

$$(j^r)^{-1} : \text{Image}\, j^r \subset C^0(\pi^r) \longrightarrow C^r(\pi). \tag{14.11}$$

14.2 Iterated (Nonholonomic) Jets and Iterated Jet Extensions

Similarly to jets as described above, iterated jets may be useful in the formulation of higher order continuum mechanics. Simply put, the iterated r-jet bundle is defined as

$$\hat{J}^r(\mathcal{X},\mathcal{Y}) := \overbrace{J^1(J^1 \cdots\cdots (J^1(\mathcal{X},\mathcal{Y}))\cdots)}^{r\text{-times}} \tag{14.12}$$

and for $B \in \hat{J}^r(\mathcal{X},\mathcal{Y})$,

$$\hat{\xi}^r(B) := \overbrace{\xi^1(\xi^1 \cdots\cdots (\xi^1(B))\cdots)}^{r\text{-times}}. \tag{14.13}$$

14.2.1 Iterated Jets

Formally, completely nonholonomic jets for the fiber bundle $\xi : \mathcal{Y} \to \mathcal{X}$ are defined inductively as follows. First, one defines the fiber bundles

$$\hat{J}^0(\mathcal{X}, \mathcal{Y}) = \mathcal{Y}, \qquad \hat{J}^1(\mathcal{X}, \mathcal{Y}) := J^1(\mathcal{X}, \mathcal{Y}) \tag{14.14}$$

and projections

$$\hat{\xi}^1 = \xi^1 : \hat{J}(\mathcal{X}, \mathcal{Y}) \longrightarrow \mathcal{X}, \qquad \hat{\xi}^1_0 = \xi^1_0 : \hat{J}^1(\mathcal{X}, \mathcal{Y}) \longrightarrow \mathcal{Y}. \tag{14.15}$$

Then, we define the *iterated r-jet bundle* as

$$\hat{J}^r(\mathcal{X}, \mathcal{Y}) := J^1(\mathcal{X}, \hat{J}^{r-1}(\mathcal{X}, \mathcal{Y})), \tag{14.16}$$

with projection

$$\hat{\xi}^r = \hat{\xi}^{r-1} \circ \xi^{1,r}_{r-1} : \hat{J}^r(\mathcal{X}, \mathcal{Y}) \longrightarrow \mathcal{X}, \tag{14.17}$$

where

$$\xi^{1,r}_{r-1} : \hat{J}^r(\mathcal{X}, \mathcal{Y}) = J^1(\mathcal{X}, \hat{J}^{r-1}(\mathcal{X}, \mathcal{Y})) \longrightarrow \hat{J}^{r-1}(\mathcal{X}, \mathcal{Y}). \tag{14.18}$$

By induction, $\hat{\xi}^r : \hat{J}^r(\mathcal{X}, \mathcal{Y}) \to \mathcal{X}$ is a well-defined fiber bundle.

Notation Equivalently, we will use the notation $\hat{J}^r\xi$ for $\hat{J}^r(\mathcal{X}, \mathcal{Y})$ and write

$$\hat{\xi}^r : \hat{J}^r\xi := J^1(\hat{\xi}^{r-1}) \longrightarrow \mathcal{X}. \tag{14.19}$$

When the projections $\xi^{1,r}_{r-1}$ are used inductively l-times, we obtain a projection

$$\hat{\xi}^r_{r-l} : \hat{J}^r(\mathcal{X}, \mathcal{Y}) \longrightarrow \hat{J}^{r-l}(\mathcal{X}, \mathcal{Y}). \tag{14.20}$$

Let $\kappa : \mathcal{X} \to \mathcal{Y}$ be a C^r-section of ξ. The iterated jet extension mapping

$$\hat{j}^r : C^r(\xi) \longrightarrow C^0(\hat{\xi}^r), \qquad \kappa \longmapsto \hat{j}^r\kappa, \tag{14.21}$$

is naturally defined by

$$\hat{j}^1 = j^1 : C^l(\xi) \longrightarrow C^{l-1}(\hat{\xi}^1), \qquad \text{and} \qquad \hat{j}^r = j^1 \circ \hat{j}^{r-1}. \tag{14.22}$$

Note that we use j^1 here as a generic jet extension mapping, omitting the indication of the domain.

There is a natural inclusion

$$\mathcal{J}^r : J^r(\mathcal{X}, \mathcal{Y}) \longrightarrow \hat{J}^r(\mathcal{X}, \mathcal{Y}), \qquad \text{given by} \qquad j^r\kappa(X) \longmapsto \hat{j}^r\kappa(X). \tag{14.23}$$

Let $\pi : W \to \mathcal{X}$ be a vector bundle; then $\hat{\pi}^1 = \pi^1 : \hat{J}^1 W = J^1 W \to \mathcal{X}$ is a vector bundle. Continuing inductively,

$$\hat{\pi}^r : \hat{J}^r W \longrightarrow \mathcal{X} \tag{14.24}$$

is a vector bundle. In this case, the inclusion $\mathcal{I}^r : J^r\pi \to \hat{J}^r\pi$ is linear.

14.2.2 Local Representation of Iterated Jets[1]

The local representatives of iterated jets are also constructed inductively. At each step, G, which we refer to as *generation*, the number of arrays is multiplied. Hence, powers of two are naturally used below. We will use multi-indices of the form $I_{\mathfrak{p}}$, where Fraktur characters, $\mathfrak{p}$, $\mathfrak{q}$, etc., are binary numbers that enumerate the various arrays included in the representation. For example, a typical element of $\hat{J}^3(\mathcal{X}, \mathcal{Y})$, in the form

$$(X^j; y^\alpha; y_{i_1}^{1\beta_1}; y_{i_2}^{2\beta_2}, y_{i_3 i_4}^{3\beta_3}; y_{i_5}^{4\beta_4}, y_{i_6 i_7}^{5\beta_5}, y_{i_8 i_9}^{6\beta_6}, y_{i_{10} i_{11} i_{12}}^{7\beta_3}). \tag{14.25}$$

For each component of the form $y_I^{q\beta_q}$, the binary representation of q contains information as to the sequence of differentiations that led to this component. Thus, such a typical element is written as

$$(X^j; y_0^{0\beta_0}; y_{I_1}^{1\beta_1}; y_{I_{10}}^{10\beta_{10}}, y_{I_{11}}^{11\beta_{11}}; y_{I_{100}}^{100\beta_{100}}, y_{I_{101}}^{101\beta_{101}}, y_{I_{110}}^{110\beta_{110}}, y_{I_{111}}^{111\beta_{111}}), \tag{14.26}$$

and for short

$$(X^j; y_{I_{\mathfrak{p}}}^{\mathfrak{p}\beta_{\mathfrak{p}}}), \quad \text{for all } \mathfrak{p} \text{ with } 0 \le G_{\mathfrak{p}} \le 3. \tag{14.27}$$

Here, G_p is the generation where the $\mathfrak{p}$-th array appears first. It is given by

$$G_{\mathfrak{p}} = \lfloor \log_2 \mathfrak{p} \rfloor + 1, \tag{14.28}$$

where $\lfloor \log_2 \mathfrak{p} \rfloor$ denotes the integer part of $\log_2 \mathfrak{p}$. In (14.25), (14.26), the generations are separated by semicolons. As indicated in the example above, with each $\mathfrak{p}$, we associate a multi-index $I_{\mathfrak{p}} = i_1 \cdots i_p$ as follows. For each binary digit 1 in $\mathfrak{p}$, there is an index i_l, $l = 1, \dots, p$. Thus, the total number of digits 1 in $\mathfrak{p}$, which is denoted by $|\mathfrak{p}|$, is the total number of indices, p, in $I_{\mathfrak{p}}$. In other words, the length, $|I_{\mathfrak{p}}|$, of the induced multi-index $I_{\mathfrak{p}}$ satisfies

$$|I_{\mathfrak{p}}| = p = |\mathfrak{p}|\,. \tag{14.29}$$

[1] This section may be skipped without interrupting the reading of most of the following.

Note also that the expression $\beta_{\mathfrak{p}}$ is not a multi-index. Here, the subscript $\mathfrak{p}$ serves for the enumeration of the β indices. If no ambiguity may arise, we will often make the notation somewhat shorter and write $y_{I_{\mathfrak{p}}}^{\beta_{\mathfrak{p}}}$ for $y_{I_{\mathfrak{p}}}^{\mathfrak{p}\beta_{\mathfrak{p}}}$. Continuing by induction, let a section $\hat{A}$ of $\hat{J}^{r-1}(\mathcal{X},\mathcal{Y})$ be represented locally by $(X^j; y_{I_{\mathfrak{p}}}^{\mathfrak{p}\alpha_{\mathfrak{p}}}(X^j))$, $G_{\mathfrak{p}} \le r-1$. Then, its 1-jet extension, a section of $\hat{J}^r(\mathcal{X},\mathcal{Y})$, is represented in the form

$$(X^j; y_{I_{\mathfrak{p}}}^{\mathfrak{p}\alpha_{\mathfrak{p}}}(X^J); y_{I_{\mathfrak{p}},k_{\mathfrak{p}}}^{\mathfrak{p}\beta_{\mathfrak{p}}}(X^j)), \qquad G_{\mathfrak{p}} \le r-1, \tag{14.30}$$

or equivalently,

$$(X^j; y_{I_{\mathfrak{p}}}^{\mathfrak{p}\alpha_{\mathfrak{p}}}(X^j); y_{I_{\mathfrak{p}},k_{\mathfrak{p}}}^{1\mathfrak{p}\alpha_{1\mathfrak{p}}}(X^j)), \qquad G_{\mathfrak{p}} \le r-1, \tag{14.31}$$

where $1\mathfrak{p}$ is the binary representation of $2^r+\mathfrak{p}$. It is noted that the array $y^{1\mathfrak{p}}$ contains the derivatives of the array $y^{\mathfrak{p}}$ and that $G_{1\mathfrak{p}} = r$. Thus indeed, the number of digits 1 that appear in $\mathfrak{q}$, i.e., $|\mathfrak{q}|$, determines the length of the index $I_{\mathfrak{q}}$.

It follows that an element of $\hat{J}^r(\mathcal{X},\mathcal{Y})$ may be represented in the form

$$(X^j; y_{I_{\mathfrak{p}}}^{\mathfrak{p}\alpha_{\mathfrak{p}}}; y_{I_{\mathfrak{p}}k_{\mathfrak{p}}}^{\mathfrak{p}\alpha_{1\mathfrak{p}}}), \qquad G_{\mathfrak{p}} \le r-1 \tag{14.32}$$

or

$$(X^j; y_{I_{\mathfrak{q}}}^{\mathfrak{q}\alpha_{\mathfrak{q}}}), \qquad \text{for all } \mathfrak{q} \text{ with } G_{\mathfrak{q}} \le r. \tag{14.33}$$

That is, for each $\mathfrak{p}$ with $G_{\mathfrak{p}} \le r-1$, we have an index $\mathfrak{q} = \mathfrak{q}(\mathfrak{p})$ such that $\mathfrak{q} = \mathfrak{q}(\mathfrak{p}) = 1\mathfrak{p}$ if $G_{\mathfrak{q}} = r$, and $\mathfrak{q} = \mathfrak{q}(\mathfrak{p}) = \mathfrak{p}$ if $G_{\mathfrak{q}} < r$.

A similar line of reasoning leads to the expression for the local representatives of the iterated jet extension mapping. For a section $\kappa : \mathcal{X} \to \mathcal{Y}$, the iterated jet extension $B = \hat{j}^r\kappa$, a section of $\hat{\xi}^r$, the local representation $(X^j; B_{I_{\mathfrak{q}}}^{\mathfrak{q}\alpha_{\mathfrak{q}}})$, $G_{\mathfrak{q}} \le r$, $|I_{\mathfrak{q}}| = |\mathfrak{q}|$, satisfies

$$B_{I_{\mathfrak{q}}}^{\mathfrak{q}\alpha_{\mathfrak{q}}} = \kappa_{,I_{\mathfrak{q}}}^{\alpha_{\mathfrak{q}}}, \quad |I_{\mathfrak{q}}| = |\mathfrak{q}|\,, \text{ independently of the particular value of } \mathfrak{q}. \tag{14.34}$$

Indeed, if $(X^j, B_{I_{\mathfrak{q}}}^{\mathfrak{q}\alpha_{\mathfrak{q}}})$, $G_{\mathfrak{q}} \le r-1$, with $B_{I_{\mathfrak{q}}}^{\mathfrak{q}\alpha_{\mathfrak{q}}} = \kappa_{,I_{\mathfrak{q}}}^{\alpha_{\mathfrak{q}}}$, represent $\hat{j}^{r-1}\kappa$, then $\hat{j}^r\kappa$ is represented locally by

$$(X^i, B_{I_{\mathfrak{q}}}^{\mathfrak{q}\alpha_{\mathfrak{q}}}; B_{I_{\mathfrak{q}},j_{\mathfrak{q}}}^{\mathfrak{q}\alpha_{1\mathfrak{q}}}), \quad G_{\mathfrak{q}} \le r-1. \tag{14.35}$$

Thus, by induction, any $\mathfrak{p} = \mathfrak{p}(\mathfrak{q})$ with $G_{\mathfrak{p}} = r$ and $G_{\mathfrak{q}} < r$ may be written as $\mathfrak{p} = 2^{r-1} + \mathfrak{q}$, $I_{\mathfrak{p}} = I_{\mathfrak{q}}j_{\mathfrak{q}}$, so that $B_{I_{\mathfrak{p}}}^{\mathfrak{p}\alpha_{\mathfrak{p}}} = B_{I_{\mathfrak{q}},j_{\mathfrak{q}}}^{\mathfrak{q}\alpha_{1\mathfrak{q}}} = \kappa_{,I_{\mathfrak{p}}}^{\alpha_{\mathfrak{p}}}$.

Let an element $A \in \hat{J}^r(\mathcal{X}, \mathcal{Y})$ be represented by $(X^j; y_{I_\mathfrak{p}}^{\mathfrak{p}\alpha_\mathfrak{p}})$, $G_\mathfrak{p} \leq r$; then $\hat{\xi}_l^r(A)$ is represented by $(X^i; y_{I_\mathfrak{p}}^{\mathfrak{p}\alpha_\mathfrak{p}})$, $G_\mathfrak{p} \leq l$.

14.2.3 The Iterated Jet Extension Mapping for a Vector Bundle

We now consider the iterated jet extension mapping

$$\hat{j}^r : C^r(\pi) \longrightarrow C^0(\hat{\pi}^r) \tag{14.36}$$

for a vector bundle $\pi : W \to \mathcal{X}$. Specializing Equation (14.33) for the case of the nonholonomic r-jet bundle

$$\hat{\pi}^r : \hat{J}^r W \longrightarrow \mathcal{X}, \tag{14.37}$$

a section B of $\hat{\pi}^r$ is represented locally in the form

$$X^i \longmapsto (X^i, B_{aI_\mathfrak{q}}^{\mathfrak{q}\alpha_\mathfrak{q}}(X^i)), \quad \text{for all } \mathfrak{q} \text{ with } G_\mathfrak{q} \leq r. \tag{14.38}$$

Thus, the induced norm on $C^0(\hat{\pi}^r)$ is given by

$$\|B\|^0 = \sup\left\{\left|B_{aI_\mathfrak{q}}^{\mathfrak{q}\alpha_\mathfrak{q}}(X^i)\right|\right\}, \tag{14.39}$$

where the supremum is taken over all $X \in \varphi_a(K_a)$, $a = 1, \ldots, A$, $\alpha_\mathfrak{q} = 1, \ldots, m$, $I_\mathfrak{q}$ with $|I| = |\mathfrak{q}|$, and $\mathfrak{q}$ with $G_\mathfrak{q} \leq r$.

Specializing (14.34) for the case of a vector bundle, it follows that if section B of $\hat{\pi}^r$ satisfies $B = \hat{j}^r w$, its local representatives satisfy

$$B_{aI_\mathfrak{q}}^{\mathfrak{q}\alpha_\mathfrak{q}} = w_{a,I_\mathfrak{q}}^{\alpha_\mathfrak{q}} \quad |I_\mathfrak{q}| = |\mathfrak{q}|\,, \text{ independently of the particular value of } \mathfrak{q}. \tag{14.40}$$

It follows that in

$$\|\hat{j}^r w\|^0 = \sup\left\{\left|w_{a,I_\mathfrak{q}}^{\alpha_\mathfrak{q}}(X^i)\right|\right\} \tag{14.41}$$

(where the supremum is taken over all $X^i \in \varphi_a(K_a)$, $a = 1, \ldots, A$, $\alpha = 1, \ldots, m$, $I_\mathfrak{q}$ with $|I| = |\mathfrak{q}|$, and $\mathfrak{q}$ with $G_\mathfrak{q} \leq r$), it is sufficient to take simply all derivatives $w_{a,I}^{\alpha}(X^i)$, for $|I| \leq r$. Hence,

$$\|\hat{j}^r w\|^0 = \sup\left\{\left|w_{a,I}^{\alpha_\mathfrak{q}}(X^i)\right|\right\}, \tag{14.42}$$

where the supremum is taken over all $X^i \in \varphi_a(K_a)$, $a = 1, \dots, A$, $\alpha = 1, \dots, m$, and I with $|I| \leq r$. It is therefore concluded that

$$\|\hat{j}^r w\|^0 = \|j^r w\|^0 = \|w\|^r. \tag{14.43}$$

In other words, one has a sequence of linear embeddings

$$C^r(\pi) \xrightarrow{\;j^r\;} C^0(\pi^r) \xrightarrow{\;C^0(\mathcal{J}^r)\;} C^0(\hat{\pi}^r), \qquad \hat{j}^r: C^r(\pi) \to C^0(\hat{\pi}^r) \tag{14.44}$$

where $\mathcal{J}^r : J^r W \to \hat{J}^r W$ is the natural inclusion (14.23), and $C^0(\mathcal{J}^r)$, defined as $C^0(\mathcal{J}^r)(A) := \mathcal{J}^r \circ A$ for every continuous section A of $J^r W$, is the inclusion of sections. These embeddings are not surjective. In particular, sections of $\hat{J}^r W$ need not have the symmetry properties that hold for sections of $J^r W$.

Chapter 15
Manifolds of Sections and Embeddings

In this chapter, the construction of a Banach manifold structure to the set of C^r-sections of a fiber bundle is outlined. This includes the set of C^r-mappings, $C^r(\mathcal{X}, \mathcal{S})$ as a special case. Moreover, the subset of C^r-embeddings, $\mathrm{Emb}^r(\mathcal{X}, \mathcal{S})$, is shown to be an open subset of $C^r(\mathcal{X}, \mathcal{S})$, for $r \geq 1$.

15.1 The Construction of Charts for the Manifold of Sections

In this section, for a fiber bundle, $\xi : \mathcal{Y} \to \mathcal{X}$, over a compact manifold with corners, $\mathcal{X}$, we outline the construction of charts for the Banach manifold structure on the collection of sections $C^r(\xi)$ as in [Pal68]. (See a detailed presentation of the subject in [Mic20, Section 5.9].)

Let κ be a C^r-section of ξ. Similar to the construction of tubular neighborhoods, the basic idea is to identify points in a neighborhood of Image κ with vectors at Image κ which are tangent to the fibers. This is achieved by defining a second-order differential equation, so that a neighboring point y in the same fiber as $\kappa(X)$ is represented through the solution $c(t)$ of the differential equations with the initial condition $\dot{c}(0) \in T_{\kappa(X)}(\mathcal{Y}_X)$, by $y = c(t = 1)$. In other words, y is the image of v under the exponential mapping.

To ensure that the image of the exponential mapping is located on the same fiber, $\mathcal{Y}_X$, the spray inducing the second-order differential equation is a vector field

$$\omega : V\mathcal{Y} \longrightarrow T(V\mathcal{Y}) \tag{15.1}$$

which is again tangent to the fiber in the sense that for

$$T\tau_{\mathcal{Y}} : T(V\mathcal{Y}) \longrightarrow T\mathcal{Y}, \quad \text{one has} \quad T\tau_{\mathcal{Y}} \circ \omega \in V\mathcal{Y}. \tag{15.2}$$

R. Segev, *Foundations of Geometric Continuum Mechanics*, Advances in Continuum Mechanics 49, https://doi.org/10.1007/978-3-031-35655-1_15

This condition, together with the analog of the standard condition for a second-order differential equation, namely,

$$T\tau_{\mathcal{Y}} \circ \omega(v) = v, \quad \text{for all} \quad v \in V\mathcal{Y}, \tag{15.3}$$

implies that ω is represented locally in the form

$$(X^i, y^\alpha, 0, \dot{y}^\beta) \longmapsto (X^i, y^\alpha, 0, \dot{y}^\beta, 0, \dot{y}^\gamma, 0, \tilde{\omega}^\alpha(X^i, y^\alpha, \dot{y}^\beta)). \tag{15.4}$$

Finally, ω is a *bundle spray* so that

$$\tilde{\omega}^\alpha(X^i, y^\alpha, a_0\dot{y}^\beta) = a_0^2\tilde{\omega}^\alpha(X^i, y^\alpha, \dot{y}^\beta). \tag{15.5}$$

Bundle sprays can always be constructed on compact manifolds using partitions of unity, and the induced exponential mappings have the required properties.

The resulting structure makes it possible to identify an open neighborhood, U, of Image κ in $\mathcal{Y}$—*a vector bundle neighborhood*—with

$$V\mathcal{Y}|_{\text{Image}\,\kappa} = \kappa^* V\mathcal{Y}. \tag{15.6}$$

(We note that a rescaling is needed if U is to be identified with the whole of $\kappa^* V\mathcal{Y}$. Otherwise, only an open neighborhood of the zero section of $\kappa^* V\mathcal{Y}$ will be used to parametrize U.)

Once the identification of U with $\kappa^* V\mathcal{Y}$ is available, the collection of sections $C^r(\mathcal{X}, U)$ may be identified with $C^r(\kappa^* V\mathcal{Y})$, $\kappa \in C^r(\xi)$. Thus, a chart valued in a Banachable space is constructed, where κ is identified with the zero section. (See a simplified illustration of the construction in Fig. 15.1.)

The construction of charts on the manifold of sections implies that curves in $C^r(\xi)$ in a neighborhood of κ are represented locally by curves in the Banachable space $C^r(\kappa^* V\mathcal{Y})$. Thus, tangent vectors $w \in T_\kappa C^r(\xi)$ may be identified with elements of $C^r(\kappa^* V\mathcal{Y})$. We therefore make the identification

$$T_\kappa C^r(\xi) = C^r(\kappa^* V\mathcal{Y}). \tag{15.7}$$

15.2 The C^r-Topology on the Space of Sections of a Fiber Bundle

The topology on the space of sections of fiber bundles is conveniently described in terms of filters of neighborhoods (e.g., [Trè67]).

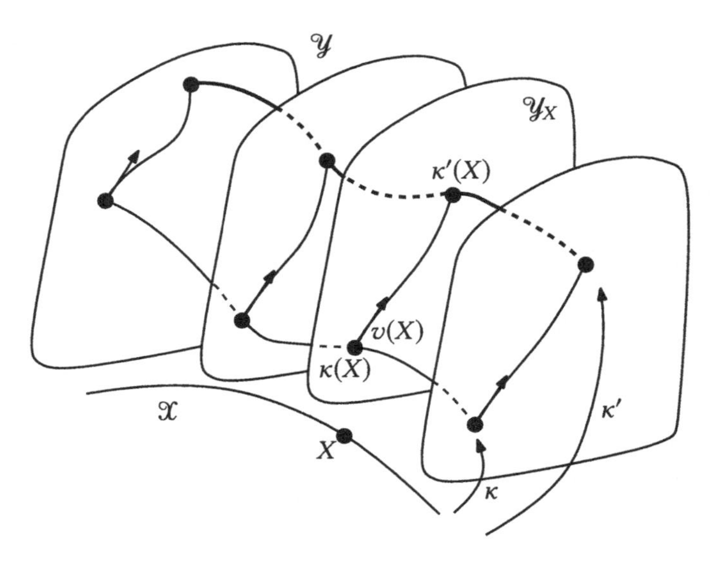

Fig. 15.1 Constructing the manifold of sections, a rough illustration

15.2.1 Finite Local Representation of a Section

We consider a fiber bundle $\xi : \mathcal{Y} \to \mathcal{X}$, where $\mathcal{X}$ is assumed to be a compact manifold with corners and the typical fiber is a manifold $\mathcal{S}$ without a boundary. Let $\{(U_b, \varphi_b, \Phi_b)\}$, $b \in B$, be fiber bundle trivialization on $\mathcal{Y}$ (Sect. 6.1.3). That is, the $\{U_b\}$ cover $\mathcal{X}$, and for each $b \in B$, $\Phi_b : \xi^{-1}(U_b) \to U_b \times \mathcal{S}$ is a diffeomorphism. Let $\kappa : \mathcal{X} \to \mathcal{Y}$ be a section, and for a generic X, let U_b be a domain of a chart such that $X \in U_b$. Then, $\mathrm{pr}_2 \circ \Phi_b \circ \kappa(X) \in \mathcal{S}$, and there is a chart $\psi_X : V_X \to \mathbb{R}^m$ in $\mathcal{S}$ such that $\mathrm{pr}_2 \circ \Phi_b \circ \kappa(X) \in V_X$. It follows that $\Phi^{-1}(U_b \times V_x)$ is open in Y, and $\kappa(X) \in \Phi_b^{-1}(U_b \times V_X)$. From the continuity of κ, it follows that there is an open subset $U_X \subset U_b$ containing X such that $\kappa(U_X) \subset \Phi_b^{-1}(U_b \times V_X)$. Using charts, and, for example, open balls in $\mathbb{R}^n$ centered at $\varphi_b(X)$, we can assume that there is a compact subset K_X contained in U_X and the interior, K_X°, is nonempty and contains X. Since $\Phi_b \circ \kappa(U_X) \subset U_b \times V_X$, we have a local representation $\boldsymbol{\kappa}_X$ of κ such that

$$\boldsymbol{\kappa}_X = \psi_X \circ \mathrm{pr}_2 \circ \Phi_b \circ \kappa \circ \varphi^{-1} : \varphi_b(U_X) \longrightarrow \mathbb{R}^m. \tag{15.8}$$

Letting X vary over $\mathcal{X}$, the sets $\{U_X\}$, $X \in \mathcal{X}$ cover $\mathcal{X}$. Due to the compactness of $\mathcal{X}$, there is a finite number A of sets that cover $\mathcal{X}$. Enumerating the finite set, for each $a = 1, \dots, A$, there is a fiber bundle chart (U_a, φ_a, Φ_a) such that U_a contains a compact set K_a with a nonempty interior, the collection $\{K_a^{\circ}\}$ cover $\mathcal{X}$. For each $a = 1, \dots, A$, there is a chart $\psi_a : V_a \to \mathbb{R}^m$ in $\mathcal{S}$, with $\Phi_a \circ \kappa \circ \varphi^{-1}|_{\varphi_a(U_a)}(U_a) \subset V_a$, and κ is represented locally by the mappings

$$\boldsymbol{\kappa}_a = \psi_a \circ \mathrm{pr}_2 \circ \Phi_a \circ \kappa \circ \varphi_a^{-1} : \varphi_a(U_a) \longrightarrow \psi_a(V_a) \subset \mathbb{R}^m. \tag{15.9}$$

15.2.2 Neighborhoods for $C^r(\xi)$ and the C^r-Topology

Let $\kappa \in C^r(\xi)$ be given and let $\{(U_a, \varphi_a, \Phi_a)\}$, $a = 1, \ldots, A$, be a finite fiber bundle atlas as above with local representatives $\boldsymbol{\kappa}_a : \varphi_a(U_a) \to \psi_a(V_a)$, where $\psi_a : V_a \to \mathbb{R}^m$ is a chart in $\mathcal{S}$. For any $\varepsilon > 0$, consider sets $U_{\kappa,\varepsilon}$ of sections having the following properties: (*i*) For each $\kappa' \in U_{\kappa,\varepsilon}$, Image $\mathrm{pr}_2 \circ \Phi_a \circ \kappa'|_{K_a} \subset V_a$ for all $a = 1, \ldots, A$, so that the local representatives $\boldsymbol{\kappa}'_a|_{K_a}$ are well-defined. (*ii*) For each $\kappa' \in U_{\kappa,\varepsilon}$, $a = 1, \ldots, A$, $\boldsymbol{X} \in \varphi_a(K_a)$, $\alpha = 1, \ldots, m$, and multi-index I, with $|I| \leq r$,

$$\left|((\boldsymbol{\kappa}'_a)^\alpha - \boldsymbol{\kappa}^\alpha_a)_{,I}(\boldsymbol{X})\right| < \varepsilon. \tag{15.10}$$

In other words, for sections κ' satisfying condition (*i*) above,

$$U_{\kappa,\varepsilon} = \left\{\kappa' \in C^r(\xi) \mid \sup\left\{\left|((\boldsymbol{\kappa}'_a)^\alpha - \boldsymbol{\kappa}^\alpha_a)_{,I}(\boldsymbol{X})\right|\right\} < \varepsilon\right\}, \tag{15.11}$$

where the supremum is taken over all

$$\boldsymbol{X} \in \varphi_a(K_a), \ \alpha = 1, \ldots, m, \ |I| \leq r, \ a = 1, \ldots, A.$$

The C^r-topology on $C^r(\xi)$ uses all such sets as a basis of neighborhoods. Using the transformation rules for the various variables, it may be shown that other choices of atlases and other choices of precompact trivializations will lead to equivalent topologies. It is noted that we use here the compactness of $\mathcal{X}$, which implies that the weak and strong C^r-topologies on the manifold of mappings (see [Hir76, p. 35]) are identical.

Remark 15.1 If one keeps the value of $a = 1, \ldots, A$, fixed, then the collection of sections

$$U_{\kappa,\varepsilon_a,a} = \left\{\kappa' \in C^r(\xi) \mid \sup\left\{\left|((\boldsymbol{\kappa}'_a)^\alpha - \boldsymbol{\kappa}^\alpha_a)_{,I}(\boldsymbol{X})\right|\right\} < \varepsilon_a\right\},$$

such that Image $\mathrm{pr}_2 \circ \Phi_a \circ \kappa'|_{K_a} \subset V_a$, and the supremum is taken over all

$$\boldsymbol{X} \in \varphi_a(K_a), \ \alpha = 1, \ldots, m, \ |I| \leq r,$$

is a neighborhood, as it contains the open neighborhood U_{κ,ε_a}. In fact, since $\mathcal{X}$ is assumed to be compact, the collection of sets of the form $\{U_{\kappa,\varepsilon_a,a}\}$ is a subbasis of neighborhoods of κ for the topology on $C^r(\xi)$.

15.2.3 Open Neighborhoods for $C^r(\xi)$ Using Vector Bundle Neighborhoods

In order to specialize the preceding constructions for the case where a vector bundle neighborhood is used, we consider again local representations of sections.

Let $\kappa \in C^r(\xi)$ be a section, and let

$$\kappa^* \tau_{\mathcal{Y}} : \kappa^* V\mathcal{Y} \longrightarrow \mathcal{X} \tag{15.12}$$

be the vector bundle identified with an open subbundle U of $\mathcal{Y}$. (We will use the two aspects of the vector bundle neighborhood, interchangeably.) Since the typical fiber of $\kappa^* V\mathcal{Y}$ may be identified with $\mathbb{R}^m$, one may choose a precompact vector bundle atlas $\{(U_a, \varphi_a, \Phi_a)\}$, $K_a \subset U_a$, $a = 1, \dots, A$, on $\kappa^* V\mathcal{Y}$, such that

$$(\varphi_a \circ \mathrm{pr}_1, \mathrm{pr}_2) \circ \Phi_a : (\kappa^* \tau_{\mathcal{Y}})^{-1}(U_a) \longrightarrow \mathbb{R}^n \times \mathbb{R}^m. \tag{15.13}$$

Thus, if we identify all open subsets V_a in Sect. 15.2.1 above with the typical fiber $\mathbb{R}^m$, the local representatives of a section are of the form

$$\kappa_a := \mathrm{pr}_2 \circ \Phi_a \circ \kappa|_{U_a} \circ \varphi_a^{-1} : \varphi_a(U_a) \longrightarrow \mathbb{R}^m. \tag{15.14}$$

A basic neighborhood of κ is given by Eq. (15.11). However, from the point of view of a vector bundle neighborhood, κ is represented by the zero section, and each κ' is viewed as a section of the vector bundle $\kappa^* \tau_{\mathcal{Y}}$, which we denote by w'. Thus, we may make the identification

$$U_{\kappa,\varepsilon} \cong \left\{ w' \in C^r(\kappa^* \tau_{\mathcal{Y}}) \mid \sup \left\{ \left| (w'_a)^\alpha_{,I}(X^i) \right| \right\} < \varepsilon \right\}. \tag{15.15}$$

In other words, using the structure of a vector bundle neighborhood, we have

$$U_{\kappa,\varepsilon} \cong \left\{ w' \in C^r(\kappa^* \tau_{\mathcal{Y}}) \mid \|w'\|^r < \varepsilon \right\}, \tag{15.16}$$

so that $U_{\kappa,\varepsilon}$ is identified with a ball of radius ε in $C^r(\kappa^* \tau_{\mathcal{Y}})$ at the zero section.

It is concluded that the charts on $C^r(\xi)$, induced by the vector bundle neighborhood, are compatible with the C^r-topology on $C^r(\xi)$.

15.3 The Manifold of Embeddings

The kinematic aspect of the Lagrangian formulation of continuum mechanics is founded on the notion of a configuration, an embedding of a body manifold $\mathcal{X}$ into the space manifold $\mathcal{S}$. The restriction of configurations to be embeddings, rather than generic C^r-mappings of the body into space, follows from the traditional principle of material impenetrability, which requires that configurations be injective and that infinitesimal volume elements are not mapped into sets of zero volume.

It is noted that any configuration $\kappa : \mathcal{X} \to \mathcal{S}$ may be viewed as a section of the trivial fiber bundle $\xi : \mathcal{X} \to \mathcal{Y} = \mathcal{X} \times \mathcal{S}$. Thus, the constructions for the manifolds of sections described above apply immediately to configurations in continuum mechanics. In this particular case, we will write $C^r(\mathcal{X}, \mathcal{S})$ for the collection of all C^r-mappings. Our objective in this section is to show that the collection of embeddings, $\text{Emb}^r(\mathcal{X}, \mathcal{S})$, is an open subset of $C^r(\mathcal{X}, \mathcal{S})$, for $r \geq 1$. In particular, it will follow that at each configuration κ, $T_\kappa \text{Emb}^r(\mathcal{X}, \mathcal{S}) = T_\kappa C^r(\mathcal{X}, \mathcal{S})$. Since the C^r-topologies, for $r > 1$, are finer than the C^1-topology, it is sufficient to prove that $\text{Emb}^1(\mathcal{X}, \mathcal{S})$ is open in $C^1(\mathcal{X}, \mathcal{S})$. This brings to light the special role that the case $r = 1$ plays in continuum mechanics.

15.3.1 The Case of a Trivial Fiber Bundle: Manifolds of Mappings

It is observed that the definitions of Sects. 15.2.1 and 15.2.2 hold with natural simplifications for the case of the trivial bundle. Thus, we use a precompact atlas $\{(U_a, \varphi_a)\}$, $a = 1, \dots, A$, and $K_a \subset U_a$, in $\mathcal{X}$ (the interiors, K_a°, cover $\mathcal{X}$). Given $\kappa \in C^1(\mathcal{X}, \mathcal{S})$, we have charts $\{(V_a, \psi_a)\}$ on $\mathcal{S}$ such that $\kappa(U_a) \subset V_a$. The local representatives of κ are of the form

$$\boldsymbol{\kappa}_a = \psi_a \circ \kappa \circ \varphi_a^{-1} : \varphi_a(U_a) \longrightarrow \psi_a(V_a) \subset \mathbb{R}^m. \tag{15.17}$$

For the case $r = 1$, Eq. (15.11) reduces to

$$U_{\kappa,\varepsilon} = \left\{\kappa' \in C^1(\mathcal{X}, \mathcal{S}) \mid \sup\left\{\left|(\boldsymbol{\kappa}'_a)^\alpha(\boldsymbol{X}) - \boldsymbol{\kappa}_a^\alpha(\boldsymbol{X})\right|, \left|(\boldsymbol{\kappa}'_a)^\alpha_{,j}(\boldsymbol{X}) - \boldsymbol{\kappa}^\alpha_{a,j}(\boldsymbol{X})\right|\right\} < \varepsilon\right\}, \tag{15.18}$$

where $\text{Image}\, \kappa'|_{K_a} \subset V_a$, and the supremum is taken over all

$$\boldsymbol{X} \in \varphi_a(K_a),\ \alpha = 1, \dots, m,\ a = 1, \dots, A.$$

Remark 15.2 It is noted that in analogy with Remark 15.1, for a fixed $a = 1, \dots, A$, a subset of mappings of the form

$$\begin{aligned} U_{\kappa,\varepsilon,a} = \Big\{\kappa' \in C^1(\mathcal{X}, \mathcal{S}) \mid \sup\{&\left|(\boldsymbol{\kappa}'_a)^\alpha(\boldsymbol{X}) - \boldsymbol{\kappa}_a^\alpha(\boldsymbol{X})\right|, \\ &\left|(\boldsymbol{\kappa}'_a)^\alpha_{,j}(\boldsymbol{X}) - \boldsymbol{\kappa}^\alpha_{a,j}(\boldsymbol{X})\right|\} < \varepsilon\Big\}, \end{aligned} \tag{15.19}$$

where $\text{Image}\, \kappa'|_{K_a} \subset V_a$, and the supremum is taken over all

$$\boldsymbol{X} \in \varphi_a(K_a),\ j = 1, \dots, n,\ \alpha = 1, \dots, m,$$

is a neighborhood of κ, since it contains a neighborhood as defined above. The collection of such sets, for various values of a and ε, forms a subbasis of neighborhoods for the topology on $C^1(\mathcal{X}, \mathcal{S})$.

15.3.2 *The Space of Immersions*

Let $\kappa \in C^1(\mathcal{X}, \mathcal{S})$ be an immersion, so that $T_X\kappa : T_X\mathcal{X} \to T_{\kappa(X)}\mathcal{S}$ is injective for every $X \in \mathcal{X}$. We show that there is a neighborhood $U_\kappa \subset C^1(\mathcal{X}, \mathcal{S})$ of κ such that all $\kappa' \in U_\kappa$ are immersions.

Note first that since the evaluation of determinants of $n \times n$ matrices is a continuous mapping, the collection of $m \times n$ matrices for which all $n \times n$ minors vanish is a closed set. Hence, the collection $L_{\mathrm{In}}(\mathbb{R}^n, \mathbb{R}^m)$ of all injective $m \times n$ matrices is open in $L(\mathbb{R}^n, \mathbb{R}^m)$. Let κ be an immersion with representatives $\boldsymbol{\kappa}_a$ as above. For each a, the derivative

$$D\boldsymbol{\kappa}_a : \varphi_a(U_a) \longrightarrow L(\mathbb{R}^n, \mathbb{R}^m), \qquad \boldsymbol{X} \longmapsto D\boldsymbol{\kappa}_a(\boldsymbol{X}), \tag{15.20}$$

is continuous; hence, $D\boldsymbol{\kappa}_a(K_a)$ is a compact set of injective linear mappings. Choosing any norm in $L(\mathbb{R}^n, \mathbb{R}^m)$, one can cover $D\boldsymbol{\kappa}_a(K_a)$ by a finite number of open balls all containing only injective mappings. In particular, setting

$$\|T\| = \max_{i,\alpha} \left\{ \left| T_i^\alpha \right| \right\}, \qquad T \in L(\mathbb{R}^n, \mathbb{R}^m), \tag{15.21}$$

let ε_a be the least radius of balls in this covering. Thus, we are guaranteed that any linear mapping T, such that $\|T - D\boldsymbol{\kappa}_a(\boldsymbol{X})\| < \varepsilon_a$ for some $\boldsymbol{X} \in \varphi_a(K_a)$, is injective. Specifically, for any $\kappa' \in C^1(\mathcal{X}, \mathcal{S})$, if

$$\sup_{\boldsymbol{X} \in \varphi_a(K_a)} \left| (\boldsymbol{\kappa}'_a)^\alpha_{,j}(\boldsymbol{X}) - \boldsymbol{\kappa}^\alpha_{a,j}(\boldsymbol{X}) \right| \leqslant \varepsilon_a, \tag{15.22}$$

then, $D\boldsymbol{\kappa}'_a$ is injective everywhere in $\varphi_a(K_a)$. Letting $\varepsilon = \min_a \varepsilon_a$, any configuration in $U_{\kappa,\varepsilon}$ as in (15.18) is an immersion.

15.3.3 *Open Neighborhoods of Local Embeddings*

Let $\kappa \in C^1(\mathcal{X}, \mathcal{S})$, and let $X \in \mathcal{X}$. It is shown below that if $T_X\kappa$ is injective, then there is a neighborhood of mappings $U_{\kappa,X}$ of κ such that every $\kappa' \in U_{\kappa,X}$ is injective in some fixed neighborhood of X. Specifically, there are a neighborhood W_X of X and a neighborhood $U_{\kappa,X}$ of κ such that for each $\kappa' \in U_{\kappa,X}$, $\kappa'|_{W_X}$ is injective.

Let (U, φ) and (V, ψ) be coordinate neighborhoods of X and $\kappa(X)$, respectively, such that $\kappa(U) \subset V$. Let $\boldsymbol{X}$ and $\boldsymbol{\kappa}$ be the local representative of X and κ relative to these charts. Thus, we are guaranteed that

$$M := \inf_{|\boldsymbol{v}|=1} |D\boldsymbol{\kappa}(\boldsymbol{X})(\boldsymbol{v})| > 0. \tag{15.23}$$

By a standard corollary of the inverse function theorem, due to the injectivity of $T_X\kappa$, we can choose U to be small enough so that the restrictions of κ and $\boldsymbol{\kappa}$ to U and its image under φ, respectively, are injective. Next, let $W_X \subset U$ be a neighborhood of X such that $\varphi(W_X)$ is convex and its closure, $\overline{W}_X$, is a compact subset of U. In addition, due to the continuity of $D\boldsymbol{\kappa}$, we may choose W_X so that

$$\left|D\boldsymbol{\kappa}(\boldsymbol{X}') - D\boldsymbol{\kappa}(\boldsymbol{X})\right| < \frac{M}{4}, \qquad \text{for all} \quad \boldsymbol{X}' \in \varphi(\overline{W}_X), \tag{15.24}$$

where $|A| = \sup_{|v|=1} |A(v)|$ is a norm on the space of matrices.

Define the neighborhood $U_{\kappa,X} \subset C^1(\mathcal{X}, \mathcal{S})$, the elements, κ', of which satisfy the conditions

$$\kappa'(\overline{W}_X) \subset V, \qquad \text{and} \qquad \left|D\boldsymbol{\kappa}'(\boldsymbol{X}') - D\boldsymbol{\kappa}(\boldsymbol{X}')\right| < \frac{M}{4}, \qquad \text{for all} \quad \boldsymbol{X}' \in \varphi(\overline{W}_X). \tag{15.25}$$

By the definition of neighborhoods in $C^1(\mathcal{X}, \mathcal{S})$ in (15.18), $U_{\kappa,X}$ contains a neighborhood of κ; hence, it is also a neighborhood. Thus, for all $\boldsymbol{X}' \in \varphi(\overline{W}_X)$,

$$\begin{aligned}
\left|D\boldsymbol{\kappa}'(\boldsymbol{X}') - D\boldsymbol{\kappa}(\boldsymbol{X})\right| &= \left|D\boldsymbol{\kappa}'(\boldsymbol{X}') - D\boldsymbol{\kappa}(\boldsymbol{X}') + D\boldsymbol{\kappa}(\boldsymbol{X}') - D\boldsymbol{\kappa}(\boldsymbol{X})\right|, \\
&\leqslant \left|D\boldsymbol{\kappa}'(\boldsymbol{X}') - D\boldsymbol{\kappa}(\boldsymbol{X}')\right| + \left|D\boldsymbol{\kappa}(\boldsymbol{X}') - D\boldsymbol{\kappa}(\boldsymbol{X})\right|, \\
&< \frac{M}{2}.
\end{aligned} \tag{15.26}$$

Now, it is shown that the fact that the values of the derivatives of representatives of elements of $U_{\kappa,X}$ are close to the injective $D\boldsymbol{\kappa}(\boldsymbol{X})$ everywhere in $\varphi(\overline{W}_X)$ implies that these mappings are close to the linear approximation using $D\boldsymbol{\kappa}(\boldsymbol{X})$, which, in turn, implies injectivity in $\overline{W}_X$ of these elements. Specifically, for and $\boldsymbol{X}_1, \boldsymbol{X}_2 \in \varphi(\overline{W}_X)$, since

$$\boldsymbol{\kappa}'(\boldsymbol{X}_2) - \boldsymbol{\kappa}'(\boldsymbol{X}_1) = \boldsymbol{\kappa}'(\boldsymbol{X}_2) - \boldsymbol{\kappa}'(\boldsymbol{X}_1) - D\boldsymbol{\kappa}(\boldsymbol{X})(\boldsymbol{X}_2 - \boldsymbol{X}_1) + D\boldsymbol{\kappa}(\boldsymbol{X})(\boldsymbol{X}_2 - \boldsymbol{X}_1), \tag{15.27}$$

the triangle inequality implies that

$$\begin{aligned}
\left|\boldsymbol{\kappa}'(\boldsymbol{X}_2) - \boldsymbol{\kappa}'(\boldsymbol{X}_1)\right| \geq{}& |D\boldsymbol{\kappa}(\boldsymbol{X})(\boldsymbol{X}_2 - \boldsymbol{X}_1)| \\
&- \left|D\boldsymbol{\kappa}(\boldsymbol{X})(\boldsymbol{X}_2 - \boldsymbol{X}_1) - (\boldsymbol{\kappa}'(\boldsymbol{X}_2) - \boldsymbol{\kappa}'(\boldsymbol{X}_1))\right|, \\
\geq{}& M\,|\boldsymbol{X}_2 - \boldsymbol{X}_1| - \left|\boldsymbol{\kappa}'(\boldsymbol{X}_2) - \boldsymbol{\kappa}'(\boldsymbol{X}_1) - D\boldsymbol{\kappa}(\boldsymbol{X})(\boldsymbol{X}_2 - \boldsymbol{X}_1)\right|.
\end{aligned} \tag{15.28}$$

Using the mean value theorem and the convexity of $\varphi(\overline{W}_X)$, there is a point $\boldsymbol{X}_0 \in \varphi(\overline{W}_X)$ such that

$$\boldsymbol{\kappa}'(\boldsymbol{X}_2) - \boldsymbol{\kappa}'(\boldsymbol{X}_1) = D\boldsymbol{\kappa}'(\boldsymbol{X}_0)(\boldsymbol{X}_2 - \boldsymbol{X}_1). \tag{15.29}$$

Hence,

$$\begin{aligned}\left|\kappa'(X_2)-\kappa'(X_1)-D\kappa(X)(X_2-X_1)\right| &= \left|(D\kappa'(X_0)-D\kappa(X))(X_2-X_1)\right|,\\ &\leq \left|D\kappa'(X_0)-D\kappa(X)\right|\left|X_2-X_1\right|,\\ &< \frac{M}{2}\left|X_2-X_1\right|.\end{aligned} \tag{15.30}$$

It follows that

$$\left|\kappa'(X_2)-\kappa'(X_1)\right| > \frac{M}{2}\left|X_2-X_1\right|, \tag{15.31}$$

which proves the injectivity.

15.3.4 Open Neighborhoods of Embeddings

Finally, it is shown how every $\kappa \in \mathrm{Emb}^1(\mathcal{X},\mathcal{S}) \subset C^1(\mathcal{X},\mathcal{S})$ has a neighborhood consisting of embeddings only. It will follow that $\mathrm{Emb}^1(\mathcal{X},\mathcal{S})$ is an open subset of $C^1(\mathcal{X},\mathcal{S})$. This has far-reaching consequences in continuum mechanics, and it explains the special role played by the C^1-topology in continuum mechanics.

Let κ be a given embedding. Using the foregoing result, for each $X \in \mathcal{X}$ there are an open neighborhood W_X of X and a neighborhood $U_{\kappa,X}$ of κ, such that for each $\kappa' \in U_{\kappa,X}$, $\kappa'|_{\overline{W}_X}$ is injective. The collection of neighborhoods $\{W_X\}$, $X \in \mathcal{X}$, is an open cover of $\mathcal{X}$, and by compactness, it has a finite sub-cover. Denote the finite collection of open sets of the form W_X as above by W_a, $a = 1,\dots,A$. Hence, $K_a := \overline{W}_a$ is a compact subset of U_a, $\kappa(U_a) \subset V_a$ for some chart (V_a,ψ_a) in $\mathcal{S}$. For each a, we have a neighborhood $U_{\kappa,a}$ of κ such that every $\kappa' \in U_{\kappa,a}$ satisfies the condition that $\kappa'|_{K_a}$ is injective. Let $\mathcal{N}_1 = \bigcap_{a=1}^{A} U_{\kappa,a}$, so that for each $\kappa' \in \mathcal{N}_1$, $\kappa'|_{K_a}$ is injective for all a. Let $\mathcal{N}_2$ be a neighborhood of κ that contains only immersions as in Sect. 15.3.2. Thus, $\mathcal{N}_0 = \mathcal{N}_1 \cap \mathcal{N}_2$ contains immersions which are locally injective in all subsets K_a.

Let κ be an embedding, and let $\mathcal{N}_0$ be the open set constructed as above. If no neighborhood of κ contains only injective mappings, then, for each $\nu = 1,2,\dots$, there is a $\kappa_\nu \in U_{\kappa,\varepsilon_\nu}$, $\varepsilon_\nu = 1/\nu$, and points $X_\nu, X'_\nu \in \mathcal{X}$, $X_\nu \neq X'_\nu$, such that $\kappa_\nu(X_\nu) = \kappa_\nu(X'_\nu)$. As $\mathcal{N}_0$ is a neighborhood of κ, we may assume that $\kappa_\nu \in \mathcal{N}_0$ for all ν. By the compactness of $\mathcal{X}$ and $\mathcal{X}\times\mathcal{X}$, we can extract a converging subsequence from the sequence $((X_\nu, X'_\nu))$ in $\mathcal{X}\times\mathcal{X}$. We keep the same notation for the converging subsequences, and let

$$(X_\nu, X'_\nu) \longrightarrow (X, X'), \qquad \text{as} \quad \nu \longrightarrow \infty. \tag{15.32}$$

We first exclude the possibility that $X = X'$. Assume $X = X' \in K_{a_0}$, for some $a_0 \in \{1, \dots, A\}$. Then, for any neighborhood $U_{\kappa,\varepsilon_\nu}$ of κ and any neighborhood of $X = X'$, there is a configuration κ_ν such that κ_ν is not injective. This contradicts the construction of local injectivity above.

Thus, one should consider the situation for which $X \neq X'$. Assume $X \in K_{a_0}$ and $X' \in K_{a_1}$ for $a_0, a_1 \in \{1, \dots, A\}$. By the definition of $U_{\kappa,\varepsilon_\nu}$, the local representatives of $\kappa_\nu|K_{a_0}$ and $\kappa_\nu|K_{a_1}$ converge uniformly to the local representatives of $\kappa|K_{a_0}$ and $\kappa|K_{a_1}$, respectively. This implies that

$$\kappa_\nu(X_\nu) \longrightarrow \kappa(X), \quad \kappa_\nu(X'_\nu) \longrightarrow \kappa(X'), \qquad \text{as} \quad \nu \longrightarrow \infty. \tag{15.33}$$

However, since for each ν, $\kappa_\nu(X_\nu) = \kappa_\nu(X'_\nu)$, it follows that $\kappa(X) = \kappa(X')$, which contradicts the assumption that κ is an embedding.

It is finally noted that the set of Lipschitz embeddings equipped with the Lipschitz topology may be shown to be open in the manifold of all Lipschitz mappings $\mathcal{X} \to \mathcal{S}$. See [FN05] for details, and see [FS15] for an application to continuum mechanics.

Chapter 16
The General Framework for Global Analytic Stress Theory

The preceding chapter implied that for the case where the kinematics of a material body $\mathcal{X}$ is described by its embeddings in a physical space $\mathcal{S}$, the collection of configurations, the *configuration space*

$$\mathbb{Q} := \mathrm{Emb}^r(\mathcal{X}, \mathcal{S}),$$

is an open subset of the manifold of mappings $C^r(\mathcal{X}, \mathcal{S})$, for $r \geq 1$. As a result, the configuration space is a Banach manifold in its own right, and

$$T_\kappa\mathbb{Q} = T_\kappa C^r(\mathcal{X}, \mathcal{S}) = T_\kappa C^r(\xi),$$

where $\xi : \mathcal{X} \times \mathcal{S} \to \mathcal{X}$ is the natural projection of the trivial fiber bundle.

In view (15.7), $T_\kappa\mathbb{Q} = C^r(\kappa^* V\mathcal{Y})$, where now

$$V\mathcal{Y} = \{v \in T\mathcal{Y} = T\mathcal{X} \times T\mathcal{S} \mid T\xi(v) = 0 \in T\mathcal{X}\}.$$

Hence, one may make the identifications

$$V\mathcal{Y} = \mathcal{X} \times T\mathcal{S},$$

and

$$(\kappa^* V\mathcal{Y})_X = (V\mathcal{Y})_{\kappa(X)} = T_{\kappa(X)}\mathcal{S}.$$

A section w of $\kappa^*\tau_\mathcal{Y} : \kappa^* V\mathcal{Y} \to \mathcal{X}$ is of the form

$$X \longmapsto w(X) \in T_{\kappa(X)}\mathcal{S}$$

R. Segev, *Foundations of Geometric Continuum Mechanics*, Advances in Continuum Mechanics 49, https://doi.org/10.1007/978-3-031-35655-1_16

and may be viewed as a *vector field along* κ, i.e., a mapping

$$w : \mathcal{X} \longrightarrow T\mathcal{S}, \qquad \text{such that,} \qquad \tau_{\mathcal{S}} \circ w = \kappa.$$

In other words, we identify the mappings w and $\tilde{w}$ in the following diagram due to the universality property of the pullback.

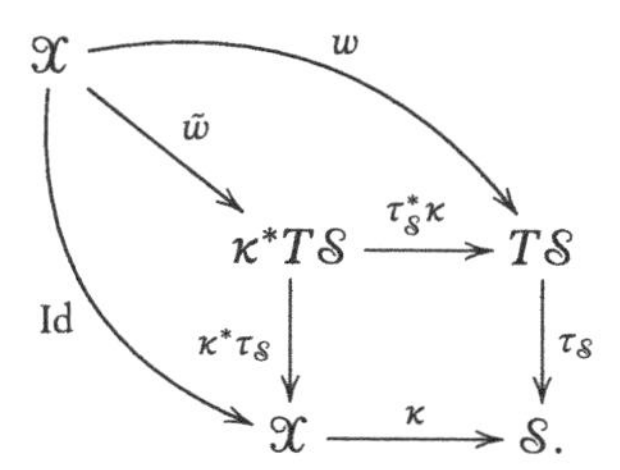

Thus, a tangent vector to the configuration space at the configuration κ may be viewed as a C^r-vector field along κ. This is a straightforward generalization of the standard notion of a virtual velocity field, and we summarize these observations by writing

$$T_\kappa \mathbb{Q} = C^r(\kappa^* V\mathcal{Y}) = \{w \in C^r(\mathcal{X}, T\mathcal{S}) \mid \tau \circ w = \kappa\}.$$

In the case of generalized continua, where $\xi : \mathcal{Y} \to \mathcal{X}$ need not be a trivial vector bundle, this simplification does not apply, of course. However, the foregoing discussion motivates the definition of the *configuration space* for a general continuum mechanical system specified by the fiber bundle $\xi : \mathcal{Y} \to \mathcal{X}$ as

$$\mathbb{Q} = C^r(\xi), \qquad r \geq 1.$$

We note that the condition that configurations are embeddings is meaningless when they are represented by sections of a general fiber bundle ξ.

The general framework for global analytic stress theory adopts the geometric structure for the statics of systems having a finite number of degrees of freedom. Once a configuration manifold $\mathbb{Q}$ is specified, *generalized* or *virtual velocities* are defined to be elements of the tangent bundle, $T\mathbb{Q}$, and *generalized forces* are defined to be elements of the cotangent bundle $T^*\mathbb{Q}$. The action, $F(w)$, of a force $F \in T^*_\kappa\mathbb{Q}$ on a virtual velocity $w \in T_\kappa\mathbb{Q}$ is interpreted as *virtual power,* and as such, the notion of power plays a fundamental role in this formulation.

The foregoing discussion implies that a force at a configuration $\kappa \in C^r(\xi)$ is an element of $C^r(\kappa^* V\mathcal{Y})^*$—a continuous and linear functional on the Banachable space of C^r-sections of the vector bundle. In the following chapters, we consider the properties of linear functionals on the space of C^r-sections of a vector bundle W. Of particular interest is the fact that our base manifold, or body manifold, is a manifold with corners rather than a manifold without boundary. The relation between such functionals, on the one hand, and Schwartz distribution and de Rham currents, on the other hand, is described. In Chap. 21, we show that the notions of stresses and hyper-stresses emerge naturally from a representation theorem for such functionals, and in Chap. 22 we study further the properties of stresses for the standard case $r = 1$.

Chapter 17
Dual Spaces Corresponding to Spaces of Differentiable Sections of a Vector Bundle: Localization of Sections and Functionals

As implied by the paradigm outlined above, generalized forces are modeled mathematically as elements of the dual space $C^r(\pi)^* = C^r(W)^*$ of the space of C^r-sections of a vector bundle $\pi : W \to \mathcal{X}$. This section reviews the basic notions corresponding to such continuous linear functionals, with particular attention to localization properties. While we assume that our base manifold $\mathcal{X}$ is compact with corners, we want to relate the nature of functionals defined on sections over $\mathcal{X}$ with the analogous objects for the case where $\mathcal{X}$ is a manifold without boundary. Thus, one can relate the properties of generalized forces with objects like distributions, de Rham currents, and generalized sections on manifolds. (See, in particular, Sect. 17.5.) As an additional motivation for considering sections over manifolds without boundaries, it is observed that in both the Eulerian formulation of continuum mechanics and classical field theories, the base manifold, either space or spacetime, is usually taken as manifold without boundary. We start with the case where $\mathcal{X}$ is a manifold without boundary and continue with the case where bodies are modeled by compact manifolds with corners.

17.1 Spaces of Differentiable Sections Over a Manifold Without Boundary and Linear Functionals

A comprehensive introduction to the subject considered here is available in the Ph.D. thesis [Ste00] and the corresponding [GKOS01, Chapter 3]. See also [GS77, Chapter VI] and [Kor08].

Consider the space of C^r-sections of a vector bundle $\pi : W \to \mathcal{X}$, for $0 \leq r \leq \infty$. For manifolds without boundary that are not necessarily compact, the setting of Sect. 14.1 will not give a norm on the space of sections. Thus, one extends the settings used for Schwartz distributions to sections of a vector bundle

R. Segev, *Foundations of Geometric Continuum Mechanics*, Advances in Continuum Mechanics 49, https://doi.org/10.1007/978-3-031-35655-1_17

(see also [Die72, Chapter XVII]). Specifically, we turn our attention to $C^r_c(\pi)$, the space of *test sections*—C^r-sections of π having compact supports in $\mathcal{X}$.

Let $\{(U_a, \varphi_a, \Phi_a)\}_{a\in A}$ be a vector bundle atlas so that

$$(\varphi_a, \Phi_a) : \pi^{-1}(U_a) \longrightarrow \mathbb{R}^n \times \mathbb{R}^m, \qquad v \longmapsto (X^i, v^\alpha) \tag{17.1}$$

and let K be a compact subset of $\mathcal{X}$. Consider the vector subspace $C^r_{c,K}(\pi) \subset C^r_c(\pi)$ of sections, the supports of which are contained in K. Let $a_l \in A$ indicate a finite collection of charts such that $\{U_{a_l}\}$ cover K, and for each U_{a_l} let $K_{a_l,\mu}$, $\mu = 1, 2, \ldots$, be a fundamental sequence of compact sets, i.e., $K_{a_l,\mu} \subset K^{\mathrm{o}}_{a_l,\mu+1}$, covering $\varphi_{a_l}(U_{a_l}) \subset \mathbb{R}^n$. Then, for a section $\chi \in C^r_{c,K}(\pi)$, the collection of semi-norms

$$\|\chi\|^r_{K,\mu} = \sup_{a_l,\alpha,|I|\leq r} \left\{ \sup_{X\in K_{a_l,\mu}} \left\{ \left|(\chi^\alpha_{a_l})_{,I}(X)\right| \right\} \right\} \tag{17.2}$$

induces a Fréchet space structure for $C^\infty_{c,K}(\pi)$. Since for each compact subset K, one has the inclusion mapping $\mathcal{J}_K : C^r_{c,K}(\pi) \to C^r_c(\pi)$, one may define the topology on $C^r_c(\pi)$ as the inductive limit topology generated by these inclusions, i.e., the strongest topology on $C^r_c(\pi)$ for which all the inclusions are continuous. A sequence of sections in $C^r_c(\pi)$ converges to zero, if there is a compact subset $K \subset \mathcal{X}$ such that the supports of all sections in the sequence are contained in K and the r-jets of the sections converge uniformly to zero in K (in the sense that all the semi-norms $\|\chi\|^k_{K,\mu}$ converge to zero for all $k \leqslant r$).

A linear functional $T \in C^r_c(\pi)^*$ is continuous when it satisfies the following condition. Let (χ_j) be a sequence of sections of π all of which are supported in a compact subset $K \subset U_a$ for some $a \in A$. In addition, assume that the local representatives of χ_j and their derivatives of all orders $k \leq r$ converge uniformly to zero in K. Then,

$$\lim_{j\to\infty} T(\chi_j) = 0. \tag{17.3}$$

Functionals in $C^r_c(\pi)^*$ for a finite value of r are referred to as functionals of order r.

For a linear functional T, the *support*, $\operatorname{supp} T$, is defined as follows. An open set $U \subset \mathcal{X}$ is termed a null set of T if $T(\chi) = 0$ for any section of π with $\operatorname{supp}\chi \subset U$. The union of all null sets, U_0, is an open set which is a null set also. Thus, one defines

$$\operatorname{supp} T := \mathcal{X} \setminus U_0. \tag{17.4}$$

17.2 Localization for Manifolds Without Boundaries

Let $\{(U_a, \varphi_a, \Phi_a)\}_{a\in A}$ be a locally finite vector bundle atlas on W, and consider

$$E_{U_a} : C^r_c(\pi|_{U_a}) \longrightarrow C^r_c(\pi), \tag{17.5}$$

the natural zero extension of sections supported in a compact subset of U_a to the space of sections that are compactly supported in $\mathcal{X}$. This is evidently a linear and continuous injection of the subspace. On its image, the subspace of sections χ with $\operatorname{supp}\chi \subset U_a$, we have a left inverse, the natural restriction

$$\rho_{U_a} : \text{Image}\, E_{U_a} \longrightarrow C^r_c(\pi|_{U_a}), \tag{17.6}$$

a surjective mapping. However, it is well-known (e.g., [Sch63, Trè67, pp. 245–246]) that the inverse ρ_{U_a} is not continuous.

The dual,

$$E^*_{U_a} : C^r_c(\pi)^* \longrightarrow C^r_c(\pi|_{U_a})^*, \tag{17.7}$$

is the restriction of functionals on $\mathcal{X}$ to sections supported on U_a, and as ρ_{U_a} is not continuous, $E^*_{U_a}$ is not surjective (*loc. cit.*). We will write

$$T|_{U_a} := \tilde{T}_a := E^*_{U_a} T. \tag{17.8}$$

We also note that the restrictions, $\{\tilde{T}_a\}$, satisfy the condition

$$\tilde{T}_a(\chi|_{U_a}) = \tilde{T}_b(\chi|_{U_b}) = T(\chi) \tag{17.9}$$

for any section χ supported in $U_a \cap U_b$.

Consider the mapping

$$s : \prod_{a\in A} C^r_c(\pi|_{U_a}) \longrightarrow C^r_c(\pi) \tag{17.10}$$

given by

$$s(\chi_1, \ldots, \chi_a, \ldots) := \sum_{a\in A} E_{U_a}(\chi_a). \tag{17.11}$$

(Note that at each point, the sum contains only a finite number of nonvanishing terms since the atlas is locally finite.) Due to overlappings between domains of definition, the mapping s is not injective. However, s is surjective because using a partition of unity, $\{u_a\}$, which is subordinate to this atlas, for each section, χ, $u_a\chi$ is compactly supported in U_a, and $\chi = \sum_a u_a\chi$. Hence, the dual mapping,

$$s^* : C^r_c(\pi)^* \longrightarrow \prod_{a\in A} C^r_c(\pi|_{U_a})^*, \tag{17.12}$$

given by,

$$(s^*T)_a := E^*_{U_a}T = T|_{U_a}, \qquad (s^*T)(\chi_1,\dots) = T\Big(\sum_{a\in A} E_{U_a}(\chi_a)\Big), \tag{17.13}$$

is injective. In other words, a functional is determined uniquely by the collection of its restrictions. Note that no compatibility condition is imposed above on the local sections $\{\chi_a\}$.

Since $\{\tilde{T}_a\} \in \operatorname{Image} s^*$ satisfy the compatibility condition (17.9), s^* is not surjective. However, it is easy to see that $\operatorname{Image} s^*$ is exactly the subspace of $\prod_{a\in A} C^r_c(\pi|_{U_a})^*$ containing the compatible collections of local functionals. Let $\{\tilde{T}_a\}$ be local functionals that satisfy (17.9) and let $\{u_a\}$ be a partition of unity. Consider the functional $T \in C^r_c(\pi)^*$ given by

$$T(\chi) = \sum_{a\in A} \tilde{T}_a(u_a\chi). \tag{17.14}$$

If χ is supported in U_b for $b \in A$, then

$$\begin{aligned} T(\chi) &= \sum_{a\in A} \tilde{T}_a(u_a\chi), && U_a \cap U_b \neq \varnothing, \\ &= \sum_{a\in A} \tilde{T}_b(u_a\chi), && \text{by (17.9)}, \\ &= \tilde{T}_b\Big(\sum_{a\in A} u_a\chi\Big), \\ &= \tilde{T}_b(\chi). \end{aligned} \tag{17.15}$$

Thus, T is a well-defined functional on $C^r_c(\pi)$, and it is uniquely determined by the collection $\{\tilde{T}_a\}$—its restrictions, independent of the partition of unity chosen.

As mentioned above, a partition of unity induces an injective right inverse to s in the form

$$p : C^r_c(\pi) \longrightarrow \prod_{a\in A} C^r_c(\pi|_{U_a}), \qquad p(\chi)_a = (u_a\chi)|_{U_a}, \tag{17.16}$$

that evidently satisfies $s \circ p = \mathrm{Id}$. It is noted that p is not a left inverse. In particular, for a section χ_a supported in U_a, with $\chi_b = 0$ for all $b \neq a$,

$$(p \circ s\{\chi_1,\dots\})_a = u_a\chi_a \tag{17.17}$$

which need not be equal to χ_a. Evidently, p depends on the partition of unity.

For the surjective dual mapping

$$p^* : \prod_{a \in A} C_c^r(\pi|_{U_a})^* \longrightarrow C_c^r(\pi)^*, \tag{17.18}$$

one has

$$p^*(T_1, \dots) = \sum_{a \in A} u_a(\rho^*_{U_a} T_a), \qquad p^*(T_1, \dots)(\chi) = \sum_{a \in A} T_a((u_a \chi)|_{U_a}), \tag{17.19}$$

where $u_a T$ denotes the functional defined by $u_a T(\chi) = T(u_a \chi)$. We note that $p^* \circ s^* = \mathrm{Id}$, while $s^* \circ p^* \neq \mathrm{Id}$, in general. The surjectivity of p^* implies that every functional T may be represented by a non unique collection $\{T_a\}$ in the form

$$T(\chi) = \sum_{a \in A} T_a((u_a \chi)|_{U_a}), \qquad T = \sum_{a \in A} u_a T_a. \tag{17.20}$$

Evidently, the relation between T and the collection $\{T_a\}$ depends on the partition of unity. The constructions are roughly illustrated in the following diagram.

$$C_c^r(\pi) \underset{s}{\overset{p}{\rightleftarrows}} \prod_{a \in A} C_c^r(\pi|_{U_a})$$

$$C_c^r(\pi)^* \underset{s^*}{\overset{p^*}{\leftrightarrows}} \prod_{a \in A} C_c^r(\pi|_{U_a})^*.$$

Nevertheless, we may restrict p^* to the subspace of compatible local functionals, Image s^*, i.e., those satisfying (17.9). Thus, the restriction

$$p^*|_{\mathrm{Image}\, s^*} : \mathrm{Image}\, s^* \longrightarrow C_c^r(\pi)^*, \tag{17.21}$$

is an isomorphism (which depends on the partition of unity). It follows that

$$s^* \circ p^* = \mathrm{Id} : \mathrm{Image}\, s^* \longrightarrow \mathrm{Image}\, s^*. \tag{17.22}$$

In other words, if the collection $\{T_a\}$, $a \in A$, satisfies the compatibility condition, then, the restrictions, $\tilde{T}_a$, satisfy $\tilde{T}_a = T_a$ for all $a \in A$. (For additional details, see [Die72, p. 244–245], which is restricted to the case of de Rham currents, and [GKOS01, pp. 234–235].)

17.3 Localization for Compact Manifolds with Corners

In analogy with Sect. 17.2, we consider the various aspects of localization relevant to the case of compact manifolds with corners. Thus, the base manifold for the vector bundle $\pi : W \longrightarrow \mathcal{X}$ is assumed to be a manifold with corners, and we are concerned with elements of $C^r(\pi)^*$ acting on sections that need not necessarily vanish together with their first r-jets on the boundary of $\mathcal{X}$.

In [Pal68, pp 10–11], Palais proves what he refers to as the "Mayer-Vietoris Theorem." Adapting the notation and specializing the theorem to the C^r-topology, the theorem may be stated as follows.

Theorem 17.1 *Let $\mathcal{X}$ be a compact, smooth n-dimensional manifold, and let $K_1, \dots, K_A$ be a finite collection of compact n-dimensional submanifolds of $\mathcal{X}$, the interiors of which cover $\mathcal{X}$ (such as in a precompact atlas). Given the vector bundle $\pi : W \to \mathcal{X}$, set*

$$\tilde{C}^r(\pi) := \left\{(\chi_1, \dots, \chi_A) \in \prod_{a=1}^{A} C^r(\pi|_{K_a}) \,\middle|\, \chi_a|_{K_b} = \chi_b|_{K_a}\right\}, \tag{17.23}$$

and define

$$\iota : C^r(\pi) \longrightarrow \tilde{C}^r(\pi), \qquad \text{by} \qquad \iota(\chi) = (\chi|_{K_1}, \dots, \chi|_{K_A}). \tag{17.24}$$

Then, ι is an isomorphism of Banach spaces.

We will refer to the condition in (17.23) as the *compatibility condition*. The most significant part of the proof is the construction of ι^{-1}. One has to construct a field χ when a collection $(\chi_1, \dots, \chi_A)$, satisfying the compatibility condition, is given. This is done using a partition of unity which is subordinate to the interiors of $K_1, \dots, K_A$.

It is noted that the situation may be viewed as "dual" to that described in Sect. 17.2. For functionals on spaces of sections with compact supports defined on a manifold without boundary, there is a natural restriction of functionals, $E^*_{U_a}$, and the images $\{\tilde{T}_a\}$ of a functional T under the restrictions satisfy the compatibility condition (17.9). The collection of restrictions determines T uniquely. Here, it follows from Theorem 17.1 that we have a natural restriction of sections, and the restricted sections satisfy the compatibility condition (17.23). The restrictions $\{\chi|_{K_a}\}$ also determine the global section χ, uniquely.

In Sect. 17.2, we observed that sections with compact supports on $\mathcal{X}$ cannot be "restricted" naturally to sections with compact supports on the various U_a. Such restrictions depend on the chosen partition of unity. The analogous situation for functionals on compact manifolds with corners is described below.

Corollary 17.1 *Let $T \in C^r(\pi)^*$, then T may be represented (non uniquely) by $(T_1, \dots, T_A)$, $T_a \in C^r(\pi|_{K_a})^*$, in the form*

$$T(\chi) = \sum_{a=1}^{A} T_a(\chi|_{K_a}). \tag{17.25}$$

Indeed, as ι in Theorem 17.1 is an embedding of $C^r(\pi)$ into a subspace of $\prod_{a=1}^{A} C^r(\pi|_{K_a})$, one has a surjective dual mapping

$$\iota^* : \prod_{a=1}^{A} C^r(\pi|_{K_a})^* \longrightarrow C^r(\pi)^*, \tag{17.26}$$

given by

$$\iota^*(T_1, \dots, T_A)(\chi) = \sum_{a=1}^{A} T_a(\chi|_{K_a}). \tag{17.27}$$

17.4 Supported Sections, Static Indeterminacy, and Body Forces

The foregoing observations are indicative of the fundamental problem of continuum mechanics—that of static indeterminacy. Consider a vector bundle $\pi : W \to \mathcal{X}$ over a compact manifold with corners $\mathcal{X}$. Given a force, F, on the body $\mathcal{X}$, as an element of $C^r(\pi)^*$, and a sub-body $\mathcal{R} \subset \mathcal{X}$, there is no unique restriction of F to a force on $\mathcal{R}$, an element of $C^r(\pi|_{\mathcal{R}})^*$. This problem is evident for standard continuum mechanics in Euclidean spaces, and it still exists for continuum mechanics of higher order on differentiable manifolds.

Remark 17.1 It is noted that for the case $r = 0$, $C^0(\pi)^*$ contains Radon measures which can be naturally restricted to measurable subsets. This fact indicates the reason why elements of $C^0(\pi)^*$ do not model force functionals of continuum mechanics.

Adopting the notation of [Mel96], denote, by $\dot{C}^r(\pi)$, the space of sections of π, the r-jet extensions of which vanish on all the components of the boundary $\partial\mathcal{X}$. Let $\tilde{\mathcal{X}}$ be a manifold without boundary extending $\mathcal{X}$ (as in Sect. 6.4), and let

$$\tilde{\pi} : \tilde{W} \longrightarrow \tilde{\mathcal{X}} \tag{17.28}$$

be an extension of π. Then, we may use zero extensions to obtain an isomorphism

$$\dot{C}^r(\pi) \cong \{\chi \in C^r_c(\tilde{\pi}) \mid \operatorname{supp} \chi \subset \mathcal{X}). \tag{17.29}$$

If $\mathcal{R}$ is a sub-body of $\mathcal{X}$, then, zero extensions induce an inclusion

$$\dot{C}^r(\pi|_{\mathcal{R}}) \hookrightarrow \dot{C}^r(\pi). \tag{17.30}$$

The dual, $\dot{C}^r(\pi)^*$, to the space of sections supported in $\mathcal{X}$ is the space of *extendable functionals.* From [Mel96, Proposition 3.3.1] it follows that the restriction

$$\rho : C^r(\pi)^* \longrightarrow \dot{C}^r(\pi)^*. \tag{17.31}$$

is surjective and its kernel is the space of functionals on $\tilde{\mathcal{X}}$ supported in $\partial\mathcal{X}$.

Thus, if we interpret $T \in C^r(\pi)^*$ as a force, $\rho(T) \in \dot{C}^r(\pi)^*$ is interpreted as the corresponding *body force*. For a sub-body $\mathcal{R}$, using the dual of (17.30), one has

$$\rho_{\mathcal{R}} : \dot{C}^r(\pi)^* \longrightarrow \dot{C}^r(\pi|_{\mathcal{R}})^*. \tag{17.32}$$

We conclude that even in this very general setting, body forces of any order may be restricted naturally to sub bodies.

17.5 Supported Functionals

Distributions on closed subsets of $\mathbb{R}^n$ have been considered by Glaeser [Gla58], Malgrange [Mal66, Chapter 7], and Oksak [Oks76]. The basic tool in the analysis of distributions on closed sets is Whitney's extension theorem [Whi34] (see also [See64, Hör90]) that guarantees that a differentiable function defined on a compact subset of $\mathbb{R}^n$ may be extended to a compactly supported differentiable function on $\mathbb{R}^n$. For the case of manifolds with corners, the extension mapping between the corresponding function spaces is continuous. (See discussion and counter-examples in [Mic20, Section 4.3].) The extension theorem implies that the restriction of functions is surjective, and so, the dual of the restriction mapping associates a unique distribution in an open subset of $\mathbb{R}^n$ with a linear functional defined on the given compact set. Distributions and functionals on a manifold with corners have been considered by Melrose [Mel96, Chapter 3], who we follow below.

Thus, for a vector bundle $\pi : W \to \mathcal{X}$, where $\mathcal{X}$ is a compact manifold with corners, let $T \in C^r(\pi)^*$, and let $\tilde{\pi} : \tilde{W} \to \tilde{\mathcal{X}}$ be an extension of π, such that $\tilde{\mathcal{X}}$ is a manifold without a boundary. The Whitney-Seeley extension

$$E : C^r(\pi) \longrightarrow C^r_c(\tilde{\pi}) \tag{17.33}$$

is a continuous injection. It follows that the natural restriction

$$\rho_{\mathcal{X}} : C^r_c(\tilde{\pi}) \longrightarrow C^r(\pi), \tag{17.34}$$

its left inverse satisfying $\rho_{\mathcal{X}} \circ E = \mathrm{Id}$, is surjective and the inclusion

$$\rho_{\mathcal{X}}^{*} : C^{r}(\pi)^{*} \longrightarrow C_{c}^{r}(\tilde{\pi})^{*} \tag{17.35}$$

is injective. In other words, each functional $T \in C^{r}(\pi)^{*}$ determines uniquely a functional $\tilde{T} = \rho_{\mathcal{X}}^{*} T$ satisfying

$$\tilde{T}(\tilde{\chi}) = T(\tilde{\chi}|_{\mathcal{X}}). \tag{17.36}$$

The last equation implies also that $\tilde{T}(\tilde{\chi}) = 0$ for any section $\tilde{\chi}$ supported in $\tilde{\mathcal{X}} \setminus \mathcal{X}$. Hence, $\tilde{T}$ is supported in $\mathcal{X}$.

Conversely, every $\tilde{T} \in C_{c}^{r}(\tilde{\pi})^{*}$, with $\operatorname{supp} \tilde{T} \subset \mathcal{X}$, represents a functional $T \in C^{r}(\pi)^{*}$, such that $\tilde{T} = \rho_{\mathcal{X}}^{*} T$. This may be deduced as follows. For any such $\tilde{T}$, consider $T = E^{*}\tilde{T}$. One needs to show that $\tilde{T} = \rho_{\mathcal{X}}^{*} \circ E^{*}(\tilde{T})$. Let $\tilde{\chi} \in C_{c}^{r}(\tilde{\pi})$, then,

$$\begin{aligned} (\tilde{T} - \rho_{\mathcal{X}}^{*} \circ E^{*}(\tilde{T}))(\tilde{\chi}) &= \tilde{T}(\tilde{\chi}) - \tilde{T}(E \circ \rho_{\mathcal{X}}(\tilde{\chi})), \\ &= \tilde{T}(\tilde{\chi} - E \circ \rho_{\mathcal{X}}(\tilde{\chi})). \end{aligned} \tag{17.37}$$

It is observed that $\tilde{\chi} - E \circ \rho_{\mathcal{X}}(\tilde{\chi})$ vanishes on $\mathcal{X}$, so that $\operatorname{supp}(\tilde{\chi} - E \circ \rho_{\mathcal{X}}(\tilde{\chi}) \subset \overline{\tilde{\mathcal{X}} \setminus \mathcal{X}}$. Since $\operatorname{supp} \tilde{T} \subset \mathcal{X}$, $\tilde{T}(\chi') = 0$ for any section χ' supported in $\tilde{\mathcal{X}} \setminus \mathcal{X}$. However, approximating the section $\tilde{\chi} - E \circ \rho_{\mathcal{X}}(\tilde{\chi})$, supported in the closure, $\overline{\tilde{\mathcal{X}} \setminus \mathcal{X}}$, by sections supported in $\tilde{\mathcal{X}} \setminus \mathcal{X}$, one concludes that $\tilde{T}(\tilde{\chi} - E \circ \rho_{\mathcal{X}}(\tilde{\chi})) = 0$, also.

Due to this construction, Melrose, [Mel96, Chapter 3], refers to such functionals (distributions) as *supported*. It is noted that such functionals of compact support are of a finite order r.

17.6 Generalized Sections and Distributions

A simple example for functionals on spaces of sections of a vector bundle $\pi : W \to \mathcal{X}$ is provided by *smooth functionals*. Consider the density dual vector bundle $W' = L\big(W, \bigwedge^{n} T^{*}\mathcal{X}\big)$ as in Sect. 6.2.5. Smooth functionals on sections with compact supports are induced by smooth sections of W'. For a section

$$S : \mathcal{X} \longrightarrow W' = L\big(W, \textstyle\bigwedge^{n} T^{*}\mathcal{X}\big),$$

and a section χ of W having a compact support, let $S \cdot \chi$ be the n-form given by

$$(S \cdot \chi)(X) = S(X)(\chi(X)). \tag{17.38}$$

The smooth functional T_S induced by S is defined by

$$T_S(\chi) := \int_{\mathcal{X}} S \cdot \chi. \tag{17.39}$$

Let $\pi_0 : W_0 \to \mathcal{X}$ be a vector bundle, and consider the case where the vector bundle $\pi : W \to \mathcal{X}$ above is set to be

$$\pi : W := W_0' = L\big(W_0, \textstyle\bigwedge^n T^*\mathcal{X}\big) \cong W_0^* \otimes \bigwedge^n T^*\mathcal{X} \longrightarrow \mathcal{X}. \tag{17.40}$$

Thus, the corresponding functionals on compactly supported sections of π are elements of

$$C_c^r(\pi)^* = C_c^r\left(L\big(W_0, \textstyle\bigwedge^n T^*\mathcal{X}\big)\right)^* \cong C_c^r\left(W_0^* \otimes \bigwedge^n T^*\mathcal{X}\right)^*. \tag{17.41}$$

In this case, smooth functionals are represented by smooth sections of

$$\begin{aligned} L\big(W, \textstyle\bigwedge^n T^*\mathcal{X}\big) &= L\big(W_0^* \otimes \textstyle\bigwedge^n T^*\mathcal{X}, \bigwedge^n T^*\mathcal{X}\big), \\ &\cong \left(W_0^* \otimes \textstyle\bigwedge^n T^*\mathcal{X}\right)^* \otimes \bigwedge^n T^*\mathcal{X}, \\ &\cong W_0 \otimes \textstyle\bigwedge^n T\mathcal{X} \otimes \bigwedge^n T^*\mathcal{X}, \\ &\cong W_0. \end{aligned} \tag{17.42}$$

One concludes that smooth functionals in $C_c^r(W_0^* \otimes \bigwedge^n T^*\mathcal{X})^*$ are represented by sections of W_0. The action of the functional T_{χ_0}, induced by a section χ_0 of W_0, on a compactly supported section S of W_0' is given by

$$T_{\chi_0}(S) = \int_{\mathcal{X}} S \cdot \chi_0. \tag{17.43}$$

In the case where $\mathcal{X}$ is a manifold without boundary, the elements of

$$C^{-r}(W_0) := C_c^r(W_0')^* \tag{17.44}$$

are referred to as *generalized sections* of W_0 (see [AB67, GS77, GKOS01], [Kor08, p. 676]).

In view of the properties of functionals on compact manifolds with corners, as considered in the preceding section, in the case where the base manifold $\mathcal{X}$ is a compact manifold with corners, functionals in $C^r(W_0')^*$ are elements of $C^{-r}(\tilde{W}_0) := C_c^r(\tilde{W}_0')^*$ that are supported in $\mathcal{X}$ (where we recall that $\tilde{\pi} : \tilde{W} \to \tilde{\mathcal{X}}$ denotes the extension of the vector bundle to the manifold without boundary $\tilde{\mathcal{X}} \supset \mathcal{X}$).

In the particular case where $W_0 = \mathcal{X} \times \mathbb{R}$ is the natural line bundle, smooth functionals are represented by real-valued functions on $\mathcal{X}$. Consequently, elements of

$$C_c^r\left(\mathbb{R} \otimes \textstyle\bigwedge^n T^*\mathcal{X}\right)^* = C_c^r\left(\textstyle\bigwedge^n T^*\mathcal{X}\right)^* \tag{17.45}$$

are referred to as *generalized functions*.

The apparent complication in the definition of generalized sections using density dual bundles is justified in the sense that each element in $C^{-r}(W_0)$ may be approximated by a sequence of smooth functionals induced by sections of W_0. (See [GKOS01, p. 241].)

In the literature, the term *section distribution* is used in different ways in this context. For example, in [GKOS01] and [Kor08, p. 676], W_0-valued distributions are defined as elements of $C^r(W_0')^*$, i.e., what are referred to here as generalized sections of W_0. (In [AB67, AS68] they are referred to as *distributional sections.*) On the other hand, in [GS77], distributions are defined as generalized sections of $\bigwedge^n T^*\mathcal{X}$—elements of $C^r(\mathcal{X})^*$. See further comments on this issue and the corresponding terms *section distributional densities* and *generalized densities* in [GS77, GKOS01, Kor08].

Chapter 18
de Rham Currents

Using the terminology introduced in the previous chapter, for a manifold without boundary $\mathcal{X}$, de Rham currents (see [dR84b, Sch73, Fed69a]) are functionals corresponding to the case of the vector bundle

$$\pi : \bigwedge^p T^*\mathcal{X} \longrightarrow \mathcal{X}, \tag{18.1}$$

so that test sections are smooth p-forms over $\mathcal{X}$ having compact supports. Thus, a *p-current of order r* on $\mathcal{X}$ is a continuous linear functional acting on elements of $C^r_c(\bigwedge^p T^*\mathcal{X})$.

A particular type of p-currents, smooth currents, are induced by differential $(n-p)$-forms. Such an $(n-p)$ form, ϕ, induces the currents ${}_\phi T$ and $T_\phi = (-1)^{p(n-p)}{}_\phi T$ by

$${}_\phi T(\psi) = \int_{\mathcal{X}} \phi \wedge \psi, \qquad T_\phi(\psi) = \int_{\mathcal{X}} \psi \wedge \phi. \tag{18.2}$$

In case a current, T, is in one of the forms above, we will say that T is a *smooth current represented by the form ϕ*.

The current ${}_\phi T$ may be interpreted geometrically as follows. Let ϕ_0 be a decomposable p-covector at a point $X \in \mathcal{X}$. It follows that one may choose a basis $\{e_i\}$, $i = 1, \dots, n$, of $T_X\mathcal{X}$ with dual basis $\{\phi^i\}$ such that $\phi_0 = \phi^1 \wedge \cdots \wedge \phi^p$. Let ψ_0 be an $(n-p)$-covector such that $\phi \wedge \psi_0 \neq 0$. Then, ψ_0 must be of the form $\psi_0 = a\phi^{p+1} \wedge \cdots \wedge \phi^n + \alpha$, with $\phi_0 \wedge \alpha = 0$, for some nonvanishing number a. Since $e_i \lrcorner \phi_0$ vanishes for $i = p+1, \dots, n$, ϕ_0 is annihilated by vectors in the subspace spanned by $\{e_{p+1}, \dots, e_n\}$. Let $\{v_1, \dots, v_n\}$ be n vectors in $T_X\mathcal{X}$ satisfying $\phi_0 \wedge \psi_0(v_1, \dots, v_n) = \phi_0 \wedge \psi_0(v_1 \wedge \cdots \wedge v_n) \neq 0$. Then, $v_1 \wedge \cdots \wedge v_n$ must be of the form

$$v_1 \wedge \cdots \wedge v_n = be_1 \wedge \cdots \wedge e_n, \tag{18.3}$$

R. Segev, *Foundations of Geometric Continuum Mechanics*, Advances in Continuum Mechanics 49, https://doi.org/10.1007/978-3-031-35655-1_18

for some real number b. The fact that the two forms annihilate vectors in the respective subspaces implies that

$$\phi \wedge \psi_0(v_1 \wedge \cdots \wedge v_n) = \phi(e_1, \ldots, e_p)\psi_0(be_{p+1} \wedge \cdots \wedge e_n) = ab. \tag{18.4}$$

This quantity, as well as the identical $((v_1 \wedge \cdots \wedge v_n) \llcorner \phi)(\psi_0)$, is interpreted as the amount of cells formed by the hyperplanes induced by the forms ϕ and ψ_0 contained in the n-parallelepiped determined by $v_1, \ldots, v_n$.

Accordingly, for a p-form ϕ and an $(n-p)$-form ψ, one may interpret ${}_\phi T(\psi)$ as the total amount of cells in $\mathcal{X}$.

Another simple p-current, $T_{\mathfrak{X}}$, is induced by an oriented p-dimensional submanifold $\mathfrak{X} \subset \mathcal{X}$. It is naturally defined by

$$T_{\mathfrak{X}}(\psi) = \int_{\mathfrak{X}} \psi. \tag{18.5}$$

These two examples illustrate the two points of view on currents. On the one hand, the example of the current ${}_\omega T$ suggests that a current in $C^r_c(\bigwedge^p T^*\mathcal{X})^*$ is viewed as a generalized $(n-p)$-form. With this point of view in mind, elements of $C^r_c(\bigwedge^p T^*\mathcal{X})$ are referred to as *currents of degree* $n-p$. In accordance with the notation in (17.44), and in view of the isomorphism (6.126) (see also (6.108)), the space of p-currents on $\mathcal{X}$ may be denoted by

$$C^{-r}(\textstyle\bigwedge^{n-p} T^*\mathcal{X}) = C^r_c(\bigwedge^p T^*\mathcal{X})^*. \tag{18.6}$$

Evidently, the notation is also suggested by the example of currents induced by differential forms as above.

On the other hand, the example of the current $T_{\mathfrak{X}}$ induced by a p-dimensional manifold $\mathfrak{X}$ suggests that currents be viewed as a geometric object of dimension p. Thus, an element of $C^r_c(\bigwedge^p T^*\mathcal{X})^*$ is referred to as a *p-dimensional current*.

18.1 Basic Operations with Currents

The contraction operations of a $(p+q)$-current T and a q-form α yield the p-currents $T \llcorner \alpha$ and $\alpha \lrcorner T$ defined by

$$(T \llcorner \alpha)(\psi) = T(\psi \wedge \alpha) \qquad \text{and} \qquad (\alpha \lrcorner T)(\psi) = T(\alpha \wedge \psi), \tag{18.7}$$

so that

$$T \llcorner \alpha = (-1)^{pq} \alpha \lrcorner T. \tag{18.8}$$

In particular, if a p-current ${}_{\omega}T$ is induced by an $(n-p)$-form ω as above, and α is a q-form with $q \leqslant p$, then, $\alpha \lrcorner {}_{\omega}T$ is the $(p-q)$-current given by

$$(\alpha \lrcorner {}_{\omega}T)(\psi) = \int_{\mathcal{X}} \omega \wedge \alpha \wedge \psi = {}_{\omega\wedge\alpha}T(\psi); \tag{18.9}$$

that is,

$$\alpha \lrcorner {}_{\omega}T = {}_{\omega\wedge\alpha}T. \tag{18.10}$$

Note that our notation is different from that of [dR84b] and different in sign from that of [Fed69a]. In particular, given a p-current T, any p-form ψ induces naturally a zero-current

$$T \cdot \psi := T \llcorner \psi, \qquad \text{so that} \qquad (T \cdot \psi)(u) := T(u\psi). \tag{18.11}$$

The p-currents T_{ω} and ${}_{\omega}T$ defined above can be expressed using contraction in the form

$$T_{\omega} = T_{\mathcal{X}} \llcorner \omega, \qquad {}_{\omega}T = \omega \lrcorner T_{\mathcal{X}}. \tag{18.12}$$

For a p-current T and a q-multi-vector field ξ, the $(p+q)$-currents $\xi \wedge T$ and $T \wedge \xi$ are defined by

$$(\xi \wedge T)(\psi) := T(\xi \lrcorner \psi), \qquad (T \wedge \xi)(\psi) := T(\psi \llcorner \xi), \tag{18.13}$$

for any $(p+q)$-form ψ. Using, $\xi \lrcorner \psi = (-1)^{qp} \psi \llcorner \xi$, one has

$$\xi \wedge T = (-1)^{pq} T \wedge \xi, \tag{18.14}$$

in analogy with the corresponding expression for multi-vectors.

Note that a real-valued function u defined on $\mathcal{X}$ may be viewed both as a zero-form and as a zero-multi-vector. Hence, we may write uT for any of the four operations defined above so that $(uT)(\psi) = T(u\psi)$.

The *boundary* operator

$$\partial : C^{-r}(\textstyle\bigwedge^{n-p} T^*\mathcal{X}) \longrightarrow C^{-(r+1)}(\textstyle\bigwedge^{n-(p-1)} T^*\mathcal{X}), \tag{18.15}$$

defined by

$$\partial T(\psi) = T(\mathrm{d}\psi), \tag{18.16}$$

is a linear and continuous operator. In other words, the boundary of a p-current of degree r is a $(p-1)$-current of degree $r+1$. In particular, for a smooth current, ${}_{\omega}T$, represented by the $(n-p)$-form ω, one has

$$\begin{aligned}\partial_{\omega}T(\psi) &= \int_{\mathcal{X}} \omega \wedge \mathrm{d}\psi, \\ &= (-1)^{n-p} \int_{\mathcal{X}} \mathrm{d}(\omega \wedge \psi) - (-1)^{n-p} \int_{\mathcal{X}} \mathrm{d}\omega \wedge \psi, \\ &= (-1)^{n-p+1} T_{\mathrm{d}\omega}(\psi).\end{aligned} \tag{18.17}$$

Here, to arrive at the last line, we used the Stokes theorem and the fact that ψ has a compact support in $\mathcal{X}$. It is concluded that

$$\partial_{\omega}T = (-1)^{n-p+1} T_{\mathrm{d}\omega}. \tag{18.18}$$

Similarly,

$$\partial T_{\omega} = (-1)^{p+1} T_{\mathrm{d}\omega}. \tag{18.19}$$

In order to strengthen further the point of view that a p-current is a generalized $(n-p)$-form, the *exterior derivative of a p-current* $\mathrm{d}T$ is defined by

$$\mathrm{d}T = (-1)^{n-p+1} \partial T. \tag{18.20}$$

Thus, in the smooth case,

$$\mathrm{d}_{\omega}T = T_{\mathrm{d}\omega}. \tag{18.21}$$

In addition, Stokes's theorem implies that for the p-current $T_{\mathfrak{T}}$, induced by the p-dimensional submanifold with boundary $\mathfrak{T}$, the boundary, a $(p-1)$-current, is given by

$$\partial T_{\mathfrak{T}} = T_{\partial \mathfrak{T}}. \tag{18.22}$$

It is quite evident, therefore, that the notion of a boundary generalizes and unites both the exterior derivative of forms and the boundaries of manifolds.

The boundary of a current satisfies the identity

$$\partial^2 := \partial \circ \partial = 0. \tag{18.23}$$

Evidently,

$$(\partial(\partial T))(\psi) = (\partial T)(\mathrm{d}\psi) = T(\mathrm{d}^2\psi) = 0. \tag{18.24}$$

The constancy theorem for currents (see de Rham [dR84b, p. 80]) asserts that on a connected n-dimensional manifold, $\mathcal{X}$, a *closed* n-current T—that is, a current satisfying $\partial T = 0$—is represented by a constant c in the form

$$T(\psi) = c \int_{\mathcal{X}} \psi. \tag{18.25}$$

18.2 Local Representation of Currents

We consider next the local representation of de Rham currents that are defined in coordinate neighborhoods.

18.2.1 Representation by 0-Currents

Let $R = E^*_{U_a} T$ be the restriction of a p-current T to forms supported in a particular coordinate neighborhood—a local representative of T. Using the summation convention for the increasing multi-index λ, we may write

$$\begin{aligned} R(\psi) &= R(\psi_\lambda \mathrm{d}x^\lambda), \qquad |\lambda| = p, \\ &= (\mathrm{d}x^\lambda \lrcorner R)(\psi_\lambda). \end{aligned} \tag{18.26}$$

(We could have used $R \llcorner \mathrm{d}x^\lambda$ just the same as the ψ_λ are real-valued functions.) Hence, one notes that locally

$$R(\psi) = R^\lambda(\psi_\lambda), \qquad \text{where,} \qquad R^\lambda := \mathrm{d}x^\lambda \lrcorner R \tag{18.27}$$

is a 0-current. Using the exterior product of a multi-vector field ξ and a current in (18.13), we have

$$R(\psi) = R^\lambda(\partial_\lambda \lrcorner \psi) = \partial_\lambda \wedge R^\lambda(\psi), \tag{18.28}$$

and so a current may be represented locally in the form

$$R = \partial_\lambda \wedge R^\lambda = R^\lambda \wedge \partial_\lambda, \tag{18.29}$$

where the second equality follows from the fact that R^λ is a zero-current.

This representation suggests that T be interpreted as a generalized multi-vector field (cf. [Whi57]).

In the sequel, when we refer to a local representative of a current T, we will often keep the same notation, T, and it will be implied that we consider the restriction of T to forms (or sections, in general) supported in a generic coordinate neighborhood.

18.2.2 Representation by n-Currents

Alternatively (cf. [dR84b, p. 36]), for a p-current R defined in a coordinate neighborhood, and an increasing multi-index $\hat{\lambda}$ with $|\hat{\lambda}| = n - p$, consider the n-currents

$$R_{\hat{\lambda}} := \partial_{\hat{\lambda}} \wedge R, \quad \text{so that} \quad R_{\hat{\lambda}}(\theta) = R(\partial_{\hat{\lambda}} \lrcorner \theta). \tag{18.30}$$

Then, for every p-form ω,

$$\begin{aligned}(\mathrm{d}x^{\hat{\lambda}} \lrcorner R_{\hat{\lambda}})(\omega) &= R_{\hat{\lambda}}(\mathrm{d}x^{\hat{\lambda}} \wedge \omega), \\ &= R(\partial_{\hat{\lambda}} \lrcorner (\mathrm{d}x^{\hat{\lambda}} \wedge \omega)), \\ &= R_{\hat{\lambda}}(\varepsilon^{\hat{\lambda}\mu} \omega_\mu \mathrm{d}x),\end{aligned} \tag{18.31}$$

where $|\mu| = p$, and we used

$$\mathrm{d}x^{\hat{\lambda}} \wedge \omega = \varepsilon^{\hat{\lambda}\mu} \omega_\mu \mathrm{d}x. \tag{18.32}$$

Also,

$$\begin{aligned}(\partial_{\hat{\lambda}} \lrcorner \mathrm{d}x)(\partial_\mu) &= \mathrm{d}x(\partial_{\hat{\lambda}} \wedge \partial_\mu), \\ &= \varepsilon_{\hat{\lambda}\mu}, \\ &= \varepsilon_{\hat{\lambda}\nu} \mathrm{d}x^\nu(\partial_\mu),\end{aligned} \tag{18.33}$$

$|\nu| = p$, implies

$$\partial_{\hat{\lambda}} \lrcorner \mathrm{d}x = \varepsilon_{\hat{\lambda}\nu} \mathrm{d}x^\nu, \tag{18.34}$$

and so

$$\partial_{\hat{\lambda}} \lrcorner (\mathrm{d}x^{\hat{\lambda}} \wedge \omega) = \omega_\lambda \mathrm{d}x^\lambda = \omega, \tag{18.35}$$

as expected. Hence,

$$(\mathrm{d}x^{\hat{\lambda}} \lrcorner R_{\hat{\lambda}})(\omega) = R(\omega). \tag{18.36}$$

We conclude that R may be represented by the n-currents

$$R_{\hat{\lambda}} := \partial_{\hat{\lambda}} \wedge R, \qquad \text{in the form} \qquad R = \mathrm{d}x^{\hat{\lambda}} \lrcorner R_{\hat{\lambda}}, \tag{18.37}$$

with

$$R(\omega) = R_{\hat{\lambda}}(\mathrm{d}x^{\hat{\lambda}} \wedge \omega) = \varepsilon^{\hat{\lambda}\mu} R_{\hat{\lambda}}(\omega_\mu \mathrm{d}x). \tag{18.38}$$

(It is recalled that in the last expression, summation is implied, where λ and $\hat{\lambda}$ are considered as distinct indices.) This representation suggests again that a p-current T be interpreted as a generalized $(n - p)$-form. In particular, an n-current is a generalized function and is often referred to as a distribution on the manifold (e.g., [Mel96, Chapter 3]).

Remark 18.1 It is noted that one may set alternatively

$$R'_{\hat{\lambda}} := R \wedge \partial_{\hat{\lambda}}. \tag{18.39}$$

Using (18.14) for the $(n-p)$-multi-vector $\partial_{\hat{\lambda}}$

$$R'_{\hat{\lambda}} = (-1)^{p(n-p)} R_{\hat{\lambda}}. \tag{18.40}$$

In addition, by (18.37) and (18.8), for the n-current $R_{\hat{\lambda}}$ and the $(n-p)$-form $\mathrm{d}x^{\hat{\lambda}}$,

$$R = R'_{\hat{\lambda}} \llcorner \mathrm{d}x^{\hat{\lambda}}, \tag{18.41}$$

and

$$R(\omega) = R'_{\hat{\lambda}}(\omega \wedge \mathrm{d}x^{\hat{\lambda}}) = \sum_{\lambda} \varepsilon^{\lambda\hat{\lambda}} R'_{\hat{\lambda}}(\omega_{\lambda}\mathrm{d}x). \tag{18.42}$$

Remark 18.2 The relation between the local representation of an r-current by 0-currents and the representation by n-currents may be obtained as follows. For a real-valued function u, compactly supported in the domain of a chart,

$$\begin{aligned} R^{\lambda}(u) &= (R \llcorner \mathrm{d}x^{\lambda})(u), \qquad |\lambda| = r, \\ &= R(u\mathrm{d}x^{\lambda}), \\ &= R'_{\hat{\mu}} \llcorner \mathrm{d}x^{\hat{\mu}}(u\mathrm{d}x^{\lambda}), \qquad |\mu| = r, \\ &= R'_{\hat{\mu}}(u\mathrm{d}x^{\lambda} \wedge \mathrm{d}x^{\hat{\mu}}), \\ &= \varepsilon^{\lambda\hat{\mu}} R'_{\hat{\mu}}(u\mathrm{d}x), \\ &= \varepsilon^{\lambda\hat{\mu}} (R'_{\hat{\mu}} \llcorner \mathrm{d}x)(u). \end{aligned} \tag{18.43}$$

Hence,

$$R^{\lambda} = \varepsilon^{\lambda\hat{\mu}} R'_{\hat{\mu}} \llcorner \mathrm{d}x. \tag{18.44}$$

In addition, since we have

$$\begin{aligned} R^{\lambda}(u) &= R^{\lambda}(u\mathrm{d}x \llcorner \partial_x), \\ &= (R^{\lambda} \wedge \partial_x)(u\mathrm{d}x), \end{aligned} \tag{18.45}$$

it follows from (18.43) that

$$R'_{\hat{\lambda}} = \varepsilon_{\lambda\hat{\lambda}} R^{\lambda} \wedge \partial_x. \tag{18.46}$$

18.2.3 Representation of the Boundary of a Current

The representation of a p-current by n-currents in Sect. 18.2.2 makes it possible to represent the boundary by $(n-1)$-currents as follows. Using (18.38), for a current R, defined in a coordinate neighborhood, the definition of ∂R implies that for any $(p-1)$-form, ω, compactly supported in the coordinate neighborhood, one has

$$\begin{aligned}
\partial R(\omega) &= R(\mathrm{d}\omega) = R_{\hat{\lambda}}(\mathrm{d}x^{\hat{\lambda}} \wedge \mathrm{d}\omega), \quad |\lambda| = p,\ |\hat{\lambda}| = n-p, \\
&= (-1)^{n-p} R_{\hat{\lambda}}(\mathrm{d}(\mathrm{d}x^{\hat{\lambda}} \wedge \omega)), \\
&= (-1)^{n-p} \partial R_{\hat{\lambda}}(\mathrm{d}x^{\hat{\lambda}} \wedge \omega), \\
&= (-1)^{n-p} (\mathrm{d}x^{\hat{\lambda}} \lrcorner\, \partial R_{\hat{\lambda}})(\omega).
\end{aligned} \tag{18.47}$$

It is concluded that ∂R is given in terms of the boundaries of the representing n-currents, $R_{\hat{\lambda}}$, as

$$\partial R = (-1)^{n-p} \mathrm{d}x^{\hat{\lambda}} \lrcorner\, \partial R_{\hat{\lambda}}. \tag{18.48}$$

Alternatively, using the notation of Remark 18.1, for a $(p-1)$-form, ω,

$$\begin{aligned}
\partial R(\omega) &= R(\mathrm{d}\omega) = R'_{\hat{\lambda}}(\mathrm{d}\omega \wedge \mathrm{d}x^{\hat{\lambda}}), \quad |\lambda| = p,\ |\hat{\lambda}| = n-p, \\
&= R'_{\hat{\lambda}}(\mathrm{d}(\omega \wedge \mathrm{d}x^{\hat{\lambda}})), \\
&= \partial R'_{\hat{\lambda}}(\omega \wedge \mathrm{d}x^{\hat{\lambda}}), \\
&= \partial R'_{\hat{\lambda}} \llcorner \mathrm{d}x^{\hat{\lambda}}(\omega).
\end{aligned} \tag{18.49}$$

Hence, ∂R is also represented by the boundaries, $\partial R'_{\hat{\lambda}}$, of the representing n-currents as

$$\partial R = \partial R'_{\hat{\lambda}} \llcorner \mathrm{d}x^{\hat{\lambda}}. \tag{18.50}$$

Just as any other current, the boundary, ∂R, of an r-current, R, may also be represented by the n-currents $(\partial R)_{\hat{\lambda}}$, or $(\partial R)'_{\hat{\lambda}}$, $|\hat{\lambda}| = r-1$, in the form

$$\partial R = (\partial R)'_{\hat{\lambda}} \llcorner \mathrm{d}x^{\hat{\lambda}} = \mathrm{d}x^{\hat{\lambda}} \lrcorner (\partial R)_{\hat{\lambda}}. \tag{18.51}$$

For example, given a real-valued differentiable function, u, which is compactly supported locally,

$$\begin{aligned}
(\partial R)'_{\hat{\lambda}}(u\mathrm{d}x) &= (\partial R \wedge \partial_{\hat{\lambda}})(u\mathrm{d}x), \\
&= (\partial R)(u\mathrm{d}x \llcorner \partial_{\hat{\lambda}}), \\
&= (\partial R)(u\varepsilon_{\mu\hat{\lambda}}\mathrm{d}x^{\mu}), \quad |\mu| = r-1, \\
&= R(\varepsilon_{\mu\hat{\lambda}}u_{,i}\mathrm{d}x^{i} \wedge \mathrm{d}x^{\mu}), \\
&= R(\varepsilon_{\mu\hat{\lambda}}\varepsilon^{i\mu}_{\nu}u_{,i}\mathrm{d}x^{\nu}), \quad |\nu| = r, \\
&= \varepsilon_{\mu\hat{\lambda}}\varepsilon^{i\mu}_{\nu}(R \llcorner \mathrm{d}x^{\nu})(u_{,i}), \\
&= \varepsilon_{\mu\hat{\lambda}}\varepsilon^{i\mu}_{\nu}\partial_i(R \llcorner \mathrm{d}x^{\nu})(u),
\end{aligned} \tag{18.52}$$

where, for a 0-current T in $\mathbb{R}^n$, the 0-current $\partial_i T$ is defined by

$$\partial_i T(u) = T(u_{,i}). \tag{18.53}$$

In addition, one can write the representation of the boundary of an r-current in terms of the 0-currents, R^{λ}, representing it as in Sect. 18.2.1. For any $(r-1)$-form, ψ, compactly supported in the domain of a chart,

$$\begin{aligned}
\partial R(\psi) &= R(\mathrm{d}\psi), \\
&= R(\mathrm{d}(\psi_{\mu}\mathrm{d}x^{\mu})), \quad |\mu| = r-1, \\
&= R(\mathrm{d}\psi_{\mu} \wedge \mathrm{d}x^{\mu}), \\
&= (R \llcorner \mathrm{d}x^{\mu})(\mathrm{d}\psi_{\mu}), \\
&= \partial(R \llcorner \mathrm{d}x^{\mu})(\psi(\partial_{\mu})), \\
&= \{[\partial(R \llcorner \mathrm{d}x^{\mu})] \wedge \partial_{\mu}\}(\psi).
\end{aligned} \tag{18.54}$$

It follows that

$$\partial R = [\partial(R \llcorner \mathrm{d}x^{\mu})] \wedge \partial_{\mu}. \tag{18.55}$$

Comparing the last equation with (18.29), it is concluded that

$$(\partial R)^{\mu} = \partial(R \llcorner \mathrm{d}x^{\mu}) \tag{18.56}$$

are the 0-currents representing ∂R.

Chapter 19
Interlude: Singular Distributions of Defects in Bodies

The theory of de Rham currents enables an extension of the geometric theory of smooth distribution of defects, as in Chap. 7, to cases where the distributions are singular. As a simple example of such a singular distribution, one can think of a single edge dislocation in a continuous body. The proposed theory of singular defects and some examples are given in this chapter.

19.1 Layering Currents and Defect Currents

We recall that the basic object in the geometric modeling of a smooth distribution of defects of layers of dimension $n - p$ in an n-dimensional body $\mathcal{X}$ is a p-layering form ϕ. For a $(p + 1)$-dimensional submanifold with boundary $\mathcal{S} \subset \mathcal{X}$, the total production (or destruction) of layers within $\mathcal{S}$ is defined to be

$$\Phi := \int_{\partial \mathcal{S}} \phi = \int_{\mathcal{S}} \mathrm{d}\phi. \tag{19.1}$$

Thus, the $(p + 1)$-form $\mathrm{d}\phi$—the defect form—describes the sources (or sinks) of layers within $\mathcal{X}$. In particular, for the case $p = 1$, $\mathrm{d}\phi = 0$ implies that locally, the $(n - 1)$-dimensional layers, generalizing the Bravais planes, may be parametrized by a real-valued function u such that $\phi = \mathrm{d}u$.

Using de Rham currents, the natural generalization of the notion of a p-layering form is an $(n - p)$-*layering current*, or a *structure current*. As mentioned in the preceding chapter, a p-layering form, ϕ, induces an $(n - p)$-layering current, ${}_{\phi}T$, by

$${}_{\phi}T(\psi) = \int_{\mathcal{X}} \phi \wedge \psi. \tag{19.2}$$

R. Segev, *Foundations of Geometric Continuum Mechanics*, Advances in Continuum Mechanics 49, https://doi.org/10.1007/978-3-031-35655-1_19

As another example, for an $(n-p)$-dimensional differentiable submanifold $\mathcal{S} \subset \mathcal{X}$, the induced the current $T_{\mathcal{S}}$, defined by

$$T_{\mathcal{S}}(\psi) := \int_{\mathcal{S}} \psi, \tag{19.3}$$

models a single material layer which is identified with the submanifold $\mathcal{S}$.

We recall that the boundary of a current generalizes both the exterior derivative of a form and the boundary of a manifold. Since the exterior derivative of a layering form represents the defects in the layering, and the boundary of a single layer, given in terms of a submanifold $\mathcal{S}$ in $\mathcal{X}$, represents the defect associated with the layer, it is natural to define the *defect current* corresponding to the layering current T to be ∂T.

Let $R = \partial T$ be the defect current associated with the structure current T. Then, Eq. (18.23) implies immediately that

$$\partial R = \partial^2 T = 0. \tag{19.4}$$

The last condition is the generalization of the Frank rule for dislocations to the continuum singular case.

Some examples of currents modeling singular dislocations and disclinations are described below. For additional examples and discussions, see [ES15, ES20].

19.2 Dislocations

Dislocation defects are associated with the case where $p = 1$, so that we consider $(n-1)$-currents. The following examples demonstrate the application of the general framework of de Rham currents to the geometric description of singular distributions of dislocations.

19.2.1 Edge Dislocations

Assume that $\mathcal{X}$ is an n-dimensional manifold without boundary, and let $\mathcal{S}$ be an $(n-1)$-submanifold with boundary of $\mathcal{X}$. We consider the $(n-1)$-structure current $R_{\mathcal{S}}$, given by

$$R_{\mathcal{S}}(\psi) = \int_{\mathcal{S}} \psi, \tag{19.5}$$

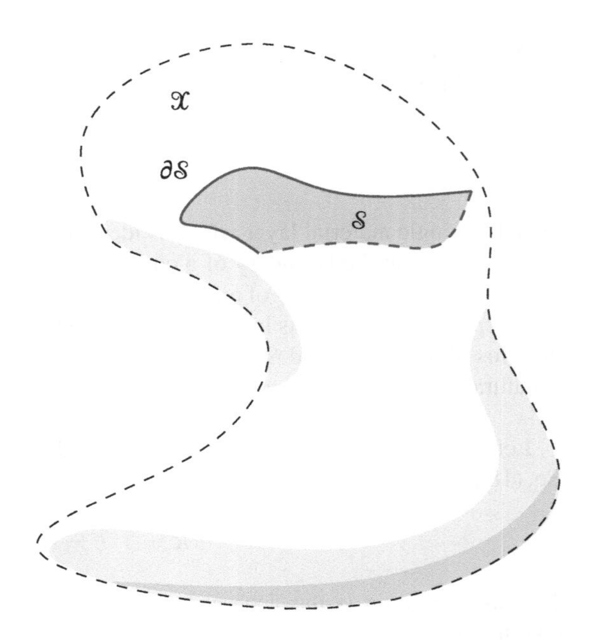

Fig. 19.1 Illustrating an edge dislocation

in analogy with (18.5). Then, as shown in (18.22), the dislocation $(n-2)$-current is given by $R_{\partial\mathcal{S}}$ (see Fig. 19.1, where the solid blue line illustrates $\partial\mathcal{S}$). As expected, $\partial(\partial R_{\mathcal{S}}) = 0$ since it acts on forms with compact support in $\mathcal{X}$.

As a concrete example, consider the case where $\mathcal{X}$ is an oriented manifold without boundary that may be covered by a single chart. Let X^i be coordinates on $\mathcal{X}$ such that their order agrees with the orientation of $\mathcal{X}$. Without a loss of generality, we may assume that for some point $X_0 \in \mathcal{X}$, the coordinates $X_0^i = 0$, for all $i = 1, \ldots, n$. Let

$$\mathcal{S} = \{X \in \mathcal{X} \mid X^1 = 0,\ X^2 \leqslant 0\} \tag{19.6}$$

equipped with the orientation induced by the form $\mathrm{d}X^2 \wedge \mathrm{d}X^3 \wedge \cdots \wedge \mathrm{d}X^n$ (see Fig. 19.2). The current $R_{\mathcal{S}}$ represents an added or a removed "half hypersurface." Then, $\partial R_{\mathcal{S}} = R_{\partial\mathcal{S}}$, where $\partial\mathcal{S} = \{X \in \mathcal{X} \mid X^1 = 0,\ X^2 = 0\}$, oriented naturally by the form $\mathrm{d}X^3 \wedge \cdots \wedge \mathrm{d}X^n$, is the singular dislocation submanifold. As expected, for the case $n = 3$, the dislocation submanifold is the X^3-curve.

19.2.2 Screw Dislocations

Let $L \subset \mathbb{R}^3$ be given by $L = \{(0, 0)\} \times \mathbb{R} = \{(0, 0, Z) \mid Z \in \mathbb{R}\}$, and let $D \subset \mathbb{R}^3$ be given by $D = \mathbb{R}^3 \setminus L = \{(X, Y, Z) \in \mathbb{R}^3 \mid (X, Y) \neq (0, 0)\}$. It is noted that on D we may use a cylindrical coordinate system (r, θ, Z), where for simplicity, we

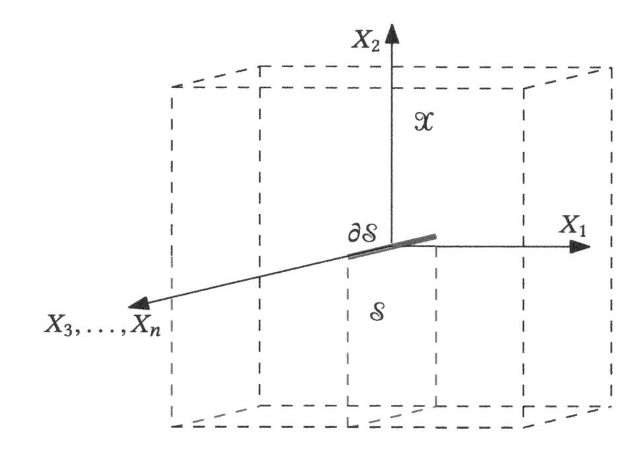

Fig. 19.2 Illustrating local representation of an edge dislocation

take the domain $[0, 2\pi)$ for θ without using a proper atlas on the unit circle (see Fig. 19.3).

Consider the layering 1-form ϕ on D defined by

$$\phi = -\frac{b}{2\pi}\mathrm{d}\theta + \mathrm{d}Z. \tag{19.7}$$

Evidently, as its components are constants, ϕ is a closed form. It thus follows from Poincare's lemma that locally ϕ is exact. Since D is not contractible to a point, ϕ is not exact globally. In fact, in the open set $D \setminus \{(r, \theta, Z) \mid \theta = 0\}$, $\phi = \mathrm{d}F$ for the real-valued

$$F(r, \theta, Z) = -\frac{b\theta}{2\pi} + Z, \tag{19.8}$$

the level sets

$$Z = \frac{b\theta}{2\pi} + C, \qquad C \in \mathbb{R} \tag{19.9}$$

of which describe spiraling screw threads of pitch b (see Fig. 19.3).

For any $r > 0$, let $S_{r,l} = \{(X, Y, Z) \in \mathbb{R}^3 \mid X^2 + Y^2 = r^2,\ Z = l\}$ be the circle of radius r situated at $Z = l$, and let $\mathcal{J} : S_{r,l} \to D$ be the inclusion. Then, for example,

$$\begin{aligned} \int_{S_{r,l}} \phi &= \int_{S_{r,l}} \mathcal{J}^*(\phi), \\ &= \int_{S_{r,l}} -\frac{b}{2\pi}\mathrm{d}\theta, \\ &= -b. \end{aligned} \tag{19.10}$$

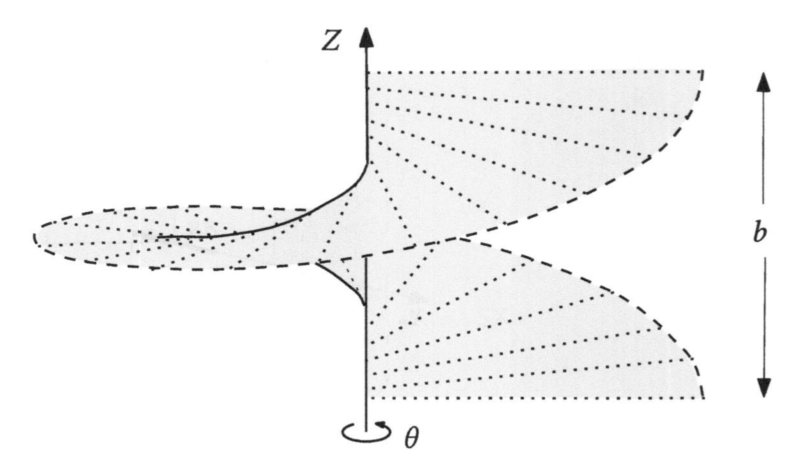

Fig. 19.3 Illustrating a screw dislocation

(It is observed that $\mathcal{J}^*(\phi)(\partial_\theta) = \phi(\mathcal{J}_*(\partial_\theta)) = \phi(\partial_\theta) = -b/2\pi$.)

We now consider the layering 2-current ${}_\phi R$ in $\mathbb{R}^3$, given by

$$ {}_\phi R(\omega) = \int_D \phi \wedge \omega \tag{19.11}$$

for any 2-form ω on $\mathbb{R}^3$ with compact support. (Note that while ${}_\phi R$ is defined for forms with compact support in $\mathbb{R}^3$, the integration is taken over D, only.) In order to determine the associated geometry of the dislocation, we examine the defect 1-current, the boundary, $\partial_\phi R$. For any 1-form α, we have

$$\begin{aligned} \partial_\phi R(\alpha) &= {}_\phi R(\mathrm{d}\alpha), \\ &= \int_D \phi \wedge \mathrm{d}\alpha, \\ &= -\int_D \mathrm{d}(\phi \wedge \alpha) + \int_D \mathrm{d}\phi \wedge \alpha. \end{aligned} \tag{19.12}$$

Since $\mathrm{d}\phi = 0$ in D, we conclude that

$$\partial_\phi R(\alpha) = -\int_D \mathrm{d}(\phi \wedge \alpha). \tag{19.13}$$

Let $C_\varepsilon = \{(X, Y, Z,) \in \mathbb{R}^3 \mid X^2 + Y^2 < \varepsilon^2\}$ be an open cylinder of radius ε about the Z-axis, and let $D_\varepsilon = \mathbb{R}^3 \setminus C_\varepsilon$. We may write

$$\partial_\phi R(\alpha) = -\int_D \mathrm{d}(\phi \wedge \alpha) = -\lim_{\varepsilon\to 0}\int_{D_\varepsilon} \mathrm{d}(\phi \wedge \alpha). \tag{19.14}$$

It is noted that D_ε is a manifold with a boundary. In fact, setting $S_\varepsilon = \{(X, Y) \in \mathbb{R}^2 \mid X^2 + Y^2 = \varepsilon^2\}$, $\partial D_\varepsilon = S_\varepsilon \times \mathbb{R}$. We may therefore use Stokes's theorem in (19.14) and obtain

$$\partial_\phi R(\alpha) = -\lim_{\varepsilon\to 0}\int_{\partial D_\varepsilon} \mathcal{J}^*(\phi \wedge \alpha), \tag{19.15}$$

where $\mathcal{J}^*(\phi \wedge \alpha)$ is the pullback under the inclusion $\mathcal{J} : \partial D_\varepsilon \to D_\varepsilon$ (which is simply the restriction of $\phi \wedge \alpha$ to vectors tangent to ∂D_ε).

A 1-form α is represented by $\alpha = \alpha_X \mathrm{d}X + \alpha_Y \mathrm{d}Y + \alpha_Z \mathrm{d}Z$ for the smooth functions α_X, α_Y, and α_Z defined on $\mathbb{R}^3$. In D, the form α may also be represented using cylindrical coordinates as $\alpha = \alpha_r \mathrm{d}r + \alpha_\theta \mathrm{d}\theta + \alpha_Z \mathrm{d}Z$. Since $\alpha_X \mathrm{d}X + \alpha_Y \mathrm{d}Y = \alpha_r \mathrm{d}r + \alpha_\theta \mathrm{d}\theta$, using $X = r\cos\theta$, $Y = r\sin\theta$ and

$$\mathrm{d}X = \frac{\partial X}{\partial r}\mathrm{d}r + \frac{\partial X}{\partial \theta}\mathrm{d}\theta, \quad \mathrm{d}Y = \frac{\partial Y}{\partial r}\mathrm{d}r + \frac{\partial Y}{\partial \theta}\mathrm{d}\theta, \tag{19.16}$$

one has

$$\alpha_\theta = r(-\alpha_X \sin\theta + \alpha_Y \cos\theta). \tag{19.17}$$

The restriction to ∂D_ε satisfies

$$\mathcal{J}^*(\phi \wedge \alpha) = (\phi_\theta \alpha_Z - \phi_Z \alpha_\theta)\mathrm{d}\theta \wedge \mathrm{d}Z = \left(-\frac{b}{2\pi}\alpha_Z - \alpha_\theta\right)\mathrm{d}\theta \wedge \mathrm{d}Z, \tag{19.18}$$

and it follows that

$$\begin{aligned}\partial_\phi R(\alpha) &= \lim_{\varepsilon\to 0}\int_{-\infty}^{\infty}\mathrm{d}Z\left[\int_{S_\varepsilon}\left(\frac{b}{2\pi}\alpha_Z + \alpha_\theta\right)\mathrm{d}\theta\right],\\ &= \int_{-\infty}^{\infty}\mathrm{d}Z\left\{\lim_{\varepsilon\to 0}\left[\int_{S_\varepsilon}\left(\frac{b}{2\pi}\alpha_Z + \alpha_\theta\right)\mathrm{d}\theta\right]\right\}.\end{aligned} \tag{19.19}$$

Examining the limit in the second line of (19.19), we first note that

$$\begin{aligned}\lim_{\varepsilon\to 0}\int_{S_\varepsilon}\alpha_\theta \mathrm{d}\theta &= \lim_{\varepsilon\to 0}\int_{S_\varepsilon}\varepsilon(-\alpha_X \sin\theta + \alpha_Y\cos\theta)\mathrm{d}\theta,\\ &= 0,\end{aligned} \tag{19.20}$$

since $\alpha_X \to \alpha_X(X = 0, Y = 0, Z)$, $\alpha_Y \to \alpha_Y(X = 0, Y = 0, Z)$, as $\varepsilon \to 0$ (and thus are independent of θ), and since the integrals of the trigonometric functions of over the circle vanish. In addition,

$$\lim_{\varepsilon\to 0}\int_{S_\varepsilon}\frac{b}{2\pi}\alpha_Z \mathrm{d}\theta = b\alpha_Z(0,0,Z), \tag{19.21}$$

and one concludes that

$$\partial_\phi R(\alpha) = b\int_{-\infty}^{\infty}\alpha_Z(0,0,Z)\mathrm{d}Z. \tag{19.22}$$

If we assign the natural orientation to $L = \{(0,0)\}\times\mathbb{R}\subset\mathbb{R}^3$, we may use R_L to denote the 1-current given by

$$T_L(\alpha) = \int_L \mathcal{I}_L^*(\alpha). \tag{19.23}$$

Here $\mathcal{I}_L : L \to \mathbb{R}^3$ is the natural inclusion, so that for any 1-form $\alpha = \alpha_X \mathrm{d}X + \alpha_Y \mathrm{d}Y + \alpha_Z \mathrm{d}Z$, $\mathcal{I}_L^*(\alpha) = \alpha_Z \mathrm{d}Z$. Finally, we may write the current as

$$\partial_\phi R = bT_L. \tag{19.24}$$

Remark 19.1 Using the same notation as above, consider the case where instead of ϕ given in (19.7), one has the 1-form ϕ' given by

$$\phi' = -\frac{b}{2\pi}\mathrm{d}\theta. \tag{19.25}$$

Since ϕ' is annihilated by the vector space spanned by the base vectors ∂_r and ∂_Z, the layers induced by ϕ' look like the pages of a book spread evenly in all directions (see Fig. 19.4). If we follow the same steps as above, we obtain

$$\phi'\wedge\alpha = -\frac{b}{2\pi}(\alpha_r \mathrm{d}\theta\wedge \mathrm{d}r + \alpha_Z \mathrm{d}\theta\wedge \mathrm{d}Z), \tag{19.26}$$

so that

$$\mathcal{I}^*(\phi'\wedge\alpha) = -\frac{b}{2\pi}\alpha_Z \mathrm{d}\theta\wedge \mathrm{d}Z. \tag{19.27}$$

If follows that $\partial_{\phi'}R = \partial_\phi R$. That is, the dislocations corresponding to these two layering geometries are identical. This observation may be viewed as follows. Let ${}_{\mathrm{d}Z}R$ be the current induced by the form $\mathrm{d}Z$. Then, since $\mathrm{d}^2 Z = 0$, $\partial_{\mathrm{d}Z}R = 0$. Since ${}_\phi R = {}_{\phi'}R - (b/2\pi)\,{}_{\mathrm{d}Z}R$, it follows that $\partial_\phi R = \partial_{\phi'}R$. Alternatively, one may envisage a smooth twist of $\mathbb{R}^3$ about the Z-axis under which the book is deformed into the screw. Since our objects are invariant under diffeomorphisms, both layering structures have the same dislocations. Thus, for example, a similar observation will hold if the pages of the book are not plane but are bent around the Z-axis forming the shape of a whirlpool.

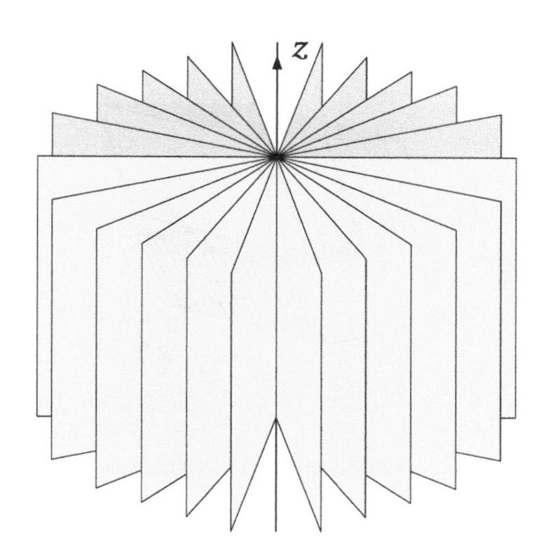

Fig. 19.4 A "book" dislocation

19.2.3 Interfaces

For the sake of simplicity, the next example is formulated in $\mathbb{R}^3$. The generalization to a general manifold is straightforward.

Let

$$\mathcal{R} = \mathbb{R}^3 \setminus \{(X, Y, Z) \mid Y = 0\}, \tag{19.28}$$

and let

$$\mathcal{R}^+ = \{(X, Y, Z) \mid Y > 0\}, \qquad \mathcal{R}^- = \{(X, Y, Z) \mid Y < 0\}. \tag{19.29}$$

Consider the 1-form ϕ defined on $\mathcal{R}$ by

$$\phi = \begin{cases} \phi^+ = \phi_X^+ \mathrm{d}X + \phi_Y^+ \mathrm{d}Y + \phi_Z^+ \mathrm{d}Z, & \text{in } \mathcal{R}^+, \\ \phi^- = \phi_X^- \mathrm{d}X + \phi_Y^- \mathrm{d}Y + \phi_Z^- \mathrm{d}X, & \text{in } \mathcal{R}^-, \end{cases} \tag{19.30}$$

where $\mathrm{d}\phi^\pm = 0$ (see Fig. 19.5).

The form ϕ induces the 2-current ${}_\phi R$ on $\mathbb{R}^3$ by

$${}_\phi R(\psi) = (\phi \lrcorner \mathcal{R})(\psi) = \int_{\mathcal{R}} \phi \wedge \psi. \tag{19.31}$$

For a compactly supported 1-form ω in $\mathbb{R}^3$, the action of the 1-current $\partial_\phi R$ satisfies

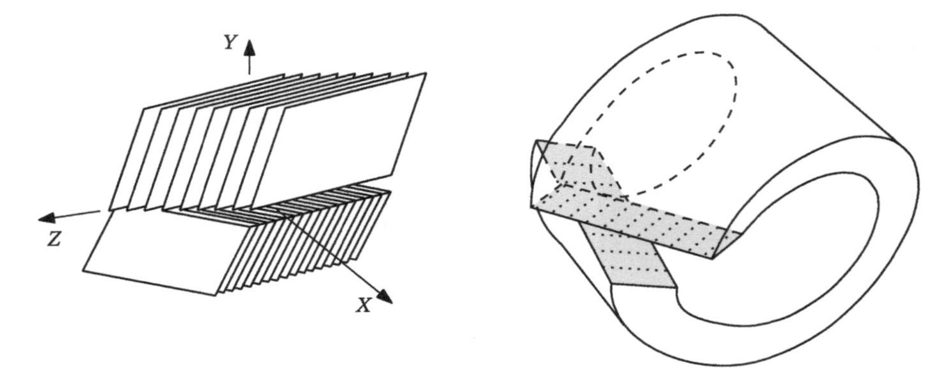

Fig. 19.5 An incoherent interface and the corresponding Volterra disclination

$$
\begin{aligned}
\partial_\phi R(\omega) &= \int_{\mathscr{R}} \phi \wedge \mathrm{d}\omega, \\
&= \int_{\mathscr{R}} (-\mathrm{d}(\phi \wedge \omega) + \mathrm{d}\phi \wedge \omega), \\
&= \int_{\mathscr{R}^+} (-\mathrm{d}(\phi^+ \wedge \omega) + \mathrm{d}\phi^+ \wedge \omega) \\
&\quad + \int_{\mathscr{R}^-} (-\mathrm{d}(\phi^- \wedge \omega) + \mathrm{d}\phi^- \wedge \omega), \\
&= -\int_{\partial\bar{\mathscr{R}}^+} \phi^+ \wedge \omega - \int_{\partial\bar{\mathscr{R}}^-} \phi^- \wedge \omega,
\end{aligned}
\tag{19.32}
$$

where we arrived at the last line by using the Stokes theorem and the fact that $\mathrm{d}\phi^\pm = 0$ in $\mathscr{R}^\pm$, respectively. In addition, $\partial\bar{\mathscr{R}}^\pm$ denote the boundaries of the respective manifolds with boundary, $\bar{\mathscr{R}}^\pm$, obtained as the closures of $\mathscr{R}^\pm$. With the representation

$$
\omega = \omega_X \mathrm{d}X + \omega_Y \mathrm{d}Y + \omega_Z \mathrm{d}Z,
\tag{19.33}
$$

we have

$$
\begin{aligned}
\phi \wedge \omega &= (\phi_X \mathrm{d}X + \phi_Y \mathrm{d}Y + \phi_Z \mathrm{d}Z) \wedge (\omega_X \mathrm{d}X + \omega_Y \mathrm{d}Y + \omega_Z \mathrm{d}Z) \\
&= (\phi_X \omega_Y - \phi_Y \omega_X) \mathrm{d}X \wedge \mathrm{d}Y + (\phi_Y \omega_Z - \phi_Z \omega_Y) \mathrm{d}Y \wedge \mathrm{d}Z \\
&\quad + (\phi_X \omega_Z - \phi_Z \omega_X) \mathrm{d}X \wedge \mathrm{d}Z.
\end{aligned}
\tag{19.34}
$$

It is noted that both restrictions of the forms $\mathrm{d}X \wedge \mathrm{d}Y$ and $\mathrm{d}Y \wedge \mathrm{d}Z$ to the interface, $Y = 0$, vanish, and using (19.32),

$$\partial_\phi R(\omega) = -\int_{\partial\bar{\mathscr{R}}^+} (\phi_X^+ \omega_Z - \phi_Z^+ \omega_X) \mathrm{d}X \wedge \mathrm{d}Z - \int_{\partial\bar{\mathscr{R}}^-} (\phi_X^- \omega_Z - \phi_Z^- \omega_X) \mathrm{d}X \wedge \mathrm{d}Z. \tag{19.35}$$

It is noted that $(\partial_X, \partial_Y, \partial_Z)$ make a positively oriented basis in $\mathbb{R}^3$, and the same holds for even permutations of this triplet. Thus, $\mathrm{d}X \wedge \mathrm{d}Y \wedge dZ$ is a positive volume element. Since $-\partial_Y$ is an outward pointing vector for $\partial\bar{\mathscr{R}}^+$,

$$-\partial_Y \lrcorner (\mathrm{d}X \wedge \mathrm{d}Y \wedge \mathrm{d}Z) = \mathrm{d}X \wedge \mathrm{d}Z \tag{19.36}$$

is the induced positive volume element on $\partial\bar{\mathscr{R}}^+$. Similarly, ∂_Y points out of $\partial\bar{\mathscr{R}}^-$, and so $\mathrm{d}Z \wedge \mathrm{d}X$ is a positive volume element on $\partial\bar{\mathscr{R}}^-$. It follows that

$$\partial_\phi R(\omega) = \int_L [(\phi_X^- - \phi_X^+)\omega_Z - (\phi_Z^- - \phi_Z^+)\omega_X] \mathrm{d}Z \mathrm{d}X, \tag{19.37}$$

where $L = \partial\bar{\mathscr{R}}^-$ is the horizontal plane with the standard orientation.

It is concluded that the distribution of dislocations is given by

$$\partial_\phi R = (\phi_X^- - \phi_X^+)(\partial_Z \wedge L) - (\phi_Z^- - \phi_Z^+)(\partial_X \wedge L). \tag{19.38}$$

We note again that the results are invariant under smooth deformations.

Next, consider the simpler case where the layers contain the Z axis so that the 1-form, ϕ, is defined in terms of closed forms, ϕ^+ and ϕ^-, by

$$\phi = \begin{cases} \phi^+ = \phi_X^+ \mathrm{d}X + \phi_Y^+ \mathrm{d}Y, & \text{in } \mathscr{R}^+, \\ \phi^- = \phi_X^- \mathrm{d}X + \phi_Y^- \mathrm{d}Y, & \text{in } \mathscr{R}^- \end{cases} \tag{19.39}$$

(see Fig. 19.6). We observe that

$$\phi(\partial_X) = \begin{cases} \phi_X^+ & \text{in } \mathscr{R}^+, \\ \phi_X^- & \text{in } \mathscr{R}^-, \end{cases} \tag{19.40}$$

so that ϕ_X^+ and ϕ_X^- are the densities in which the layers intersect the interface on the positive and negative sides of the horizontal plane, respectively.

The form ϕ induces the 2-current ${}_\phi R$ on $\mathbb{R}^3$ by

$${}_\phi R(\psi) = (\phi \lrcorner \mathscr{R})(\psi) = \int_{\mathscr{R}} \phi \wedge \psi. \tag{19.41}$$

Inserting $\phi_Z^+ = \phi_Z^- = 0$, above, we obtain

$$\partial_\phi R(\omega) = \int_L (\phi_X^- - \phi_X^+)\omega_Z \mathrm{d}Z \mathrm{d}X; \tag{19.42}$$

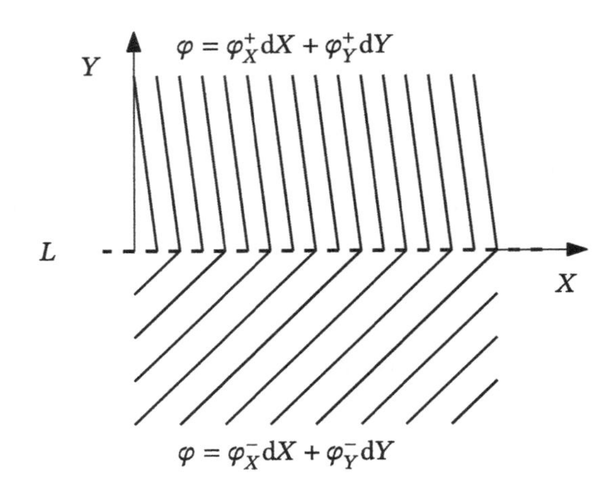

Fig. 19.6 An incoherent interface

the distribution of dislocations is given by

$$\partial_\phi R = (\phi_X^- - \phi_X^+) L \wedge \partial_Z. \tag{19.43}$$

Remark 19.2 This result simply means that the defect is induced by the jump in the density of layers intersecting the horizontal plane across the interface. A mere change in direction of the layers across the interface causes no defects. Note that in the situation illustrated in Fig. 19.7, where $\phi_X^+ = -\phi_X^-$, one obtains

$$\partial_\phi R = 2\phi_X^- L \wedge \partial_Z. \tag{19.44}$$

Remark 19.3 The case $\phi_X^+ = -\phi_X^-$ as in Fig. 19.7 indicates a continuous sink of layers on the upper side of the interface and a continuous source of layers on the upper side of the interface. Thus, we have continuous sources of edge dislocations distributed on either side of the interface. These continuous distributions of dislocations are interpreted as a disclination. On the other hand, the situation where $\phi_X^+ = \phi_X^-$ as illustrated on the left of Fig. 19.8 does not represent defects in the layering, but rather, singular bending of the layers. While this example may be rejected as physically irrelevant, it is valid from the geometric point of view. Thus, in the representation of the layering structure using forms and currents, additional information is provided. This information may not be available in traditional illustrations of lattice defects.

Remark 19.4 Consider, for example, the situation illustrated in Fig. 19.8. Since there are no defects in the layers shown on the left, there are no defects if the layers are smoothly deformed as illustrated on the right. We view the diffeomorphism as a transformation of coordinates, and the statement $\partial_\phi R = 0$ is invariant.

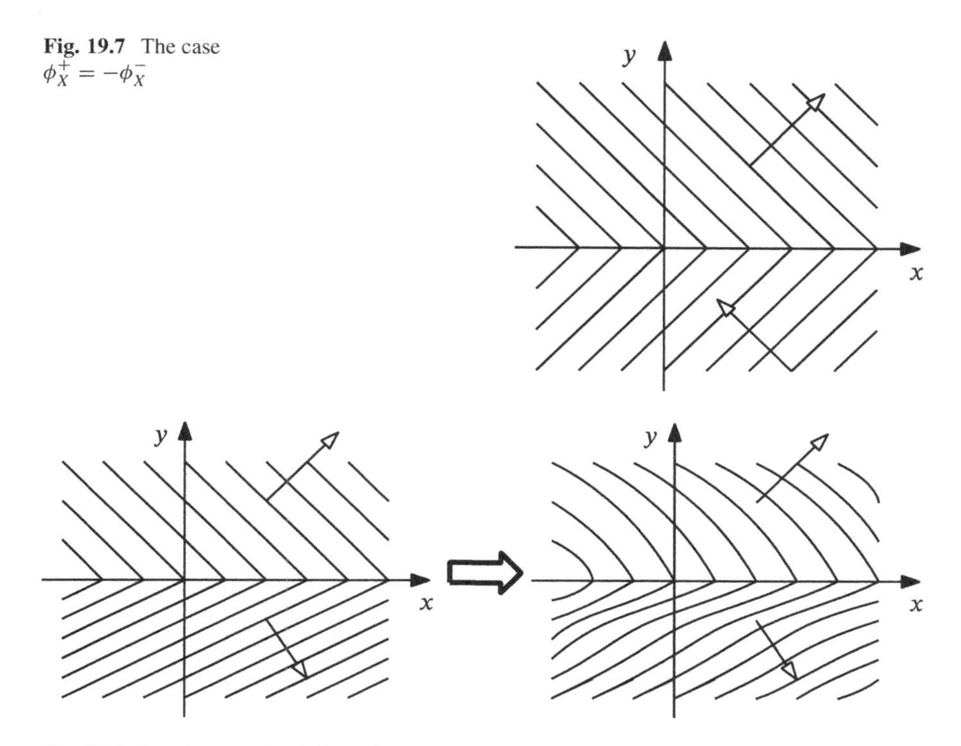

Fig. 19.7 The case $\phi_X^+ = -\phi_X^-$

Fig. 19.8 Invariance under deformations

19.3 Singular Distributions of Inclinations and Disclinations

It is recalled (see Sect. 7.5) that a smooth distribution of inclinations is modeled as an $(n-1)$-form, ϕ, and the corresponding distribution of disclinations is given by $\mathrm{d}\phi$. Thus, an inclination ϕ induces a de Rham 1-current ${}_\phi R$ as in (18.2). In the non-smooth case, we replace the inclination 1-form ϕ and the current it induces by a general *inclination* 1-*current* R. Inclination currents that are not given in terms of smooth $(n-1)$-forms represent singular, or concentrated, director fields, as the examples below illustrate. The boundary, ∂R, of the inclination current represents the distribution of disclinations.

Example 19.1 (A Non-coherent Interface 1) Consider the locally integrable $(n-1)$-form ϕ in $\mathbb{R}^n$ given by

$$\phi(X) = \begin{cases} \phi^+ = \mathrm{d}X^1 \wedge \cdots \wedge \mathrm{d}X^{n-1}, & \text{for } X \in \overline{\mathbb{R}}^{n+}, \\ \phi^- = a\mathrm{d}X^1 \wedge \cdots \wedge \mathrm{d}X^{n-1}, & \text{for } X \in \mathbb{R}^{n-}, \end{cases} \tag{19.45}$$

where $a \in \mathbb{R}$, $\mathbb{R}^{n-} = \{X \in \mathbb{R}^n \mid X^n < 0\}$, and $\overline{\mathbb{R}}^{n+} = \{X \in \mathbb{R}^n \mid X^n \geqslant 0\}$. Evidently, $\mathrm{d}\phi^+ = \mathrm{d}\phi^- = 0$. The inclination form ϕ induces a 1-current ${}_\phi R$ by

$$ {}_\phi R(\omega) = \int_{\mathbb{R}^n} \phi \wedge \omega. \tag{19.46} $$

Clearly, the 1-dimensional subspace spanned by $\partial/\partial X^n$ annihilates $\phi(X)$ for all X for which $X^n \neq 0$. Thus, the directors are aligned in the X^n direction.

For any smooth compactly supported 0-form α in $\mathbb{R}^n$,

$$ \begin{aligned} \partial_\phi R(\alpha) &= \int_{\mathbb{R}^n} \phi \wedge \mathrm{d}\alpha, \\ &= \int_{\mathbb{R}^{n-}} \phi \wedge \mathrm{d}\alpha + \int_{\overline{\mathbb{R}}^{n+}} \phi \wedge \mathrm{d}\alpha, \\ &= (-1)^{n-1} \left[\int_{\mathbb{R}^{n-}} \mathrm{d}(\alpha\phi) - \int_{\mathbb{R}^{n-}} \alpha \mathrm{d}\phi + \int_{\mathbb{R}^{n+}} \mathrm{d}(\alpha\phi) - \int_{\mathbb{R}^{n+}} \alpha \mathrm{d}\phi \right], \\ &= (-1)^{n-1} \left[\int_{\partial\mathbb{R}^{n-}} \alpha\phi + \int_{\partial\mathbb{R}^{n+}} \alpha\phi \right] \end{aligned} \tag{19.47} $$

where in the third line we used (7.7). Let P be the hyperplane in $\mathbb{R}^n$ defined by $X^n = 0$ oriented such that $P = \partial\mathbb{R}^{n-} = -\partial\mathbb{R}^{n+}$. (We view $\mathbb{R}^{n-}$ and $\mathbb{R}^{n+}$ as currents.) Thus, $\theta_P = \mathrm{d}X^1 \wedge \cdots \wedge \mathrm{d}X^{n-1}$ is the natural volume element on P. Let R_P be the 0-current given by

$$ R_P(\alpha) = \int_P \alpha\theta_P. \tag{19.48} $$

We conclude that

$$ \partial_\phi R = (-1)^{n-1}(a-1)R_P, \tag{19.49} $$

which is interpreted as a concentrated source of directors of magnitude $a-1$ which is distributed over the $X^1, \ldots, X^{n-1}$ hyperplane.

In the slightly more general case, let ϕ be defined in terms of closed forms, ϕ^+, ϕ^-, by

$$ \phi(X) = \begin{cases} \phi^+(X), & \text{for } X \in \overline{\mathbb{R}}^{n+}, \\ \phi^-(X), & \text{for } X \in \mathbb{R}^{n-}. \end{cases} \tag{19.50} $$

In analogy with the foregoing computations, one obtains

$$ \partial_\phi R(\alpha) = (-1)^{n-1} \int_P \alpha(\phi^- - \phi^+). \tag{19.51} $$

We conclude that the disclination current vanishes if ϕ^- and ϕ^+ have the same restriction to P, i.e., both forms have the same component relative to $\mathrm{d}X^1 \wedge \cdots \wedge \mathrm{d}X^{n-1}$.

Example 19.2 (An Edge Disclination) Let L be a connected and oriented 1-dimensional submanifold with a boundary of $\mathcal{X}$. Then, L induces a 1-current R_L by

$$R_L(\omega) = \int_L \omega, \tag{19.52}$$

for all compactly supported smooth 1-forms ω in $\mathcal{X}$. Using Stokes's theorem, one has

$$\partial R_L(\alpha) = \int_L \mathrm{d}\alpha = \int_{\partial L} \alpha. \tag{19.53}$$

Evidently, as ∂L is a 0-dimensional submanifold, and assuming it is not empty, it may contain one or two points. It is noted that the orientation of a 0-dimensional vector space may be given the value of either $+1$ or -1. This applies to ∂L if it contains one point. If ∂L contains two points, each may have either a $+1$-orientation or a -1-orientation, while the other point, if exists, has the opposite orientation.

In the case where ∂L contains one point X_1, and assuming its orientation is positive, one has $\partial L(\alpha) = \alpha(X_0)$, representing an edge disclination originating at X_0. This will be the situation if $\mathcal{X} = (-1, 1)^3 \subset \mathbb{R}^3$ and $L = \{(0, 0, Z) \mid -1 < Z \leqslant 0\}$ so that $X_1 = (0, 0, 0)$. In this case, the disclination does not terminate inside the body. In the case where ∂L contains also the additional point X_2 having a negative orientation, $\partial L(\alpha) = \alpha(X_1) - \alpha(X_2)$ and the disclination terminates at X_2.

Example 19.3 (Directors Emanating from a Singular Line) Using the notation introduced in Sect. 19.2.2 on screw dislocations, consider the inclination $n - 1 = 2$-form ϕ defined on $D \subset \mathbb{R}^3$ by

$$\phi = \mathrm{d}\theta \wedge \mathrm{d}Z. \tag{19.54}$$

This inclination form induces an inclination 1-current, R, on $\mathbb{R}^3$ by

$${}_{\phi}R(\omega) = \int_D \phi \wedge \omega, \tag{19.55}$$

for any 1-form ω. It is noted that in its domain of definition, $\mathrm{d}\phi = 0$.

To compute the disclination 0-current ∂R, one observes that for any smooth function α, compactly supported in $\mathbb{R}^3$, (7.7) implies that

$$
\begin{aligned}
\partial_\phi R(\alpha) &= \int_D \mathrm{d}(\phi\alpha) - \int_D \alpha \mathrm{d}\phi, \\
&= \lim_{\varepsilon\to 0} \int_{D_\varepsilon} \mathrm{d}(\alpha\phi), \\
&= \lim_{\varepsilon\to 0} \int_{\partial D_\varepsilon} \mathcal{J}^*(\alpha\phi), \\
&= \lim_{\varepsilon\to 0} \int_{\partial D_\varepsilon} \alpha \mathrm{d}\theta \wedge \mathrm{d}Z.
\end{aligned}
\tag{19.56}
$$

In analogy with the computations of Sect. 19.2.2, one obtains

$$
\partial_\phi R(\alpha) = 2\pi \int_{Z=-\infty}^{\infty} \alpha(0, 0, Z)\mathrm{d}Z,
\tag{19.57}
$$

which we may write as

$$
\partial_\phi R = 2\pi T_L \llcorner \mathrm{d}Z.
\tag{19.58}
$$

Thus, we have a uniform distribution of sources of radial directors along the Z-axis

Chapter 20
Vector-Valued Currents

A natural extension of the notions of generalized sections and de Rham currents yields vector-valued currents that will be used to model stresses. Vector-valued currents and their local representations will be considered in this chapter. These may be viewed as the singular counterparts of vector-valued forms considered in Sect. 6.2.4. For a comprehensive treatment of the subject of vector-valued distributions on manifolds, see Chapter 3 of [GKOS01].

20.1 Vector-Valued Currents

We start by substituting the vector bundle $W^* \otimes \bigwedge^p T^*\mathcal{X}$ for the vector bundle W_0 in definition (17.44) of generalized sections. Thus, we obtain

$$C^{-r}(W^* \otimes \bigwedge^p T^*\mathcal{X}) = C^r_c((W^* \otimes \bigwedge^p T^*\mathcal{X})')^* \cong C^r_c(W \otimes \bigwedge^p T\mathcal{X} \otimes \bigwedge^n T^*\mathcal{X})^*. \tag{20.1}$$

Using the isomorphism

$$\wedge_\lrcorner : \bigwedge^{n-p} T^*\mathcal{X} \longrightarrow L(\bigwedge^p T^*\mathcal{X}, \bigwedge^n T^*\mathcal{X}) \cong \bigwedge^p T\mathcal{X} \otimes \bigwedge^n T^*\mathcal{X}, \tag{20.2}$$

whereby $\wedge_\lrcorner(\omega)(\psi) = \omega \wedge \psi$ (see (6.108)), it is concluded that we may make the identifications

$$C^{-r}(W^* \otimes \bigwedge^p T^*\mathcal{X}) = C^r_c\big(W \otimes \bigwedge^{n-p} T^*\mathcal{X}\big)^* \tag{20.3}$$

(see [Sch73, p. 340]).

R. Segev, *Foundations of Geometric Continuum Mechanics*, Advances in Continuum Mechanics 49, https://doi.org/10.1007/978-3-031-35655-1_20

Remark 20.1 It is noted that the choice to use the isomorphism $\wedge_\lrcorner$ to relate $\bigwedge^{n-p} T^*\mathcal{X}$ with $\bigwedge^p T\mathcal{X} \otimes \bigwedge^n T^*\mathcal{X}$ is somewhat arbitrary. One could also use the isomorphism $\wedge_\llcorner$ as in (6.106). Such a choice would lead to objects and results that may differ by sign.

Remark 20.2 Let $W \otimes U$ be the tensor product of two vector bundles, W and U, over $\mathcal{X}$. Then, the mapping

$$\widehat{\otimes} : C^\infty(W) \times C^\infty(U) \longrightarrow C^\infty(W \otimes U), \quad (w, u) \longmapsto w \otimes u \tag{20.4}$$

is bilinear for the modules over the ring of smooth functions on $\mathcal{X}$. In addition, one has a natural isomorphism (see [GHV72, p. 80] and [Con01, p. 232])

$$C^\infty(W) \otimes C^\infty(U) \cong C^\infty(W \otimes U). \tag{20.5}$$

Remark 20.3 In order to determine a vector-valued current uniquely, it is sufficient to know its action on sections of the tensor product $W \otimes \bigwedge^{n-p} T^*\mathcal{X}$ of the form $w \otimes \omega$, where w is a section of W, and ω is an $(n-p)$-form. Locally, if a section, χ, of $W \otimes \bigwedge^{n-p} T^*\mathcal{X}$ is represented by $\chi^\alpha_\lambda g_\alpha \otimes \mathrm{d}x^\lambda$, one may consider the $(n-p)$-forms, $\omega^\alpha = \chi^\alpha_\lambda \mathrm{d}x^\lambda$. Then, evidently, χ is also represented by $g_\alpha \otimes \omega^\alpha$. This will be done frequently in what follows.

Comparing Equation (20.3) to vector-valued forms, as in Sect. 6.2.4, we may refer to elements of these spaces as *generalized vector-valued p-forms* or as *vector-valued $(n-p)$-currents*.

A smooth vector-valued $(n-p)$-current is represented by a $W^* \otimes \bigwedge^{n-p} T\mathcal{X}$-valued n-form—a smooth section S of $W^* \otimes \bigwedge^{n-p} T\mathcal{X} \otimes \bigwedge^n T^*\mathcal{X}$ as

$$\chi \longmapsto \int_{\mathcal{X}} S \cdot \chi, \tag{20.6}$$

where it is noted that $S \cdot \chi$ is an n-form. Locally, for an increasing multi-index μ with $|\mu| = n - p$, using the notation of (6.79), we may write

$$S = S^\mu_\alpha g^\alpha \otimes \partial_\mu \otimes \mathrm{d}x, \qquad S \cdot \chi = S^\mu_\alpha \chi^\alpha_\mu \mathrm{d}x. \tag{20.7}$$

Alternatively, and similarly to the case of smooth scalar-valued currents, smooth elements, ${}_{\widehat{S}}T$ and $T_{\widehat{S}}$, of $C^{-r}\left(W^*, \bigwedge^p T^*\mathcal{X}\right)$ are induced by a section $\widehat{S}$ of $W^* \otimes \bigwedge^p T^*\mathcal{X}$ in the forms

$${}_{\widehat{S}}T(\chi) = \int_{\mathcal{X}} \widehat{S} \dot{\wedge} \chi, \qquad T_{\widehat{S}}(\chi) = \int_{\mathcal{X}} \chi \dot{\wedge} \widehat{S}, \tag{20.8}$$

for every compactly supported C^r-section χ of $W \otimes \bigwedge^{n-p} T^*\mathcal{X}$. In the definitions, we used the notation introduced in Sect. 6.2.4, and in particular (6.116). Locally, for $|\hat{\mu}| = p$, $|\lambda| = n - p$, the vector-valued form $\widehat{S}$ is represented by $\widehat{S}_{\alpha\hat{\mu}} g^\alpha \otimes \mathrm{d}x^{\hat{\mu}}$, and

$$
\begin{aligned}
\widehat{S}\dot{\wedge}\chi &= \widehat{S}_{\alpha\hat{\mu}}\chi^{\alpha}_{\lambda}\mathrm{d}x^{\hat{\mu}} \wedge \mathrm{d}x^{\lambda} = \varepsilon^{\hat{\lambda}\lambda}\widehat{S}_{\alpha\hat{\lambda}}\chi^{\alpha}_{\lambda}\mathrm{d}x, \\
\chi\dot{\wedge}\widehat{S} &= \widehat{S}_{\alpha\hat{\mu}}\chi^{\alpha}_{\lambda}\mathrm{d}x^{\lambda} \wedge \mathrm{d}x^{\hat{\mu}} = \varepsilon^{\lambda\hat{\lambda}}\widehat{S}_{\alpha\hat{\lambda}}\chi^{\alpha}_{\lambda}\mathrm{d}x = (-1)^{p(n-p)}\widehat{S}\dot{\wedge}\chi.
\end{aligned}
\tag{20.9}
$$

(Note that as both λ and $\hat{\lambda}$ are increasing multi-indices, an expression such as $\varepsilon^{\hat{\mu}\lambda}\widehat{S}_{\alpha\hat{\mu}} = \varepsilon^{\hat{\lambda}\lambda}\widehat{S}_{\alpha\hat{\lambda}}$, for a given λ, contains only one nonvanishing term.)

Let $\widehat{S}_{\lrcorner}$ and $\widehat{S}_{\llcorner}$ be two smooth vector-valued forms so that for every section χ,

$$
{}_{\widehat{S}_{\lrcorner}}T(\chi) = \int_{\mathcal{X}} \widehat{S}_{\lrcorner}\dot{\wedge}\chi = \int_{\mathcal{X}} \chi\dot{\wedge}\widehat{S}_{\llcorner} = T_{\widehat{S}_{\llcorner}}(\chi); \tag{20.10}
$$

then,

$$
\widehat{S}_{\lrcorner} = (-1)^{p(n-p)}\widehat{S}_{\llcorner}. \tag{20.11}
$$

Comparing (20.7), (20.9), and (20.11), one concludes that if the sections S, $\widehat{S}_{\lrcorner}$, and $\widehat{S}_{\llcorner}$ represent the same current, we have

$$
\widehat{S}_{\lrcorner\alpha\hat{\lambda}} = \varepsilon_{\hat{\lambda}\lambda}S^{\lambda}_{\alpha}, \qquad \widehat{S}_{\llcorner\alpha\hat{\lambda}} = \varepsilon_{\lambda\hat{\lambda}}S^{\lambda}_{\alpha}. \tag{20.12}
$$

Globally, it follows that

$$
\widehat{S}_{\lrcorner} = \mathsf{C}_{\llcorner}(S), \qquad \widehat{S}_{\llcorner} = \mathsf{C}_{\lrcorner}(S), \tag{20.13}
$$

where the contractions

$$
\mathsf{C}_{\llcorner}, \mathsf{C}_{\lrcorner} : W^* \otimes \bigwedge\nolimits^{n-p} T\mathcal{X} \otimes \bigwedge\nolimits^{n} T^*\mathcal{X} \longrightarrow W^* \otimes \bigwedge\nolimits^{p} T^*\mathcal{X} \tag{20.14}
$$

are induced by the right and left contractions $\theta\llcorner\eta_1(\eta_2) = \theta(\eta_2\wedge\eta_1)$ and $\eta_1\lrcorner\theta(\eta_2) = \theta(\eta_1 \wedge \eta_2)$, respectively (see Sect. 6.2.3).

The foregoing analysis suggests the definition of contraction operations, $S \lrcorner R$ and $R \llcorner S$, between a scalar-valued $(p+q)$-current R and a co-vector-valued q-form S (a section of $W^* \otimes \bigwedge^q T^*\mathcal{X}$). The contractions are vector-valued p-currents given by

$$
\begin{aligned}
(S \lrcorner R)(\chi) &:= R(S\dot{\wedge}\chi), \\
(R \llcorner S)(\chi) &:= R(\chi\dot{\wedge}S).
\end{aligned}
\tag{20.15}
$$

for vector-valued p-forms χ.

The definitions in (20.15) may be used to define the smooth vector-valued currents of (20.8) by

$$
{}_{\widehat{S}}T = \widehat{S} \lrcorner T_{\mathcal{X}}, \qquad T_{\widehat{S}} = T_{\mathcal{X}} \llcorner \widehat{S}, \tag{20.16}
$$

where it is recalled that $T_{\mathcal{X}}$ is the n-current induced naturally by the manifold $\mathcal{X}$.

20.2 Various Operations for Vector-Valued Currents and Local Representation

We introduce first some basic operations on vector-valued currents, in analogy with Sect. 18.1. Next, we consider the local representation of the restriction of a vector-valued p-current to vector-valued forms supported in some given vector bundle chart.

20.2.1 The Inner Product of a Vector-Valued Current and a Vector Field

Given a vector-valued current T in $C^r_c(W \otimes \bigwedge^p T^*\mathcal{X})^*$ and a C^r-section w of W, we define the (scalar) p-current $w \cdot T$ by

$$w \cdot T(\omega) = T(w \otimes \omega). \tag{20.17}$$

The bilinear product defined above induces a linear mapping

$$\circledcirc : C^r_c(W \otimes \textstyle\bigwedge^p T^*\mathcal{X})^* \longrightarrow L[C^r_c(W), C^r_c(\bigwedge^p T^*\mathcal{X})^*] \tag{20.18}$$

by

$$\circledcirc\,(T)(w) := w \cdot T \quad \text{and we set,} \quad T_{\circledcirc} := \circledcirc(T). \tag{20.19}$$

It is noted that $\circledcirc$ is invertible. For if $T_{\circledcirc} \in L[C^r_c(W), C^r_c(\bigwedge^p T^*\mathcal{X})^*]$ is given, we set $T = \circledcirc^{-1}(T_{\circledcirc})$ by

$$T(w \otimes \omega) := (T_{\circledcirc}(w))(\omega). \tag{20.20}$$

By Remark 20.3, the last condition determines T uniquely.

For the local representation of a vector-valued current, T, defined in the domain of a chart, one may consider the (scalar-valued) p-currents,

$$T_\alpha := g_\alpha \cdot T, \qquad T_\alpha(\omega) = T(g_\alpha \otimes \omega). \tag{20.21}$$

Thus, in analogy with (18.26), (18.27), we have

$$\begin{aligned} T(w \otimes \omega) &= T(w^\alpha g_\alpha \otimes \omega), \\ &= (g_\alpha \cdot T)(w^\alpha \omega), \\ &= T_\alpha(w^\alpha \omega). \end{aligned} \tag{20.22}$$

20.2.2 The Tensor Product of a Current and a Co-vector Field

A (scalar) p-current, T, and a C^r-section of W^*, g, induce a vector-valued current $g \otimes T \in C^r_c\big(W \otimes \bigwedge^p T^*\mathcal{X}\big)^*$ by setting

$$(g \otimes T)(w \otimes \omega) := T((g \cdot w)\omega). \tag{20.23}$$

In particular, when T is defined in the domain of a chart, locally,

$$(g^\alpha \otimes T)(w \otimes \omega) := T(w^\alpha \omega). \tag{20.24}$$

Utilizing this definition, one may write for the local representatives,

$$\begin{aligned} g^\alpha \otimes T_\alpha(w \otimes \omega) &= T_\alpha(w^\alpha \omega), \\ &= T(w^\alpha g_\alpha \otimes \omega), \end{aligned} \tag{20.25}$$

and so, complementing (20.22), one has

$$T = g^\alpha \otimes T_\alpha. \tag{20.26}$$

20.2.3 Representation by 0-Currents

Proceeding as in Sect. 18.2, the local p-current T may be represented by the 0-currents

$$T^\lambda_\alpha := (T_\alpha)^\lambda = \mathrm{d}x^\lambda \lrcorner T_\alpha = \mathrm{d}x^\lambda \lrcorner (g_\alpha \cdot T), \quad |\lambda| = p, \tag{20.27}$$

and by (18.29)

$$T_\alpha = \partial_\lambda \wedge T^\lambda_\alpha. \tag{20.28}$$

Hence, for a section w and a p-form ω, such that $w \otimes \omega$ is compactly supported in the domain of a chart,

$$\begin{aligned} T^\lambda_\alpha(w^\alpha \omega_\lambda) &= [\mathrm{d}x^\lambda \lrcorner (g_\alpha \cdot T)](w^\alpha \omega_\lambda), \\ &= (g_\alpha \cdot T)(w^\alpha \omega_\lambda \mathrm{d}x^\lambda), \\ &= T(w^\alpha \omega_\lambda g_\alpha \mathrm{d}x^\lambda), \\ &= T(w \otimes \omega). \end{aligned} \tag{20.29}$$

Therefore, the action of the current T is given in the form

$$T(\chi) = T^{\lambda}_{\alpha}(\chi^{\alpha}_{\lambda}). \tag{20.30}$$

Moreover, using (20.26) and (18.29), we have

$$T = g^{\alpha} \otimes (\partial_{\lambda} \wedge T^{\lambda}_{\alpha}). \tag{20.31}$$

In the case where the 0-currents T^{λ}_{α} are represented locally by smooth n-forms $S^{\lambda}_{\alpha}\mathrm{d}x$, one has

$$T(\chi) = \int_{U} S^{\lambda}_{\alpha}\chi^{\alpha}_{\lambda}\mathrm{d}x \tag{20.32}$$

in accordance with (20.7).

20.2.4 Representation by n-Currents

In analogy with the representation of a vector-valued p-current by (scalar-valued) 0-currents as above, we can use the constructions of Sect. 18.2.2 to represent the scalar-valued local currents T_{α} by scalar-valued n-currents. Thus, we set

$$T_{\alpha\hat{\lambda}} := (T_{\alpha})_{\hat{\lambda}} = \partial_{\hat{\lambda}} \wedge T_{\alpha}, \qquad T'_{\alpha\hat{\lambda}} = (T_{\alpha})'_{\hat{\lambda}} = T_{\alpha} \wedge \partial_{\hat{\lambda}}. \tag{20.33}$$

It follows immediately from Sect. 18.2.2 and (20.26) that

$$T_{\alpha} = \mathrm{d}x^{\hat{\lambda}} \lrcorner T_{\alpha\hat{\lambda}} = T'_{\alpha\hat{\lambda}} \llcorner \mathrm{d}x^{\hat{\lambda}} \tag{20.34}$$

and

$$T = g^{\alpha} \otimes (\mathrm{d}x^{\hat{\lambda}} \lrcorner T_{\alpha\hat{\lambda}}) = g^{\alpha} \otimes (T'_{\alpha\hat{\lambda}} \llcorner \mathrm{d}x^{\hat{\lambda}}). \tag{20.35}$$

For example, as expected,

$$\begin{aligned}
\{g^{\alpha} \otimes [((g_{\alpha} \cdot T) \wedge \partial_{\hat{\lambda}}) \llcorner \mathrm{d}x^{\hat{\lambda}}]\}(w \otimes \omega) &= [((g_{\alpha} \cdot T) \wedge \partial_{\hat{\lambda}}) \llcorner \mathrm{d}x^{\hat{\lambda}}]((g^{\alpha} \cdot w) \otimes \omega), \\
&= [(g_{\alpha} \cdot T) \wedge \partial_{\hat{\lambda}}](w^{\alpha}\omega \wedge \mathrm{d}x^{\hat{\lambda}}), \\
&= [(g_{\alpha} \cdot T) \wedge \partial_{\hat{\lambda}}](w^{\alpha}\omega_{\lambda}\mathrm{d}x\varepsilon^{\lambda\hat{\lambda}}), \\
&= (g_{\alpha} \cdot T)(w^{\alpha}\omega_{\lambda}\mathrm{d}x \llcorner \partial_{\hat{\lambda}}\varepsilon^{\lambda\hat{\lambda}}), \\
&= T(w^{\alpha}\omega_{\lambda}g_{\alpha} \otimes \mathrm{d}x^{\mu}\varepsilon_{\mu\hat{\lambda}}\varepsilon^{\lambda\hat{\lambda}}), \\
&= T(w^{\alpha}\omega_{\lambda}g_{\alpha} \otimes \mathrm{d}x^{\lambda}), \\
&= T(w \otimes \omega).
\end{aligned} \tag{20.36}$$

20.2.5 The Exterior Product of a Vector-Valued Current and a Multi-Vector Field

Next, in analogy with Sect. 18.2, for a vector-valued p-current T and a q-multi-vector field, η, $q \leq n-p$, consider the vector-valued $(p+q)$-current $\eta \wedge T$ defined by

$$(\eta \wedge T)(w \otimes \omega) := T(w \otimes (\eta \lrcorner \omega)). \tag{20.37}$$

In particular, for multi-indices $\hat{\lambda}$, $|\hat{\lambda}| = n-p$, we define locally the vector-valued n-currents

$$T_{\hat{\lambda}} := \partial_{\hat{\lambda}} \wedge T, \qquad T_{\hat{\lambda}}(w \otimes \omega) = T(w \otimes (\partial_{\hat{\lambda}} \lrcorner \omega)), \tag{20.38}$$

so that

$$T_{\hat{\lambda}}(w \otimes \mathrm{d}x) = T(w \otimes (\partial_{\hat{\lambda}} \lrcorner \mathrm{d}x)). \tag{20.39}$$

20.2.6 The Contraction of a Vector-Valued Current and a Form

Furthermore, for a vector-valued p-current T and a q-form ψ, $q \leq p$, define the vector-valued $(p-q)$-currents $\psi \lrcorner T$ and $T \llcorner \psi$ as follows. For each compactly supported $(p-q)$-form ω,

$$(\psi \lrcorner T)(w \otimes \omega) := T(w \otimes (\psi \wedge \omega)) \tag{20.40}$$

and

$$(T \llcorner \psi)(w \otimes \omega) := T(w \otimes (\omega \wedge \psi)), \tag{20.41}$$

with $T \llcorner \psi = (-1)^{p(p-q)} \psi \lrcorner T$. In the case where $q = p$, we obtain an element $\psi \lrcorner T \in C_c^r(W)^*$, a vector-valued 0-current, satisfying

$$(\psi \lrcorner T)(w) = T(w \otimes \psi). \tag{20.42}$$

Locally, one may consider the local functionals—local vector-valued 0-currents,

$$T^{\lambda} := \mathrm{d}x^{\lambda} \lrcorner T, \quad |\lambda| = p, \qquad T^{\lambda}(w) = T(w \otimes \mathrm{d}x^{\lambda}). \tag{20.43}$$

Hence,

$$\begin{aligned} T(w \otimes \omega) &= T(w \otimes (\omega_\lambda \mathrm{d}x^\lambda)), \\ &= \omega_\lambda T(w \otimes \mathrm{d}x^\lambda), \\ &= \omega_\lambda T^\lambda(w), \end{aligned} \tag{20.44}$$

and so,

$$T(w \otimes \omega) = T^\lambda(\omega_\lambda w) \tag{20.45}$$

is a local representation of the action of T using vector-valued 0-currents. It is implied by the identity $T^\lambda(\omega_\lambda w) = \partial_\lambda \wedge T^\lambda(w \otimes \omega)$, that

$$T = \partial_\lambda \wedge T^\lambda. \tag{20.46}$$

20.2.7 Alternative Representation by n-Currents

Next, for a local basis $\mathrm{d}x^{\hat{\lambda}}$ of $\bigwedge^{n-p} T^*\mathcal{X}$, using (20.38) and (20.42),

$$\begin{aligned} (\mathrm{d}x^{\hat{\lambda}} \lrcorner T_{\hat{\lambda}})(w \otimes \omega) &= T_{\hat{\lambda}}(w \otimes (\mathrm{d}x^{\hat{\lambda}} \wedge \omega)), \\ &= T(w \otimes (\partial_{\hat{\lambda}} \lrcorner (\mathrm{d}x^{\hat{\lambda}} \wedge \omega))). \end{aligned} \tag{20.47}$$

Following the same procedure as that leading to (18.37) and (18.38), one concludes that the local vector-valued p-current T may be represented by the vector-valued n-currents $T_{\hat{\lambda}}$ in the form

$$T = \mathrm{d}x^{\hat{\lambda}} \lrcorner T_{\hat{\lambda}}, \tag{20.48}$$

and

$$T(w \otimes \omega) = T_{\hat{\lambda}}(w \otimes (\mathrm{d}x^{\hat{\lambda}} \wedge \omega)) = \varepsilon^{\hat{\lambda}\lambda} T_{\hat{\lambda}}(\omega_\lambda w \otimes \mathrm{d}x). \tag{20.49}$$

Using (20.17) and (20.26), we may define the (scalar) n-currents

$$(T_{\hat{\lambda}})_\alpha := g_\alpha \cdot T_{\hat{\lambda}}, \qquad \text{so that} \qquad T_{\hat{\lambda}} = g^\alpha \otimes (T_{\hat{\lambda}})_\alpha, \tag{20.50}$$

and

$$(T_{\hat{\lambda}})_\alpha(\theta) = T_{\hat{\lambda}}(g_\alpha \otimes \theta) = T(g_\alpha \otimes (\partial_{\hat{\lambda}} \lrcorner \theta)). \tag{20.51}$$

Consider the local p-currents T_α in (20.21) and their components $T_{\alpha\hat{\lambda}} := (T_\alpha)_{\hat{\lambda}}$ defined in (20.33). Since for an n-form, θ, locally supported in the domain of a chart,

$$\begin{aligned} T_{\alpha\hat{\lambda}}(\theta) = (T_\alpha)_{\hat{\lambda}}(\theta) &= \partial_{\hat{\lambda}} \wedge (g_\alpha \cdot T)(\theta), \\ &= (g_\alpha \cdot T)(\partial_{\hat{\lambda}} \lrcorner \theta), \\ &= T(g_\alpha \otimes (\partial_{\hat{\lambda}} \lrcorner \theta)), \end{aligned} \tag{20.52}$$

we conclude that

$$(T_{\hat{\lambda}})_\alpha = (T_\alpha)_{\hat{\lambda}} = T_{\alpha\hat{\lambda}}. \tag{20.53}$$

Remark 20.4 In the foregoing discussion, we have made special choices, and used, for example, the definitions $T_{\hat{\lambda}} := \partial_{\hat{\lambda}} \wedge T$ and $T^\lambda := \mathrm{d}x^\lambda \lrcorner T$ rather than $T'_{\hat{\lambda}} := T \wedge \partial_{\hat{\lambda}}$ and $T'^\lambda := T \llcorner \mathrm{d}x^\lambda$, defined in (20.38), respectively. The correspondence between the two schemes is a natural extension of Remark 18.1. In particular, $\varepsilon_{\hat{\lambda}\lambda}$ will be replaced by $\varepsilon_{\lambda\hat{\lambda}}$. Thus, in view of Eqs. (20.35) and (20.50),

$$T = \mathrm{d}x^{\hat{\lambda}} \lrcorner (g^\alpha \otimes T_{\alpha\hat{\lambda}}) = (g^\alpha \otimes T'_{\alpha\hat{\lambda}}) \llcorner \mathrm{d}x^{\hat{\lambda}} = g^\alpha \otimes (\mathrm{d}x^{\hat{\lambda}} \lrcorner T_{\alpha\hat{\lambda}}) = g^\alpha \otimes (T'_{\alpha\hat{\lambda}} \llcorner \mathrm{d}x^{\hat{\lambda}}). \tag{20.54}$$

Remark 20.5 The relations between the representation of a current by 0-currents and the representation in terms of n-currents in Remark 18.2 imply that

$$\mathrm{d}x \lrcorner T_{\alpha\hat{\lambda}} = \varepsilon_{\hat{\lambda}\lambda} T^\lambda_\alpha, \quad T_{\alpha\hat{\lambda}} = \varepsilon_{\hat{\lambda}\lambda} \partial_x \wedge T^\lambda_\alpha, \quad T^\lambda_\alpha = \varepsilon^{\lambda\hat{\mu}} T'_{\alpha\hat{\mu}} \llcorner \mathrm{d}x, \quad T'_{\alpha\hat{\lambda}} = \varepsilon_{\lambda\hat{\lambda}} T^\lambda_\alpha \wedge \partial_x. \tag{20.55}$$

Remark 20.6 In the smooth case, the n-currents $T_{\alpha\hat{\lambda}}$ are represented by functions $\widehat{S}_{\alpha\hat{\lambda}}$ that make up the vector-valued $(n-p)$-form

$$\widehat{S} = \widehat{S}_{\alpha\hat{\mu}} g^\alpha \otimes \mathrm{d}x^{\hat{\mu}}, \tag{20.56}$$

as in (20.8), (20.9).

Remark 20.7 In summary, the representation by zero currents (e.g., (20.32)) corresponds to viewing the vector-valued current as an element of $C^r_c(W \otimes \bigwedge^p T^*\mathcal{X})^* =: C^{-r}(W^* \otimes \bigwedge^p T\mathcal{X} \otimes \bigwedge^n T^*\mathcal{X})$. On the other hand, the representation as in (20.32), (20.56) uses the isomorphism of $\bigwedge^p T\mathcal{X} \otimes \bigwedge^n T^*\mathcal{X}$ with $\bigwedge^{n-p} T^*\mathcal{X}$, to make the identification

$$C^r_c(W \otimes \textstyle\bigwedge^p T^*\mathcal{X})^* \cong C^{-r}(W^* \otimes \textstyle\bigwedge^{n-p} T^*\mathcal{X}). \tag{20.57}$$

Chapter 21
The Representation of Forces by Stresses and Hyperstresses

In this chapter, the existence of stress is proved from an entirely different perspective, in comparison with Chap. 10. We no longer start with systems of forces, the analogs of Cauchy's postulates, and proceed with geometric constructions based on those of Chap. 4. Rather, using the infinite dimensional analog of the paradigm introduced in Chap. 2 and having the Banach manifold structure of the configuration space, $\mathrm{Emb}^r(\mathcal{X}, \mathcal{S})$, $r \geqslant 1$, at our disposal (Chap. 15), we simply define a force to be an element of the cotangent bundle of the configuration space. From this point onward, mathematical analysis takes over. A representation theorem for forces, based on the Hahn-Banach theorem, leads to significant results without using any additional assumption of physical nature. These results may be summarized roughly as follows.

- Forces are represented by Radon measures valued in the dual of the jet bundles of the vector bundle of virtual velocity fields.
- These measures are generalizations of the stress object of continuum mechanics for the case $r = 1$. Thus, henceforth we will refer to these measures as *variational stress measures* or simply as *stresses*.
- The representation of forces is not unique, that is, a collection of stress measures represent a single force.
- The origin of the non uniqueness is the fact that the jet extension mapping of vector fields is not surjective (nor does it have a dense image). For a given force, all stress fields representing it will have the same action on compatible (integrable) jet fields. However, their actions on incompatible jet fields are, in general, different. In other words, while the force object determines a definite action on compatible jet fields, the stress object extends this action to incompatible jet fields.
- The condition that a given stress represents a force is a generalization of the principle of virtual work and the "equilibrium" equation of continuum mechanics. This holds even though no additional assumption of physical nature has been made.

R. Segev, *Foundations of Geometric Continuum Mechanics*, Advances in Continuum Mechanics 49, https://doi.org/10.1007/978-3-031-35655-1_21

- The condition that a variational stress measure ς represents a force F is given simply by

$$F = j^{1*}(\varsigma), \tag{21.1}$$

where j^{1*} is the dual of the jet extension mapping of sections.
- All the above, apply respectively also to hyper-stresses and nonholonomic stresses. All one has to do is follow the same procedure for $r > 1$. To obtain nonholonomic stresses, one has to replace the jet extension mapping by the iterated, nonholonomic, jet extension corresponding to iterated jet bundles.
- As already mentioned, the stress object may be as irregular as a Radon measure. For example, one may consider a "force dipole," which is just a stress field distributed as Dirac measure.
- In case the variational stress measures are represented by smooth densities, body force fields, surface force fields, and their relations to stresses (as in Chap. 10) are recovered. The relation between the force fields and stress fields is obtained.

This chapter considers the general case of C^r-sections of a fiber bundle. For the special case of simple stresses, $r = 1$, further details are presented in Chap. 22.

21.1 Representation of Forces by Hyper-Stresses

We recall that the tangent space, $T_\kappa C^r(\xi)$, to the Banach manifold of C^r-sections of the fiber bundle $\xi : \mathcal{Y} \to \mathcal{X}$, at the section $\kappa : \mathcal{X} \to \mathcal{Y}$, may be identified with the Banachable space, $C^r(\kappa^* V\mathcal{Y})$, of sections of the pullback vector bundle $\kappa^* \tau_{\mathcal{Y}} : \kappa^* V\mathcal{Y} \longrightarrow \mathcal{X}$ (cf. [Mic20, Section 5.8]). Elements of the tangent space at κ to the configuration manifold represent generalized velocities of the continuous mechanical system. Consequently, a generalized force is modeled mathematically by an element $F \in C^r(\kappa^* V\mathcal{Y})^*$. The central message of this section is that although such functionals cannot be restricted naturally to sub bodies of $\mathcal{X}$, as discussed in Sect. 17.3, forces may be represented, non uniquely, by stress objects that enable restriction of forces to sub bodies. In order to simplify the notation, we will replace $\kappa^* V\mathcal{Y}$ by a general vector bundle $\pi : W \to \mathcal{X}$, as in Chap. 14, and the notation introduced there will be used throughout. The construction is analogous to the representation theorem for distributions of finite order (e.g., [Sch73, p. 91] or [Trè67, p. 259]).

Consider the jet extension linear mapping

$$j^r : C^r(\pi) \longrightarrow C^0(\pi^r) \tag{21.2}$$

as in Sect. 14.1. As noted, j^r is an embedding, and under the appropriate norms induced by atlases, it is even isometric. Evidently, due to the compatibility constraint, Image j^r is a proper subset of $C^0(\pi^r)$, which is not dense. Hence, the inverse

$$(j^r)^{-1} : \mathrm{Image}\, j^r \longrightarrow C^r(\pi) \tag{21.3}$$

is a well-defined linear homeomorphism. Given a force $F \in C^r(\pi)^*$, the linear functional

$$F \circ (j^r)^{-1} : \mathrm{Image}\, j^r \longrightarrow \mathbb{R} \tag{21.4}$$

is a continuous and linear functional on $\mathrm{Image}\, j^r$. Hence, by the Hahn-Banach theorem, it may be extended to a linear functional $\varsigma \in C^0(\pi^r)^*$. It follows that for every $w \in C^r(\pi)$,

$$\varsigma(j^r w) = F(w). \tag{21.5}$$

In other words, the linear mapping

$$j^{r*} : C^0(\pi^r)^* \longrightarrow C^r(\pi)^* \tag{21.6}$$

is surjective, and ς represents a force F if and only if

$$j^{r*}\varsigma = F. \tag{21.7}$$

The object $\varsigma \in C^0(\pi^r)^*$ is interpreted as a generalization of the notion of a hyper-stress in higher-order continuum mechanics and will be so referred to. For $r = 1$, ς is a generalization of the standard stress tensor field. The condition (21.5), resulting from the representation theorem, is a generalization of the principle of virtual work, or virtual power, as it states that the power expended by the force F for a virtual velocity field w is equal to the power expended by the hyper-stress for $j^r w$—containing the first r derivatives of the velocity field. Accordingly, Eq. (21.7) is a generalization of the equilibrium equation of continuum mechanics.

In terms of vector-valued currents, the representation theorem states that the 0-current F, which is valued in W and is dual to C^r-sections, is represented by some 0-current, ς, which is valued in $J^r W$ and is dual to C^0-sections.

It is noted that ς is not unique. The non uniqueness originates from the fact that the image of the jet extension mapping, containing the compatible jet fields, is not a dense subset of $C^0(\pi^r)$. Thus, static indeterminacy of continuum mechanics follows naturally from the representation theorem.

21.2 The Representation of Forces by Nonholonomic Stresses

In view of (14.43), the same procedure applies if we use the nonholonomic jet extension $\hat{j}^r : C^r(\pi) \to C^0(\hat{\pi}^r)$ as in Eq. (14.23). A force F may then be represented by a non unique, nonholonomic stress $\hat{\varsigma} \in C^0(\hat{\pi})^*$ in the form

$$F = \hat{j}^{r*}\hat{\varsigma}. \tag{21.8}$$

The mapping $C^0(\mathcal{J}^r)$ of Sect. 14.2.3 is an embedding. Hence, a hyper-stress ς may be represented by some non unique, nonholonomic stress $\hat{\varsigma}$ in the form

$$\varsigma = C^0(\mathcal{J}^r)^*(\hat{\varsigma}), \tag{21.9}$$

and in the following commutative diagram, all mappings are surjective.

$$C^r(\pi)^* \xleftarrow{j^{r*}} C^0(\pi^r)^* \xleftarrow{C^0(\mathcal{J}^r)^*} C^0(\hat{\pi}^r)^*, \qquad (\hat{j}^r)^*: C^0(\hat{\pi}^r)^* \to C^r(\pi)^* \tag{21.10}$$

21.3 Smooth Stresses

In view of the discussion in Sect. 17.6, hyper-stresses are elements of $C^0(\pi^r)^*$, and so, in particular, smooth sections of $\pi^{r*} \otimes \bigwedge^n T^*\mathcal{X}$, i.e., n-forms valued in the dual of the r-jet bundle, induce hyper-stresses. Thus, a *variational hyper-stress density*, a smooth section, S, of $\pi^{r*} \otimes \bigwedge^n T^*\mathcal{X}$, represents a force functional, F, by

$$F(w) = \int_{\mathcal{X}} S \cdot j^r w. \tag{21.11}$$

The integration is meaningful as $S \cdot j^r w$ is an n-form over $\mathcal{X}$.

Similarly, nonholonomic stresses are elements of $C^0(\hat{\pi}^r)^*$, and smooth nonholonomic stresses are n-forms valued in the dual of the r-iterated jet bundle. A *nonholonomic stress density* , a smooth section, $\hat{S}$, of $\hat{\pi}^{r*} \otimes \bigwedge^n T^*\mathcal{X}$ represents a force, F, by

$$F(w) = \int_{\mathcal{X}} \hat{S} \cdot \hat{j}^r w. \tag{21.12}$$

Locally, such a section is represented by the collection $(S_\alpha^{\boldsymbol{I}})$, $0 \leqslant |\boldsymbol{I}| \leqslant r$, $\alpha = 1, \dots, m$ so that the integrand, $\hat{S} \cdot \hat{j}^r w$, is represented in the form

$$S_\alpha^{\boldsymbol{I}} w_{\boldsymbol{I}}^\alpha \mathrm{d}x. \tag{21.13}$$

21.4 Stress Measures

Analytically, stresses and hyper-stresses of the types described above are vector-valued zero-currents that are representable by integration, or simply, measures. (See [Fed69a, Section 4.1], for the scalar case.)

As noted in Sect. 17.2, given a vector bundle atlas $\{(U_a, \varphi_a, \Phi_a)\}_{a\in A}$, a linear functional is uniquely determined by its restrictions to sections supported in the various domains $\varphi_\alpha(U_\alpha)$—its local representatives. In particular, for the case of an m-dimensional vector bundle $\pi : W \to \mathcal{X}$, and a functional $T \in C^0(\pi)^*$, a typical local representative is an element $T_a \in \left(C^0_c(\varphi_a(U_a))^*\right)^m$. Thus, each component $(T_a)_\alpha \in C^0_c(\varphi_a(U_a))^*$ is a Radon measure, or a distribution representable by integration, defined on $\varphi_a(U_a)$. We will use the same notation T_a, $T_{a\alpha} := (T_a)_\alpha$, for the measure. Consequently, for a section, w_a, compactly supported in $\varphi_a(U_a)$, we may write

$$T_a(w_a) = \int_{\varphi_a(U_a)} w_a \cdot \mathrm{d}T_a := \int_{\varphi_a(U_a)} w^\alpha_a \mathrm{d}T_{a\alpha} \tag{21.14}$$

(where summation over α, but not over a, is implied). Given a partition of unity $\{u_a\}$ subordinate to the atlas, one has

$$T(w) = \int_{\mathcal{X}} w \cdot \mathrm{d}T := \sum_{a\in A} \int_{\varphi_a(U_a)} \Phi_a(u_a w) \cdot \mathrm{d}T_a = \sum_{a\in A} \int_{\varphi_a(U_a)} u_a w^\alpha_a \mathrm{d}T_{a\alpha}. \tag{21.15}$$

(Alternatively, one could use the representation by local representatives of T as in Corollary 17.1, Sect. 17.3.)

For the case of hyper-stresses, one has to replace W above by $J^r W$, w^α_a by χ^α_{aI}, $|I| \leq r$, T by ς, and $T_{a\alpha}$ by $\varsigma^I_{a\alpha}$. In addition, since $\mathcal{X}$ is a manifold with corners, representing measures may be viewed as measures on the extension $\widetilde{\mathcal{X}}$, which are supported in $\mathcal{X}$. Thus, for a section χ of $J^r W$, represented locally by χ^α_{aI},

$$\varsigma(\chi) = \int_{\mathcal{X}} \chi \cdot \mathrm{d}\varsigma := \sum_{a\in A} \int_{\varphi_a(U_a)} \Phi_a(u_a \chi) \cdot \mathrm{d}\varsigma_a = \sum_{a\in A} \int_{\varphi_a(U_a)} u_a \chi^\alpha_{aI} \mathrm{d}\varsigma^I_{a\alpha}. \tag{21.16}$$

We note that the components $\varsigma^I_{a\alpha}$ have the same symmetry under permutations of I as sections of the jet bundle. If w is a section of a vector bundle W_0, then,

$$\varsigma(j^r w) = \int_{\mathcal{X}} j^r w \cdot \mathrm{d}\varsigma = \sum_{a\in A} \int_{\varphi_a(U_a)} u_a w^\alpha_{a,I} \mathrm{d}\varsigma^I_{a\alpha}. \tag{21.17}$$

The same reasoning applies to the representation by nonholonomic stresses; only here, we consider sections $\hat{\chi}$ of the iterated jet bundle represented locally by $\hat{\chi}^{\mathrm{p}\alpha_{\mathrm{p}}}_{aI_{\mathrm{p}}}$, $G_{\mathrm{p}} \leq r$. The local nonholonomic stress measures have components $\hat{\varsigma}^{\mathrm{p}I_{\mathrm{p}}}_{a\alpha_{\mathrm{p}}}$ and

$$\hat{\varsigma}(\hat{\chi}) = \int_{\mathcal{X}} \hat{\chi} \cdot \mathrm{d}\hat{\varsigma} := \sum_{a\in A} \int_{\varphi_a(U_a)} \Phi_a(u_a \hat{\chi}) \cdot \mathrm{d}\hat{\varsigma}_a = \sum_{a\in A} \int_{\varphi_a(U_a)} u_a \hat{\chi}^{\mathrm{p}\alpha_{\mathrm{p}}}_{aI_{\mathrm{p}}} \mathrm{d}\hat{\varsigma}^{\mathrm{p}I_{\mathrm{p}}}_{a\alpha_{\mathrm{p}}}, \tag{21.18}$$

$$\hat{\varsigma}(\hat{j}^r w) = \int_{\mathcal{X}} \hat{j}^r w \cdot \mathrm{d}\hat{\varsigma} = \sum_{a \in A} \int_{\varphi_a(U_a)} u_a w_{a, I_\mathfrak{p}}^{\mathfrak{p}\alpha_\mathfrak{p}} \mathrm{d}\hat{\varsigma}_{a\alpha_\mathfrak{p}}^{\mathfrak{p}I_\mathfrak{p}}, \tag{21.19}$$

where summation is implied on all values of $\alpha_\mathfrak{p}$, $I_\mathfrak{p}$, for all values of $\mathfrak{p}$ such that $G_\mathfrak{p} \leq r$.

It is concluded that for a given force F, there is some non unique vector-valued hyper-stress measure ς and some non unique nonholonomic stress measure $\hat{\varsigma}$, such that

$$F(w) = \int_{\mathcal{X}} j^r w \cdot \mathrm{d}\varsigma = \int_{\mathcal{X}} \hat{j}^r w \cdot \mathrm{d}\hat{\varsigma}. \tag{21.20}$$

21.5 Force System Induced by Stresses

It was noted in Sect. 17.3 that given a force on a body $\mathcal{X}$—an n-dimensional manifold with corners—there is no unique way to restrict it to an n-dimensional submanifold with corners, a sub-body $\mathcal{R} \subset \mathcal{X}$. We view this as the fundamental problem of continuum mechanics—static indeterminacy.

A stress distribution, although not determined uniquely by a force, provides a means for determining a *force system,* the assignment of a force $F_{\mathcal{R}}$ to each sub-body $\mathcal{R}$. Indeed, once a stress measure is given, be it a hyper-stress or a nonholonomic stress, integration theory makes it possible to consider the force system given by

$$F_{\mathcal{R}}(w) = \int_{\mathcal{R}} j^r w \cdot \mathrm{d}\varsigma = \int_{\mathcal{R}} \hat{j}^r w \cdot \mathrm{d}\hat{\varsigma}, \tag{21.21}$$

for any section w of $\pi|_{\mathcal{R}}$.

Further details on the relation between hyper-stresses and force systems are available in [Seg86, SD91]. Specifically, it is shown that given a force system that is induced by a stress measure, then the stress measure is unique. In addition, the conditions of boundedness, additivity and continuity, are shown in [SD91] to be sufficient (they are evidently necessary conditions) so that a force system $\{F_{\mathcal{R}}\}$, $\mathcal{R} \subset \mathcal{X}$ is induced by a stress measure.

It is our opinion that the foregoing line of reasoning captures the essence of stress theory in continuum mechanics accurately and elegantly.

21.6 On the Kinematic Mapping

In Chap. 2 we proposed a setting for the problem of statics for systems whose configuration spaces were finite-dimensional. We had a kinematic mapping

$$h : \mathbb{Q} \to \mathcal{P}, \tag{21.22}$$

between the two configuration spaces. For $q \in \mathbb{Q}$, the equilibrium condition for a generalized force, $g \in T^*_{h(q)}\mathcal{P}$, and a force, $F \in T^*_q Q$, was given by

$$F = (T_p h)^*(g), \tag{21.23}$$

where $(T_p h)^*$ is the dual of the tangent mapping to h at p.

For the case of continuum mechanics, we have identified so far the configuration space, $\mathbb{Q} = C^r(\mathcal{Y})$, or $Q = \mathrm{Emb}^r(\mathcal{B}, \mathcal{S})$, where the latter is an open submanifold of $C^r(\mathcal{B}, \mathcal{S})$. In this section we complete the analogy with the finite dimensional situation, and we show how the representation of forces by stresses is a natural part of this setting.

The setting outlined below uses a particular simple case of constructions and results of Palais [Pal68, Chapters 15, 17] for what he refers to as *nonlinear differential operators* and their *linearization*. The general framework for continuum statics is related to the notion of *local model* used in [ES80, Seg81, SE83] (see also [Seg86, KOS17b]).

Thus, we set

$$\mathcal{P} := C^0(J^r\mathcal{Y}), \tag{21.24}$$

that is, $\mathcal{P}$ is the Banach manifold of all continuous sections of the r-th jet bundle of $\xi : \mathcal{Y} \to \mathcal{X}$. Elements of $\mathcal{P}$ are *r-jet fields*, a generalization of the jet fields of Sect. 12.1 (see also [KOS17a]). As such, the elements of $\mathcal{P}$ generalize deformation gradient fields of order r, where for the case of manifolds, no invariant decomposition of jets into derivatives of distinct orders is available.

Naturally, *r-jet velocities* at the jet field $\chi \in \mathcal{P}$ are conceived as elements $\eta \in T_\chi\mathcal{P}$. These generalize the jet velocities of Sect. 12.1. As implied by (15.7), there is a natural isomorphism

$$T_\chi\mathcal{P} := T_\chi C^0(J^r\mathcal{Y}) \cong C^0(\chi^*(VJ^r\mathcal{Y})). \tag{21.25}$$

Correspondingly, elements of the cotangent space,

$$T^*_\chi\mathcal{P} \cong C^0(\chi^*(VJ^r\mathcal{Y}))^*, \tag{21.26}$$

may be referred to as *r-jet forces*, but are naturally identified with stress distributions.

Consequently, the role of the kinematic mapping h above is assumed by the jet extension mapping,

$$j^r_{\mathcal{Y}} : \mathbb{Q} = C^r(\mathcal{Y}) \longrightarrow \mathcal{P} = C^0(J^r\mathcal{Y}). \tag{21.27}$$

Hence, velocities are related by the tangent mapping,

$$Tj^r_{\mathcal{Y}} : T\mathbb{Q} = TC^r(\mathcal{Y}) \longrightarrow T\mathcal{P} = TC^0(J^r\mathcal{Y}). \tag{21.28}$$

In particular, for $\kappa \in C^r(\mathcal{Y})$, and $\chi = j^r_{\mathcal{Y}}\kappa$,

$$\begin{aligned} T_\kappa j^r_{\mathcal{Y}} : T_\kappa\mathbb{Q} = T_\kappa C^r(\mathcal{Y}) \cong C^r(\kappa^* V\mathcal{Y}) &\longrightarrow T_\chi\mathcal{P} \\ &= T_\chi C^0(J^r\mathcal{Y}) \cong C^0(\chi^* V J^r\mathcal{Y}). \end{aligned} \tag{21.29}$$

By Palais [Pal68, Theorem 17.1, p. 82], in analogy with (12.51), for $\chi = j^r_{\mathcal{Y}}\kappa$, there is a natural isomorphism,

$$\chi^* V J^r\mathcal{Y} \cong J^r(\kappa^* V\mathcal{Y}). \tag{21.30}$$

Thus, we may write the previous equation as

$$T_\kappa j^r_{\mathcal{Y}} : T_\kappa\mathbb{Q} \cong C^r(\kappa^* V\mathcal{Y}) \longrightarrow T_\chi\mathcal{P} \cong C^0(J^r(\kappa^* V\mathcal{Y})). \tag{21.31}$$

One is led, therefore, to consider the tangent map $T_\kappa j^r_{\mathcal{Y}}$. In the terminology of [Pal68, Chapter 17], the tangent map is the "linearization" of the jet extension mapping $j^r_{\mathcal{Y}} : C^r(\mathcal{Y}) \to C^0(J^r\mathcal{Y})$. In view of Chap. 15, we may use sections of a vector bundle neighborhood of κ to represent a neighborhood of κ in $C^r(\mathcal{Y})$ and apply the jet extension to them. Thus, observing that the jet extension mapping for sections of a vector bundles is linear, it is naturally identified with its tangent at κ. (See [Pal68, Chapter 17] for a thorough exposition of the subject.) It is concluded that

$$T_\kappa j^r_{\mathcal{Y}} = j^r : C^r(\kappa^* V\mathcal{Y}) \longrightarrow C^0(J^r(\kappa^* V\mathcal{Y})). \tag{21.32}$$

Hence, the condition that an element

$$\varsigma \in T^*_{j^r_{\mathcal{Y}}\kappa}\mathcal{P} \cong C^0(J^r(\kappa^* V\mathcal{Y}))^* \tag{21.33}$$

is in equilibrium with a given force

$$F \in T^*_\kappa\mathbb{Q} \cong C^r(\kappa^* V\mathcal{Y})^*, \tag{21.34}$$

is

$$F = j^{r*}(\varsigma), \tag{21.35}$$

where

$$j^{r*} = (T_\kappa j^r_{\mathcal{Y}})^* : C^0(J^r(\kappa^* V\mathcal{Y}))^* \longrightarrow C^r(\kappa^* V\mathcal{Y})^* \tag{21.36}$$

is the dual to the tangent of the jet extension mapping. This is evidently in complete agreement with the representation procedure for forces by stresses as presented in Sect. 21.1.

21.7 Global Elastic Constitutive Equations and the Problem of Elasticity

We now generalize the definitions of Sect. 12.2.5 to the global (non-local) non-smooth case.

A *global loading* for a body $\mathcal{X}$ is a section, a 1-form,

$$\Phi_{\mathbb{Q}} : \mathbb{Q} = C^r(\mathcal{Y}) \longrightarrow T^*\mathbb{Q} = T^*C^r(\mathcal{Y}). \tag{21.37}$$

As expected, a loading form associates generalized forces with the various configurations of the body.

In analogy, a *global constitutive relation* for $\mathcal{X}$ is a section, a 1-form

$$\Phi_{\mathcal{P}} : \mathcal{P} = C^0(J^r\mathcal{Y}) \longrightarrow T^*\mathcal{P} = T^*C^0(J^r\mathcal{Y}). \tag{21.38}$$

A global constitutive relation associates stress distributions to jet fields, where the latter need not necessarily be integrable.

Remark 21.1 It is noted that by definition, for a fixed base manifold $\mathcal{X}$, a global constitutive relation, assigning a stress distribution to a jet field, is non-local. The value, $\chi(y)$, $\chi \in \mathcal{P}$, $y \in J^r\mathcal{Y}$, may, in general, influence the stress distribution over the whole body. However, in continuum mechanics it is customary to postulate the principle of determinism of the stress (see [TN65, p. 56]) implying, for elastic materials, that the stress in each body be determined by the configuration of the body. This implies, in particular, that for every sub-body $\mathcal{X}' \subset \mathcal{X}$, there is a global constitutive relation

$$\Phi_{\mathcal{P}'} : \mathcal{P}' = C^0(J^r\mathcal{Y}|_{\mathcal{X}'}) \longrightarrow T^*\mathcal{P}' = T^*C^0(J^r\mathcal{Y}|_{\mathcal{X}'}). \tag{21.39}$$

In addition, since the stress distribution, as a measure, may be restricted to sub-bodies, the principle of determinism implies that

$$\Phi_{\mathcal{P}'}(\chi|_{\mathcal{X}'}) = \Phi_{\mathcal{P}}(\chi)|_{\mathcal{X}'}, \quad \text{for all} \quad \chi \in \mathcal{P},\ \mathcal{X}' \subset \mathcal{X}. \tag{21.40}$$

In other words, the stress distribution in an arbitrarily small open set is determined by the values of χ in this small set. We refer to this property as *germ locality*. In particular, it is sufficient to determine the constitutive relation for sub bodies in the domains of fiber bundle charts.

Stronger results follow from further regularity assumptions. Assume that the stress measures resulting from continuous sections in $\mathcal{P}$ are represented by continuous sections of $L(VJ^r\mathcal{Y}, \pi^{r*}\bigwedge^d T^*\mathcal{X})$ (see Sect. 12.2.4). It follows that the value of a stress distribution at a point $x \in \mathcal{X}$ is meaningful. Moreover, endow Image$\Phi_{\mathcal{P}} \subset C^0(L(VJ^r\mathcal{Y}, \pi^{r*}\bigwedge^d T^*\mathcal{X}))$ with the C^0-topology, and assume that for this topology, the mapping $\Phi_{\mathcal{P}}$ is continuous. Then, one can show that the constitutive relation $\Phi_{\mathcal{P}}$ is actually local. That is, for each point $x \in \mathcal{X}$, there is a continuous mapping,

$$\Phi_x : J^r_x\mathcal{Y} \longrightarrow L(VJ^1\mathcal{Y}, \pi^{1*}\bigwedge^d T^*\mathcal{X})|_x, \quad \text{such that} \quad (\Phi_{\mathcal{P}}(\chi))(x) = \Phi_x(\chi(x)). \tag{21.41}$$

For the proof (using Whitney's extension theorem), and other results on locality of constitutive relations, see [Seg88].

The jet extension kinematic mapping $j^r_{\mathcal{Y}}$ of (21.27) may be used to pull back the 1-form $\Phi_{\mathcal{P}}$ onto $\mathbb{Q}$. Specifically, we have the one-form

$$(Tj^r_{\mathcal{Y}})^*\Phi_{\mathcal{P}} : \mathbb{Q} \longrightarrow T^*\mathbb{Q} \tag{21.42}$$

given by

$$\begin{aligned}(Tj^r_{\mathcal{Y}})^*\Phi_{\mathcal{P}}(\kappa)(w) &= \Phi_{\mathcal{P}}(j^r_{\mathcal{Y}}\kappa)(T_\kappa j^r_{\mathcal{Y}}(w)),\\ &= \Phi_{\mathcal{P}}(j^r_{\mathcal{Y}}\kappa)(j^r w),\end{aligned} \tag{21.43}$$

for all $\kappa \in \mathbb{Q}$, $w \in T_\kappa\mathbb{Q}$. Clearly, the pullback of $\Phi_{\mathcal{P}}$ is a 1-form on $\mathbb{Q}$, an object of the same type as a loading on the body.

Thus, the problem of elasticity of Sect. 12.2.5 is now generalized naturally as follows. Given a loading $\Phi_{\mathbb{Q}}$ and a constitutive relation $\Phi_{\mathcal{P}}$, find a configuration $\kappa \in \mathbb{Q}$ such that

$$(Tj^r_{\mathcal{Y}})^*\Phi_{\mathcal{P}}(\kappa) = \Phi_{\mathbb{Q}}(\kappa), \tag{21.44}$$

specifically,

$$\Phi_{\mathcal{P}}(j^r_{\mathcal{Y}}\kappa)(j^r w) = \Phi_{\mathbb{Q}}(\kappa)(w), \quad \text{for all} \quad w \in T_\kappa\mathbb{Q}. \tag{21.45}$$

Chapter 22
Simple Forces and Stresses

We restrict ourselves now to the most natural setting for continuum mechanics, the case $r = 1$—the first value for which the set of C^r-embeddings is open in the manifold of mappings. (See [FS15] for consideration of configurations modeled as Lipschitz mappings.) Evidently, hyper-stresses and nonholonomic stresses become identical now, and therefore, it is natural in this case to use the terminology *simple forces* and *stresses*.

Consequently, this chapter presents the weak, non-smooth analog, of Sects. 10.7 and 10.8.

22.1 Simple Variational Stresses

A simple stress ς on a body $\mathcal{X}$ is an element of $C^0(J^1W)^*$, which implies that smooth stress distributions are sections of

$$(J^1W)^* \otimes \bigwedge^n T^*\mathcal{X} = L\big(J^1W, \bigwedge^n T^*\mathcal{X}\big). \tag{22.1}$$

Following the discussion in Sect. 17.5, ς may be viewed as a generalized section of $(J^1\tilde{W})^* \otimes \bigwedge^n T^*\tilde{\mathcal{X}}$, which is supported in $\mathcal{X}$, where we use the extension of the vector bundle $\pi : W \to \mathcal{X}$ to a vector bundle $\tilde{\pi} : \tilde{W} \to \tilde{\mathcal{X}}$ over a compact manifold without boundary $\tilde{\mathcal{X}}$.

We conclude from Sect. 21.1 that simple variational stresses are vector-valued measures, ς, that represent forces in the C^1-case, in the form

$$F = j^{1*}(\varsigma), \quad \text{or,} \quad F_{\mathcal{R}}(w) = \int_{\mathcal{R}} j^1 w \cdot \mathrm{d}\varsigma. \tag{22.2}$$

In the second equation above, we emphasized the fact that a stress induces a force system on the collections of subbodies of a given body.

R. Segev, *Foundations of Geometric Continuum Mechanics*, Advances in Continuum Mechanics 49, https://doi.org/10.1007/978-3-031-35655-1_22

A section, χ, supported in the domain of a chart of the jet bundle, is represented locally in the form

$$\chi^{\alpha} g_{\alpha} + \chi_i^{\alpha} \mathrm{d}x^i \otimes g_{\alpha}, \tag{22.3}$$

so that locally,

$$\begin{aligned} \varsigma(\chi) &= \varsigma(\chi^{\alpha} g_{\alpha} + \chi_i^{\alpha} \mathrm{d}x^i \otimes g_{\alpha}), \\ &= \varsigma_{\alpha}(\chi^{\alpha}) + \varsigma_{\alpha}^i(\chi_i^{\alpha}). \end{aligned} \tag{22.4}$$

Here, ς_{α} and ς_{α}^i are 0-currents, represented by Radon measures, defined by

$$\varsigma_{\alpha}(u) := (g_{\alpha} \cdot \varsigma)(u) = \varsigma(u g_{\alpha}), \tag{22.5}$$

$$\varsigma_{\alpha}^i(u) := ((\mathrm{d}x^i \otimes g_{\alpha}) \cdot \varsigma)(u) = \varsigma(u \mathrm{d}x^i \otimes g_{\alpha}), \tag{22.6}$$

where u is any continuous function that is compactly supported in the domain of the chart.

In accordance with Sect. 10.7, in the smooth case, ς is represented by a section S of $(J^1 W)^* \otimes \bigwedge^n T^* \mathcal{X}$ in the form

$$\varsigma(\chi) = \int_{\mathcal{X}} S \cdot \chi. \tag{22.7}$$

Locally, such a vector-valued form is represented as

$$S = (R_{\alpha} g^{\alpha} + S_{\alpha}^i \partial_i \otimes g^{\alpha}) \otimes \mathrm{d}x, \tag{22.8}$$

and for the domain of a chart, U, and a section χ with $\operatorname{supp}\chi \subset U$,

$$\varsigma(\chi) = \int_U (R_{\alpha} \chi^{\alpha} + S_{\alpha}^i \chi_i^{\alpha}) \mathrm{d}x. \tag{22.9}$$

22.2 The Vertical Projection

As in Sect. 10.7, we make use of the vertical sub-bundle,

$$V_0^1 \pi^1 : V_0^1 J^1 W \longrightarrow \mathcal{X} \tag{22.10}$$

—the kernel of the natural projection

$$\pi_0^1 : J^1 W \longrightarrow W. \tag{22.11}$$

It is recalled that the vertical sub-bundle may be identified with the vector bundle $T^*\mathcal{X} \otimes W$, and denoting the natural inclusion by

$$\mathcal{I}_{V_0^1} : V_0^1 J^1 W \longrightarrow J^1 W, \tag{22.12}$$

one has the induced inclusion

$$C^0(\mathcal{I}_{V_0^1}) : C^0(V_0^1 \pi^1) \longrightarrow C^0(\pi^1), \qquad \chi \longmapsto \mathcal{I}_{V_0^1} \circ \chi. \tag{22.13}$$

Clearly, $C^0(\mathcal{I}_{V_0^1})$ is injective and a homeomorphism onto its image. Hence, its dual

$$C^0(\mathcal{I}_{V_0^1})^* : C^0(\pi^1)^* \longrightarrow C^0(V_0^1 \pi^1)^* \cong C^0(T^*\mathcal{X} \otimes W)^* \tag{22.14}$$

is a well-defined surjection. Simply put, $C^0(\mathcal{I}_{V_0^1})^*(\varsigma)$ is the restriction of the stress ς to sections of the vertical sub-bundle. Accordingly, we will refer to an element $\varsigma^+ \in C^0(T^*\mathcal{X} \otimes W)^*$ as a *vertical stress* and to $C^0(\mathcal{I}_{V_0^1})^*$ as the *vertical projection*.

In case the stress ς is represented locally by the 0-currents $(\varsigma_\alpha, \varsigma_\beta^i)$ as in (22.4), then $C^0(\mathcal{I}_{V_0^1})^*(\varsigma)$ is represented by (ς_β^i).

Let $\varsigma^+ \in C^0(T^*\mathcal{X} \otimes W)^*$ be a vertical stress and let $w \in C^0(\pi)$. Then, $\varsigma^+ \cdot w$, defined by

$$(\varsigma^+ \cdot w)(\varphi) := \varsigma^+(\varphi \otimes w), \qquad \varphi \in C^0(T^*\mathcal{X}) \tag{22.15}$$

is a (scalar-valued) 1-current. This is an indication of the fact that ς^+ may be viewed as a vector-valued 1-current (assuming we ignore the transposition in comparison with (20.1) and (20.17)).

We may use the procedures for the local representation of currents as in Sect. 20.2 to represent ς^+ by the scalar 1-currents $\varsigma_\alpha^+ := (\varsigma^+ \cdot g_\alpha)$ defined locally. For a form φ, compactly supported in a coordinate neighborhood, we set

$$\varsigma_\alpha^+(\varphi) = (\varsigma^+ \cdot g_\alpha)(\varphi) := \varsigma^+(\varphi \otimes g_\alpha). \tag{22.16}$$

As

$$\begin{aligned}
\varsigma^+(\varphi \otimes w) &= \varsigma^+(\varphi \otimes g_\alpha w^\alpha), \\
&= (\varsigma^+ \cdot g_\alpha)(\varphi w^\alpha), \\
&= (\varsigma^+ \cdot g_\alpha)(\varphi(g^\alpha \cdot w)), \\
&= (\varsigma_\alpha^+ \otimes g^\alpha)(\varphi \otimes w),
\end{aligned}$$

we infer that

$$\varsigma^+ = \varsigma^+_\alpha \otimes g^\alpha. \tag{22.17}$$

Note that

$$\begin{aligned}
\{\partial_i \wedge [(\mathrm{d}x^i \lrcorner (\varsigma^+ \cdot g_\alpha)) \otimes g^\alpha]\}(\varphi \otimes w) &= [(\mathrm{d}x^i \lrcorner (\varsigma^+ \cdot g_\alpha)) \otimes g^\alpha]((\partial_i \lrcorner \varphi) \otimes w),\\
&= [(\mathrm{d}x^i \lrcorner (\varsigma^+ \cdot g_\alpha)) \otimes g^\alpha](\varphi_i w),\\
&= [\mathrm{d}x^i \lrcorner (\varsigma^+ \cdot g_\alpha)](\varphi_i g^\alpha \cdot w),\\
&= [\mathrm{d}x^i \lrcorner (\varsigma^+ \cdot g_\alpha)](\varphi_i w^\alpha),\\
&= [(\varsigma^+ \cdot g_\alpha)](\mathrm{d}x^i \wedge \varphi_i w^\alpha),\\
&= \varsigma^+(\varphi \otimes g_\alpha w^\alpha),\\
\{[\partial_i \wedge ((\mathrm{d}x^i \lrcorner \varsigma^+) \cdot g_\alpha)] \otimes g^\alpha\}(\varphi \otimes w) &= \varsigma^+(\varphi \otimes w).
\end{aligned} \tag{22.18}$$

Thus, setting

$$\varsigma^{+i}_\alpha := \mathrm{d}x^i \lrcorner (\varsigma^+ \cdot g_\alpha) = (\mathrm{d}x^i \lrcorner \varsigma^+) \cdot g_\alpha, \tag{22.19}$$

we may write

$$\varsigma^+ = \partial_i \wedge (\varsigma^{+i}_\alpha \otimes g^\alpha) = (\partial_i \wedge \varsigma^{+i}_\alpha) \otimes g^\alpha. \tag{22.20}$$

Finally, it is observed that in case $\varsigma^+ = C^0(\mathcal{J}_{V^1_0})^*(\varsigma)$, then $\varsigma^{+i}_\alpha = \varsigma^i_\alpha$.

Alternatively, one may represent ς^+ by n-currents. We first note that

$$\begin{aligned}
\{\mathrm{d}x^{\hat{\imath}} \lrcorner [(\partial_{\hat{\imath}} \wedge (\varsigma^+ \cdot g_\alpha)) \otimes g^\alpha]\}(\varphi \otimes w) &= [(\partial_{\hat{\imath}} \wedge (\varsigma^+ \cdot g_\alpha)) \otimes g^\alpha]((\mathrm{d}x^{\hat{\imath}} \wedge \varphi) \otimes w),\\
&= [(\partial_{\hat{\imath}} \lrcorner (\varsigma^+ \cdot g_\alpha)) \otimes g^\alpha](\varepsilon^{\hat{\imath}j}\varphi_j \mathrm{d}x \otimes w),\\
&= (\partial_{\hat{\imath}} \lrcorner (\varsigma^+ \cdot g_\alpha))(\varepsilon^{\hat{\imath}j}\varphi_j \mathrm{d}x(g^\alpha \cdot w)),\\
&= [(\varsigma^+ \cdot g_\alpha)](\varepsilon^{\hat{\imath}j}\varphi_j \partial_{\hat{\imath}} \lrcorner \mathrm{d}x w^\alpha),\\
&= [(\varsigma^+ \cdot g_\alpha)](\varepsilon^{\hat{\imath}j}\varphi_j \varepsilon_{\hat{\imath}k} \mathrm{d}x^k w^\alpha),\\
&= \varsigma^+(\varphi \otimes g_\alpha w^\alpha),\\
\{[\mathrm{d}x^{\hat{\imath}} \lrcorner ((\partial_{\hat{\imath}} \wedge \varsigma^+) \cdot g_\alpha)] \otimes g^\alpha\}(\varphi \otimes w) &= \varsigma^+(\varphi \otimes w).
\end{aligned} \tag{22.21}$$

Hence, in analogy with Sect. 20.2.4, we set

$$\varsigma^+_{\hat{\imath}\alpha} := (\varsigma^+_\alpha)_{\hat{\imath}} = \partial_{\hat{\imath}} \wedge \varsigma^+_\alpha, \qquad \varsigma^{+\prime}{}_{\hat{\imath}\alpha} = (\varsigma^+_\alpha)'_{\hat{\imath}} = \varsigma^+_\alpha \wedge \partial_{\hat{\imath}}, \tag{22.22}$$

so that

$$\varsigma^+_\alpha = \mathrm{d}x^{\hat{\imath}} \lrcorner \varsigma^+_{\hat{\imath}\alpha} = \varsigma^{+\prime}{}_{\hat{\imath}\alpha} \llcorner \mathrm{d}x^{\hat{\imath}}. \tag{22.23}$$

As in (20.35), one has

$$\varsigma^+ = (\mathrm{d}x^{\hat{\imath}} \lrcorner \varsigma^+_{\hat{\imath}\alpha}) \otimes g^\alpha = (\varsigma^{+\prime}{}_{\hat{\imath}\alpha} \llcorner \mathrm{d}x^{\hat{\imath}}) \otimes g^\alpha. \tag{22.24}$$

In the smooth case, in agreement with Sect. 10.7, the vertical projection of the stress is represented by a section S^+ of $T\mathcal{X} \otimes W^* \otimes \bigwedge^n T^*\mathcal{X}$. Its action satisfies

$$C^0(\mathcal{J}_{V^1_0})^*(\varsigma)(\chi) = \int_{\mathcal{X}} S^+ \cdot \chi, \tag{22.25}$$

where the n-form $S^+ \cdot \chi$ is given by

$$(S^+ \cdot \chi)(v_1, \ldots, v_n)(X) = S^+(X)(\chi(X) \otimes (v_1(X) \wedge \cdots \wedge v_n(X))). \tag{22.26}$$

For a vertical stress ς^+ that is represented locally by smooth S^i_α as above and a field w, the 1-current $\varsigma^+ \cdot w$ is given locally by

$$\varphi \longmapsto \int_U S^i_\alpha w^\alpha \varphi_i \mathrm{d}x, \tag{22.27}$$

for a 1-form φ supported in U. In other words, if the vertical stress ς^+ is represented by the section

$$S^+ = S^i_\alpha \partial_i \otimes g^\alpha \otimes \mathrm{d}x \tag{22.28}$$

of $T\mathcal{X} \otimes W^* \otimes \bigwedge^n T^*\mathcal{X}$, then, the 1-current $\varsigma^+ \cdot w$ is represented by the density $S^+ \cdot w$, a section of $T\mathcal{X} \otimes \bigwedge^n T^*\mathcal{X}$, given by

$$(S^+ \cdot w)(\varphi) = S^+(\varphi \otimes w). \tag{22.29}$$

22.3 Traction Stresses

Using the transposition $\mathrm{tr} : W \otimes T^*\mathcal{X} \to T^*\mathcal{X} \otimes W$, one has a mapping on the space of vertical stresses

$$C^0(\mathrm{tr})^* : C^0(T^*\mathcal{X} \otimes W)^* \longrightarrow C^0(W \otimes T^*\mathcal{X})^*. \tag{22.30}$$

We define *traction stress distributions* to be elements of $C^0(W \otimes T^*\mathcal{X})^*$. It is noted that a traction stress is not much different than a vertical stress distribution, but transposition enables its representation as a vector-valued current.

Using local representation in accordance with Sects. 20.2.4 and 20.2.7, a traction stress σ is represented locally by n-currents $\sigma_{\alpha\hat{\imath}}$, $|\hat{\imath}| = n-1$, in the form

$$\sigma = \mathrm{d}x^{\hat{\imath}} \lrcorner\, \sigma_{\hat{\imath}} = g^{\alpha} \otimes \sigma_{\alpha} = \mathrm{d}x^{\hat{\imath}} \lrcorner\, (g^{\alpha} \otimes \sigma_{\alpha\hat{\imath}}) = g^{\alpha} \otimes (\mathrm{d}x^{\hat{\imath}} \lrcorner\, \sigma_{\alpha\hat{\imath}}), \tag{22.31}$$

where

$$\sigma_{\hat{\imath}} := \partial_{\hat{\imath}} \wedge \sigma, \qquad \sigma_{\alpha} := g_{\alpha} \cdot \sigma, \qquad \sigma_{\alpha\hat{\imath}} := g_{\alpha} \cdot \sigma_{\hat{\imath}} = (\sigma_{\alpha})_{\hat{\imath}} = g_{\alpha} \cdot (\partial_{\hat{\imath}} \wedge \sigma). \tag{22.32}$$

Hence, in analogy with (18.38),

$$\begin{aligned} \sigma(w \otimes \varphi) &= [\mathrm{d}x^{\hat{\imath}} \lrcorner\, (g^{\alpha} \otimes \sigma_{\alpha\hat{\imath}})](w \otimes \varphi), \\ &= \sigma_{\alpha\hat{\imath}}(w^{\alpha} \mathrm{d}x^{\hat{\imath}} \wedge \varphi) \\ &= \varepsilon^{\hat{\imath}i} \sigma_{\alpha\hat{\imath}}(w^{\alpha} \varphi_i \mathrm{d}x), \\ &= \sum_i (-1)^{n-i} \sigma_{\alpha\hat{\imath}}(w^{\alpha} \varphi_i \mathrm{d}x). \end{aligned} \tag{22.33}$$

We can also write

$$\sigma(w \otimes \varphi) = \mathrm{d}x^{\hat{\imath}} \lrcorner\, (\sigma \cdot w)_{\hat{\imath}}, \qquad \text{where,} \quad (\sigma \cdot w)_{\hat{\imath}} := \partial_{\hat{\imath}} \wedge (\sigma \cdot w) \tag{22.34}$$

are scalar n-currents.

Alternatively, as in Sect. 20.2.3, we can represent σ by the scalar local 0-currents σ^i_{α} as

$$\sigma = g^{\alpha} \otimes (\partial_i \wedge \sigma^i_{\alpha}), \tag{22.35}$$

where

$$\sigma^i_{\alpha} := (\sigma_{\alpha})^i = \mathrm{d}x^i \lrcorner\, \sigma_{\alpha} = \mathrm{d}x^i \lrcorner\, (g_{\alpha} \cdot \sigma). \tag{22.36}$$

In addition,

$$(\sigma \cdot w)^i := \mathrm{d}x^i \lrcorner\, (\sigma \cdot w), \qquad \text{so that} \quad \sigma(w \otimes \varphi) = \partial_i \wedge (\sigma \cdot w)^i. \tag{22.37}$$

The main conclusion of this section follows from the foregoing analysis.

Corollary 22.1 *Introducing the notation*

$$p_{\sigma} := C^0(\mathrm{tr})^* \circ C^0(\mathcal{J}_{V^1_0})^* : C^0(\pi^1)^* \longrightarrow C^0(W \otimes T^*\mathcal{X})^*, \tag{22.38}$$

a simple stress distribution ς induces a unique traction stress distribution σ by

$$\sigma = p_\sigma(\varsigma). \tag{22.39}$$

For the non-smooth case, the relation between variational stresses and traction stresses may also be introduced naturally via the following observation.

Proposition 22.1 *Let u be a C^1-function on $\mathcal{X}$, let ς be a variational stress measure, and let $\sigma = p_\sigma(\varsigma)$. Then, for each $w \in C^1(\pi)$,*

$$F(uw) = \varsigma(j^1(uw)) = \varsigma(uj^1w) + \sigma(w \otimes \mathrm{d}u). \tag{22.40}$$

Proof We first note that if u is a differentiable function on a body, then, the product rule for the jet of the product, uw, with a vector field w—the analog of the Leibniz rule—is

$$j^1(uw) = uj^1w + \mathcal{I}_{V_0^1}(\mathrm{d}u \otimes w), \tag{22.41}$$

where $\mathcal{I}_{V_0^1}$ is the inclusion of the vertical subbundle as in (22.12). Since the last equation is formulated invariantly, it suffices to prove it using local representation. Indeed, if w is represented locally by w^α, then, $j^1(uw)$ is represented locally by

$$\begin{aligned}(uw^\alpha, (uw^\beta)_{,i}) &= (uw^\alpha, u_{,i}w^\beta + uw^\beta_{,i}),\\ &= u(w^\alpha, w^\beta_{,i}) + (0, (\mathrm{d}u \otimes w)^\beta_{,i}).\end{aligned} \tag{22.42}$$

Since $(0, (\mathrm{d}u \otimes w)^\beta_{,i})$ is the local representative of $\mathcal{I}_{V_0^1}(\mathrm{d}u \otimes w)$, (22.41) follows.

Linearity implies that

$$\begin{aligned}F(uw) = \varsigma(j^1(uw)) &= \varsigma(uj^1w) + \varsigma[\mathcal{I}_{V_0^1}(\mathrm{d}u \otimes w)],\\ &= \varsigma(uj^1w) + \varsigma^+(\mathrm{d}u \otimes w),\\ &= \varsigma(uj^1w) + \sigma(w \otimes \mathrm{d}u),\end{aligned} \tag{22.43}$$

as asserted.

Let $\sigma = p_\sigma(\varsigma)$; then, comparing the last equation with (22.18), it is concluded that, in accordance with (20.55), locally,

$$(-1)^{n-i}\sigma_{\alpha\hat{i}}(u\mathrm{d}x) = \varsigma^i_\alpha(u) \tag{22.44}$$

for any function u, and so

$$\mathrm{d}x \lrcorner \sigma_{\alpha\hat{i}} = (-1)^{n-i}\varsigma^i_\alpha, \qquad \sigma_{\alpha\hat{i}} = (-1)^{n-i}\partial_x \wedge \varsigma^i_\alpha. \tag{22.45}$$

Finally, by (22.44), (22.33),

$$\sigma(w \otimes \varphi) = \varsigma_{\alpha}^{i}(w^{\alpha}\varphi_i). \tag{22.46}$$

Remark 22.1 Continuing Remark 20.4, it is observed that one may consider

$$\sigma'_{\hat{\imath}} := \sigma \wedge \partial_{\hat{\imath}}, \qquad \sigma'_{\alpha\hat{\imath}} := g_{\alpha} \cdot \sigma'_{\hat{\imath}} = (\sigma_{\alpha})'_{\hat{\imath}} = g_{\alpha} \cdot (\sigma \wedge \partial_{\hat{\imath}}), \tag{22.47}$$

so that

$$\sigma = \sigma'_{\hat{\imath}} \llcorner \mathrm{d}x^{\hat{\imath}} = (g^{\alpha} \otimes \sigma'_{\alpha\hat{\imath}}) \llcorner \mathrm{d}x^{\hat{\imath}}. \tag{22.48}$$

Hence,

$$\begin{aligned} \sigma(w \otimes \varphi) &= (g^{\alpha} \otimes \sigma'_{\alpha\hat{\imath}}) \llcorner \mathrm{d}x^{\hat{\imath}}(w \otimes \varphi), \\ &= \sigma'_{\alpha\hat{\imath}}(w^{\alpha}\varphi \wedge \mathrm{d}x^{\hat{\imath}}), \\ \varsigma_{\alpha}^{+i}(w^{\alpha}\varphi_i) &= \sum_i (-1)^{i-1}\sigma'_{\alpha\hat{\imath}}(\varphi_i w^{\alpha}\mathrm{d}x), \end{aligned} \tag{22.49}$$

where we used (22.18) in the last line. The last computation and comparison with (22.45) imply that

$$\sigma'_{\alpha\hat{\imath}} \llcorner \mathrm{d}x = (-1)^{i-1}\varsigma_{\alpha}^{i}, \qquad \sigma'_{\alpha\hat{\imath}} = (-1)^{i-1}\varsigma_{\alpha}^{i} \wedge \partial_x = (-1)^{n-1}\sigma_{\alpha\hat{\imath}}. \tag{22.50}$$

It is noted that as ς_{α}^{i} are zero currents, the order of the contraction and wedge product in the last equation is immaterial.

In summary, overlooking the transposition, traction stresses and vertical components of simple variational stresses are essentially identical objects, represented by either n-currents, for traction stresses, or 0-currents, for vertical components.

22.4 Smooth Traction Stresses

In the smooth case, we adapt the constructions of Sect. 20 to the current context. A traction stress σ may be represented by sections $s_{\lrcorner}$ and $s_{\llcorner}$ of $W^* \otimes \bigwedge^{n-1} T^*\mathcal{X}$, so that

$$\sigma = s_{\lrcorner} \lrcorner T_{\mathcal{X}} = T_{\mathcal{X}} \llcorner s_{\llcorner}, \tag{22.51}$$

in accordance with (20.16), (20.10), (20.11), and

$$\sigma(\chi) = \int_{\mathcal{X}} s_{\lrcorner} \dot{\wedge} \chi = \int_{\mathcal{X}} \chi \dot{\wedge} s_{\llcorner}, \quad s_{\lrcorner} = (-1)^{(n-1)} s_{\llcorner}. \tag{22.52}$$

Locally,

$$
\begin{aligned}
s_{\lrcorner} &= s_{\lrcorner\alpha\hat{\imath}} g^{\alpha} \otimes \mathrm{d}x^{\hat{\imath}} = \varepsilon^{\hat{\imath}i} s_{\lrcorner\alpha\hat{\imath}} g^{\alpha} \otimes (\mathrm{d}x \llcorner \partial_i), \qquad & s_{\lrcorner} \dot{\wedge} \chi &= \varepsilon^{\hat{\imath}i} s_{\lrcorner\alpha\hat{\imath}} \chi_i^{\alpha} \mathrm{d}x, \\
s_{\llcorner} &= s_{\llcorner\alpha\hat{\imath}} g^{\alpha} \otimes \mathrm{d}x^{\hat{\imath}} = \varepsilon^{\hat{\imath}i} s_{\llcorner\alpha\hat{\imath}} g^{\alpha} \otimes (\mathrm{d}x \llcorner \partial_i), \qquad & \chi \dot{\wedge} s_{\llcorner} &= \varepsilon^{i\hat{\imath}} s_{\llcorner\alpha\hat{\imath}} \chi_i^{\alpha} \mathrm{d}x.
\end{aligned}
\tag{22.53}
$$

Let ς be a smooth stress represented by the vector-valued form S, a section of $(J^1 W)^* \otimes \bigwedge^n T^*\mathcal{X}$ as in (22.8), (22.9), and let S^+ be its vertical component as in (22.28). Then, using (20.12),

$$
\begin{aligned}
s_{\lrcorner\alpha\hat{\imath}} &= \varepsilon_{\hat{\imath}i} S_{\alpha}^{i} = (-1)^{n-i} S_{\alpha}^{i}, \\
s_{\llcorner\alpha\hat{\imath}} &= \varepsilon_{i\hat{\imath}} S_{\alpha}^{i} = (-1)^{i-1} S_{\alpha}^{i}, \\
s_{\llcorner\alpha\hat{\imath}} &= (-1)^{n-1} s_{\lrcorner\alpha\hat{\imath}}
\end{aligned}
\tag{22.54}
$$

(cf. (10.62)).

22.5 Further Aspects of Stress Representation

We consider further aspects of non-smooth variational stress distributions and their relations to forces and traction stresses.

Consider a force, F, and a generalized velocity field, $w \in C^1(W)$. Then, viewing F as a W-valued 0-current acting on C^1-sections, and using the notation of Sect. 20.2.1, we may consider the scalar-valued 0-current, a distribution,

$$
w \cdot F \in C^1(\mathcal{X})^*, \quad \text{defined as} \quad (w \cdot F)(u) = F(uw). \tag{22.55}
$$

We can also use the notation introduced in (20.18), (20.19) and consider the associated mapping

$$
F_{\circledcirc} = \circledcirc(F) : C^1(W) \longrightarrow C^1(\mathcal{X})^*, \qquad F_{\circledcirc}(w) := w \cdot F. \tag{22.56}
$$

Similarly, for a variational stress measure, ς, and a section, $\chi \in C^0(J^1 W)$, we consider

$$
\chi \cdot \varsigma \in C^0(\mathcal{X})^*, \qquad (\chi \cdot \varsigma)(u) = \varsigma(u\chi), \tag{22.57}
$$

and the corresponding

$$
\varsigma_{\circledcirc} = \circledcirc(\varsigma) : C^0(J^1 W) \longrightarrow C^0(\mathcal{X})^*, \qquad \varsigma_{\circledcirc}(\chi) := \chi \cdot \varsigma. \tag{22.58}
$$

In particular, for a generalized velocity $w \in C^1(W)$,

$$(j^1 w \cdot \varsigma)(u) = \varsigma(u j^1 w). \tag{22.59}$$

Remark 22.2 It is observed that, in general, $uj^1 w$ is not necessarily integrable, that is, there need not be a generalized velocity w' such that $j^1 w' = u j^1 w$. Thus, the functional $j^1 w \cdot \varsigma = \varsigma_{\circledcirc}(j^1 w)$ depends on ς and not only on the force $F = j^{1*}\varsigma$ that it represents. In other words, the action of the functional $j^1 w \cdot \varsigma$ on the various functions u enables one to evaluate power for incompatible sections of the jet bundle.

In the sequel, we will refer to scalar-valued 0-currents, such as $w \cdot F$ and $j^1 w \cdot \varsigma$, as *induced functionals* or *induced distributions.*

Let R be a scalar-valued 0-current acting on C^r-functions. Define the *integration operation* by setting (see [GKOS01, pp. 249-250] and [Mel96, p. III.4])

$$\text{int}(R) = \int R := R(\mathbf{1}_{\mathcal{X}}), \tag{22.60}$$

where $\mathbf{1}_{\mathcal{X}}(X) = 1$, for all $X \in \mathcal{X}$. Note that the definition is an extension of the usual definition of integration to currents that are not necessarily measures (representable by integration). In order to make a clear distinction between this operation and standard integration, we have introduced the non-conventional notation "int".

In particular, for 0-current, T, acting on C^r-sections of a vector bundle W_0, and a section $w_0 \in C^r(W_0)$, one has

$$T(w_0) = \text{int}(w_0 \cdot T) = \int w_0 \cdot T. \tag{22.61}$$

With these definitions at hand, the assertion of Proposition 22.1, in the form

$$(w \cdot F)(u) = \varsigma(j^1(uw)) = (j^1 w \cdot \varsigma)(u) + \partial(w \cdot \sigma)(u), \tag{22.62}$$

may be expressed alternatively as

$$F_{\circledcirc}(w)(u) = \varsigma(j^1(uw)) = \varsigma_{\circledcirc}(j^1 w)(u) + \partial(\sigma_{\circledcirc}(w))(u). \tag{22.63}$$

Hence, we obtain

$$F_{\circledcirc}(w) = \varsigma_{\circledcirc}(j^1 w) + \partial(\sigma_{\circledcirc}(w)). \tag{22.64}$$

It is emphasized that all the terms in the foregoing equations are viewed as 0-currents.

Let ς be a stress, let $w \in C^1(W)$, and set $\sigma = p_\sigma(\varsigma)$. We define the operator

$$\tilde{\partial} : C^0(W \otimes T^*\mathcal{X})^* \longrightarrow L(C^1(W), C^1(\mathcal{X})^*) \tag{22.65}$$

by the condition that the 0-current $(\tilde{\partial}\sigma)_{\circledcirc}(w) = (w \cdot \tilde{\partial}\sigma)$ satisfies

$$[(\tilde{\partial}\sigma)_{\circledcirc}(w)](u) = (w \cdot \tilde{\partial}\sigma)(u) := \sigma(w \otimes \mathrm{d}u) = \partial(w \cdot \sigma)(u), \tag{22.66}$$

so that

$$(\tilde{\partial}\sigma)_{\circledcirc}(w) = \partial(\sigma_{\circledcirc}(w)). \tag{22.67}$$

We conclude that, as 0-currents,

$$F_{\circledcirc} = \varsigma_{\circledcirc} \circ j^1 + (\tilde{\partial}\sigma)_{\circledcirc}. \tag{22.68}$$

Let T be a $(p+q)$-current and let ψ be a differentiable q-form. We recall the definition of $T \llcorner \psi$ as the p-current

$$(T \llcorner \psi)(\omega) = T(\omega \wedge \psi). \tag{22.69}$$

Consider the case where T is a $(p+q+1)$-current, so that $T \llcorner \psi$ is a $(p+1)$-current, and $\partial(T \llcorner \psi)$ is a p-current as before. We obtain an expression for $\partial(T \llcorner \psi)$ as follows.

$$\begin{aligned}
\partial(T \llcorner \psi)(\omega) &= (T \llcorner \psi)(\mathrm{d}\omega), \\
&= T(\mathrm{d}\omega \wedge \psi), \\
&= T[\mathrm{d}(\omega \wedge \psi) - (-1)^p \omega \wedge \mathrm{d}\psi], \\
&= \partial T(\omega \wedge \psi) - (-1)^p T(\omega \wedge \mathrm{d}\psi), \\
&= (\partial T \llcorner \psi)(\omega) - (-1)^p (T \llcorner \mathrm{d}\psi)(\omega).
\end{aligned} \tag{22.70}$$

Consequently,

$$\partial(T \llcorner \psi) = \partial T \llcorner \psi - (-1)^p T \llcorner \mathrm{d}\psi. \tag{22.71}$$

For the current T above, we now substitute the n-current $T_{\mathfrak{X}}$ as in (18.5). Assuming that the traction stress σ is represented by a differentiable vector-valued form $s_{\llcorner}$ as in (22.51), we substitute for the form ψ above the $(n-1)$-form $w \cdot s_{\llcorner}$, where $w \in C^1(W)$. Then, we have for the 1-current $T_{\mathfrak{X}} \llcorner (w \cdot s_{\llcorner})$,

$$\partial[T_{\mathfrak{X}} \llcorner (w \cdot s_{\llcorner})] = \partial T_{\mathfrak{X}} \llcorner (w \cdot s_{\llcorner}) - T_{\mathfrak{X}} \llcorner \mathrm{d}(w \cdot s_{\llcorner}), \tag{22.72}$$

which holds for such a differentiable traction stress density $s_{\llcorner}$.

For the case where the variational stress is represented by a differentiable vector-valued form, S, the 0-current (scalar) $j^1 w \cdot \varsigma$ and the 1-current $w \cdot \sigma$ may be expressed in terms of smooth densities as

$$j^1 w \cdot \varsigma = T_{\mathcal{X}} \llcorner (j^1 w \cdot S), \qquad w \cdot \sigma = T_{\mathcal{X}} \llcorner (w \cdot s_{\llcorner}). \tag{22.73}$$

Thus, with the help of (22.72), Eq. (22.62) may be written as

$$\begin{aligned}(w \cdot F)(u) &= (T_{\mathcal{X}} \llcorner (j^1 w \cdot S))(u) + \partial(T_{\mathcal{X}} \llcorner (w \cdot s_{\llcorner}))(u),\\ &= (T_{\mathcal{X}} \llcorner (j^1 w \cdot S))(u) + [\partial T_{\mathcal{X}} \llcorner (w \cdot s_{\llcorner}) - T_{\mathcal{X}} \llcorner \mathrm{d}(w \cdot s_{\llcorner})](u).\end{aligned} \tag{22.74}$$

As scalar-valued currents,

$$w \cdot F = T_{\mathcal{X}} \llcorner (j^1 w \cdot S - \mathrm{d}(w \cdot s_{\llcorner})) + \partial T_{\mathcal{X}} \llcorner (w \cdot s_{\llcorner}) \tag{22.75}$$

is the invariant decomposition of the force into a body force and a surface force. The fact that the body force, the term with $T_{\mathcal{X}} \llcorner$, is independent of the derivatives of w follows from the computations of Sect. 10.7.

22.6 Example: Non-smooth p-Form Electrodynamics on Manifolds

As an example for the framework introduced above, we present below the non-smooth counterpart for p-forms electrodynamics, viewed as a special case of geometric continuum mechanics. In other words, we consider here the non-smooth global counterpart of Sect. 11.2. It is noted that for equations written in terms of differential forms, such as the p-form Maxwell equations, one may directly obtain the non-smooth counterparts by replacing a p-form with an $(n - p)$-current and replacing exterior differentiation with the boundary operator for currents. However, in this section, we would like to obtain the non-smooth formulation, afresh.

Accordingly, we interpret the manifold $\mathcal{X}$ as spacetime and assume it is an n-dimensional without boundary. The bundle $\bigwedge^p T^*\mathcal{X}$ is substituted for the vector bundle W, and a typical section of $\bigwedge^p T^*\mathcal{X}$, a p-form, will be denoted by $\mathfrak{a}$. Such a differential form is interpreted as a generalized vector potential of electromagnetism or a variation thereof. For the sake of simplicity, we will consider p-forms supported in the interior of a compact n-dimensional submanifold (possibly with a boundary), $\mathcal{R}$ of $\mathcal{X}$. The Banachable space of such p-forms will be denoted as $C^1_{\mathcal{R}}(\bigwedge^p T^*\mathcal{X})$, so that a force $F \in C^1_{\mathcal{R}}(\bigwedge^p T^*\mathcal{X})^*$ is a p-current or order 1.

The representation procedure of forces by stress measures applies in this case. Proposition 22.1, as well as its representation in Eq. (22.62), still hold. Thus, one has

$$(\mathfrak{a} \cdot F)(u) = (j^1 \mathfrak{a} \cdot \varsigma)(u) + (\mathfrak{a} \cdot \sigma)(\mathrm{d}u) \tag{22.76}$$

for any differentiable function u over $\mathcal{X}$ supported in $\mathcal{R}$. Here,

$$\varsigma \in C^0_{\mathcal{R}}(J^1(\textstyle\bigwedge^p T^*\mathcal{X}))^* = C^{-0}_{\mathcal{R}}(J^1(\bigwedge^p T^*\mathcal{X})^*, \bigwedge^n T^*\mathcal{X}) \tag{22.77}$$

is a vector-valued measure over $\mathcal{R}$ (which may be viewed as a generalized vector-valued form), and

$$\sigma \in C^0_{\mathcal{R}}(\textstyle\bigwedge^p T^*\mathcal{X} \otimes T^*\mathcal{X})^* = C^{-0}_{\mathcal{R}}(\bigwedge^p T\mathcal{X} \otimes \bigwedge^{n-1} T^*\mathcal{X}). \tag{22.78}$$

We recall that $\mathfrak{a} \cdot \sigma$ is a 1-current given by $(\mathfrak{a} \cdot \sigma)(\phi) = \sigma(\mathfrak{a} \otimes \phi)$, for any 1-form ϕ.

We now make the assumption that the traction stress, σ, has a particularly simple form. Namely, there is a $(p+1)$-current of order zero,

$$\mathfrak{g} \in C^{-0}_{\mathcal{R}}(\textstyle\bigwedge^{n-p-1} T^*\mathcal{X}) = C^0_{\mathcal{R}}(\bigwedge^{p+1} T^*\mathcal{X})^*, \tag{22.79}$$

such that

$$\sigma(\mathfrak{a} \otimes \phi) = \mathfrak{g}(\phi \wedge \mathfrak{a}), \tag{22.80}$$

or equivalently,

$$\mathfrak{a} \cdot \sigma = \mathfrak{g} \llcorner \mathfrak{a}. \tag{22.81}$$

It follows that

$$\begin{aligned} (\mathfrak{a} \cdot \sigma)(\mathrm{d}u) &= \mathfrak{g}(\mathrm{d}u \wedge \mathfrak{a}), \\ &= \mathfrak{g}(\mathrm{d}(u\mathfrak{a}) - u\mathrm{d}\mathfrak{a}), \\ &= \partial\mathfrak{g}(u\mathfrak{a}) - (\mathfrak{g} \llcorner \mathrm{d}\mathfrak{a})(u). \end{aligned} \tag{22.82}$$

Hence,

$$\partial(\mathfrak{a} \cdot \sigma) = (\partial\mathfrak{g}) \llcorner \mathfrak{a} - \mathfrak{g} \llcorner \mathrm{d}\mathfrak{a}. \tag{22.83}$$

Next, we note that since we consider sections that vanish of $\partial\mathcal{R}$, surface forces become irrelevant and, in fact, the force functional is a body force as in Sect. 17.4. Thus, assumption (11.33) is simply replaced now by the assumption that the force vanishes, that is,

$$F(\mathfrak{a}) = 0 \tag{22.84}$$

for all test forms $\mathfrak{a}$. As a result, it follows from (22.76) that the analog of (11.34) is

$$j^1\mathfrak{a} \cdot \varsigma = -\partial(\mathfrak{a} \cdot \sigma) = -(\partial\mathfrak{g}) \llcorner \mathfrak{a} + \mathfrak{g} \llcorner \mathrm{d}\mathfrak{a}. \tag{22.85}$$

One may introduce the notation

$$\mathfrak{f} := -\mathrm{d}\mathfrak{a}, \qquad \mathfrak{J} := -\partial\mathfrak{g}. \tag{22.86}$$

Thus,

$$j^1\mathfrak{a}\cdot\varsigma = \mathfrak{J} \llcorner \mathfrak{a} - \mathfrak{g} \llcorner \mathfrak{f}. \tag{22.87}$$

Evidently,

$$\mathrm{d}\mathfrak{f} = 0, \qquad \partial\mathfrak{J} = 0. \tag{22.88}$$

It is noted that different signs will result in if we defined $\mathfrak{g}$ to act on $\mathfrak{a} \wedge \phi$ in (22.80) (contraction on the left in (22.81)).

22.7 Flat Forces

As another special case related to the forces that are conjugate to p-forms, we consider the non-smooth version of the example in Remark 11.1. Thus, instead of making the assumptions in (22.80) and (22.84), we continue as follows.

Similarly to Remark 11.1, we set

$$\mathfrak{A}_1 : J^1\left(\bigwedge^p T^*\mathcal{X}\right) \longrightarrow \bigwedge^p T^*\mathcal{X} \times \bigwedge^{p+1} T^*\mathcal{X} \tag{22.89}$$

for the natural mapping given by

$$\mathfrak{A}_1(j^1_x\mathfrak{a}) := (\mathfrak{a}(x), \mathrm{d}\mathfrak{a}(x)). \tag{22.90}$$

The mapping $\mathfrak{A}_1$ may be applied pointwise to sections in $C^0_{\mathcal{R}}(J^1\left(\bigwedge^p T^*\mathcal{X}\right))$, and keeping the same notation, we obtain

$$\mathfrak{A}_1 : C^0_{\mathcal{R}}(J^1\left(\bigwedge^p T^*\mathcal{X}\right)) \longrightarrow C^0_{\mathcal{R}}(\bigwedge^p T^*\mathcal{X}) \times C^0_{\mathcal{R}}(\bigwedge^{p+1} T^*\mathcal{X}), \tag{22.91}$$

whereby

$$j^1\mathfrak{a} \longmapsto (\mathfrak{a}, \mathrm{d}\mathfrak{a}). \tag{22.92}$$

Using the C^0-topology on $C^0_{\mathcal{R}}(\bigwedge^p T^*\mathcal{X}) \times C^0_{\mathcal{R}}(\bigwedge^{p+1} T^*\mathcal{X})$, it follows that an element of the dual space, $C^0_{\mathcal{R}}(\bigwedge^p T^*\mathcal{X})^* \times C^0_{\mathcal{R}}(\bigwedge^{p+1} T^*\mathcal{X})^*$, is represented by a pair of currents of order zero, measures, (G, H), such that G is a p-current and H is a $(p+1)$-current, in the form

$$(G, H)(\alpha, \beta) = G(\alpha) + H(\beta). \tag{22.93}$$

Moreover, it is evident that the mapping $\mathfrak{A}_1$ is linear, continuous, and injective. Composing with the jet extension mapping, which is also linear, continuous, and injective, we get

$$\mathfrak{d} = \mathfrak{A}_1 \circ j^1 : C^1_{\mathcal{R}}\left(\bigwedge^p T^*\mathcal{X}\right) \longrightarrow C^0_{\mathcal{R}}(\bigwedge^p T^*\mathcal{X} \times \bigwedge^{p+1} T^*\mathcal{X}), \quad \mathfrak{a} \longmapsto (\mathfrak{a}, \mathrm{d}\mathfrak{a}). \tag{22.94}$$

The dual mapping

$$\begin{aligned} \mathfrak{d}^* = (\mathfrak{A}_1 \circ j^1)^* = j^{1*} \circ \mathfrak{A}_1^* : C^0_{\mathcal{R}}(\bigwedge^p T^*\mathcal{X})^* \times C^0_{\mathcal{R}}(\bigwedge^{p+1} T^*\mathcal{X})^* \\ \longrightarrow C^1_{\mathcal{R}}\left(\bigwedge^p T^*\mathcal{X}\right)^*, \end{aligned} \tag{22.95}$$

satisfies

$$((j^{1*} \circ \mathfrak{A}_1^*)(G, H))(\mathfrak{a}) = G(\mathfrak{a}) + H(\mathrm{d}\mathfrak{a}) = (G + \partial H)(\mathfrak{a}). \tag{22.96}$$

A force $F \in C^1_{\mathcal{R}}(\bigwedge^p T^*\mathcal{X})^*$ will be referred to as *flat* if it belongs to Image$(\mathfrak{A}_1 \circ j^1)^*$. Thus, for a flat force, F, there are currents $G \in C^0_{\mathcal{R}}(\bigwedge^p T^*\mathcal{X})^*$ and $H \in C^0_{\mathcal{R}}(\bigwedge^{p+1} T^*\mathcal{X})^*$, such that

$$F = G + \partial H \tag{22.97}$$

(cf. Sect. 23.3 describing Federer's definition of flat chains). Evidently, the current H is not unique. If G and H represent F as above, so would G and $H + \partial H'$ for any $(p+2)$-current H'. It is further noticed that if ς represents a flat force F, given in terms of the currents G and H of order zero, then,

$$\varsigma(j^1\mathfrak{a}) = (G + \partial H)(\mathfrak{a}). \tag{22.98}$$

Chapter 23
Whitney's Geometric Integration Theory and Non-smooth Bodies

Whitney's geometric integration theory, [Whi57], offers another perspective on the subject of non-smooth fluxes, and in particular, non-smooth bodies, in a Euclidean space. In this chapter, based on [RS03b], we mainly follow Whitney's approach which we have found to be in harmony with the traditional continuum mechanics approach to fluxes. In fact, as presented in [Fed69a, Chapter 4], Whitney's geometric integration theory may be introduced as a special case of the theory of de Rham currents.

In Whitney's theory, general geometric objects, referred to as *flat chains* and *sharp chains*, are defined. Chains may be used to model generalized non-smooth bodies of continuum mechanics and non-smooth fields defined on such bodies. In particular, chains can model bodies having fractal boundaries, in particular, boundaries of infinite measure, and boundaries to which normal vectors cannot be defined. The collections of chains have structures of Banach spaces. Elements of the dual Banach spaces, referred to as *cochains*, are represented by forms defined on the Euclidean space, and the action of a cochain on a chain is represented by integration.

23.1 The Flat Norm: Motivation

For continuum mechanics in a three-dimensional Euclidean space, the assumptions leading to the Cauchy flux theorem, as represented in Eqs. (8.3) and (8.5), imply that for each region $\mathscr{R}$,

$$|\Phi_{\partial\mathscr{R}}| \leqslant \int_{\partial\mathscr{R}} |\tau_{\mathscr{R}}| \,\mathrm{d}A, \qquad |\Phi_{\partial\mathscr{R}}| \leqslant \int_{\mathscr{R}} |\varsigma - \beta| \,\mathrm{d}V, \tag{23.1}$$

where now we view the integrands as scalar fields rather than differential forms.

R. Segev, *Foundations of Geometric Continuum Mechanics*, Advances in Continuum Mechanics 49, https://doi.org/10.1007/978-3-031-35655-1_23

In analogy with Chap. 4, we view the total flux $\Phi_{\partial\mathscr{R}}$ as the result of applying a real-valued function ω defined on the collection of two-dimensional oriented surfaces so that

$$\Phi_{\partial\mathscr{R}} = \omega(\partial\mathscr{R}). \tag{23.2}$$

It also follows from the boundary integral above that for $\mathscr{V} \subset \partial\mathscr{R}$, we may consider the total flux,

$$\Phi_{\mathscr{V}} = \int_{\mathscr{V}} \tau_{\mathscr{R}} \mathrm{d}A, \tag{23.3}$$

such that

$$|\Phi_{\mathscr{V}}| := |\omega(\mathscr{V})| \leqslant \int_{\mathscr{V}} |\tau_{\mathscr{R}}| \, \mathrm{d}A. \tag{23.4}$$

Let $|\mathscr{R}|$ denote the volume of $\mathscr{R}$. Then, in analogy with (8.6), we assume that there is a positive constant M, such that for any region $\mathscr{R}$.

$$|\omega(\partial\mathscr{R})| \leqslant M\,|\mathscr{R}|\,. \tag{23.5}$$

In addition, for $\mathscr{V} \subset \partial\mathscr{R}$, let $|\mathscr{V}|$ denote the area of $\mathscr{V}$. Thus, assuming that $\tau_{\mathscr{R}}$ is bounded, it follows from (23.4) that

$$|\omega(\mathscr{V})| \leqslant N\,|\mathscr{V}|\,. \tag{23.6}$$

It follows that for any $\mathscr{V}$ and any region $\mathscr{R}$,

$$\begin{aligned} |\omega(\mathscr{V})| &\leqslant |\omega(\mathscr{V}) - \omega(\partial\mathscr{R}) + \omega(\partial\mathscr{R})|\,, \\ &\leqslant |\omega(\mathscr{V}) - \omega(\partial\mathscr{R})| + |\omega(\partial\mathscr{R})|\,. \end{aligned} \tag{23.7}$$

Next, in order to view the flux function as a linear mapping, the collection of surfaces is chosen to be the collection of 2-polyhedral cells, $\mathbf{C}_2(\mathbb{E}^3)$, and the collection of regions is chosen as $\mathbf{C}_3(\mathbb{E}^3)$. Hence, once linearity is well defined, we may write,

$$\begin{aligned} |\omega(\mathscr{V})| &\leqslant |\omega(\mathscr{V} - \partial\mathscr{R})| + |\omega(\partial\mathscr{R})|\,, \\ &\leqslant N\,|\mathscr{V} - \partial\mathscr{R}| + M\,|\mathscr{R}|\,, \end{aligned} \tag{23.8}$$

for an arbitrary polyhedral 3-chain $\mathscr{R}$. It is concluded that there is a positive constant, C_ω, such that

$$|\omega(\mathscr{V})| \leqslant C_\omega(|\mathscr{V} - \partial\mathscr{R}| + |\mathscr{R}|), \quad \text{for all} \quad \mathscr{R} \in \mathbf{C}_3(\mathbb{E}^3). \tag{23.9}$$

Moreover, since (23.9) holds for an arbitrary polyhedral 3-chain, $\mathcal{R}$, one has

$$|\omega(\mathcal{V})| \leqslant C_\omega \inf\{|\mathcal{V} - \partial\mathcal{R}| + |\mathcal{R}|) \mid \mathcal{R} \in \mathbf{C}_3(\mathbb{E}^3)\}. \tag{23.10}$$

Since ω is a linear functional defined on the infinite-dimensional space $\mathbf{C}_2(\mathbb{E}^3)$, one would like to view the preceding equation as a continuity condition for ω. This suggests that the definition of a norm of a polyhedral 2-cell, the *flat norm*, $\|\mathcal{V}\|^\flat$, be given by

$$\|\mathcal{V}\|^\flat := \inf\{|\mathcal{V} - \partial\mathcal{R}| + |\mathcal{R}|) \mid \mathcal{R} \in \mathbf{C}_3(\mathbb{E}^3)\}. \tag{23.11}$$

With the definition of the flat norm, Eq. (23.10) may be written as

$$|\omega(\mathcal{V})| \leqslant C_\omega \|\mathcal{V}\|^\flat \tag{23.12}$$

—a continuity condition for the flux linear functional.

Whitney, [Whi57], proves that (23.11) indeed defines a norm for the general case of $\mathbf{C}_r(\mathbb{E}^n)$. Once the norm has been defined, the space of polyhedral r-chains in $\mathbb{E}^n$ may be completed relative to the flat norm. The resulting Banach space will be denoted by $\mathbf{C}_r^\flat(\mathbb{E}^n)$, and its elements are referred to as *flat chains*. Flat r-chains model a rich collection of geometric objects, including r-dimensional submanifolds of $\mathbb{E}^n$ on the one hand, and fractals on the other hand. It is emphasized that the class of geometric objects obtained and the properties of these geometric objects reflect the norm used, which, in turn, has been chosen based on the classical principles of continuum mechanics.

Furthermore, the flat norm also determines the nature of elements of the dual space $\mathbf{C}_r^\flat(\mathbb{E}^n)^*$, the elements of which are referred to as *flat cochains*. Indeed, we view flat cochains as generalized flux functionals. As one could have hoped, Whitney proved that flat cochains may be represented by differential forms and that their action on flat chains can be described as integration of the corresponding forms over the chains.

These properties, as well as additional properties to be considered below, serve as evidences of the elegance of Whitney's geometric integration theory and its relevance to continuum mechanics.

23.2 Flat Chains

We first present the flat norm in further detail. The *mass* of a polyhedral r-chain $A = \sum a_i s_i$ in $\mathbb{E}^n$ is defined to be $|A| = \sum |a_i||s_i|$, where $|s_i|$ denotes the r-dimensional volume of $|s_i|$. Thus, in case the interiors of the cells s_i of a polyhedral r-chain do not intersect, and if $a_i = 1$, then, the mass of the polyhedral chain is exactly its r-dimensional volume. It can be shown that the mass operation $A \mapsto |A|$ is a norm on $\mathbf{C}_r(\mathbb{E}^n)$.

As a straightforward extension of the arguments above, the flat norm, $\|A\|^\flat$, of a polyhedral r-chain A in $\mathbb{E}^n$, is defined by

$$\|A\|^\flat = \inf\{|A - \partial D| + |D| \mid D \in \mathbf{C}_{r+1}(\mathbb{E}^n)\}, \tag{23.13}$$

using all polyhedral $(r + 1)$-chains, D.

We note that it is not immediate that $\| \cdot \|^\flat$ is indeed a norm (see Whitney [Whi57, p. 173]). Furthermore, the actual calculation of the flat norm may be quite complicated even for simple r-chains. (For example, consider a 1-chain in the plane consisting of two oriented line segments.) It is observed also that by taking $D = 0$ in the definition above,

$$\|A\|^\flat \leqslant |A| . \tag{23.14}$$

Completing the space $\mathbf{C}_r(\mathbb{E}^n)$ with respect to the flat norm gives a Banach space denoted by $\mathbf{C}^\flat_r(\mathbb{E}^n)$. That is, $\mathbf{C}^\flat_r(\mathbb{E}^n)$ contains the formal limits of all the sequences of polyhedral r-chains (A_i), such that $\lim_{i,j\to\infty} \|A_j - A_i\|^\flat = 0$. Elements of $\mathbf{C}^\flat_r(\mathbb{E}^n)$ that are formal limits of such sequences are sometimes denoted by $\lim^\flat A_i$. Elements of $\mathbf{C}^\flat_r(\mathbb{E}^n)$ are *flat r-chains* in $\mathbb{E}^n$. If there are no intersections between cells and all coefficients have the value of 1, we roughly identify the flat chain with the set that contains its points.

Example 23.1 (A 1-Dipole) Consider the sequence of 1-chains (A_i) in $\mathbb{E}^2$ such that $A_i = L_{1i} + L_{2i}$ where L_{1i} and L_{i2} are 1-simplices associated with two parallel line segments having the same length L, having opposite orientations, and the line segment corresponding to L_{2i} is obtained from the line segment corresponding to L_{1i} by a translation of distance d_i perpendicularly to its direction (see Fig. 23.1a). If we take the rectangle generated by the two line segments as D_i (shown in yellow) for the definition of the flat norm, it follows that $\|A_i\|^\flat \leqslant (L+2)d_i$. Thus, if $d_i \to 0$, the sequence (A_i) converges to the zero chain in the flat norm. On the other hand, in the mass norm, for $d_i \neq d_j$, another dipole remains and we have $|A_i - A_j| = 2L$. Thus, the sequence does not converge in the mass norm. Roughly speaking, the geometrical significance of the flat norm in this case is that, unlike the mass norm, it takes into account how closely the two segments are situated.

Example 23.2 (A Square) Consider the oriented boundary, A_i, of the square of side d_i as shown in black in Fig. 23.1b. Let D_i be the square of the same side length. Then, $A_i - \partial D_i = 0$, and it follows that

$$\|A_i\|^\flat \leqslant d_i^2. \tag{23.15}$$

Example 23.3 (The Staircase) We consider the "staircase" sequence (B_i) illustrated in Fig. 23.2. Here, B_0 is the sum of two oriented segments of unit lengths as shown, and

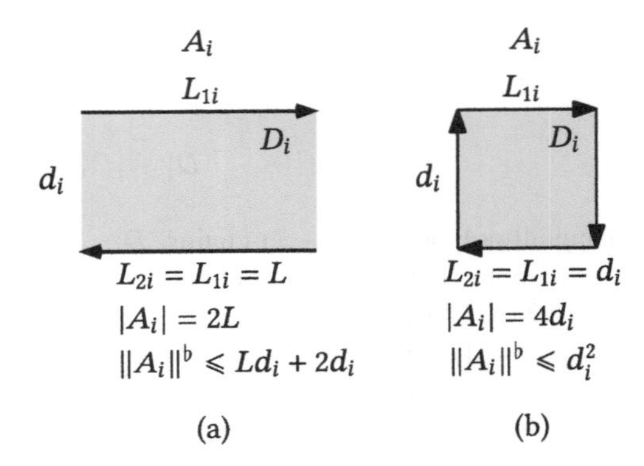

Fig. 23.1 A 1-dipole (**a**) and a square (**b**)

$$B_i = B_0 + \sum_{j=1}^{i} A_j, \tag{23.16}$$

where

$$A_j = \sum_{l=1}^{2^{j-1}} A_{jl} \tag{23.17}$$

is the sum of 2^{j-1} oriented 1-squares of size $d_j = 1/2^j$. We consider the limit of the sequence (B_i), as $i \to \infty$. For each square A_{jl}, let D_{jl} be the cell such that $A_{jl} = \partial D_{jl}$. Then, using D_{jl} in the definition of the flat norm, we get $\|A_{jl}\|^\flat \leqslant d_j^2 = 2^{-2j}$. Hence,

$$\begin{aligned} \|A_j\|^\flat &= \left\| \sum_{l=1}^{2^{j-1}} A_{jl} \right\|^\flat, \\ &\leqslant \sum_{l=1}^{2^{j-1}} \|A_{jl}\|^\flat \\ &\leqslant 2^{j-1} 2^{-2j}, \\ &= 2^{-j-1}, \end{aligned} \tag{23.18}$$

and it follows that for $k > i$,

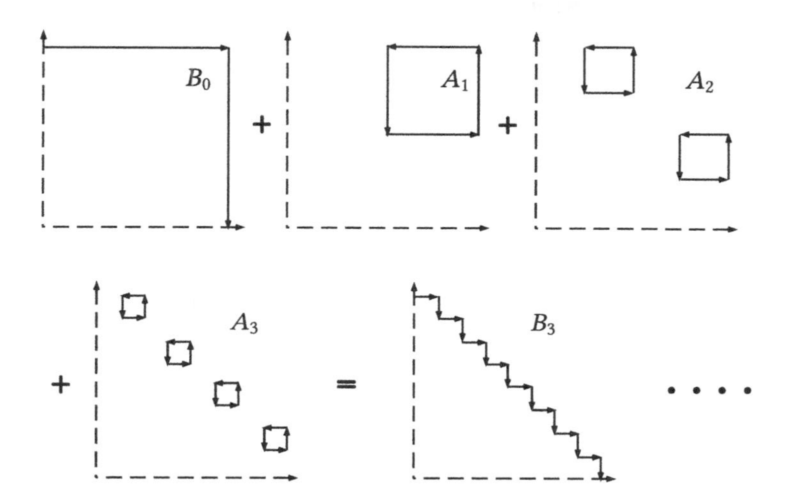

Fig. 23.2 The staircase

$$
\begin{aligned}
\|B_k - B_i\|^\flat &= \left\| \sum_{j=i+1}^{k} A_j \right\|^\flat, \\
&\leqslant \sum_{j=i+1}^{k} \|A_j\|^\flat, \\
&\leqslant \sum_{j=i+1}^{k} 2^{-j-1}, \\
&\leqslant \sum_{j=i+1}^{\infty} 2^{-j-1},
\end{aligned}
\tag{23.19}
$$

which tends to zero for $i \rightarrow \infty$. It is concluded that the sequence (B_i) is a Cauchy sequence, and the staircase is a flat 1-chain. This example demonstrates how non-smooth geometric objects may be represented by flat chains. In particular, the normal vector to the staircase is not defined.

Example 23.4 (The van Koch Snowflake) The next example considers a popular fractal. In particular, the normal vector to the snowflake is not defined. In addition, the mass norm of the fractal, its length, is infinite. Figure 23.3 illustrates the construction of the van Koch snowflake (actually one third of it) as the limit of a sequence (B_i) of polyhedral 1-chains. One starts with the 1-simplex B_0 of unit length, and at each step one sets

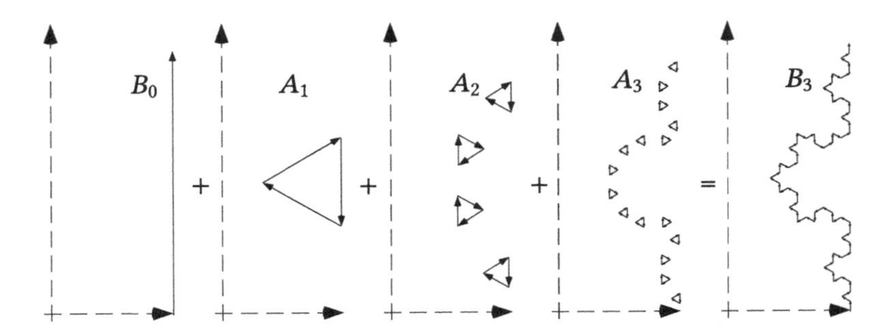

Fig. 23.3 The van Koch snowflake

$$B_i = B_0 + \sum_{j=1}^{i} A_j, \tag{23.20}$$

as illustrated. Now,

$$A_j = \sum_{l=1}^{4^{j-1}} A_{jl}, \tag{23.21}$$

where A_{jl} is the oriented boundary of a triangle of sides 3^{-i}. The triangles are situated in such a way that at each step, the middle third of the edge of the previous step is canceled. Since the flat norm of each A_{jl} is bounded by the area of the corresponding triangle,

$$\|A_{jl}\|^\flat \leqslant \frac{\sqrt{3}}{2} 3^{-2j}. \tag{23.22}$$

Thus,

$$\begin{aligned} \|A_j\|^\flat &= \left\| \sum_{l=1}^{4^{j-1}} A_{jl} \right\|^\flat, \\ &\leqslant \sum_{l=1}^{4^{j-1}} \|A_{jl}\|^\flat \\ &\leqslant 4^{j-1} \frac{\sqrt{3}}{2} 3^{-2j}, \\ &= \frac{\sqrt{3}}{8} \left(\frac{4}{9}\right)^j. \end{aligned} \tag{23.23}$$

It follows that for $k > i$,

$$\begin{aligned}\|B_k - B_i\|^{\flat} &= \left\| \sum_{j=i+1}^{k} A_j \right\|^{\flat}, \\ &\leqslant \sum_{j=i+1}^{k} \|A_j\|^{\flat}, \\ &\leqslant \sum_{j=i+1}^{k} \frac{\sqrt{3}}{8} \left(\frac{4}{9}\right)^j \\ &\leqslant \sum_{j=i+1}^{\infty} \frac{\sqrt{3}}{8} \left(\frac{4}{9}\right)^j,\end{aligned} \tag{23.24}$$

which is a convergent geometric series, the first term of which tends to zero as $i \to \infty$. We conclude that the snowflake is a flat chain.

On the other hand, the mass norm of the elements of the sequence (B_i) diverges. Since at each triangle adds two segments of size 3^{-j}, the total length added at the j-th step is

$$4^{j-1} \cdot 2 \cdot 3^{-j} = \frac{1}{2}\left(\frac{4}{3}\right)^j. \tag{23.25}$$

While above we exhibited the non-smooth character of flat chains, Whitney proved that smooth r-dimensional submanifolds of $\mathbb{E}^n$ may also be represented by flat r-chains.

Let (A_i) be a sequence of polyhedral r-chains; then, the boundaries ∂A_i comprise a sequence of $(r-1)$-polyhedral chains. Whitney shows ([Whi57, pp. 154–155]) that if the sequence (A_i) converges in the flat norm, then the sequence (∂A_i) also converges in the flat norm. Thus, if $A = \lim^{\flat} A_i$, its boundary is defined by

$$\partial A := \lim{}^{\flat} \partial A_i. \tag{23.26}$$

It is concluded that the boundary of a flat r-chain always exists as a flat $(r-1)$-chain. In addition for each flat chain A

$$\|\partial A\|^{\flat} \leqslant \|A\|^{\flat}. \tag{23.27}$$

It is implied that the boundary operator

$$\partial : \mathbf{C}^{\flat}_r(\mathbb{E}^n) \longrightarrow \mathbf{C}^{\flat}_{r-1}(\mathbb{E}^n), \qquad A \longmapsto \partial A, \tag{23.28}$$

is a continuous and linear mapping.

The Riemann integral of a continuous r-form ω over a flat r-chain $A = \lim A_i$ is defined to be

$$\int_A \omega = \lim \int_{A_i} \omega, \tag{23.29}$$

if the limit exists.

23.3 Federer's Definition of Flat Chains

In this section, we review briefly the definition of flat chains by Federer [Fed69b, Chapter 4]. Federer views r-flat chains as r-currents in $\mathbb{R}^n$ having some specific properties.

Simply put, a current is said to be representable by integration if it is of order zero, that is, if it is continuous in the C^0-topology so that its action may be represented by integration relative to a (vector-valued) measure. A current T is said to be normal if it is representable by integration, its boundary ∂T is representable by integration, and supp$T$ is compact.

Let $U \subset \mathbb{R}^n$ be an open subset and $K \subset U$ be compact. For any smooth r-form ϕ defined on U, consider the flat seminorm

$$\|\phi\|_K^\flat := \sup_{x \in K}\{|\phi(x)|, |\mathrm{d}\phi(x)|\}, \tag{23.30}$$

where one uses the natural metric structure of $\mathbb{R}^n$ to evaluate $|\phi(x)|$ and $|\mathrm{d}\phi(x)|$. We note that the construction is similar to that of Sect. 22.7. If T is an r-current, the flat, dual seminorm (actually, a norm) is given by

$$\|T\|_K^\flat := \sup\{T(\phi) \mid \|\phi\|_K^\flat = 1\}. \tag{23.31}$$

Assume that for an r-current T, $\|T\|_K^\flat < \infty$. Then, for any smooth r-form ϕ,

$$T(\phi) \leqslant \|T\|_K^\flat \|\phi\|_K^\flat. \tag{23.32}$$

For a pair of smooth forms, an r-form α and an $(r+1)$-form β, we define the seminorm

$$\|(\alpha, \beta)\|_K^0 := \sup_{x \in K}\{|\alpha(x)|, |\beta(x)|\}. \tag{23.33}$$

Consider the mapping (cf. Sect. 22.7)

$$\mathfrak{d} : C_c^\infty(\bigwedge^r \mathbb{R}^n) \longrightarrow C_c^\infty(\bigwedge^r \mathbb{R}^n) \times C_c^\infty(\bigwedge^{r+1} \mathbb{R}^n), \qquad \mathfrak{d}(\phi) := (\phi, \mathrm{d}\phi). \tag{23.34}$$

Evidently, $\mathfrak{d}$ is linear, injective, and

$$\|\mathfrak{d}\phi\|_K^0 = \|\phi\|_K^\flat. \tag{23.35}$$

Let

$$\mathfrak{d}^{-1} : \mathrm{Image}\mathfrak{d} \longrightarrow C_c^\infty(\bigwedge^r \mathbb{R}^n) \tag{23.36}$$

be the left inverse. For an r-current T with $\|T\|_K^\flat < \infty$,

$$T \circ \mathfrak{d}^{-1} : \mathrm{Image}\mathfrak{d} \longrightarrow \mathbb{R}, \tag{23.37}$$

and

$$\begin{aligned} T \circ \mathfrak{d}^{-1}(\phi, \mathrm{d}\phi) &= T \circ \mathfrak{d}^{-1}(\mathfrak{d}\phi), \\ &= T(\phi), \\ &\leqslant \|T\|_K^\flat \|\phi\|_K^\flat, \\ &= \|T\|_K^\flat \|(\phi, \mathrm{d}\phi)\|_K^0. \end{aligned} \tag{23.38}$$

Thus, $T \circ \mathfrak{d}^{-1}$ is bounded by the seminorm $\|(\phi, \mathrm{d}\phi)\|_K^0$, and by the Hahn-Banach theorem, it may be extended to a linear functional S on $C_c^\infty(\bigwedge^r \mathbb{R}^n) \times C_c^\infty(\bigwedge^{r+1} \mathbb{R}^n)$ which is bounded by the $\|(\alpha, \beta)\|_K^0$ seminorm. As in Sect. 22.7, the functional S may be represented by an r-current G and an $(r+1)$-current H, both represented by integration, so that

$$T(\phi) = G(\phi) + H(\mathrm{d}\phi). \tag{23.39}$$

Hence, T may be represented in the form

$$T = G + \partial H. \tag{23.40}$$

Let $\mathbf{C}_r^\flat(K)$ be the closure of the collection of normal currents supported in K relative to the $\|T\|_K^\flat$-norm. Finally, the space of flat chains in U is defined to be the union

$$\mathbf{C}_r^\flat(U) = \bigcup_{K \subset U} \mathbf{C}_r^\flat(K). \tag{23.41}$$

Federer proves that this definition is equivalent to Whitney's definition of flat chains.

23.4 Sharp Chains

Whitney obtained chains that are even less regular than the flat chains by introducing a possibly smaller norm. Thus, more Cauchy sequences will converge and one ends up with a larger completed space.

The *sharp norm* $\|A\|^{\sharp}$ of a polyhedral r-chain $A = \sum a_i s_i$ is defined by

$$\|A\|^{\sharp} = \inf\left\{\frac{\sum |a_i||s_i||v_i|}{r+1} + \left\|\sum a_i \text{trans}_{v_i} s_i\right\|^{\flat}\right\}, \tag{23.42}$$

using all collections of vectors $v_i \in \mathbb{E}^n$. Here trans_v is a translation operator that moves each point p of s to $p + v$, giving a translated cell $\text{trans}_v s$ having the same orientation as s.

Clearly, setting all $v_i = 0$, we conclude that $\|A\|^{\sharp} \leqslant \|A\|^{\flat}$, so the sharp norm defines a coarser topology. Completing the space $\mathbf{C}_r(\mathbb{E}^n)$ with respect to the sharp norm gives a Banach space denoted by $\mathbf{C}_r^{\sharp}(\mathbb{E}^n)$ whose elements are referred to as *sharp chains*. It follows that $\mathbf{C}_r^{\flat}(\mathbb{E}^n)$ is a Banach subspace of $\mathbf{C}_r^{\sharp}(\mathbb{E}^n)$).

Example 23.5 (The Staircase Strainer and Staircase Mixer) Consider first the sequence of pairs of 1-vectors in $\mathbb{E}^2$ as in Example 23.1. Here, we consider the case where each segment is of length d_i and the two segments are a distance d_i apart in Fig. 23.4. Take $v_1 = 0$, and take v_2 as the vector such that trans_{v_2} will cause the two line segments to overlap, so $|v_2| = d_i$. In this case, $\sum a_i \text{trans}_{v_i} s_i = 0$ and $\sum |a_i||s_i||v_i| = d_i^2$. Therefore, the sharp norm of the dipole is bounded from above by $d_i^2/2$. For $d_i \to 0$, the sharp norm of the shrinking pairs tends to zero faster than the flat norm.

The "staircase strainer" is constructed using the sequence (B_i) as shown in Fig. 23.5. Here, A_j is the sum of 2^{j-1} dipoles of size $d_j = 1/2^j$, $B_i = B_0 + \sum_{j=1}^{i} A_j$, and we take the limit as $i \to \infty$. For the flat norm we have

$$\|B_i - B_{i-1}\|^{\flat} = \|A_i\|^{\flat} \leqslant 2^{i-1} 2/2^i = 1, \tag{23.43}$$

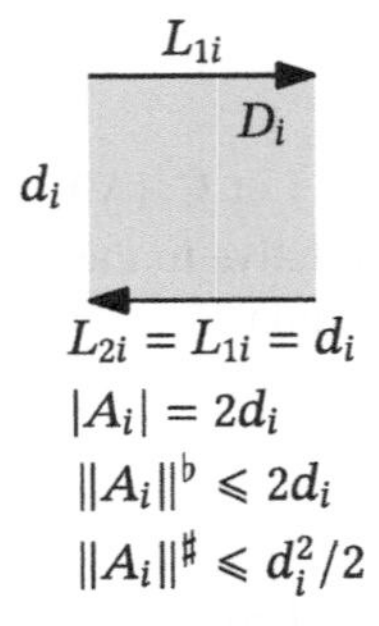

Fig. 23.4 Dipoles using the sharp norm

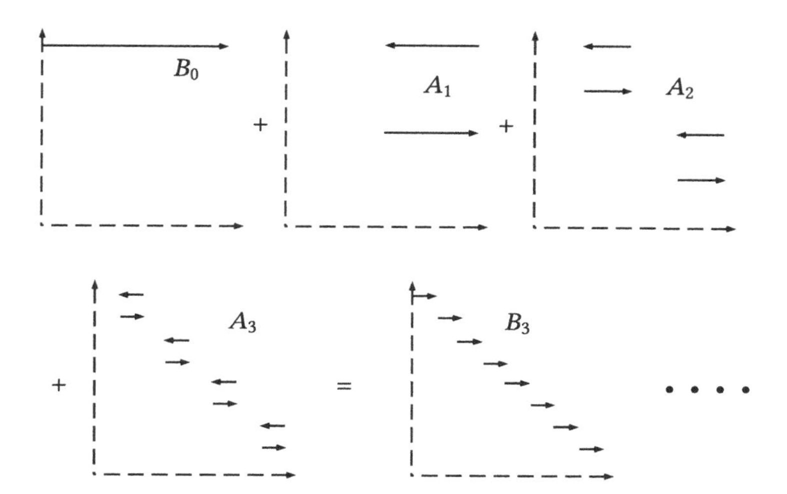

Fig. 23.5 Constructing the staircase strainer

so the sequence (B_i) does not converge. On the other hand, using the sharp norm, for any $k > i$,

$$\begin{aligned} \|B_k - B_i\|^\sharp &= \left\| \sum_{j=i+1}^{k} A_j \right\|^\sharp, \\ &\leqslant \sum_{j=i+1}^{k} \|A_j\|^\sharp, \\ &\leqslant \sum_{j=i+1}^{k} 2^{j-1} \left(\frac{1}{2}\right)^{2j+1}, \\ &\leqslant \sum_{j=i+1}^{\infty} 2^{-j-2}, \end{aligned} \tag{23.44}$$

and the sequence converges. The staircase strainer (as it allows flux only through the horizontal segments) is therefore a sharp 1-chain.

The staircase mixer (as it reverses the orientation in each step) can be constructed as in Fig. 23.6. The staircase mixer can also be viewed as a combination of two staircase strainers with opposite orientations. As such, it is also a sharp chain, but not a flat chain.

Roughly speaking, the difference in behavior between the flat norm and the sharp norm may be described as follows. Consider a sequence (A_i) of shrinking r-polyhedral chains of typical size $a_i \to 0$. If A_i is the boundary ∂B_i of a shrinking

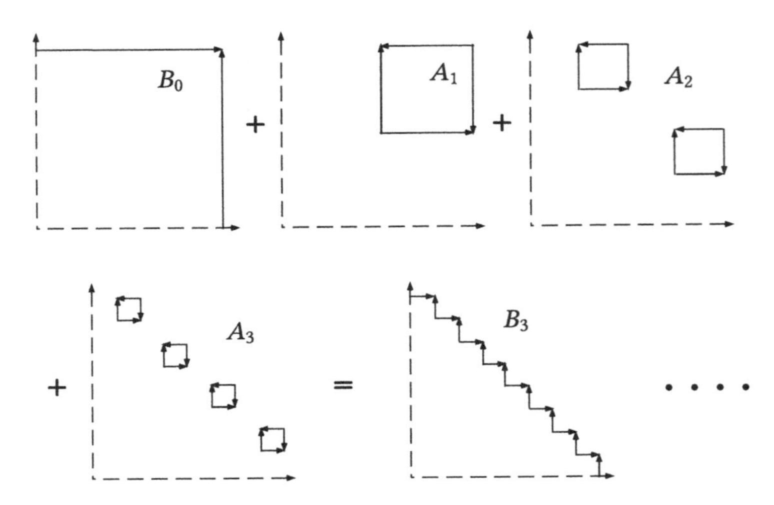

Fig. 23.6 Constructing the staircase mixer

$(r + 1)$-chain B_i, then taking $D = B_i$ in the definition of the flat norm, $\|A_i\|^\flat$ shrinks like a_i^{r+1}. If A_i cannot be represented as the boundary of an $(r + 1)$-chain, the r-dimensional mass of some subset of A_i will always be present in the definition of the flat norm, and hence, $\|A_i\|^\flat$ will shrink like a_i^r only. On the other hand, for the sharp norm, one can cancel the flat norm of a chain by translating simplices using vectors of the same order of magnitude as a_i. The price to pay, according to the definition of the sharp norm, is bounded by a_i^{r+1} whether A_i is the boundary of another chain or not.

It is noted that being less regular than a flat chain, the boundary of a sharp chain need not exist as a sharp chain.

The Riemann integral of a continuous r-form τ over a sharp r-chain $A = \lim A_i$ is defined to be $\int_A \tau = \lim \int_{A_i} \tau$, if the limit exists.

23.5 Cochains

Cochains are elements of the dual spaces to the Banach spaces of flat and sharp chains. The basic idea of the application of Whitney's geometric integration theory to the analysis of Cauchy fluxes is that cochains in the various dual spaces are abstract counterparts of total flux operators. Specifically, for classical continuum mechanics, we regard the total flux Φ_A of a certain extensive property, P, through a two-dimensional domain A in $\mathbb{E}^3$, as the action $T \cdot A$ of a 2-cochain T on the 2-chain A associated with the domain. For the sake of simplicity of the notation, we used here the same notation for both the oriented domain and the representing chain. It is noted that chains contain more information than just the domain where they are supported. For example, any continuous function defined on a submanifold of the Euclidean space may be represented as a chain. Obviously, the coefficients for the simplices that make up the chain will be different from 1 and will represent the values of the function. In such a case, one may interpret the value of the function as some potential associated with the extensive property, so that the total flux is interpreted as the flux of power. Alternatively, if we interpret the value of the function as a component of a velocity field, the action of a cochain on the chain may be interpreted as the calculation of mechanical power. Thus, geometric integration theory combines the classical approaches to flux theory and the variational weak approach. An immediate benefit of using geometric integration theory is that the analysis holds for r-chains in $\mathbb{E}^n$ for all values of $r \leqslant n$.

The properties of cochains that make them suitable mathematical models for Cauchy fluxes follow firstly from the linearity of their action on chains, which is common to both Banach spaces considered above. Linearity of the action of cochains implies both the additivity and the action-interaction-antisymmetry properties assumed in continuum mechanics as described in some detail in Chaps. 4 and 8. For example, given a cochain T, we have $T \cdot (-A) = -T \cdot A$. Secondly, the properties of the various cochains are determined by the continuity of their action on chains. The continuity condition is directly linked to the norm on the respective space of chains. Basic observations regarding the relations between the various norms, and the properties we expect fluxes to have, will be described below.

23.5.1 Flat Cochains

Flat r-cochains in $\mathbb{E}^n$ are the elements of $\mathbf{C}_r^\flat(\mathbb{E}^n)^*$, the dual space of $\mathbf{C}_r^\flat(\mathbb{E}^n)$.

Consider norms, $\|\cdot\|^x$, on the space of polyhedral chains and wish to consider the action of a continuous flux functional T; then, $|T \cdot A| \leqslant C_T\|A\|^x$. If one requires that $\|A\|^x \leqslant |A|$ and $\|\partial D\|^x \leqslant |D|$, for any r-chain A and $(r+1)$-chain D, then the boundedness conditions are implied by continuity because

$$|T \cdot A| \leqslant C_T\|A\|^x \leqslant C_T\,|A|\,, \quad \text{and} \quad |T \cdot \partial D| \leqslant C_T\|\partial D\|^x \leqslant C_T\,|D|\,. \tag{23.45}$$

In order to admit the most general flux operators that satisfy these conditions, we need the largest norm such that $\|A\|^x \leqslant |A|$ and $\|\partial D\|^x \leqslant |D|$. Indeed, Whitney shows that the flat norm is the largest norm satisfying these two conditions.

23.5.2 Sharp Cochains

Sharp r-cochains in E^n are elements of $\mathbf{C}_r^\sharp(\mathbb{E}^n)^*$, the dual the space of $\mathbf{C}_r^\sharp(\mathbb{E}^n)$. Since flat chains form a Banach subspace of $\mathbf{C}_r^\sharp(\mathbb{E}^n)$, every sharp cochain may be restricted to flat chains. In other words, any sharp cochain is also flat.

The additional property of sharp cochains that distinguishes them from generic flat cochains is the boundedness under translation. Given a sharp cochain T, consider for an r-simplex s and a vector v, the difference in the flux due to the translation by v, i.e., $|T \cdot s - T \cdot \text{trans}_v s|$. The continuity of T implies that

$$\begin{aligned} |T \cdot s - T \cdot \text{trans}_v s| &\leqslant C_T \|s - \text{trans}_v s\|^\sharp, \\ &\leqslant C_T \frac{|s|\,|v|}{r+1}, \end{aligned} \tag{23.46}$$

by choosing $v_1 = 0$ and $v_2 = -v$ in the definition of the sharp norm. Thus, continuity implies that there is a positive N such that

$$|T \cdot s - T \cdot \text{trans}_v s| \leqslant N\,|s|\,|v|\,. \tag{23.47}$$

In particular, the difference tends to zero if so does the magnitude of v. Clearly, this imposes a regularity restriction on sharp cochains.

Consider norms $\|\cdot\|^x$ on the space of polyhedral chains and wish to consider the action of a continuous flux functional T; then, $|T \cdot A| \leqslant C_T \|A\|^x$. If one requires that $\|A\|^x \leqslant |A|$, $\|\partial D\|^x \leqslant |D|$ and $\|s - \text{trans}_v s\|^x \leqslant |s|\,|v|$, for every r-chain A, $(r+1)$-chain D, r-simplex s, and vector v, then the boundedness conditions are implied by continuity. For example, boundedness under translation is implied by

$$|T \cdot s - T \cdot \text{trans}_v s| \leqslant C_T\,|s - \text{trans}_v s|^x \leqslant C_T\,|s|\,|v|\,. \tag{23.48}$$

To admit the most general flux operators that satisfy these three conditions, we need the largest norm satisfying them. Indeed, Whitney shows that the sharp norm is the largest norm satisfying the conditions.

23.5.3 The Cauchy Mapping

The Cauchy mapping of r-directions induced by an r-cochain is completely analogous to the mapping that gives the dependence of the flux density on the unit normal in classical continuum mechanics, hence the terminology we have chosen. Let the *r-direction* α of an r-simplex s be the r-vector $\{s\}/\,|s|$. (It is recalled that in Sect. 5.3, $\{s\}$ is defined as the r-vector associated with the simplex s.) The *Cauchy mapping* D_T associated with the cochain T is defined to be the function of points, x, and r-directions, α, such that

$$D_T(x, \alpha) = \lim_{i \to \infty} T \cdot \frac{s_i}{|s_i|}, \tag{23.49}$$

where s_i is a sequence of r-simplices containing x with r-direction α, such that

$$\lim_{i\to\infty} \operatorname{diam}(s_i) = 0. \tag{23.50}$$

Since the r-direction α is a generalization of the unit normal $\mathbf{n}$ used in continuum mechanics, the analog of Cauchy's flux theorem will be the assertion that the restriction of the Cauchy mapping to each point x may be extended to a linear mapping of r-vectors. In other words, D_T is an r-form in $\mathbb{E}^n$.

23.5.4 The Representation of Sharp Cochains by Forms

The analog to Cauchy's flux theorem in Whitney's geometric integration theory for sharp cochains states the following.

Proposition 23.1 *For each sharp r-cochain T, the Cauchy mapping D_T may be extended to a unique r-form that represents T by*

$$T \cdot A = \int_A D_T, \tag{23.51}$$

for every polyhedral chain A.

Clearly, the proposition defines the integral of a form over a sharp chain by continuity.

Whitney's theory determines exactly the forms that represent sharp cochains—the *sharp forms*. Firstly, the norm $|\cdot|_0$ is defined on the space of alternating r-tensors, $\bigwedge^r \mathbf{V}^*$, by

$$|\tau|_0 = \sup\bigl\{|\tau(\mathfrak{v})| \bigm| \mathfrak{v} \text{ a simple multivector, } |\mathfrak{v}| = 1\bigr\}, \tag{23.52}$$

where the Euclidean structure is used to evaluate the norm of multivectors. The sharp norm of the form τ is defined by

$$\|\tau\|^\sharp = \sup_{x,y\in\mathbb{E}^n} \left\{|\tau(x)|_0\,,\,(r+1)\frac{|\tau(y)-\tau(x)|_0}{|y-x|}\right\}. \tag{23.53}$$

Then, a *sharp form* is defined to be a form whose sharp norm is finite. Thus, sharp forms are bounded Lipschitz forms. Using the norm topology on the space of cochains, where

$$\|T\|^\sharp = \sup_{\|A\|^\sharp=1} |T \cdot A|\,,$$

Whitney showed that the previous proposition defines an isomorphism of the Banach space of sharp cochains and the Banach space of sharp forms.

23.5.5 The Representation of Flat Cochains by Forms

While sharp r-cochains are regular enough to be represented uniquely by sharp r-forms, flat r-cochains are less regular, and each flat r-cochain is represented by an equivalence class of r-forms which satisfy certain regularity conditions. Sharp forms representing sharp cochains are continuous, and Riemann integration may be used. The representation of flat cochains by forms requires Lebesgue integration.

An r-form in $\mathbb{E}^n$ is said to be bounded and measurable if all its components relative to a basis of $\mathbf{V}$ are bounded and measurable. The Lebesgue integral of an r-form τ over an r-simplex s is defined by

$$\int_s \tau = \int_s \tau(x) \cdot \frac{\{s\}}{|s|} \, \mathrm{d}x, \tag{23.54}$$

where $|s|$ is the r-dimensional volume of s and the integral on the right is a Lebesgue integral of a real-valued function. Lebesgue integration is extended by linearity to an arbitrary polyhedral chain, $A = \sum_i a_i s_i$, by

$$\int_A \tau = \sum_i a_i \int_{s_i} \tau. \tag{23.55}$$

The Lebesgue integral of an r-form over a flat r-chain $A = \lim^\flat A_j$ is defined by

$$\int_A \tau = \lim \int_{A_j} \tau, \tag{23.56}$$

if the limit exists.

The analysis of the representation of flat cochains by forms requires more attention than the sharp counterpart. For example, in the definition of the Cauchy mapping

$$D_T(x, \alpha) = \lim_{i \to \infty} T \cdot \frac{s_i}{|s_i|}, \tag{23.57}$$

it is required that in the converging sequence (s_i), each of the simplices will contain x as a vertex. It turns out that for each r-direction α, $D_T(x, \alpha)$ is defined almost everywhere. Wolfe's representation theorem as formulated by Whitney [Whi57, p. 261] for flat cochains states as follows.

Proposition 23.2 *Let T be a flat r-cochain in an open set $R \subset \mathbb{E}^n$. Then, there is a set $Q \subset R$, with $|R - Q| = 0$, such that for each $x \in Q$, $D_T(x, \alpha)$ is defined for all r-directions α, and is extendable to all r-vectors, giving an r-covector $D_T(x)$. The r-form D_T is bounded and measurable in R. For any r-simplex s in R, D_T is a measurable r-form relative to the plane of s and*

$$T \cdot s = \int_s D_T. \tag{23.58}$$

In [Whi57, pp. 263–266], Whitney identifies exactly the space of flat forms—those forms that represent flat cochains. Identifying forms that are equal almost everywhere, and choosing an appropriate norm, the flat norm of forms, Whitney showed that the Banach space of flat forms is isomorphic to the space of flat cochains.

23.6 Coboundaries and the Differential Balance Equations

Coboundaries generalize exterior differentiation, and their definition is purely algebraic. The *coboundary* $\mathrm{d}T$ of an r-cochain T is the $(r+1)$-cochain defined by

$$\mathrm{d}T \cdot A = T \cdot \partial A, \tag{23.59}$$

i.e., it is the dual of the boundary operator for chains. As $\partial(\partial A) = 0$, one has $\mathrm{d}(\mathrm{d}T) = 0$. The basic result concerning coboundaries is that the coboundary of a flat cochain is flat and the coboundary of a sharp cochain is flat. This implies a very general formulation of the balance equation. For a cochain T that is either sharp or flat, the coboundary exists as a flat cochain, and we may define an $(r+1)$-cochain S, satisfying $\mathrm{d}T + S = 0$, so the balance equation $S \cdot A + T \cdot \partial A = 0$ holds. Here, S is interpreted as the cochain giving the rate of change of the total amount of the property, P, in the flat $(r+1)$-chain A (assuming there is no source term).

If the form D_T, representing the cochain T, is differentiable, then the flat form $D_{\mathrm{d}T}$ representing $\mathrm{d}T$ is given as the exterior derivative of D_T. That is,

$$D_{\mathrm{d}T} = \mathrm{d}D_T, \tag{23.60}$$

as one would expect. Thus, using τ for D_T, the abstract balance equation above assumes the form

$$\mathrm{d}\tau + b = 0, \qquad \int_A b + \int_{\partial A} \tau = 0. \tag{23.61}$$

In the more general case where τ is an arbitrary flat form representing the flat cochain T, $\mathrm{d}T$ is a flat cochain, and hence it may be represented by any flat form $\mathrm{d}_0\tau$ in the equivalence class of $D_{\mathrm{d}T}$. Thus, one may write the "differential" balance in the general situation of flat cochains. In fact,

$$\|T\|^\flat = \sup_x \{|D_T(x)|\,,\, |D_{\mathrm{d}T}(x)|\}. \tag{23.62}$$

The right-hand side of this identity is the flat norm of the form D_T.

In the particular case where T is a sharp cochain represented by the sharp form $\tau = D_T$, the functions giving the components of τ are Lipschitz mappings. Hence, it has an analytic exterior derivative, $\mathrm{d}\tau$, almost everywhere. Furthermore, it turns out that $\mathrm{d}_0\tau = \mathrm{d}\tau$ almost everywhere.

Chapter 24
Optimal Fields and Load Capacity of Bodies

In this chapter, based on [Seg07b], we return to the classical geometric settings of continuum mechanics in $\mathbb{R}^3$, and so it may be read independently of the foregoing material. Using analytic methods, which are analogous to those of Chaps. 2 and 21, we present a framework for the analysis of optimization problems associated with balance equations on a given region $\Omega \subset \mathbb{R}^3$. As balance equations do not have unique solutions, we look for solutions of minimum L^p-norms, in particular, minimum L^∞-norm. Letting the data fields vary, we also look for the largest ratio, K, between the norm of the optimal solution and the norm of the data fields. This largest ratio is a purely geometric property of the region Ω. Among other examples, we show that for an elastic perfectly plastic body, there is a maximal positive number, C, the *load capacity ratio,* such that the body will not collapse plastically under any distribution of loading $(\mathbf{t}, \mathbf{b})$, containing an external traction field $\mathbf{t}$ and a body force field $\mathbf{b}$, if the norm of the loading is bounded by CY_0, where Y_0 is the yield stress. We also give expressions for K and C in terms of quantities that are analogous to the norm of the trace mapping for Sobolev spaces.

24.1 Introduction

It is assumed that Ω is a connected, bounded, and open subset of $\mathbb{R}^n$ ($\mathbb{R}^3$ or $\mathbb{R}^2$ in practical cases) and that $\partial\Omega$ is C^2. (In many cases weaker smoothness conditions on the boundary are sufficient.)

The balance equations of various quantities in continuum physics may be written as

$$\nabla \cdot \sigma + \mathbf{b} = 0 \quad \text{in}\, \Omega, \quad \sigma(\mathbf{n}) = \mathbf{t}, \quad \text{on}\, \partial\Omega, \tag{24.1}$$

R. Segev, *Foundations of Geometric Continuum Mechanics*, Advances in Continuum Mechanics 49, https://doi.org/10.1007/978-3-031-35655-1_24

where $\mathbf{b}$ is given in a region Ω, $\mathbf{n}$ is the unit normal to the boundary, and the boundary condition $\mathbf{t}$ is given on $\partial\Omega$. Note that the exact tensorial character of $\mathbf{b}$, $\mathbf{t}$ and σ is not specified yet, so, for example, σ may be a flux vector field or a stress tensor field, etc. The tensor field σ is the unknown, and without having any additional data, such as constitutive relations, the system is under-determined. Thus, for the given data $(\mathbf{t}, \mathbf{b})$, there is a family $\Sigma_{(\mathbf{t},\mathbf{b})}$ of solutions to the problem. The weak version of balance equations may be written in the form

$$\int_\Omega \sigma(D(w)) = \int_{\partial\Omega} \mathbf{t}\cdot w + \int_\Omega \mathbf{b}\cdot w \tag{24.2}$$

for all vector fields w on Ω belonging to a certain function space, where D is some differential operator. Writing the integral operators as linear functionals on the space of fields w and keeping the same notation for the linear functionals (with abuse of notation), the balance equation above may be rewritten as

$$\sigma(D(w)) = (\mathbf{t}, \mathbf{b})(\delta(w)), \qquad \text{for all} \quad w, \tag{24.3}$$

where $\delta(w) = (w|_{\partial\Omega}, w)$. This may be rewritten using the definition of dual mappings as

$$D^*(\sigma) = \delta^*(\mathbf{t}, \mathbf{b}). \tag{24.4}$$

The last form of the balance equation for σ emphasizes the point of view whereby the solutions are given in terms of a generalized inverse, D^{*+}, of the dual differential operator D^*.

Since the solution is not unique, one may consider an optimization problem where, for given loading data $(\mathbf{t}, \mathbf{b})$, we look, among all $\sigma \in \Sigma_{(\mathbf{t},\mathbf{b})}$, for the solution that minimizes a certain physically meaningful norm, $\|\cdot\|$. Thus, we set

$$s^{\text{opt}}_{(\mathbf{t},\mathbf{b})} = \inf_{D^*(\sigma)=\delta^*(\mathbf{t},\mathbf{b})} \|\sigma\|. \tag{24.5}$$

Usually, generalized inverses are constructed so that they minimize the L^2-norm. For many applications in engineering, the L^∞-norm, interpreted as the maximum of the field σ, is more relevant. Thus, we consider optimization of the L^∞-norm in the examples below.

In traditional engineering stress analysis, the term stress concentration factor, K_σ, is used for the ratio between the maximum of the stress field and the maximum of the "nominal" stress field computed by oversimplifying the elasticity problem. The term may be borrowed and used for the ratio between the maximum of the stress field and the maximum of the applied loading fields that the stress equilibrates. Replacing the maximum of the various fields by possibly more general norms, we use the term *stress concentration factor* for the ratio,

$$K_\sigma = \frac{\|\sigma\|}{\|(\mathbf{t}, \mathbf{b})\|}, \tag{24.6}$$

for some norm on the space of data $\{(\mathbf{t}, \mathbf{b})\}$.

Next, we consider the ratio

$$K_{(\mathbf{t},\mathbf{b})} = \frac{S^{\text{opt}}_{(\mathbf{t},\mathbf{b})}}{\|(\mathbf{t}, \mathbf{b})\|}. \tag{24.7}$$

We refer to $K_{(\mathbf{t},\mathbf{b})}$ as the *optimal stress concentration factor* (see [Seg06, Seg05]). Finally, we consider the *generalized stress concentration factor*

$$K = \sup_{(\mathbf{t},\mathbf{b})} K_{(\mathbf{t},\mathbf{b})} = \sup_{(\mathbf{t},\mathbf{b})} \frac{S^{\text{opt}}_{(\mathbf{t},\mathbf{b})}}{\|(\mathbf{t}, \mathbf{b})\|}. \tag{24.8}$$

It is noted that by its definition, K is a purely geometric property of Ω.

Summarizing our work [Seg06, Seg05, PS07, Seg07a, FS09], we present a framework where the analysis of the quantities defined above may be carried out. We also exhibit the results for some particular cases. The following result, which we derive in the last section, is relevant to engineering stress analysis.

Theorem 24.1 *Let Ω be a region occupied by an elastic perfectly plastic body whose yield stress is Y_0, and such that the yield criterion is given in terms of a norm $|\cdot|$ applied to the stress deviator matrix. Assume that the body is supported on an open subset of its boundary. Then, there is a maximal positive number, C, to which we refer as the* load capacity ratio *of Ω, such that the body will not collapse under any pair of essentially bounded force fields $(\mathbf{t}, \mathbf{b})$ as long as the essential suprema of their magnitudes are less than or equal to CY_0. The number C satisfies*

$$\frac{1}{C} = \sup_{w \in LD(\Omega)_{P\perp}} \frac{\int_{\partial\Omega} |\gamma(w)| + \int_\Omega |w|}{\int_\Omega |\varepsilon(w)|}. \tag{24.9}$$

Here, γ is the trace mapping, i.e., $\gamma(w)$ is the boundary value of the vector field w, so for a continuous vector field u defined on $\overline{\Omega}$, $\gamma(u|_\Omega) = u|_{\partial\Omega}$; $LD(\Omega)_{P\perp}$ is the collection of incompressible integrable vector fields w for which $\varepsilon(w) = \frac{1}{2}(\nabla w + (\nabla w)^T)$, the corresponding linear strain, is integrable; and $|\varepsilon(w)(x)|$ is evaluated using the dual norm to that used for the yield criterion.

The expression above makes it evident again that the load capacity of a body, or a structure, is a purely geometric property. In some cases, $1/C$ simply reduces to the norm of the trace mapping. The results of this chapter make it possible to compute the load capacity and worst-case loading distributions of a structure, based only on the geometry of the structure (see [FS09]).

24.2 Balance Equations

24.2.1 Sobolev Spaces of Sections of a Trivial Vector Bundle

As in previous chapters, it is convenient to view vector fields w as sections of a vector bundle W, which in the current setting will be trivial. Thus, we consider a trivial vector bundle $\pi : W = \Omega \times \mathbf{V} \to \Omega$ where the fiber $\mathbf{V}$ is m-dimensional, and the base manifold Ω is a bounded open subset of $\mathbb{R}^n$ having a smooth boundary. Obviously, a section w of W is represented by a mapping $\Omega \to \mathbf{V}$. It is assumed that $\mathbf{V}$ has a Riemannian structure $\langle \cdot\, , \cdot \rangle$. In fact, in all the examples we present, $\mathbf{V}$ is taken as $\mathbb{R}^m$.

For a vector field, w, a section of W, the jet extension $j^k w$ is represented by the collection of tensor fields, $(w, \nabla w, \cdots, \nabla^k w)$, defined on Ω. It is assumed that a norm $|\cdot|$ is given on $J^k_x W$. For example, using a basis in $\mathbf{V}$ and utilizing multi-index notation, one may set,

$$\left|j^k_x w\right| := \sum_{|I|\leqslant k, j} |w^j_{,I}(x)|, \quad \text{or alternatively,} \quad \left|j^k_x w\right| := \left(\sum_{|I|\leqslant k, j} |w^j_{,I}(x)|^2\right)^{1/2}. \tag{24.10}$$

The dual vector space $(J^k_x W)^*$ is the fiber of the dual jet bundle $(J^k W)^*$. The same notation, $|\cdot|$, is used for the dual norm on $(J^k_x W)^*$.

For $1 < p < \infty$, we set as usual $p' = p/(p-1)$ with the extension $p' = \infty$ for $p = 1$. Since Ω is an open subset of $\mathbb{R}^n$, we can use the Lebesgue measure. Therefore, a measurable section χ of $J^k W$ is L^p if

$$\|\chi\|_p = \left(\int_\Omega |\chi(x)|^p\right)^{1/p} < \infty. \tag{24.11}$$

For L^p-sections, $\|\cdot\|_p$ is indeed a norm. The space of L^p-sections of $J^k W$ is denoted as $L^p(J^k W)$, and it is a Banach space with the norm above. We have the duality relation $L^p(J^k W)^* \cong L^{p'}((J^k W)^*)$. An $L^{p'}$-section $\widehat{\sigma}$ of $(J^k W)^*$ represents the continuous linear functional $\sigma \in L^p(J^k W)^*$ in the form

$$\sigma(\chi) = \int_\Omega \widehat{\sigma}(x)(\chi(x)), \tag{24.12}$$

and in the sequel we will identify σ and $\widehat{\sigma}$ and omit the "$\widehat{\ }$".

The Sobolev space $W^p_k(W)$ contains the sections w of W such that $j^k w \in L^p(J^k W)$. Note that since a section of W may be represented by a collection of real-valued functions on Ω, distributional derivatives of w make sense, and the k-jet of w is defined accordingly. For a comprehensive treatment of the subject on manifolds, see [Pal68]. For any $w \in W^p_k(W)$, $\|w\| = \|j^k(w)\|_p$ is a norm, for which $W^p_k(W)$ is a Banach space. Clearly, this choice of norm makes the jet mapping

$$j^k\colon W_k^p(W) \longrightarrow L^p(J^k W) \tag{24.13}$$

a linear, isometric injection.

24.2.2 W_k^p-Forces, (p, k)-Stresses, and Stress Field Optimization

We will refer to an element F of $W_k^p(W)^*$ as a W_k^p-force and to an element σ of $L^{p'}((J^k W)^*)$ as a (p, k)-stress. The standard representation of elements of the dual spaces to the Sobolev spaces (e.g., [Maz85, pp. 25–26]) is interpreted here physically as a representation theorem of forces by stresses, in complete analogy with Sect. 21.1. For each $F \in W_k^p(W)^*$, $\sigma_0 = F \circ (j^k)^{-1}\colon \text{Image}\, j^k \to \mathbb{R}$ is a bounded linear functional with $\sigma_0 \circ j^k = F$. As the jet extension mapping is isometric, $\|\sigma_0\| = \|F\|$. Thus, using the Hahn-Banach theorem, one may extend σ_0 from Image j^k to $L^p(J^k(W))$ giving some element $\sigma \in L^{p'}((J^k W)^*)$, with $\sigma \circ j^k = F$, or equivalently,

$$F = (j^k)^*(\sigma). \tag{24.14}$$

Moreover, by the Hahn-Banach theorem, there is a stress field σ^{hb}, such that

$$\begin{aligned} \|\sigma^{\text{hb}}\| &= \|\sigma_0\|, \\ &= \sup_{\chi \in \text{Image}\, j^k} \frac{\sigma_0(\chi)}{\|\chi\|}, \\ &= \sup_{w \in W_k^p(W)} \frac{F(w)}{\|w\|}, \\ &= \|F\|. \end{aligned} \tag{24.15}$$

Since for an arbitrary stress field $\sigma' \in L^{p'}((J^k W)^*)$ that extends σ_0,

$$\begin{aligned} \|\sigma'\| &= \sup_{\chi \in L^p(J^k W)} \frac{\sigma'(\chi)}{\|\chi\|}, , \\ &\geqslant \sup_{\chi \in \text{Image}\, j^k} \frac{\sigma'(\chi)}{\|\chi\|}, \\ &= \sup_{\chi \in \text{Image}\, j^k} \frac{\sigma_0(\chi)}{\|\chi\|}, \\ &= \|\sigma_0\|, \\ &= \|F\|, \end{aligned} \tag{24.16}$$

we conclude that

$$\|\sigma^{\mathrm{hb}}\| = \|F\| = \inf\left\{\|\sigma'\| \mid F = (j^k)^*(\sigma'), \sigma' \in L^{p'}((J^k W)^*)\right\}. \tag{24.17}$$

It follows that σ^{hb} is an optimal stress field and that $\|F\|$ is the least norm that a stress that represents the force F may have.

24.2.3 Traces and Loading Distributions

Consider the boundary $\partial\Omega$ of the region under consideration. The smoothness assumptions we have made for $\partial\Omega$ imply that the Lebesgue measure on $\mathbb{R}^n$ induces a measure on $\partial\Omega$ (see Adams [AF03, p. 163]). As a result, the Sobolev spaces of functions prescribed on the boundary, $W_l^p(W|_{\partial\Omega})$, are well-defined. (It is noted that as W is a trivial bundle over $\Omega \subset \mathbb{R}^n$, $W|_{\partial\Omega}$ is well-defined.) The trace mapping, $\gamma\colon W_k^p(W) \to W_{k_\partial}^p(W|_{\partial\Omega})$, where $k_\partial = k - 1/p$ is the reduced differentiability, assigns the boundary values corresponding to Sobolev sections. It is a well-defined bounded linear mapping (see, e.g., Adams [AF03, p. 163], Palais [Pal68, p. 27]), and for any $u \in C^\infty(\overline{\Omega}, \mathbb{R}^3)$, it satisfies $\gamma(u|_\Omega) = u|_{\partial\Omega}$.

Since we have the inclusion $\mathcal{I}_{k_\partial}: W_k^p(W) \hookrightarrow W_{k_\partial}^p(W)$, we may set

$$\delta = (\gamma, \mathcal{I}_{k_\partial})\colon W_k^p(W) \longrightarrow W_{k_\partial}^p(W|_{\partial\Omega}) \times W_{k_\partial}^p(W). \tag{24.18}$$

Clearly, δ is a bounded linear mapping, where for $(u, v) \in W_{k_\partial}^p(W|_{\partial\Omega}) \times W_{k_\partial}^p(W)$, $\|(u, v)\| = \|u\| + \|v\|$. We will refer to an element $\mathbf{t} \in W_{k_\partial}^p(W|_{\partial\Omega})^*$ as a surface force, to an element $\mathbf{b} \in W_{k_\partial}^p(W)^*$ as a body force, and to a pair $(\mathbf{t}, \mathbf{b}) \in \left(W_{k_\partial}^p(W|_{\partial\Omega}) \times W_{k_\partial}^p(W)\right)^*$ as a loading pair.

We conclude that a loading pair, $(\mathbf{t}, \mathbf{b})$, represents a force, $F = \delta^*(\mathbf{t}, \mathbf{b})$, where $\delta^*: \left(W_{k_\partial}^p(W|_{\partial\Omega}) \times W_{k_\partial}^p(W)\right)^* \to W_k^p(W)^*$ is the dual mapping. Specifically,

$$F(w) = \mathbf{t}(\gamma(w)) + \mathbf{b}(w), \quad \text{for all} \quad w \in W_k^p(W). \tag{24.19}$$

24.2.4 Loading Distributions and Stresses

The framework we presented above can be described by the following diagram

$$
\begin{array}{ccccc}
W^p_{k_\partial}(W|_{\partial\Omega}) \times W^p_{k_\partial}(W) & \xleftarrow{\ \delta\ } & W^p_k(W) & \xrightarrow{\ j^k\ } & L^p(J^kW) \\
\left(W^p_{k_\partial}(W|_{\partial\Omega}) \times W^p_{k_\partial}(W)\right)^* & \xrightarrow{\ \delta^*\ } & W^p_k(W)^* & \xleftarrow{\ (j^k)^*\ } & L^{p'}((J^kW)^*).
\end{array} \tag{24.20}
$$

From the foregoing analysis, we conclude the following

Theorem 24.2 *Let* $(\mathbf{t}, \mathbf{b}) \in \left(W^p_{k_\partial}(W|_{\partial\Omega}) \times W^p_{k_\partial}(W)\right)^*$ *be a loading pair; then, the following assertions hold.*

(*i*) Existence of stresses. *There is a subset* $\Sigma_{(\mathbf{t},\mathbf{b})} \subset L^{p'}((J^kW)^*)$ *of stress distributions that represent the loading pair, such that for* $\sigma \in \Sigma_{(\mathbf{t},\mathbf{b})}$, $\delta^*(\mathbf{t}, \mathbf{b}) = (j^k)^*(\sigma)$.

(*ii*) Optimal stresses. *Let* $s^{\text{opt}}_{(\mathbf{t},\mathbf{b})} = \inf_{\sigma\in\Sigma_{(\mathbf{t},\mathbf{b})}} \{\|\sigma\|\}$. *Then, there is a stress* $\widehat{\sigma} \in \Sigma_{(\mathbf{t},\mathbf{b})}$ *such that* $s^{\text{opt}}_{(\mathbf{t},\mathbf{b})} = \|\widehat{\sigma}\|$.

(*iii*) The expression for $s^{\text{opt}}_{(\mathbf{t},\mathbf{b})}$. *The optimum satisfies*

$$
s^{\text{opt}}_{(\mathbf{t},\mathbf{b})} = \|\delta^*(\mathbf{t}, \mathbf{b})\| = \sup_{w\in W^p_k(W)} \frac{|\mathbf{t}(\gamma(w)) + \mathbf{b}(w)|}{\|j^k(w)\|_p}. \tag{24.21}
$$

(*iv*) The generalized stress concentration factor. *Let the generalized stress concentration factor be defined by*

$$
K := \sup_{(\mathbf{t},\mathbf{b})} \frac{s^{\text{opt}}_{(\mathbf{t},\mathbf{b})}}{\|(\mathbf{t}, \mathbf{b})\|}, \qquad (\mathbf{t}, \mathbf{b}) \in \left(W^p_{k_\partial}(W|_{\partial\Omega}) \times W^p_{k_\partial}(W)\right)^*. \tag{24.22}
$$

Then,

$$
K = \|\delta\|. \tag{24.23}
$$

Proof It remains to prove (*iv*). Using Eq. (24.21), we have

$$
\begin{aligned}
K &= \sup_{(\mathbf{t},\mathbf{b})} \frac{s^{\text{opt}}_{(\mathbf{t},\mathbf{b})}}{\|(\mathbf{t}, \mathbf{b})\|}, \\
&= \sup_{(\mathbf{t},\mathbf{b})} \frac{\|\delta^*(\mathbf{t}, \mathbf{b})\|}{\|(\mathbf{t}, \mathbf{b})\|}, \\
&= \|\delta^*\|, \\
&= \|\delta\|,
\end{aligned} \tag{24.24}
$$

by a standard result of functional analysis (e.g., [Tay58, p. 214]). □

In the case where no body forces are applied, γ replaces δ in the analysis. Therefore, the preceding equations reduce to the simpler

$$s_{\mathbf{t}}^{\mathrm{opt}} = \|\gamma^*(\mathbf{t})\| = \sup_{w \in W_k^p(W)} \frac{|\mathbf{t}(\gamma(w))|}{\|j^k(w)\|_p}, \tag{24.25}$$

and

$$K := \sup_{(\mathbf{t},\mathbf{b})} \frac{s_{\mathbf{t}}^{\mathrm{opt}}}{\|\mathbf{t}\|}, \qquad \mathbf{t} \in W_{k_\partial}^p(W|_{\partial\Omega})^*, \tag{24.26}$$

satisfying

$$K = \|\gamma\| \tag{24.27}$$

—the norm of the trace mapping.

24.2.5 The Junction Problem for Fluxes

In this simplest example, we consider the case where W is the trivial line bundle $\Omega \times \mathbb{R}$ and the Sobolev space is $W_1^1(\Omega)$. The various spaces are shown in the diagram below. Sections of $J^1(\Omega \times \mathbb{R})$ in $W_1^1(\Omega)$ are identified with mappings $\Omega \to \mathbb{R}^{n+1}$ regarded as pairs $(\phi, v) \in L^1(\Omega, \mathbb{R} \times \mathbb{R}^n)$. For $\phi \in W_1^1(\Omega)$, $j^1\phi$ may be identified with $(\phi, \nabla\phi)$.

$$\begin{array}{ccccc}
L^1(\partial\Omega) \times L^1(\Omega) & \xleftarrow{\ \delta\ } & W_1^1(\Omega) & \xrightarrow{\ j^1\ } & L^1(\Omega, \mathbb{R}^{n+1}) \\
 & & & & \\
\left(L^1(\partial\Omega) \times L^1(\Omega)\right)^* & \xrightarrow{\ \delta^*\ } & W_1^1(\Omega)^* & \xleftarrow{\ (j^1)^*\ } & L^1(\Omega, \mathbb{R}^{n+1})^* \\
\| & & & & \| \\
L^\infty(\partial\Omega) \times L^\infty(\Omega) & & & & L^\infty(\Omega, \mathbb{R}^{n+1}).
\end{array} \tag{24.28}$$

One may think of this framework as a continuous logistics problem in the region Ω, possibly conceived as a junction. For some extensive property, e.g., the mass of a certain material or the thermal energy, we regard the loading pair $(\mathbf{t}, \mathbf{b}) \in L^\infty(\partial\Omega) \times L^\infty(\Omega)$ as prescribed boundary flux distribution and rate of change of density. These are balanced by a stress object, a pair $(\sigma_0, \sigma_1) \in L^\infty(\Omega) \times L^\infty(\Omega, \mathbb{R}^n)$, where σ_1 is interpreted as a flow field and σ_0 is interpreted as the source distribution of the property in Ω. Thus, for an optimal pair, (σ_0, σ_1), one wishes to minimize the magnitude of the maximal value of the flow and the maximal value of the source (production rate) density.

Note that for a differentiable $\sigma_1 : \Omega \to \mathbb{R}^n$, and fields $\mathbf{b}, \sigma_0 \in L^\infty(\Omega)$, $\mathbf{t} \in L^\infty(\partial\Omega)$, the condition $\delta^*(\mathbf{t}, \mathbf{b}) = (j^1)^*(\sigma_0, \sigma_1)$ implies that for each $\phi \in W_1^1(\Omega)$,

$$\int_\Omega \mathbf{b}\phi + \int_{\partial\Omega} \mathbf{t}\phi = \int_\Omega \sigma_0\phi + \int_{\partial\Omega} \sigma_{1i}\phi_{,i}. \tag{24.29}$$

(We use the summation convention with Cartesian tensors as we work in $\mathbb{R}^n$.) Evidently, using integration by parts, it follows that

$$\sigma_{1i,i} + \mathbf{b} = \sigma_0 \quad \text{in } \Omega, \quad \text{and} \quad \sigma_{1i}\mathbf{n}_i = \mathbf{t} \quad \text{on } \partial\Omega, \tag{24.30}$$

as expected.

We conclude that Theorem 24.2 implies the following.

(*i*) *Existence of flow objects.* Given a pair $(\mathbf{t}, \mathbf{b}) \in L^\infty(\partial\Omega) \times L^\infty(\Omega)$, consisting of a boundary flux and a density rate distributions, there is a subspace $\Sigma_{(\mathbf{t},\mathbf{b})} \subset L^\infty(\Omega) \times L^\infty(\Omega, \mathbb{R}^n)$ containing pairs of essentially bounded production rates and flux vector fields, for which the weak version of the balance equations (24.30) holds. It is noticed that we did not use any form of the existence theorem for the Cauchy flux, or stress, other than Theorem (24.2,*i*).

(*ii*) *Optimal stresses.* Let

$$s^{\text{opt}}_{(\mathbf{t},\mathbf{b})} = \inf_{\sigma \in \Sigma_{(\mathbf{t},\mathbf{b})}} \left\{ \operatorname*{ess\,sup}_{x\in\Omega} \{|\sigma_0(x)|, |\sigma_1(x)|\} \right\}, \tag{24.31}$$

where $|\sigma_1(x)|$ is any particular norm on $\mathbb{R}^n$ used for values of the flux vector field. Then, there is a pair $(\widehat{\sigma}_0, \widehat{\sigma}_1) \in \Sigma_{(\mathbf{t},\mathbf{b})}$ such that

$$s^{\text{opt}}_{(\mathbf{t},\mathbf{b})} = \operatorname*{ess\,sup}_{x\in\Omega} \{|\widehat{\sigma}_0(x)|, |\widehat{\sigma}_1(x)|\}. \tag{24.32}$$

(*iii*) *The expression for* $s^{\text{opt}}_{(\mathbf{t},\mathbf{b})}$. The optimum satisfies

$$s^{\text{opt}}_{(\mathbf{t},\mathbf{b})} = \sup_\phi \frac{\left|\int_{\partial\Omega} t\phi + \int_\Omega \mathbf{b}\phi\right|}{\int_\Omega (|\phi| + |\nabla\phi|)}, \quad \phi \in C^\infty(\overline{\Omega}), \tag{24.33}$$

where for computing $|\nabla\phi|$, we use the norm on $\mathbb{R}^n$ that is dual to the one used for the flow vector $\sigma_1(x)$. It is noted that since $C^\infty(\overline{\Omega})$ is a dense subset of $W_1^1(\Omega)$, it is sufficient to take the supremum over all smooth functions.

(*iv*) *The generalized stress concentration factor.* The generalized stress concentration factor, or rather, the flow amplification factor for our interpretation, is defined as

$$K = \sup_{(\mathbf{t},\mathbf{b})} \frac{s^{\text{opt}}_{(\mathbf{t},\mathbf{b})}}{\operatorname{ess\,sup}_{x,y} \{|\mathbf{t}(y)|\,,\, |\mathbf{b}(x)|\}}, \tag{24.34}$$

$(\mathbf{t}, \mathbf{b}) \in L^\infty(\partial\Omega) \times L^\infty(\Omega)$. Then, as $C^\infty(\overline{\Omega})$ is dense in the Sobolev spaces,

$$K = \|\delta\| = \sup_{\phi \in C^\infty(\overline{\Omega})} \frac{\int_\Omega |\phi| + \int_{\partial\Omega} |\phi|}{\int_\Omega (|\phi| + |\nabla\phi|)}. \tag{24.35}$$

24.3 Preliminaries on Rigid Velocity Fields

This section presents some properties of rigid velocity fields $\Omega \subset \mathbb{R}^3 \to \mathbb{R}^3$. In particular, we consider a projection that associates a rigid velocity field with any integrable velocity field. Such a projection will be of use below.

24.3.1 The Subspace of Rigid Velocities

A *rigid* velocity (or displacement) field is of the form

$$w(x) = a + \omega \times x, \quad x \in \Omega \tag{24.36}$$

where a and ω are fixed in $\mathbb{R}^3$ and $\omega \times x$ is the vector product. We can replace $\omega \times x$ with $\tilde{\omega}(x)$, where $\tilde{\omega}$ is the associated skew symmetric matrix, so $w(x) = a + \tilde{\omega}(x)$. We will denote the six-dimensional space of rigid body velocities by $\mathbf{R}$. For a rigid motion

$$\tilde{\omega}_{im} = \frac{1}{2}(w_{i,m} - w_{m,i}), \tag{24.37}$$

an expression that is extended to the non-rigid situation as the *vorticity* vector field. Therefore,

$$w_{i,m} = \varepsilon(w)_{im} + \tilde{\omega}_{im}. \tag{24.38}$$

Evidently, $\mathbf{R} \subset LD(\Omega)$ is a subspace, and we have an inclusion

$$\mathcal{I}_{\mathbf{R}} : \mathbf{R} \longrightarrow LD(\Omega). \tag{24.39}$$

Considering the kernel of the stretching mapping $\varepsilon \colon LD(\Omega) \longrightarrow L^1(\Omega, \mathbb{R}^6)$, a theorem whose classical version is due to Liouville states (see [Tem85, pp. 18–19]) that

$$\operatorname{Kernel}\varepsilon = \operatorname{Image}\mathcal{J}_{\mathbf{R}}. \tag{24.40}$$

Thus, we have the sequence

$$0 \longrightarrow \mathbf{R} \xrightarrow{\mathcal{J}_{\mathbf{R}}} LD(\Omega) \xrightarrow{\varepsilon} L^1(\Omega, \mathbb{R}^6) \longrightarrow 0. \tag{24.41}$$

24.3.2 Approximation by Rigid Velocities

We now wish to consider approximations of a velocity field by a rigid velocity field. Let ρ be a Radon measure on $\overline{\Omega}$ and $1 \leqslant p \leqslant \infty$. We consider the Banach space $L^{p,\rho}(\overline{\Omega}, \mathbb{R}^3)$ containing vector fields that are p-integrable relative to the measure ρ. For a given $w \in L^{p,\rho}(\overline{\Omega}, \mathbb{R}^3)$, we wish to find the rigid velocity $r = a + b \times x \in \mathbf{R}$, for which

$$(\|w - r\|_{L^{p,\rho}})^p = \inf_{r' \in \mathbf{R}} \left(\|w - r'\|_{L^{p,\rho}}\right)^p = \inf_{r' \in \mathbf{R}} \int_{\overline{\Omega}} \sum_i \left|w_i - r'_i\right|^p \mathrm{d}\rho \tag{24.42}$$

is attained. Setting

$$e_i(x) := w_i(x) - r'_i(x) = w_i(x) - a_i - \varepsilon_{ijk} b_j x_k \tag{24.43}$$

(where ε_{ijk} is the Levi-Civita alternating tensor), it is noted that

$$\frac{\partial e_i}{\partial a_l} = -\delta_{il}, \qquad \frac{\partial e_i}{\partial b_l} = -\varepsilon_{ilk} x_k, \qquad l = 1, 2, 3. \tag{24.44}$$

We are looking for vectors a and b that minimize

$$e = \int_{\overline{\Omega}} \sum_i |e_i|^p \mathrm{d}\rho. \tag{24.45}$$

For $l = 1, 2, 3$, we have

$$\begin{aligned}
\frac{\partial e}{\partial a_l} &= \int_{\overline{\Omega}} p \sum_i |e_i|^{p-1} \frac{e_i}{|e_i|}(-\delta_{il})\, \mathrm{d}\rho, \\
\frac{\partial e}{\partial b_l} &= \int_{\overline{\Omega}} p \sum_i |e_i|^{p-1} \frac{e_i}{|e_i|}(-\varepsilon_{ilk} x_k)\, \mathrm{d}\rho,
\end{aligned} \tag{24.46}$$

and we obtain the following six equations with the six unknowns, a_l, b_m, for the minimum,

$$\begin{aligned} 0 &= \int_{\overline{\Omega}} |e_l|^{p-2}\, e_l \,\mathrm{d}\rho, \\ 0 &= \int_{\overline{\Omega}} \sum_i |e_i|^{p-2}\, e_i \varepsilon_{ilk} x_k \,\mathrm{d}\rho, \end{aligned} \tag{24.47}$$

$l = 1, 2, 3$.

Particularly simple are the equations for $p = 2$. In this case we obtain

$$\int_{\overline{\Omega}} w \,\mathrm{d}\rho = \int_{\overline{\Omega}} r \,\mathrm{d}\rho, \quad \text{and} \quad \int_{\overline{\Omega}} x \times w \,\mathrm{d}\rho = \int_{\overline{\Omega}} x \times r \,\mathrm{d}\rho. \tag{24.48}$$

If we interpret ρ as a mass distribution on $\overline{\Omega}$, these two conditions simply state that the best rigid velocity approximations should give the same momentum and angular momentum as the original field.

Of particular interest (see [Tem85, p. 120]) is the case where ρ is the volume measure on Ω. Set $\bar{x}$ to be the center of volume of Ω, i.e.,

$$\bar{x} = \frac{1}{|\Omega|} \int_{\Omega} x. \tag{24.49}$$

Without loss of generality, we will assume that $\bar{x} = 0$ (otherwise, we may replace x by $x - \bar{x}$ in the sequel).

Let $\bar{w}$ be the mean of the field w and I the inertia matrix relative to the center of volume, that is,

$$\bar{w} = \frac{1}{|\Omega|} \int_{\Omega} w, \qquad I_{im} = \int_{\Omega} (x_k x_k \delta_{im} - x_i x_m), \tag{24.50}$$

and

$$I(\omega) = \int_{\Omega} x \times (\omega \times x). \tag{24.51}$$

The inertia matrix is symmetric and positive definite, and so the solution for r gives

$$r(x) = \bar{w} + \omega \times x \tag{24.52}$$

with $\bar{w}$ as above and

$$\omega = I^{-1}\left(\int_{\Omega} x \times w\right). \tag{24.53}$$

Thus, $x \longmapsto \bar{w} + \omega \times x$, with $\bar{w}$ and ω as above, is well-defined for integrable velocity fields, and we obtain a mapping

$$\mathrm{pr}_{\mathbf{R}} \colon L^1(\Omega, \mathbb{R}^3) \longrightarrow \mathbf{R}. \tag{24.54}$$

It is straightforward to show that $\mathrm{pr}_{\mathbf{R}}$ is indeed a linear projection onto $\mathbf{R}$.

Also of interest below will be the case where $p = 1$, and the measure ρ is given by

$$\rho(D) = V(D \cap \Omega) + A(D \cap \partial\Omega). \tag{24.55}$$

The conditions for best approximations $r = a + b \times x$ assume the form

$$\int_{\Omega} \frac{e_l}{|e_l|} + \int_{\partial\Omega} \frac{e_l}{|e_l|} = 0, \tag{24.56}$$

$$\int_{\Omega} \sum_i \frac{e_i}{|e_i|} \varepsilon_{ilk} x_k + \int_{\partial\Omega} \sum_i \frac{e_i}{|e_i|} \varepsilon_{ilk} x_k = 0, \tag{24.57}$$

where $z/|z|$ is taken as 0 for $z = 0$. (For an analysis of L^1-approximations, see [Pin89] and reference cited therein.)

24.4 Other Differential Operators

In our principal example, the differential operator considered is not the jet extension mapping. As a consequence, the space we consider is not a Sobolev space. Thus, we consider the following situation.

24.4.1 General Structure

Let W be a vector bundle over Ω as above, and let S be another trivial vector bundle over Ω. Consider the spaces of generalized sections $\mathcal{D}'(W) := C_c^\infty(W')^*$ and $\mathcal{D}'(S) := C_c^\infty(S')^*$ as in Sect. 17.6. Using a basis in $\mathbf{V}$, a section of W may be identified with a collection of real-valued functions on Ω, and an element of $\mathcal{D}'(W)$ may be identified with a collection of m distributions on Ω. Thus, distributional derivatives and jets of generalized sections are well-defined.

A linear mapping,

$$\tilde{D} : \mathcal{D}'(J^k W) \longrightarrow \mathcal{D}'(S), \tag{24.58}$$

induces a distributional linear k-th order (generally, non-local) differential operator

$$D : \mathscr{D}'(W) \longrightarrow \mathscr{D}'(S) \tag{24.59}$$

by setting (cf. (11.43))

$$D := \tilde{D} \circ j^k. \tag{24.60}$$

Consider the vector space

$$\mathbf{U} := D^{-1}(L^p(S)). \tag{24.61}$$

By its definition, $\mathbf{U}$ contains generalized sections of W, such that for each $w \in \mathbf{U}$, $D(w)$ is an L^p-section of S. Next, assuming that $\|w\| := \|D(w)\|_p$, $w \in \mathbf{U}$, is a norm on $\mathbf{U}$, we endow $\mathbf{U}$ with this norm. Restricting D to $\mathbf{U}$, while keeping the same notation for the restriction,

$$D\colon \mathbf{U} \longrightarrow L^p(S) \tag{24.62}$$

is a linear isometric injection (as injectivity is implied by isometry).

For the space $\mathbf{U}$ of velocity fields, a generalized force will be an element of $\mathbf{U}^*$. The properties of D in (24.62) imply that the representation and optimization properties of forces in $\mathbf{U}^*$ are completely analogous to those described in Sect. 24.2.2.

24.4.2 Example: The Space $LD(\Omega)$

Let $\Omega \subset \mathbb{R}^3$ be as above. We consider the *linear strain* or *stretching* mapping the differential operator

$$\varepsilon : \mathscr{D}'(W) = \mathscr{D}'(\Omega, \mathbb{R}^3) \longrightarrow \mathscr{D}'(\Omega, \mathbb{R}^6), \qquad \varepsilon(w) = \frac{1}{2}(\nabla w + (\nabla w)^{\mathsf{T}}), \tag{24.63}$$

(where we use distributional derivatives) for any section distribution w of W. Here, the vector space of symmetric 3×3 matrices is identified with $\mathbb{R}^6$, and the vector space of all 3×3 matrices is identified with $\mathbb{R}^9$. Similarly, we may identify the vector space $J^1_x(W)$ with $\mathbb{R}^3 \times \mathbb{R}^9 = \mathbb{R}^{12}$, and view the stretching mapping as a differential operator $\mathscr{D}'(W) \longrightarrow \mathscr{D}'(J^1 W)$. Note that ε may be factored as above so that

$$\varepsilon = \tilde{\varepsilon} \circ j^1, \tag{24.64}$$

where

$$\tilde{\varepsilon} : \mathcal{D}'(J^1W) \cong \mathcal{D}'(\Omega, \mathbb{R}^{12}) \longrightarrow \mathcal{D}'(\Omega, \mathbb{R}^6), \qquad (v_i, v_{jk}) \longmapsto \tfrac{1}{2}(v_{jk} + v_{kj}). \tag{24.65}$$

We may further consider the linear differential operator

$$(\mathrm{Id}, \varepsilon) : \mathcal{D}'(W) \longrightarrow \mathcal{D}'(\Omega, \mathbb{R}^9), \qquad w \longmapsto (w, \varepsilon(w)) \tag{24.66}$$

(which may be factored similarly to ε) that will play the role of the differential operator D above. In accordance with the foregoing structure, the role of the vector space $\mathbf{U}$ is assumed by

$$LD(\Omega) := (\mathrm{Id}, \varepsilon)^{-1}(L^1(\Omega, \mathbb{R}^9)). \tag{24.67}$$

The space $LD(\Omega)$ will be referred to as the space of *integrable stretchings*. Thus, a vector field has an integrable stretching if its components and the components of the induced stretching are L^1-integrable.

The induced norm on $LD(\Omega)$ is therefore

$$\|w\|_{LD} = \sum_i \|w_i\|_{L^1} + \sum_{i,m} \|\varepsilon(w)_{im}\|_{L^1}. \tag{24.68}$$

Clearly, we have a continuous linear inclusion $LD(\Omega) \longrightarrow L^1(\Omega, \mathbb{R}^3)$. In addition, $w \mapsto \varepsilon(w)$ is given by a continuous linear mapping

$$\varepsilon \colon LD(\Omega) \longrightarrow L^1(\Omega, \mathbb{R}^6). \tag{24.69}$$

The space $LD(\Omega)$ is shown in Temam [Tem85] to have the following properties, some of which are crucial for our analysis (see also [TS80b, Tem81]).

1. *Equivalent norm:* Let p be a continuous seminorm on $LD(\Omega)$ which is a norm on the space of rigid fields, $\mathbf{R}$. Then,

$$p(w) + \|\varepsilon(w)\|_{L^1} \tag{24.70}$$

is a norm on $LD(\Omega)$ that is equivalent to the original norm defined above. In particular, the expression

$$\|w\| = \|\mathrm{pr}_{\mathbf{R}}(w)\|_1 + \|\varepsilon(w)\|_1, \tag{24.71}$$

defines a norm on $LD(\Omega)$ that is equivalent to the norm (24.68). As a result,

$$(\mathrm{pr}_{\mathbf{R}}, \varepsilon) \colon LD(\Omega) \longrightarrow \mathbf{R} \times L^1(\Omega, \mathbb{R}^6) \tag{24.72}$$

is a norm-preserving linear injection.

2. *Trace mapping:* There is a bounded, linear trace mapping,

$$\gamma : LD(\Omega) \longrightarrow L^1(\partial\Omega, \mathbb{R}^3), \tag{24.73}$$

defined on $LD(\Omega)$, such that for $w \in C^\infty(\overline{\Omega}, \mathbb{R}^3)$, $\gamma(w|_\Omega) = w|_{\partial\Omega}$. Furthermore, the trace mapping is surjective. As an immediate implication, we can use the bounded linear injection

$$\delta = (\gamma, \mathcal{J}) : LD(\Omega) \to L^1(\partial\Omega, \mathbb{R}^3) \times L^1(\Omega, \mathbb{R}^3), \tag{24.74}$$

as in the setting described in Sect. 24.2.3. Here, $\mathcal{J} : LD(\Omega) \to L^1(\Omega)$ is the natural inclusion.

3. *Approximations:* $C^\infty(\overline{\Omega}, \mathbb{R}^3)$ is dense in $LD(\Omega)$. Thus, in particular

$$\|\gamma\| := \sup_{w \in LD(\Omega)} \frac{\|\gamma(w)\|_{L^1}}{\|w\|_{LD}} = \sup_{w \in C^\infty(\overline{\Omega}, \mathbb{R}^3)} \frac{\|w|_{\partial\Omega}\|_{L^1}}{\|w|_\Omega\|_{LD}}. \tag{24.75}$$

4. *Extensions:* There is a continuous linear extension operator

$$E : LD(\Omega) \longrightarrow LD(\mathbb{R}^3)$$

such that $E(w)(y) = w(y)$ for almost all $y \in \Omega$.

5. *Regularity:* If w is any distribution on Ω for which the corresponding stretching is L^1, then $w \in L^1(\Omega, \mathbb{R}^3)$. Consequently, $w \in LD(\Omega)$.

24.5 Quotient Spaces

In the case where the differential operator, D, is not injective, $w \mapsto \|D(w)\|$, $w \in \mathbf{U} = D^{-1}(L^p(S))$, is not a norm on $\mathbf{U}$. However, we may form the quotient space $\mathbf{U}/\operatorname{Kernel} D$ and consider

$$E = D/\operatorname{Kernel} D : \mathbf{U}/\operatorname{Kernel} D \longrightarrow L^p(S), \quad E([w]) := D(w). \tag{24.76}$$

Thus, E is an injection, and

$$\|w\| = \|E(w)\| \quad \text{is a norm on} \quad \mathbf{U}/\operatorname{Kernel} D. \tag{24.77}$$

It follows that we may extend the definition of a force, defined in Sect. 24.4.1 to be an element $F \in \mathbf{U}^*$, to elements of $(\mathbf{U}/\operatorname{Kernel} D)^*$. The properties of the differential operation E imply that the representation of force by stresses as in Sect. 24.2.2 and the optimization result in Sect. 24.2.2 apply here, also. Hence, any given force,

$$F \in (\mathbf{U}/\operatorname{Kernel} D)^* \tag{24.78}$$

is represented in the form

$$F = E^*(\sigma), \quad \text{for some stress} \quad \sigma \in L^{p'}(S^*), \tag{24.79}$$

and

$$\|F\| = \inf\{\|\sigma\|\}, \quad \text{for} \quad \sigma \in L^{p'}(S^*),\ F = E^*(\sigma). \tag{24.80}$$

24.5.1 Distortions

Let $\mathbf{U}$ be a vector space of velocities on Ω containing the rigid velocities $\mathbf{R}$. We will say that the two velocity fields $w_1, w_2 \in \mathbf{U}$ have the same *distortion* if $w_2 = w_1 + r$ for some rigid motion $r \in \mathbf{R}$. This clearly induces an equivalence relation on $\mathbf{U}$, and the corresponding quotient space $\mathbf{U}/\mathbf{R}$ will be referred to as the space of distortions. If χ is an element of $\mathbf{U}/\mathbf{R}$, then $\varepsilon(w)$ is the same for all members of $w \in \chi$. The natural projection,

$$\text{pr}\colon \mathbf{U} \longrightarrow \mathbf{U}/\mathbf{R}, \tag{24.81}$$

associates with each element $w \in \mathbf{U}$ its equivalence class $[w] = \{w + r | r \in \mathbf{R}\}$.

If $\mathbf{U}$ is a normed space, then, the induced norm on $\mathbf{U}/\mathbf{R}$ is given by

$$\|[w]\| = \inf_{w' \in [w]} \|w'\| = \inf_{r \in \mathbf{R}} \|w - r\|. \tag{24.82}$$

Thus, the norm is given in terms of the best approximation by a rigid velocity as described above.

Let $\mathbf{U}$ be a vector space of velocities contained in $L^1(\Omega, \mathbb{R}^3)$. Then, $\text{pr}_{\mathbf{R}}$ defined above induces an additional projection,

$$\text{pr}_0(w) = w - \text{pr}_{\mathbf{R}}(w) \in \mathbf{U}. \tag{24.83}$$

For $w \in \mathbf{R}$,

$$\begin{aligned} \text{pr}_0(w) &= w - \text{pr}_{\mathbf{R}}(w), \\ &= 0, \end{aligned} \tag{24.84}$$

and one has

$$\text{pr}_0 \circ \text{pr}_{\mathbf{R}} = 0. \tag{24.85}$$

Conversely, if $\text{pr}_0(w) = 0$, then, $w = \text{pr}_{\mathbf{R}}(w) \in \mathbf{R}$. Hence,

$$\text{Kernel}\,\text{pr}_0 = \mathbf{R}. \tag{24.86}$$

It is noted that

$$\begin{aligned}\text{pr}_\mathbf{R} \circ \text{pr}_0(w) &= \text{pr}_\mathbf{R}(w) - \text{pr}_\mathbf{R}(\text{pr}_\mathbf{R}(w)),\\ &= \text{pr}_\mathbf{R}(w) - \text{pr}_\mathbf{R}(w),\\ &= 0.\end{aligned} \tag{24.87}$$

Also,

$$\begin{aligned}\text{pr}_0 \circ \text{pr}_0(w) &= \text{pr}_0(w - \text{pr}_\mathbf{R}(w)),\\ &= w - \text{pr}_\mathbf{R}(w) - \text{pr}_\mathbf{R}(w - \text{pr}_\mathbf{R}(w)),\\ &= w - \text{pr}_\mathbf{R}(w),\\ &= \text{pr}_0(w).\end{aligned} \tag{24.88}$$

If $w \in \text{Image}\,\text{pr}_0$, there is a $u \in \mathbf{U}$ such that $w = \text{pr}_0(u) = u - \text{pr}_\mathbf{R}(u)$. Hence,

$$\begin{aligned}\text{pr}_\mathbf{R}(w) &= \text{pr}_\mathbf{R}(u) - \text{pr}_\mathbf{R}(\text{pr}_\mathbf{R}(u))\\ &= 0,\end{aligned} \tag{24.89}$$

and $w \in \text{Kernel}\,\text{pr}_\mathbf{R}$. Conversely, if $w \in \text{Kernel}\,\text{pr}_\mathbf{R}$, then

$$\begin{aligned}\text{pr}_0(w) &= w - \text{pr}_\mathbf{R}(w),\\ &= w,\end{aligned} \tag{24.90}$$

and so $w \in \text{Image}\,\text{pr}_0$. It is concluded that

$$\mathbf{U}_0 := \text{Kernel}\,\text{pr}_\mathbf{R} = \text{Image}\,\text{pr}_0. \tag{24.91}$$

In other words, the space $\mathbf{U}_0$ is the subspace of $\mathbf{U}$ containing velocity fields having zero approximating rigid velocities.

Note that

$$\text{pr}_0/\mathbf{R} : \mathbf{U}/\mathbf{R} \longrightarrow \mathbf{U}_0, \qquad \text{pr}_0/\mathbf{R}([w]) := \text{pr}_0(w), \tag{24.92}$$

is a bijection such that

$$(\text{pr}_0/\mathbf{R})^{-1}(w) = [w]. \tag{24.93}$$

On $\mathbf{U}_0$ we have two equivalent norms: the norm it has as a subspace $\text{Kernel}\,\text{pr}_\mathbf{R}$ of $\mathbf{U}$ and the norm that makes the bijection $\text{pr}_0/\mathbf{R}\colon \mathbf{U}/\mathbf{R} \to \mathbf{U}_0$ an isometry. With the projections pr_0 and $\text{pr}_\mathbf{R}$, $\mathbf{U}$ has a Whitney sum structure

$$\mathbf{U} = \mathbf{U}_0 \oplus \mathbf{R}. \tag{24.94}$$

The resulting structure is illustrated by the following diagram

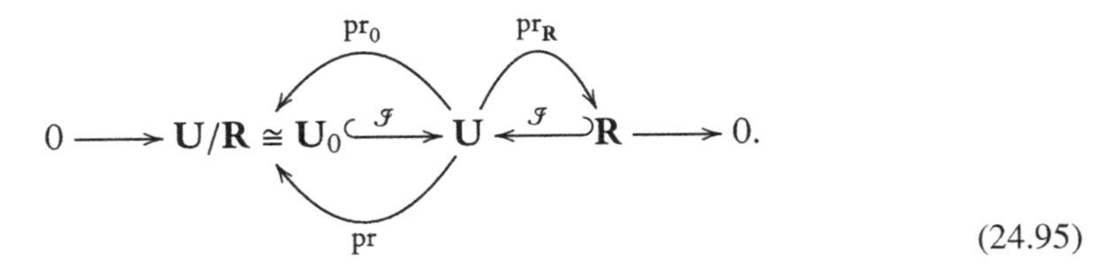

(24.95)

24.5.2 Example: The Space of LD-Distortions

We consider the case where the space, $\mathbf{U}$, of virtual velocity fields, is the space of integrable stretchings, $LD(\Omega)$. Setting $LD(\Omega)_0 := \operatorname{Kernel} \mathrm{pr}_{\mathbf{R}}$, the preceding diagram assumes the form

$$0 \longrightarrow LD(\Omega)/\mathbf{R} \cong LD(\Omega)_0 \xhookrightarrow{\mathcal{I}} LD(\Omega) \xhookleftarrow{\mathcal{I}} \mathbf{R} \longrightarrow 0. \tag{24.96}$$

(with arrows $\mathrm{pr}_0 : LD(\Omega) \to LD(\Omega)_0$, $\mathrm{pr}_{\mathbf{R}} : LD(\Omega) \to \mathbf{R}$, $\mathrm{pr} : LD(\Omega) \to LD(\Omega)/\mathbf{R}$)

In addition, instead of the differential operator $(\mathrm{Id}, \varepsilon)$, used in the definition of $LD(\Omega)$, we consider now the differential operator $(0, \varepsilon) \colon LD(\Omega) \to L^1(\mathbb{R}^9)$, which we will often write simply as $\varepsilon \colon LD(\Omega) \to L^1(\mathbb{R}^6)$.

Clearly, ε is not injective. One can easily verify that for a rotation $r \in \mathbf{R}$, $\varepsilon(r) = 0$. Conversely, it follows from a theorem by Liouville (see [Tem85, pp. 18–19]) that if $\varepsilon(w) = 0$, then, $w \in \mathbf{R}$. One concludes that

$$\operatorname{Kernel} \varepsilon = \mathbf{R} \tag{24.97}$$

—the space of rigid velocity fields. Consequently, we have the sequence as in the following diagram.

$$0 \longrightarrow \mathbf{R} \xhookrightarrow{\mathcal{I}} LD(\Omega) \xrightarrow{\varepsilon} L^1(\Omega, \mathbb{R}^6) \longrightarrow 0. \tag{24.98}$$

(with arrow $\mathrm{pr}_{\mathbf{R}} : LD(\Omega) \to \mathbf{R}$)

Taking the quotient space relative to $\mathbf{R}$, one obtains the diagram

$$
\begin{array}{ccc}
LD(\Omega) & \xrightarrow{\varepsilon} & L^1(\Omega, \mathbb{R}^6) \\
\text{pr} \downarrow & & \parallel \\
LD(\Omega)/\mathbf{R} & \xrightarrow{\varepsilon/\mathbf{R}} & L^1(\Omega, \mathbb{R}^6),
\end{array}
\tag{24.99}
$$

where pr is the natural projection on the quotient. Clearly, $\varepsilon/\mathbf{R}$ is an isometric injection. We refer to an element $u \in LD(\Omega)/\mathbf{R}$ as an *LD-distortion*, or just as a distortion, in case the context is clear.

As mentioned above for the general case, on the space of LD-distortions, $LD(\Omega)/\mathbf{R}$, there is a natural norm

$$
\|\chi\| = \inf_{w \in \chi} \|w\|_{LD}. \tag{24.100}
$$

This norm is equivalent to

$$
\|\varepsilon(\chi)\| = \sum_{i,m} \|\varepsilon(w)_{im}\|_{L^1} \tag{24.101}
$$

where w is any member of χ, and we can use any equivalent norm on the space of symmetric tensors.

Considering the decomposition,

$$
(\text{pr}_0, \text{pr}_{\mathbf{R}}) \colon LD(\Omega) \longrightarrow LD(\Omega)_0 \oplus \mathbf{R},
$$

Temam shows that there is a constant C depending only on Ω such that

$$
\|\text{pr}_0(w)\|_{L^1} = \|w - \text{pr}_{\mathbf{R}}(w)\|_{L^1} \leqslant C\|\varepsilon(w)\|_{L^1}. \tag{24.102}
$$

24.5.3 Total Forces and Equilibrated Forces

Let $\mathbf{U}$ be a vector space of velocities that contains the rigid velocities. Let $\mathcal{I}_{\mathbf{R}} \colon \mathbf{R} \to \mathbf{U}$ be the inclusion of the rigid velocities. Then,

$$
\mathcal{I}_{\mathbf{R}}^* \colon \mathbf{U}^* \longrightarrow \mathbf{R}^*, \tag{24.103}
$$

is a continuous and surjective mapping. For $F \in \mathbf{U}^*$, the value $\mathcal{I}_{\mathbf{R}}^*(F) \in \mathbf{R}^*$ will be referred to as the *total of the force*. This is, evidently, a special case of group action as discussed in Sect. 13.1. In the current situation, we have the action of the Euclidean group.

Consider the example where $\mathbf{U}$ is a subspace of $L^1(\Omega, \mathbb{R}^3)$, and $\text{pr}_{\mathbf{R}} \colon L^1(\Omega, \mathbb{R}^3) \longrightarrow \mathbf{R}$ is used as in Eqs. (24.52), (24.50), (24.53). Then, the component of $\mathcal{I}_{\mathbf{R}}^*(F)$ which is conjugate to the means, $\bar{w}$, of the velocity fields, will be referred to as the *resultant* *force*, and the component conjugate to the angular velocities, ω, will be referred to as the *resultant moment*.

A force $F \in \mathbf{U}^*$ is *equilibrated* if $F(r) = 0$ for all $r \in \mathbf{R}$. This is of course equivalent to $F(w) = F(w + r)$ for all $r \in \mathbf{R}$, so that an equilibrated force F induces a unique element of $(\mathbf{U}/\mathbf{R})^*$. Conversely, any element $G \in (\mathbf{U}/\mathbf{R})^*$ induces an equilibrated force F by $F(w) = G([w])$, where $[w]$ is the equivalence class of w. Recalling that pr^* is a norm-preserving injection relative to the quotient norm (see Taylor [Tay58, p. 227]), we may identify the equilibrated forces with elements of $(\mathbf{U}/\mathbf{R})^*$.

In other words, as the quotient projection, pr, is surjective, the dual mapping $\mathrm{pr}^* \colon (\mathbf{U}/\mathbf{R})^* \to \mathbf{U}^*$ is injective; it is simply the inclusion of all equilibrated forces, which, in turn, are orthogonal to rigid fields in Kernel pr. By definition,

$$\operatorname{Image} \mathcal{I}_{\mathbf{R}} = \operatorname{Kernel} \mathrm{pr}, \tag{24.104}$$

and in addition,

$$\operatorname{Image} \mathrm{pr}^* = \operatorname{Kernel} \mathcal{I}_{\mathbf{R}}^* \tag{24.105}$$

is orthogonal to the kernel of pr. The situation is illustrated by the sequences in the following diagram.

$$\begin{array}{c} 0 \longrightarrow \mathbf{R} \xrightarrow{\mathcal{I}_{\mathbf{R}}} \mathbf{U} \xrightarrow{\mathrm{pr}} \mathbf{U}/\mathbf{R} \longrightarrow 0 \\ \\ 0 \longleftarrow \mathbf{R}^* \xleftarrow{\mathcal{I}_{\mathbf{R}}^*} \mathbf{U}^* \xleftarrow{\mathrm{pr}^*} (\mathbf{U}/\mathbf{R})^* \longleftarrow 0. \end{array} \tag{24.106}$$

24.5.4 Stresses for Unsupported Bodies Under Equilibrated Loadings

The dual counterpart of the diagram (24.99) is

$$\begin{array}{ccc} LD(\Omega)^* & \xleftarrow{\epsilon^*} & L^\infty(\Omega, \mathbb{R}^6) \\ \uparrow{\scriptstyle \mathrm{pr}^*} & & \Vert \\ (LD(\Omega)/\mathbf{R})^* & \xleftarrow{(\epsilon/\mathbf{R})^*} & L^\infty(\Omega, \mathbb{R}^6), \end{array} \tag{24.107}$$

where $(LD(\Omega)/\mathbf{R})^*$ represents the collection of equilibrated forces.

On the other hand, since the trace mapping is well-defined for $LD(\Omega)$, we have

$$\begin{array}{ccc} L^1(\partial\Omega,\mathbb{R}^3)\times L^1(\Omega,\mathbb{R}^3) & \xleftarrow{\;\;\delta\;\;} & LD(\Omega) \\ {\scriptstyle \mathrm{pr}}\downarrow & & \downarrow{\scriptstyle \mathrm{pr}} \\ (L^1(\partial\Omega,\mathbb{R}^3)\times L^1(\Omega,\mathbb{R}^3))/\mathbf{R} & \xleftarrow{\;\delta/\mathbf{R}\;} & LD(\Omega)/\mathbf{R}, \end{array} \tag{24.108}$$

where $\delta/\mathbf{R}$ makes the diagram commutative. The dual diagram, when combined with diagram (24.107), gives

$$\begin{array}{ccccc} L^\infty(\partial\Omega,\mathbb{R}^3)\times L^\infty(\Omega,\mathbb{R}^3) & \xrightarrow{\;\;\delta^*\;\;} & LD(\Omega)^* & \xleftarrow{\;\;\varepsilon^*\;\;} & L^\infty(\Omega,\mathbb{R}^6) \\ {\scriptstyle \mathrm{pr}^*}\uparrow & & {\scriptstyle \mathrm{pr}^*}\uparrow & & \| \\ ((L^1(\partial\Omega,\mathbb{R}^3)\times L^1(\Omega,\mathbb{R}^3))/\mathbf{R})^* & \xrightarrow{(\delta/\mathbf{R})^*} & (LD(\Omega)/\mathbf{R})^* & \xleftarrow{(\varepsilon/\mathbf{R})^*} & L^\infty(\Omega,\mathbb{R}^6). \end{array} \tag{24.109}$$

Regarding elements of $((L^1(\partial\Omega,\mathbb{R}^3)\times L^1(\Omega,\mathbb{R}^3))/\mathbf{R})^*$ as equilibrated loading pairs, we may apply the basic result to this situation as follows.

(*i*) *Existence of stresses.* Given an equilibrated loading pair $(\mathbf{t},\mathbf{b}) \in L^\infty(\partial\Omega,\mathbb{R}^3)\times L^\infty(\Omega,\mathbb{R}^3)$, there is a subspace $\Sigma_{(\mathbf{t},\mathbf{b})} \subset L^\infty(\Omega,\mathbb{R}^6)$ containing essentially bounded symmetric stress tensor fields, σ, that represent the loading pair by $\delta^*(\mathbf{t},\mathbf{b}) = \mathrm{pr}^*\circ(\varepsilon/\mathbf{R})^*(\sigma)$. The condition that σ represents the loading pair $(\mathbf{t},\mathbf{b})$ is the principle of virtual work

$$\int_{\partial\Omega}\mathbf{t}\cdot\gamma(w)+\int_\Omega \mathbf{b}\cdot w=\int_\Omega \sigma_{ij}\varepsilon(w)_{ij}. \tag{24.110}$$

Again, it is noted that we did not use any form of Cauchy's stress existence theorem.

(*ii*) *Optimal stresses.* Let

$$s^{\mathrm{opt}}_{(\mathbf{t},\mathbf{b})} = \inf_{\sigma\in\Sigma_{(\mathbf{t},\mathbf{b})}}\left\{\operatorname*{ess\,sup}_{x\in\Omega}\{|\sigma(x)|\}\right\}, \tag{24.111}$$

where $|\sigma(x)|$ is any particular norm on $\mathbb{R}^6$ used for the values of the stress field. Then, there is a stress field $\widehat{\sigma}\in\Sigma_{(\mathbf{t},\mathbf{b})}$ such that

$$s^{\mathrm{opt}}_{(\mathbf{t},\mathbf{b})} = \operatorname*{ess\,sup}_{x\in\Omega}\{|\widehat{\sigma}(x)|\}. \tag{24.112}$$

The optimum satisfies

$$s^{\mathrm{opt}}_{(\mathbf{t},\mathbf{b})} = \sup_w \frac{\left|\int_{\partial\Omega} t\cdot w+\int_\Omega \mathbf{b}\cdot w\right|}{\int_\Omega|\varepsilon(w)|},\qquad w\in C^\infty(\overline{\Omega},\mathbb{R}^3),$$

where for computing $|\varepsilon(w)|$ we use the norm on $\mathbb{R}^9$ that is dual to the one used for the stress matrix $\sigma(x)$.

(*iii*) *The generalized stress concentration factor*. Let the generalized stress concentration factor be defined as

$$K = \sup_{(\mathbf{t},\mathbf{b})} \frac{s^{\text{opt}}_{(\mathbf{t},\mathbf{b})}}{\text{ess sup}_{x,y} \{|\mathbf{t}(y)|, |\mathbf{b}(x)|\}}, \tag{24.113}$$

where the loading pair $(\mathbf{t}, \mathbf{b}) \in L^\infty(\partial\Omega, \mathbb{R}^3) \times L^\infty(\Omega, \mathbb{R}^3)$ is equilibrated. Then, $K = \|\delta/\mathbf{R}\|$.

24.6 Subspaces

We now return to the case where $D\colon \mathbf{U} \to L^p(S)$ is injective. Let $\mathbf{U}_\mathrm{c}$ be a subspace of $\mathbf{U}$. Then, $D_\mathrm{c} := D|_{\mathbf{U}_\mathrm{c}}\colon \mathbf{U}_\mathrm{c} \to L^p(S)$ is again an isometric injection, and the representation of elements of $\mathbf{U}_\mathrm{c}^*$ by stresses as well as the rule about optimal stresses still holds. The next example uses this fact in the case where the subspace $\mathbf{U}_\mathrm{c}$ contains velocity fields that satisfy homogeneous boundary conditions.

24.6.1 Supported Bodies and the Space $LD(\Omega)_\mathrm{c}$

Consider the situation where the body Ω is supported on an open subset $\Gamma_0 \subset \partial\Omega$, and let $LD(\Omega)_\mathrm{c}$ be the subspace of $LD(\Omega)$ containing velocity fields that satisfy the boundary conditions, i.e.,

$$LD(\Omega)_\mathrm{c} = \left\{ w \in LD(\Omega) \mid \gamma(w)|_{\Gamma_0} = 0, \text{ a.e.} \right\}. \tag{24.114}$$

Since both γ and the restriction operator are continuous and linear, $LD(\Omega)_\mathrm{c}$ is a closed subspace of $LD(\Omega)$. Thus, we have the isometric injection

$$(\mathrm{pr}_\mathbf{R}, \varepsilon)_\mathrm{c} := (\mathrm{pr}_\mathbf{R}, \varepsilon)|_{LD(\Omega)_\mathrm{c}}\colon LD(\Omega)_\mathrm{c} \longrightarrow \mathbf{R} \times L^1(\Omega, \mathbb{R}^6). \tag{24.115}$$

Next, it is observed that the approximation procedure of integrable velocity fields by rigid velocity fields, as described in Sect. 24.3.2, applies with minor modification for vector fields defined on the open set $\Gamma_0 \subset \partial\Omega$. The integration should be carried out over Γ_0. Thus, let $\mathrm{pr}^{\Gamma_0}_\mathbf{R}\colon L^1(\Gamma_0, \mathbb{R}^3) \to \mathbf{R}$ be a projection on the space of rigid motions for fields defined on Γ_0. Define the projection

$$\mathrm{pr}_\mathbf{R}\colon LD(\Omega) \longrightarrow \mathbf{R}, \qquad \text{by} \qquad \mathrm{pr}_\mathbf{R}(w) = \mathrm{pr}^{\Gamma_0}_\mathbf{R}(\gamma(w)|_{\Gamma_0}).$$

Since all operators on the right are continuous in w, so is $\mathrm{pr}_{\mathbf{R}}$. Thus, one may use this projection mapping in Eq. (24.72). Evidently, $LD(\Omega)_{\mathrm{c}} \subset \operatorname{Kernel} \mathrm{pr}_{\mathbf{R}}$.

With this choice of a projection mapping on the space of rigid motions, the restriction of Eq. (24.72) to $LD(\Omega)_{\mathrm{c}}$ implies that

$$\varepsilon_0 = \varepsilon|_{LD(\Omega)_{\mathrm{c}}} \colon LD(\Omega)_{\mathrm{c}} \longrightarrow L^1(\Omega, \mathbb{R}^6) \tag{24.116}$$

is an isometric injection. In addition, for each $w \in LD(\Omega)_0$,

$$\|w\| = \|\varepsilon(w)\|_1. \tag{24.117}$$

24.6.2 Stress Analysis for Supported Bodies

As the body is supported on Γ_0, traction may be applied on a disjoint part $\Gamma_{\mathrm{t}} \subset \partial\Omega$. Thus, it is natural to assume that Γ_0 and Γ_{t} are nonempty and disjoint open sets, $\overline{\Gamma_0} \cup \overline{\Gamma_{\mathrm{t}}} = \partial\Omega$, and $\Lambda = \partial\Gamma_0 = \partial\Gamma_{\mathrm{t}}$ is a differentiable one-dimensional submanifold of $\partial\Omega$.

Boundary traction fields are naturally represented as elements of $L^\infty(\Gamma_{\mathrm{t}}, \mathbb{R}^3) = L^1(\Gamma_{\mathrm{t}}, \mathbb{R}^3)^*$. Let

$$L^1(\partial\Omega, \mathbb{R}^3)_{\mathrm{c}} := \left\{u \in L^1(\partial\Omega, \mathbb{R}^3) \mid u(y) = 0 \quad \text{a.e. on } \Gamma_0\right\}; \tag{24.118}$$

then, $L^1(\partial\Omega, \mathbb{R}^3)_{\mathrm{c}}$ is a closed subspace of $L^1(\partial\Omega, \mathbb{R}^3)$. The restriction mapping

$$\rho_{\mathrm{t}} \colon L^1(\partial\Omega, \mathbb{R}^3)_{\mathrm{c}} \longrightarrow L^1(\Gamma_{\mathrm{t}}, \mathbb{R}^3), \qquad \rho_{\mathrm{t}}(u) = u|_{\Gamma_{\mathrm{t}}} \tag{24.119}$$

is also linear and continuous. In addition, as $\partial\Gamma_0 = \partial\Gamma_{\mathrm{t}} = \Lambda$ has zero area measure,

$$\int_{\partial\Omega} |u| \; \mathrm{d}A = \int_{\Gamma_{\mathrm{t}}} |\rho_{\mathrm{t}}(u)| \; \mathrm{d}A, \qquad u \in L^1(\partial\Omega, \mathbb{R}^3)_{\mathrm{c}}, \tag{24.120}$$

so ρ_t is a norm-preserving injection.

Consider the zero extension mapping,

$$e_0 \colon L^1(\Gamma_{\mathrm{t}}, \mathbb{R}^3) \longrightarrow L^1(\partial\Omega, \mathbb{R}^3)_{\mathrm{c}}, \tag{24.121}$$

defined by

$$e_0(u)(y) = \begin{cases} u(y) & \text{for } y \in \Gamma_{\mathrm{t}}, \\ 0 & \text{for } y \notin \Gamma_{\mathrm{t}}. \end{cases} \tag{24.122}$$

Clearly, $\rho_{\mathrm{t}} \circ e_0$ is the identity on the space $L^1(\Gamma_{\mathrm{t}}, \mathbb{R}^3)$. Moreover, for any $u \in L^1(\partial\Omega, \mathbb{R}^3)_{\mathrm{c}}$, $e_0(\rho_{\mathrm{t}}(u))(y) = u(y)$ almost everywhere (except for $y \in \Lambda$), so $e_0 \circ \rho_{\mathrm{t}}$ is the identity on $L^1(\partial\Omega, \mathbb{R}^3)_{\mathrm{c}}$. Consequently, there are natural isometric isomorphisms

$$L^1(\Gamma_{\mathrm{t}}, \mathbb{R}^3) \cong L^1(\partial\Omega, \mathbb{R}^3)_{\mathrm{c}}, \qquad \text{and} \qquad L^\infty(\Gamma_{\mathrm{t}}, \mathbb{R}^3) \cong L^1(\partial\Omega, \mathbb{R}^3)^*_{\mathrm{c}}.$$

The dual mappings e_0^* and ρ_{t}^* induce an isometric isomorphism of the spaces $L^1(\Gamma_{\mathrm{t}}, \mathbb{R}^3)^*$ and $L^1(\partial\Omega, \mathbb{R}^3)^*_{\mathrm{c}}$. Every element $\mathbf{t}_0 \in L^1(\partial\Omega, \mathbb{R}^3)^*_{\mathrm{c}}$ is represented uniquely by an essentially bounded $\mathbf{t} \in L^\infty(\Gamma_{\mathrm{t}}, \mathbb{R}^3)$ in the form

$$\mathbf{t}_0(u) = \int_{\Gamma_{\mathrm{t}}} \mathbf{t} \cdot u \, \mathrm{d}A. \tag{24.123}$$

The mapping

$$\gamma_0 = \rho_{\mathrm{t}} \circ \gamma|_{LD(\Omega)_{\mathrm{c}}} : LD(\Omega)_{\mathrm{c}} \longrightarrow L^1(\Gamma_{\mathrm{t}}, \mathbb{R}^3) \tag{24.124}$$

is a linear and continuous surjection. Dually,

$$\gamma_0^* = \left(\gamma|_{LD(\Omega)_{\mathrm{c}}}\right)^* \circ \rho_{\mathrm{t}}^* : L^\infty(\Gamma_{\mathrm{t}}, \mathbb{R}^3) \longrightarrow LD(\Omega)^*_{\mathrm{c}} \tag{24.125}$$

The restriction of the mapping δ to $LD(\Omega)_{\mathrm{c}}$ is

$$\delta_0 : LD(\Omega)_{\mathrm{c}} \longrightarrow L^1(\Gamma_{\mathrm{t}}, \mathbb{R}^3) \times L^1(\Omega, \mathbb{R}^3). \tag{24.126}$$

The resulting structure is described by the following commutative diagrams, where $\mathcal{J}$ denotes a generic inclusion of a subspace.

$$\begin{array}{ccccc}
L^1(\partial\Omega, \mathbb{R}^3) \times L^1(\Omega, \mathbb{R}^3) & \xleftarrow{\delta} & LD(\Omega) & \xrightarrow{(\mathrm{pr}_{\mathbf{R}}, \varepsilon)} & \mathbf{R} \times L^1(\Omega, \mathbb{R}^6) \\
\uparrow \mathcal{J} & & \uparrow \mathcal{J} & & \uparrow \mathcal{J} \\
L^1(\Gamma_{\mathrm{t}}, \mathbb{R}^3) \times L^1(\Omega, \mathbb{R}^3) & \xleftarrow{\delta_0} & LD(\Omega)_{\mathrm{c}} & \xrightarrow{\varepsilon_0} & L^1(\Omega, \mathbb{R}^6).
\end{array} \tag{24.127}$$

and dually,

$$\begin{array}{ccccc}
L^\infty(\partial\Omega, \mathbb{R}^3) \times L^\infty(\Omega, \mathbb{R}^3) & \xrightarrow{\delta^*} & LD(\Omega)^* & \xleftarrow{(\mathrm{pr}_{\mathbf{R}}, \varepsilon)^*} & \mathbf{R}^* \times L^\infty(\Omega, \mathbb{R}^6) \\
\downarrow \mathcal{J}^* & & \downarrow \mathcal{J}^* & & \downarrow \mathcal{J}^* \\
L^\infty(\Gamma_{\mathrm{t}}, \mathbb{R}^3) \times L^\infty(\Omega, \mathbb{R}^3) & \xrightarrow{\delta_0^*} & LD(\Omega)^*_{\mathrm{c}} & \xleftarrow{\varepsilon_0^*} & L^\infty(\Omega, \mathbb{R}^6).
\end{array} \tag{24.128}$$

We conclude that the following can be stated for the stress analysis of supported bodies.

(*i*) *Existence of stresses*. Given a loading pair $(\mathbf{t}, \mathbf{b}) \in L^\infty(\Gamma_\mathbf{t}, \mathbb{R}^3) \times L^\infty(\Omega, \mathbb{R}^3)$, there is a subspace $\Sigma_{(\mathbf{t},\mathbf{b})} \subset L^\infty(\Omega, \mathbb{R}^6)$ such that for each $\sigma \in \Sigma_{(\mathbf{t},\mathbf{b})}$, $\delta_0^*(\mathbf{t}, \mathbf{b}) = \varepsilon_0^*(\sigma)$.

(*ii*) *Optimal stresses*. The optimum $s^{\text{opt}}_{(\mathbf{t},\mathbf{b})}$ satisfies

$$s^{\text{opt}}_{(\mathbf{t},\mathbf{b})} = \sup_w \frac{\left|\int_{\partial\Omega} t \cdot w + \int_\Omega \mathbf{b} \cdot w\right|}{\int_\Omega |\varepsilon_0(w)|}, \qquad w \in LD(\Omega)_c. \tag{24.129}$$

The optimum is attainable by some stress field $\widehat{\sigma} \in \Sigma_{(\mathbf{t},\mathbf{b})}$.

(*iii*) *The generalized stress concentration factor*. The generalized stress concentration factor is given by $K = \|\delta_0\|$. For the case where body forces are not applied, $K = \|\gamma_0\|$.

24.7 Product Structures

In this section we arrive at one of the main results of this chapter, the existence of the load capacity ratio for perfectly plastic bodies, and the expression for it, as described in the introduction to this chapter.

24.7.1 Product Structures on Subbundles of $J^k W$

It is assumed now that S has a direct sum structure

$$S = S_1 \oplus S_2 \tag{24.130}$$

for complementary subbundles S_1 and S_2 and that a norm $|\cdot|$ is given on S_1. Thus, $L^p(S) = L^p(S_1) \oplus L^p(S_2)$, and denoting the natural projections as $(\text{pr}_1, \text{pr}_2)\colon S \to S_1 \times S_2$, we have for $\alpha = 1, 2$, $D = (D_1, D_2)$, where,

$$D_\alpha\colon \mathbf{U} \longrightarrow L^p(S_\alpha), \qquad \text{is given by} \qquad D_\alpha(w)(x) = \text{pr}_\alpha(D(w)(x)). \tag{24.131}$$

Equivalently, setting

$$\text{pr}^\circ_\alpha\colon L^p(S) \longrightarrow L^p(S_\alpha), \qquad \text{pr}^\circ_\alpha(\chi) = \text{pr}_\alpha \circ \chi, \tag{24.132}$$

we have $D_\alpha = \text{pr}^\circ_\alpha \circ D$.

A seminorm, $|\cdot|_Y$, is defined on S by $|\chi|_Y = \left|\text{pr}_1(\chi)\right|$. In addition, a seminorm $\|\cdot\|_Y$ is induced on $\mathbf{U}$ by $\|w\|_Y = \|D_1(w)\|$, where the norm $|\cdot|$, defined above on S_1, is used for computing $\|D_1(w)\|$. Note that $\|\cdot\|_Y$ is a norm on

$\mathbf{U}_Y = D^{-1}(L^p(S_1) \times \{0\})$. In the sequel, we will often identify $L^p(S_1) \times \{0\}$ with $L^p(S_1)$ and view $L^p(S_1)$ as a subspace of $L^p(S)$.

Consider

$$D_Y = D|_{\mathbf{U}_Y} : \mathbf{U}_Y \to L^p(S_1). \tag{24.133}$$

The fact that D is injective implies that D_Y is a linear injection, which, by the definition of the norm, is an isometry.

We focus our attention on forces that are elements of $\mathbf{U}_Y^*$. The foregoing structure implies that a force $F \in \mathbf{U}_Y^*$ is represented by some stress $\sigma \in L^{p'}(S_1^*)$ in the form $F = D_Y^*(\sigma)$ and

$$\|F\| = \inf\left\{\|\sigma\| \mid F = D_Y^*(\sigma),\ \sigma \in L^{p'}(S_1^*)\right\}. \tag{24.134}$$

The situation is illustrated in the following diagrams.

$$\begin{array}{ccc} \mathbf{U} & \xrightarrow{D} & L^p(S) \\ \uparrow \mathcal{J} & & \mathcal{J} \uparrow \downarrow \mathrm{pr}_1^\circ \\ \mathbf{U}_Y & \xrightarrow{D_Y} & L^p(S_1), \end{array} \qquad \begin{array}{ccc} \mathbf{U}^* & \xleftarrow{D^*} & L^{p'}(S^*) \\ \downarrow \mathcal{J}^* & & \mathcal{J}^* \downarrow \uparrow \mathrm{pr}_1^{\circ *} \\ \mathbf{U}_Y^* & \xleftarrow{D_Y^*} & L^{p'}(S_1^*). \end{array} \tag{24.135}$$

24.7.2 Stress Analysis for Elastic Plastic Bodies

The foregoing constructions lead us to our main result concerning stress analysis for elastic, perfectly plastic bodies. The relevance of perfect plasticity corresponds to the following question. Is there a class of bodies, or materials, for which optimal stress fields are the resulting stress fields under certain loading situations? As shown below, it turns out that for a perfectly plastic body, the stress distribution is optimal in the limit state. This fact enables us also to define the notion of a load capacity ratio for a plastic body. For a given geometry of the body, or a structure, the load capacity ratio induces a bound on the maximum of the applied loading, which will guarantee that plastic collapse will not occur (see [Seg07a]).

Using the notation of Sects. 24.6.1–24.6.2, the differential operator we consider is $\varepsilon_0 : LD(\Omega)_c \to L^1(\Omega, \mathbb{R}^6)$. The analysis follows from the fact that yield criteria in the theory of plasticity are usually seminorms on the space of stress matrices rather than norms. Specifically, the yield criteria may be expressed usually as the application of a norm to the deviatoric component of the stress matrix. Here, the direct sum decomposition is of the form $\mathbb{R}^6 = P^\perp \oplus P$, where $P = \{aI \mid a \in \mathbb{R}\}$ is the space of spherical matrices, and the complement, $P^\perp$, is the space of trace-less or deviatoric symmetric matrices. We denote by π_P the usual projection of the space of matrices on the subspace P, i.e., $\pi_P(m) = \frac{1}{3} m_{ii} I$, and by $\pi_{P^\perp}$ the projection on $P^{\perp}$ so $\pi_{P^{\perp}}(m) = m_{P^{\perp}} = m - \pi_P(m)$. Thus, the pair $(\pi_{P^{\perp}}, \pi_P)$ induces an isomorphism of the space of symmetric matrices with $P^{\perp} \oplus P$. Again, we use the same notation $|\cdot|$ for both the norm on $\mathbb{R}^6$, the elements of which are interpreted as strain values, and for the dual norm on $\mathbb{R}^{6*}$, the elements of which are interpreted as stress values. Consequently, the yield function is of the form $|m|_Y = \left|\pi_{P^{\perp}}(m)\right|$. For example, if we take $|\cdot|$ to be the Frobenius norm on the space of deviatoric matrices, $P^{\perp}$, we get the von Mises yield criterion.

For the space of stress fields, we will therefore have the seminorm $\|\cdot\|_Y$ defined by $\|\sigma\|_Y = \|\pi_{P^{\perp}} \circ \sigma\|_{\infty}$. Let Y_0 denote the yield stress. Then, yielding will not occur when $\|\sigma\|_Y < Y_0$.

We may now combine the structures described in Sects. 24.7.1 and 24.6. The subspace under consideration is $LD(\Omega)_{\mathrm{c}} \subset LD(\Omega)$. The decomposition $S = S_1 \oplus S_2$ is taken as $P^{\perp} \oplus P$. The space $\mathbf{U}_Y$ of (24.133) will be the space of LD-vector fields that satisfy the boundary conditions on Γ_0 and are isochoric (incompressible). This space will be denoted, in what follows, as $LD(\Omega)_{P^{\perp}}$, and the restriction, D_Y of the differential operator, the linear strain mapping in our case, will be denoted by $\varepsilon_{P^{\perp}}$. In particular, $\varepsilon_{P^{\perp}}$ is a linear isometry. Finally, the restriction of the mapping δ_0 to the isochoric fields will be denoted by $\delta_{P^{\perp}}$.

The situation is depicted in the following diagram

$$
\begin{array}{ccccc}
L^1(\partial\Omega,\mathbb{R}^3)\times L^1(\Omega,\mathbb{R}^3) & \xleftarrow{\ \delta\ } & LD(\Omega) & \xrightarrow{(\mathrm{pr}_{\mathbf{R}},\varepsilon)} & \mathbf{R}\times L^1(\Omega,\mathbb{R}^6) \\
\Big\uparrow \mathcal{J} & & \Big\uparrow \mathcal{J} & & \Big\uparrow \mathcal{J} \\
L^1(\Gamma_{\mathrm{t}},\mathbb{R}^3)\times L^1(\Omega,\mathbb{R}^3) & \xleftarrow{\ \delta_0\ } & LD(\Omega)_{\mathrm{c}} & \xrightarrow{\ \varepsilon_0\ } & L^1(\Omega,\mathbb{R}^6) \\
\Big\| & & \Big\uparrow \mathcal{J} & & \mathcal{J}\Big\uparrow \Big\downarrow \mathrm{pr}^{\circ}_{P^{\perp}} \\
L^1(\Gamma_{\mathrm{t}},\mathbb{R}^3)\times L^1(\Omega,\mathbb{R}^3) & \xleftarrow{\ \delta_{P^{\perp}}\ } & LD(\Omega)_{P^{\perp}} & \xrightarrow{\ \varepsilon_{P^{\perp}}\ } & L^1(\Omega,P^{\perp}),
\end{array}
\tag{24.136}
$$

and its dual

$$
\begin{array}{ccccc}
L^\infty(\partial\Omega,\mathbb{R}^3)\times L^\infty(\Omega,\mathbb{R}^3) & \xrightarrow{\ \delta^*\ } & LD(\Omega)^* & \xleftarrow{(\mathrm{pr}_{\mathbf{R}},\varepsilon)^*} & \mathbf{R}^*\times L^\infty(\Omega,\mathbb{R}^6) \\
\mathcal{J}^*\Big\downarrow & & \Big\downarrow \mathcal{J}^* & & \Big\downarrow \mathcal{J}^* \\
L^\infty(\Gamma_{\mathrm{t}},\mathbb{R}^3)\times L^\infty(\Omega,\mathbb{R}^3) & \xrightarrow{\ \delta_0^*\ } & LD(\Omega)_{\mathrm{c}}^* & \xleftarrow{\ \varepsilon_0^*\ } & L^\infty(\Omega,\mathbb{R}^6) \\
\Big\| & & \Big\downarrow \mathcal{J}^* & & \mathcal{J}^*\Big\downarrow \Big\uparrow \mathrm{pr}^{\circ *}_{P^{\perp}} \\
L^\infty(\Gamma_{\mathrm{t}},\mathbb{R}^3)\times L^\infty(\Omega,\mathbb{R}^3) & \xrightarrow{\ \delta_{P^{\perp}}^*\ } & LD(\Omega)_{P^{\perp}}^* & \xleftarrow{\ \varepsilon_{P^{\perp}}^*\ } & L^\infty(\Omega,P^{\perp}).
\end{array}
\tag{24.137}
$$

It is noted that the dual mapping $\mathcal{J}^*$ for an inclusion $\mathcal{J}$ is simply a restriction. Thus, the last lines in the previous two diagrams are analogous to the diagram preceding Theorem 24.2. We conclude that the theorem applies in this case upon making the relevant substitutions.

For a given loading pair $(\mathbf{t}, \mathbf{b}) \in L^\infty(\Gamma_\mathbf{t}, \mathbb{R}^3) \times L^\infty(\Omega, \mathbb{R}^3)$, the expression defining the optimal stress is

$$s^{\text{opt}}_{(\mathbf{t},\mathbf{b})} = \inf_{\delta^*_{P\perp}(\mathbf{t},\mathbf{b})=\varepsilon^*_{P\perp}(\sigma)} \|\sigma\|_Y, \qquad \sigma \in L^\infty(\Omega, P^\perp). \tag{24.138}$$

We observe that the equilibrium condition

$$\delta^*_{P\perp}(\mathbf{t}, \mathbf{b}) = \varepsilon^*_{P\perp}(\sigma), \qquad \sigma \in L^\infty(\Omega, P^\perp), \tag{24.139}$$

applies only to the deviatoric stress, and the hydrostatic component of the stress is not restricted. The expression for the generalized stress concentration factor is

$$K = \sup_{(\mathbf{t},\mathbf{b})} \frac{s^{\text{opt}}_{(\mathbf{t},\mathbf{b})}}{\|(\mathbf{t}, \mathbf{b})\|_\infty}. \tag{24.140}$$

In analogy with the results for the earlier examples, one has

$$s^{\text{opt}}_{(\mathbf{t},\mathbf{b})} = \sup_{w\in LD(\Omega)_{P\perp}} \frac{\left|\int_{\partial\Omega} \mathbf{t}\cdot w + \int_\Omega \mathbf{b}\cdot w\right|}{\int_\Omega \left|\varepsilon_{P\perp}(w)\right|} = \|\delta^*_{P\perp}(\mathbf{t}, \mathbf{b})\|. \tag{24.141}$$

It is observed that for any multiplier $\lambda \in \mathbb{R}^+$,

$$s^{\text{opt}}_{\lambda(\mathbf{t},\mathbf{b})} = \|\delta^*_{P\perp}(\lambda(\mathbf{t}, \mathbf{b}))\| = \lambda s^{\text{opt}}_{(\mathbf{t},\mathbf{b})}. \tag{24.142}$$

For the generalized stress concentration factor, the theorem gives the expression

$$K = \|\delta_{P\perp}\| = \sup_{w\in LD(\Omega)_{P\perp}} \frac{\int_{\partial\Omega} |w| + \int_\Omega |w|}{\int_\Omega \left|\varepsilon_{P\perp}(w)\right|}. \tag{24.143}$$

(It is recalled that smooth vector fields on $\overline{\Omega}$ are dense in $LD(\Omega)$, and so, it is sufficient to consider only smooth vector fields in the suprema.) Again, if no body forces are applied, the stress concentration factor is equal to the norm of the trace mapping for isochoric LD-vector fields that vanish on Γ_0.

We now present the mechanical interpretation of $s^{\text{opt}}_{(\mathbf{t},\mathbf{b})}$ and K in the framework of plasticity theory. For an elastic, perfectly plastic body, the condition for unavoidable collapse under the loading pair $(\mathbf{t}, \mathbf{b})$ is $s^{\text{opt}}_{(\mathbf{t},\mathbf{b})} > Y_0$. Let $\mathscr{A}$ be the collection of all loading pairs for which collapse does not necessarily occur, that is, those pairs for which $s^{\text{opt}}_{(\mathbf{t},\mathbf{b})} \leqslant Y_0$. The boundary of $\mathscr{A}$ is the *collapse manifold* Ψ, i.e., $\Psi = \left\{(\mathbf{t}, \mathbf{b}) \mid s^{\text{opt}}_{(\mathbf{t},\mathbf{b})} = Y_0\right\}$. One may write $\sigma = \sigma_1/\lambda$, $\|\sigma_1\|_Y = Y_0$, and noting that $\|\sigma/\|\sigma\|_Y\|_Y = 1$, the expression for the optimal stress becomes

$$s^{\text{opt}}_{(\mathbf{t},\mathbf{b})} = \inf_{\substack{\varepsilon^*_{P\perp}(\sigma_1/\lambda)=\delta^*_{P\perp}(\mathbf{t},\mathbf{b}),\\ \lambda\in\mathbb{R}^+,\,\sigma_1\in\partial B}} \|\sigma_1/\lambda\|_Y, \tag{24.144}$$

where B is the ball in $L^\infty(\Omega, P^\perp)$ of radius Y_0 . Thus,

$$s^{\text{opt}}_{(\mathbf{t},\mathbf{b})} = \inf_{\substack{\varepsilon^*_{P\perp}(\sigma_1/\lambda)=\delta^*_{P\perp}(\mathbf{t},\mathbf{b}),\\ \lambda\in\mathbb{R}^+,\,\sigma_1\in\partial B}} \frac{Y_0}{\lambda}, \tag{24.145}$$

$$\frac{Y_0}{s^{\text{opt}}_{(\mathbf{t},\mathbf{b})}} = \sup\left\{\lambda \mid \exists\sigma_1 \in \partial B,\ \varepsilon^*_{P\perp}(\sigma_1) = \delta^*_{P\perp}(\lambda(\mathbf{t},\mathbf{b}))\right\}. \tag{24.146}$$

The similarity of this construction to the definition of the Minkowski functional is noted.

Clearly, in the last equation, ∂B may be replaced by $\overline{B}$ because if we consider σ with $\|\sigma\|_Y < 1$, then, $\sigma_1 = \sigma/\|\sigma\|_Y$ is on ∂B, and the corresponding λ will be multiplied by $\|\sigma\|_Y < 1$. The unit ball B contains the stress fields that are essentially bounded by the yield stress, and we are looking for the largest multiplier of the loading pair for which there is an equilibrating stress field that is essentially bounded by Y_0. Thus, we are looking for

$$\lambda^* = \frac{Y_0}{s^{\text{opt}}_{(\mathbf{t},\mathbf{b})}} = \sup\left\{\lambda \mid \exists\sigma \in B,\ \varepsilon^*_{P\perp}(\sigma) = \delta^*_{P\perp}(\lambda(\mathbf{t},\mathbf{b}))\right\}, \tag{24.147}$$

which is the limit analysis factor (e.g., Christiansen [Chr80, Chr86] and Teman and Strang [TS80a]). Thus, the optimal stress problem for elastic perfectly plastic bodies is equivalent to the limit analysis problem for perfectly plastic bodies. In particular, at the limit state, the stress distribution in a perfectly plastic body is optimal.

For the application to plasticity, we use the term *load capacity ratio* for $C := 1/K$. Hence,

$$C = \frac{1}{\sup_{(\mathbf{t},\mathbf{b})}(s^{\text{opt}}_{(\mathbf{t},\mathbf{b})}/\|(\mathbf{t},\mathbf{b})\|_\infty)} = \inf_{(\mathbf{t},\mathbf{b})} \frac{\|(\mathbf{t},\mathbf{b})\|_\infty}{s^{\text{opt}}_{(\mathbf{t},\mathbf{b})}}. \tag{24.148}$$

For every loading pair $(\mathbf{t},\mathbf{b})$, we set

$$(\mathbf{t},\mathbf{b})_\psi := \frac{(\mathbf{t},\mathbf{b})}{s^{\text{opt}}_{(\mathbf{t},\mathbf{b})}/Y_0}. \tag{24.149}$$

Using (24.142), for any $\lambda > 0$, one has,

$$s^{\text{opt}}_{(\mathbf{t},\mathbf{b})_\psi} = Y_0, \qquad \frac{\|(\mathbf{t},\mathbf{b})\|_\infty}{s^{\text{opt}}_{(\mathbf{t},\mathbf{b})}} = \frac{\|(\mathbf{t},\mathbf{b})_\psi s^{\text{opt}}_{(\mathbf{t},\mathbf{b})}/Y_0\|_\infty}{s^{\text{opt}}_{(\mathbf{t},\mathbf{b})}} = \frac{\|(\mathbf{t},\mathbf{b})_\psi\|_\infty}{Y_0}. \tag{24.150}$$

It follows that for any loading pair $(\mathbf{t}, \mathbf{b})$, $(\mathbf{t}, \mathbf{b})_\Psi$ belongs to the collapse manifold Ψ and that the operation above is a projection onto the collapse manifold. Thus,

$$C = \inf_{(\mathbf{t},\mathbf{b})} \frac{\|(\mathbf{t}, \mathbf{b})\|_\infty}{S^{\mathrm{opt}}_{(\mathbf{t},\mathbf{b})}} = \inf_{(\mathbf{t},\mathbf{b})_\Psi \in \Psi} \|(\mathbf{t}, \mathbf{b})_\Psi\|_\infty / Y_0, \tag{24.151}$$

and indeed, $CY_0 = \inf_{(\mathbf{t},\mathbf{b})_\Psi \in \Psi} \|(\mathbf{t}, \mathbf{b})_\Psi\|_\infty$ is the largest radius of a ball containing only loading pairs for which collapse does not occur. In other words, no collapse will occur for any loading pair $(\mathbf{t}, \mathbf{b})$ satisfying

$$\|(\mathbf{t}, \mathbf{b})\|_\infty \leqslant CY_0. \tag{24.152}$$

Consequently, the load capacity ratio, $C = 1/K$, which depends only on the geometry of the body, and is computed using (24.143), provides a bound on the maximum of the applied load which will guarantee that the body will not collapse, independent of the distribution of the loading. For examples of computations for bodies and structures, including the computation of worst-case loading distributions, see [FS09].

References

[AB67] M. Atiyah and R. Bott. A Lefschetz fixed point formula for elliptic complexes: I. *Annals of Mathematics*, 86:374–407, 1967.

[AF03] R.A. Adams and J.J.F. Fournier. *Sobolev Spaces*. Academic Press, 2003.

[AM77] L. Auslander and R.E. MacKenzie. *Introduction to Differentiable Manifolds*. Dover, 1977.

[AMR88] R. Abraham, J.E. Marsden, and T. Ratiu. *Manifolds, Tensor Analysis. and Applications*. Springer, 1988.

[AS68] M. Atiyah and I. Singer. The index of elliptic operators: I. *Annals of Mathematics*, 87:484–530, 1968.

[BC70] F. Brickell and R.S. Clark. *Differentiable Manifolds, an Introduction*. Van Nostrand Reinhold, 1970.

[Cap89] G. Capriz. *Continua with Microstructure*. Springer, 1989.

[Cer99] P. Cermelli. Material symmetry and singularities in solids. *Proceedings of the Royal Society of London*, A 455:299–322, 1999.

[Cha77] S. Chandrasekhar. *Liquid Crystals*. Cambridge University Press, 1977.

[Chr80] E. Christiansen. Limit analysis in plasticity as a mathematical programming problem. *Calcolo*, 17:41–65, 1980.

[Chr86] E. Christiansen. On the collapse solutions of limit analysis. *Archive for Rational Mechanics and Analysis*, 91:119–135, 1986.

[CM02] G. Capriz and P.M. Mariano, editors. *Advances in Multifield Theories of Continua with Substructure*. Birkhauser, 2002.

[Con01] L. Conlon. *Differentiable Manifolds*. Birkhauser, 2nd edition, 2001.

[DH73] A. Douady and L. Hérault. Arrondissement des variétés à coins in Appendix of A. Borel J-P. Serre: Corners and arithmetic groups. *Commentarii Mathematici Helvetici*, 48:484–489, 1973.

[Die72] J. Dieudonné. *Treatise on Analysis*, volume III. Academic Press, 1972.

[DMM99] M. Degiovanni, A. Marzocchi, and A. Musesti. Cauchy fluxes associated with tensor fields having divergence measure. *Archive for Rational Mechanics and Analysis*, 147:197–223, 1999.

[DO05] L. Dorfmann and R.W. Ogden. Nonlinear electroelasticity. *Acta Mechanica*, 174:167–183, 2005.

[DO09] L. Dorfmann and R.W. Ogden. Non-linear electro- and magnetoelastic interactions, in Continuum Mechanics, [Eds. J. Merodio and G. Saccomandi]. In *Encyclopedia of Life Support Systems*. EOLSS, 2009.

R. Segev, *Foundations of Geometric Continuum Mechanics*, Advances in Continuum Mechanics 49, https://doi.org/10.1007/978-3-031-35655-1

[DP91] C. Davini and G. Parry. A complete list of invariants for defective crystals. *Proceedings of the Royal Society of London*, 432:341–365, 1991.

[dR84a] G. de Rham. *Differentiable Manifolds*. Springer, 1984.

[dR84b] G. de Rham. *Differentiable Manifolds*. Springer, 1984.

[EC70] A.C. Eringen and W.D. Claus. A micromorphic approach to dislocation theory and its relation to several existing theories. In J.A. Simmons, R. de Wit, and R. Bullough, editors, *Fundamental Aspects of Dislocation Theory*, pages 1023–1040. U.S. National Bureau of Standards, 1970.

[EE07] M. Elzanowski and M. Epstein. *Material Inhomogeneities and their Evolution*. Springer, 2007.

[ES80] M. Epstein and R. Segev. Differentiable manifolds and the principle of virtual work in continuum mechanics. *Journal of Mathematical Physics*, 21:1243–1245, 1980.

[ES14] M. Epstein and R. Segev. Geometric aspects of singular dislocations. *Mathematics and Mechanics of Solids*, 19:335–347, 2014. https://doi.org/10.1177/1081286512465222.

[ES15] M. Epstein and R. Segev. Differential geometry and continuum mechanics. volume 137 of *Springer Proceedings in Mathematics and Statistics*, chapter 7: On the geometry and kinematics of smoothly distributed and singular defects, pages 203–234. Springer, 2015.

[ES20] M. Epstein and R. Segev. Regular and Singular Dislocations. In R. Segev and M. Epstein, editors, *Geometric Continuum Mechanics*, Advances in Mechanics and Mathematics, pages 223–265. Springer, 2020.

[Fed69a] H. Federer. *Geometric Measure Theory*. Springer, 1969.

[Fed69b] H. Federer. *Geometric Measure Theory*. Springer, 1969.

[FN05] K. Fukui and T. Nakamura. A topological property of Lipschitz mappings. *Topology and its Applications*, 148:143–152, 2005.

[Fra58] F.C. Frank. 1. Liquid crystals. On the theory of liquid crystals. *Discussions of the Faraday Society*, 25:19–28, 1958.

[Fra04] T. Frankel. *The Geometry of Physics*. Cambridge university press, 2004.

[FS09] L. Falach and R. Segev. Load capacity ratios for structures. *Computer Methods in Applied Mechanics and Engineering*, 199:77–93, 2009.

[FS15] L. Falach and R. Segev. The configuration space and principle of virtual power for rough bodies. *Mathematics and Mechanics of Solids*, 20:1049–1072, 2015. https://doi.org/10.1177/1081286513514244.

[FV89] R. Fosdick and E. Virga. A variational proof of the stress theorem of Cauchy. *Archive for Rational Mechanics and Analysis*, 105:95–103, 1989.

[GHV72] W. Greub, S. Halperin, and R Vanstone. *Connections, Curvature, and Cohomology*, volume I. Academic Press, 1972.

[GKOS01] M. Grosser, M. Kunzinger, M. Oberguggenberger, and R. Steinbauer. *Geometric Theory of Generalized Functions with Applications to General Relativity*. Springer, 2001.

[Gla58] G. Glaeser. Etude de quelques algebres tayloriennes. *Journal d'Analyse Mathématique*, 6:1–124, 1958.

[GM75] M. E. Gurtin and L. C. Martins. Cauchy's theorem in classical physics. *Archive for rational mechanics and analysis*, 60:305–324, 1975.

[GS77] V. Guillemin and S. Sternberg. *Geometric Asymptotics*. American Mathematical Society, 1977.

[GWZ86] M. E. Gurtin, W. O. Williams, and W. P. Ziemer. Geometric measure theory and the axiom of continuum thermodynamics. *Archive for Rational Mechanics and Analysis*, 92:1–22, 1986.

[HIO06] F.W. Hehl, Y. Itin, and Y.N. Obukhov. Recent developments in premetric electrodynamics. arXiv:physics/0610221v1 [physics.class-ph], October 2006.

[Hir76] M. Hirsch. *Differential Topology*. Springer, 1976.

[HO] F.W. Hehl and W.N. Obukhov. *Foundations of Classical Electrodynamics: Charge, Flus, and Metric*. Birkhauser.

[HO03] F. W. Hehl and W. N. Obukhov. *Foundations of Classical Electrodynamics: Charge, Flux, and Metric*. Birkhauser, 2003.

[Hör90] L. Hörmander. *The analysis of linear partial differential operators. I. Distribution theory and Fourier analysis*. Springer, 1990.

[HT86] M. Henneaux and C. Teitelboim. p-Form electrodynamics. *Foundations of Physics*, 16(7):593–617, 1986.

[HT88] M. Henneaux and C. Teitelboim. Dynamics of chiral (self-dual) p-forms. *Physics Letters B*, 206:650–654, 1988.

[KA75] E. Kroner and K.H. Anthony. Dislocations and disclinations in material structures: The basic topological concepts. *Annual Review of Material Science*, 5:43–72, 1975.

[Kai04] G. Kaiser. Energy-momentum conservation in pre-metric electrodynamics with magnetic charges. *Journal of Physics A: Mathematical and General*, 37:7163–7168, 2004.

[Kon55] K. Kondo. *Geometry of Elastic Deformation and incompatibility*. Tokyo Gakujutsu Benken Fukyu-Kai, IC, 1955.

[Kor08] J. Korbas. *Handbook of Global Analysis*, chapter Distributions, vector distributions, and immersions of manifolds in Euclidean spaces, pages 665–724. Elsevier, 2008.

[KOS17a] R. Kupferman, E. Olami, and R. Segev. Stress theory for classical fields. *Mathematics and Mechanics of Solids*, 2017. https://doi.org/10.1177/1081286517723697.

[KOS17b] R. Kupferman, E. Olami, and R. Segev. Stress theory for classical fields. *Mathematics and Mechanics of Solids*, 25:1472–1503, 2017. https://doi.org/10.1177/1081286517723697.

[Kot22] F. Kottler. Maxwell'sche gleichungen und metrik. *Sitzungsberichte Akademie der Wissenschaften in Wien (IIa)*, 131:119–146, 1922.

[KSM93] I. Kolár, J. Slovák, and P.W. Michor:. *Natural operations in differential geometry*. Springer, 1993.

[Kur72] K. Kuratowski. *Introduction to Set Theory and Topology*. Pergamon, 1972.

[Lee02] J.M. Lee. *Introduction to Smooth Manifolds*. Springer, 2002.

[LK06] S.A. Lurie and A.L. Kalamkarov. General theory of defects in continous media. *International Journal of Solids and Structures*, 43:91–111, 2006.

[LL95] L.D. Landau and E.M. Lifshitz. *The Classical Theory of Fields. Course of Theoretical Physics.*, volume 2. Butterworth-Heinemann, 1995.

[Mal66] B. Malgrange. *Ideals of Differentiable functions*, volume 3 of *Tata Institute of Fundamental Research*. Oxford University Press, 1966.

[Maz85] V.G. Maz'ja. *Sobolev Spaces*. Springer, 1985.

[Mel96] R.B. Melrose. *Differential Analysis on Manifolds with Corners*. www-math.mit.edu/~rbm/book.html, 1996.

[Mic80] P. W. Michor. *Manifolds of Differentiable Mappings*. Shiva, 1980.

[Mic20] P. Michor. Manifolds of mappings for continuum mechanics. chapter 3 of Geometric Continuum Mechanics, R. Segev and M. Epstein (Eds.), pages 3–75. Springer, 2020.

[MRD08] J. Margalef-Roig and E Outerelo Domíngues. *Handbook of Global Analysis*, chapter Topology of manifolds with corners, pages 983–1033. Elsevier, 2008.

[MTW73] C.W. Misner, K.S. Thorne, and J.A. Wheeler. *Gravitation*. Freeman, 1973.

[Mun84] J.R. Munkres. *Elements of Algebraic Topology*. Addison-Wesley, 1984.

[Nol59] W. Noll. the foundations of classical mechanics in the light of recent advances in continuum mechanics. In Leon Henkin, Patrick Suppes, and Alfred Tarski, editors, *The Axiomatic Method, with Special Reference to Geometry and Physics*, pages 266–281. North-Holland, 1959. Proceedings of an international symposium held at the University of California, Berkeley, December 26, 1957-January 4, 1958.

[Nol67] W. Noll. Materially uniform bodies with inhomogeneities. *Archive for Rational mechanics and Analysis*, 27:1–32, 1967.

[NS12] J. Navarro and J.B. Sancho. Energy and electromagnetism of a differential k-form. *Journal of Mathematical Physics*, 53:102501, 2012.

[NV88] W. Noll and E. G. Virga. Fit regions and functions of bounded variation. *Archive for Rational Mechanics and Analysis*, 102:1–21, 1988.

[Oks76] A.I. Oksak. On invariant and covariant Schwartz distributions in the case of a compact linear Groups. *Communications in Mathematical Physics*, 46:269–287, 1976.

[Pal68] R. S. Palais. *Foundations of Global Non-Linear Analysis*. Benjamin, 1968.

[Pin89] A. Pinkus. *On L^1-Approximation*. Cambridge tracts in mathematics Vol. 93. Cambridge University Press, 1989.

[PS07] R. Peretz and R. Segev. Bounds on the trace mapping of LD-fields. *Computers and Mathematics with Applications*, 53:665–684, 2007.

[RS03a] G. Rodnay and R. Segev. Cauchy's flux theorem in light of geometric integration theory. *Journal of Elasticity*, 71:183–203, 2003.

[RS03b] G. Rodnay and R. Segev. Cauchy's flux theorem in light of geometric integration theory. *Journal of Elasticity*, 71 (a special volume in memory of C. Truesdell):183–203, 2003.

[Sah84] D. Sahoo. Elastic continuum theories of lattice defects: a review. *Bulletin of Materials Science*, 6:775–798, 1984.

[Sau89] D. J. Saunders. *The Geometry of Jet Bundles*, volume 142 of *London Mathematical Society: Lecture Notes Series*. Cambridge University Press, 1989.

[Sch63] L. Schwartz. *Lectures on Modern Mathematics (Volume 1)*, chapter Some applications of the theory of distributions, pages 23–58. Wiley, 1963.

[Sch73] L. Schwartz. *Théorie des Distributions*. Hermann, 1973.

[SD91] R. Segev and G. DeBotton. On the consistency conditions for force systems. *International Journal of Nonlinear Mechanics*, 26:47–59, 1991.

[SE80] R. Segev and M. Epstein. Some geometrical aspects of continuum mechanics. Technical Report 158, 1980.

[SE83] R. Segev and M. Epstein. An invariant theory of stress and equilibrium. in *Mathematical Foundations of Elasticity* by J. Marsden and T.J. Hughes, 1983.

[SE22] R. Segev and M. Epstein. Proto-Galilean dynamics of a particle and a continuous body. *Journal of Elasticity*, 2022. to appear.

[See64] R.T. Seeley. Extension of C^∞ functions defined in a half space. *Proceeding of the American Mathematical Society*, 15:625–626, 1964.

[Seg81] R. Segev. *Differentiable Manifolds and Some Basic Notions of Continuum Mechanics*. PhD thesis, University of Calgary, 1981.

[Seg86] R. Segev. Forces and the existence of stresses in invariant continuum mechanics. *Journal of Mathematical Physics*, 27:163–170, 1986.

[Seg88] R. Segev. Locality and continuity in constitutive theory. *Archive for Rational Mechanics and Analysis*, 101:29–37, 1988.

[Seg02] R. Segev. Metric-independent analysis of the stress-energy tensor. *Journal of Mathematical Physics*, 43:3220–3231, 2002.

[Seg05] R. Segev. Generalized stress concentration factors for equilibrated forces and stresses. *Journal of Elasticity*, 81:293–315, 2005.

[Seg06] R. Segev. Generalized stress concentration factors. *Mathematics and Mechanics of Solids*, 11:479–493, 2006.

[Seg07a] R. Segev. Load capacity of bodies. *International Journal of Non-Linear Mechanics*, 42 (A special volume in memory of R. Rivlin):250–257, 2007.

[Seg07b] R. Segev. Optimization for the balance equations. In M. Silhavy, editor, *Mathematical Modeling of Bodies with Complicated Bulk and Boundary Behavior*, Quaderni di Matematica. University of Naples, 2007. in Mathematical Modeling of Bodies with Complicated Bulk and Boundary Behavior, M. Silhavy, ed., *Quaderni di Matematica*, **20**, 197–216.

[Sil85] M. Silhavy. The existance of the flux vector and the divergence theorem for general Cauchy flux. *Archive for Rational Mechanics and Analysis*, 90:195–212, 1985.

[Ste83] S. Sternberg. *Lectures on Differential Geometry*. American Mathematical Society, 1983.

[Ste00] R. Steinbauer. *Distributional Methods in General Relativity*. PhD thesis, Universität Wien, 2000.

[Tay58] A.E. Taylor. *Introduction to Functional Analysis*. Wiley, 1958.

[Tem81] R. Temam. On the continuity of the trace of vector functions with bounded deformation. *Applicable Analysis*, 11:291–302, 1981.

[Tem85] R. Temam. *Mathematical Problems in Plasticity*. Gauthier-Villars, 1985.

[TN65] C. A. Truesdell and W. Noll. *The nonlinear field theories of mechanics*. Springer, 1965.

[Tou68] R.A. Toupin. Dislocated and oriented media. In *Continuum Theory of Inhomogeneities in Simple Bodies*, pages 9–24. Springer, 1968.

[Trè67] F. Trèves. *Topological Vector Spaces, Distributions and Kernels*. Academic Press, 1967.

[Tru77] C. A. Truesdell. *A First Course in Rational Continuum Mechanics*. Academic Press, 1977.

[TS80a] R. Temam and G. Strang. Duality and relaxation in the variational problems of plasticity. *Journal de Mécanique*, 19:493–527, 1980.

[TS80b] R. Temam and G. Strang. Functions of bounded deformations. *Archive for Rational Mechanics and Analysis*, 75:7–21, 1980.

[TT60] C.A. Truesdell and R. Toupin. *The Classical Field Theories*, volume III/1 of *Handbuch der Physik*. Springer, 1960.

[vD34] D. van Dantzig. The fundamental equations of electromagnetism, independent of metrical geometry. *Proceedings of the Cambridge Philosophical Society*, 30:421–427, 1934.

[Wan67] C.-C. Wang. On the geometric structure of simple bodies, a mathematical foundation for the theory of continuous distributions of dislocations. *Archive for Rational Mechanics and Analysis*, 27:33–94, 1967.

[War83] F.W. Warner. *Foundations of Differentiable Manifolds and Lie Groups*. Springer, 1983.

[Whi34] H. Whitney. Analytic extensions of functions defined in closed sets. *Transactions of the American Mathematical Society*, 36:63–89, 1934.

[Whi53] E.T. Whittaker. *A History of the Theories of Aether and Electricity*, volume 2. Nelson, 1953.

[Whi57] H. Whitney. *Geometric Integration Theory*. Princeton University Press, 1957.

Index

R. Segev, *Foundations of Geometric Continuum Mechanics*, Advances in Continuum Mechanics 49, https://doi.org/10.1007/978-3-031-35655-1

Printed and bound by CPI Group (UK) Ltd, Croydon, CR0 4YY
28/11/2024
01796467-0001